AF556674

INTERNATIONAL

REVIEW OF CYTOLOGY

Volume 40

ADVISORY EDITORS

INTERNATIONAL

Review of Cytology

EDITED BY

G. H. BOURNE
Yerkes Regional Primate Research Center
Emory University
Atlanta, Georgia

J. F. DANIELLI
Center for Theoretical Biology
State University of New York at Buffalo
Buffalo, New York

ASSISTANT EDITOR
K. W. JEON
Department of Zoology
University of Tennessee
Knoxville, Tennessee

VOLUME 40

ACADEMIC PRESS New York San Francisco London 1975
A Subsidiary of Harcourt Brace Jovanovich, Publishers

ACADEMIC PRESS, INC.
111 Fifth Avenue, New York, New York 10003

United Kingdom Edition published by
ACADEMIC PRESS, INC. (LONDON) LTD.
24/28 Oval Road, London NW1

LIBRARY OF CONGRESS CATALOG CARD NUMBER: 52-5203

ISBN 0-12-364340-6

PRINTED IN THE UNITED STATES OF AMERICA

Contents

Fine Structure of the Thyroid Gland

HISAO FUJITA

Postnatal Gliogenesis in the Mammalian Brain

A. PRIVAT

Three-Dimensional Reconstruction from Serial Sections

RANDLE W. WARE AND VINCENT LOPRESTI

List of Contributors

Numbers in parentheses indicate the pages on which the authors' contributions begin.

HISAO FUJITA (197), *Hiroshima University, School of Medicine, Hiroshima, Japan*

HIDEO HAYASHI (101), *Department of Pathology, Kumamoto University Medical School, Kumamoto, Japan*

R. N. JONES (1), *Department of Agricultural Botany, University College of Wales, Aberystwyth, Wales, United Kingdom*

VINCENT LOPRESTI (325), *Columbia University, New York, New York*

VLADIMIR R. PANTIĆ (153), *Serbian Academy of Sciences and Arts, Belgrade, Yugoslavia*

A. PRIVAT (281), *Laboratoire de cultures de tissu nerveux, I.N.S.E.R.M., Paris, France*

RANDLE W. WARE (325), *California Institute of Technology, Pasadena, California*

B-Chromosome Systems in Flowering Plants and Animal Species

R. N. JONES

Department of Agricultural Botany, University College of Wales, Aberystwyth, Wales, United Kingdom

I. Introduction

In eukaryotes each chromosome of the basic complement makes a vital and unique contribution to the genome. Physiologically, a haploid set is the genetic minimum for an organism to function. But for mechanical reasons those species that reproduce sexually require at least a diploid complement in order to accomplish orderly segregation and recombination at meiosis. With few exceptions (e.g., hyacinths), any form of numerical variation other than some exact multiple of the basic complement, as in trisomics or other aneuploids, is not compatible with normal development and reproduction. There is a requirement for the individual chromosomes of the basic complement to be present in a strict numerical relationship to one another. This requirement is not mandatory though for certain nonessential chromosomes which are different from and additional to those of the basic set. Many species of higher plants and animals have what appear to be "optional extras," called *B chromosomes*, which are outside the laws of classic genetics and a perpetual source of embarrassment to cytologists.

One of the first reports of the occurrence of B chromosomes was given by Stevens (1908), who described the presence of small additional chromosomes appearing in variable numbers in about 50% of random collections of the animal species *Diabrotica soror* and *Diabrotica 12-punctata* (Coleoptera). Stevens adopted the term supernumerary which Wilson (1907) had earlier used to describe certain extra chromosomes in *Metapodius* (Hemiptera). In plants, Kuwada (1915, 1925), Reeves (1925), and Fisk (1925) commented on the variable chromosome number of maize, but it was Longley (1927) who first distinguished these extra chromosomes in maize as supernumerary and presented a major work on their character and behavior. Randolph (1928) classified the extra chromosomes of maize into three types: A—duplicates of members of the typical set; B—chromosomes of approximately the same size as the smallest members of the typical set but differing from them in form and behavior; and C—diminutive chromosomes (derivatives of the B types). Thus the term B chromosome was introduced to describe extra chromosomes which have little if any effect on visible characters of the plant, which are not homologous with the A *chromosomes* of the normal complement, and which are extremely irregular in their meiotic distribution. B Chromosomes are now known under a variety of different names (see Battaglia, 1964b), including supernumerary and accessory, which are the most common alternatives. This confusion in the nomenclature reflects the struggle that successive generations of cytologists have had to define them unambiguously.

B Chromosomes are dispensable and nonhomologous with A chromosomes. These two characteristics above all others most sharply differentiate them from members of the basic A-chromosome complement. For practical purposes of definition and description, however, they are best considered from a much broader standpoint and with reference to a unique set of attributes:

1. Bs may be present or absent within individuals of a species.
2. When present they vary in number among individuals and among different populations.
3. Their inheritance is non-Mendelian. B Chromosomes are unstable during somatic cell division and/or meiosis. Their maintenance in a population depends on an equilibrium between forces of elimination and accumulation.
4. They are generally smaller than A chromosomes and devoid of major genes.

5. Bs adversely affect vigor and fertility, especially when present in high numbers. The genetic effects are of a polygenic nature, and they are phenotypically undetectable in low numbers.

6. They alter the nuclear phenotype. Changes occur in such fundamental processes as cell cycle time, gene activity, and A-chromosome behavior at meiosis.

These exceptional characteristics are of themselves, interesting enough, but the main reason why B chromosomes are presently attracting so much attention is on account of their widespread distribution and their potential as an experimental system.

II. Distribution

Information relating to numbers of B-chromosome-containing species has been given by Melander (1950), Makino (1951), and White (1973) for animals, by Darlington and Wylie (1955) and Federov (1969) for plants, and by Battaglia (1964b) for plants and animals. For animals the number is small, 116 in the present survey (Table I). They occur mainly in the Coleoptera and Orthoptera. The Orthoptera are especially suitable for cytological studies, and this no doubt is reflected in the extent to which B chromosomes have been found in this group. Their discovery in mammalian species is recent, as indeed is the method of chromosome analysis based on blood culture, and only six species are known to carry them. In plants, B chromosomes occur much more extensively. Darlington and Wylie (1955) list the chromosome numbers of over 17,000 species of flowering plants, of which 139 (0.8%) have Bs. Federov (1969) lists chromosome numbers of more than 35,000 species of flowering plants of which 381 (1.1%) contain Bs. In Battaglia's survey, over 200 species of flowering plants with Bs were listed, together with a few animals and some mosses. The present survey (Tables I and II) records the occurrence of B chromosomes in 591 species from 219 genera of flowering plants, of which 6 are gymnosperms and the rest angiosperms. They are about equally divided between the dicotyledons (297) and the monocotyledons (294). The number is slightly overestimated in so far as polyploid members of some species have been recorded separately. However, we have to remember that many chromosome counts are based on but a single individual of a species, and thus many B chromosomes have probably been overlooked. Add to this the fact that a large number of both plant and animal species have not as yet been examined cytologically at all, and it becomes apparent that the true number of B-chromosome-containing species is probably far in excess of that listed in Table I. In

TABLE I

B CHROMOSOMES IN FLOWERING PLANT AND ANIMAL SPECIES[a]

Flowering plants

Family, genus, and species	Ploidy	2*n*	Number of Bs[b]	Reference
Gymnosperms				
Cupressaceae				
Cupressus glabra	2×	22	1	Hunziker (1961)
Ephedraceae				
Ephedra foliata	2×	14	+B	Rao (1968)
Pinaceae				
Picea obvata	2×	24	0–3	Kruklis (1971)
Picea sitchensis	2×	24	0–2	Moir and Fox (1972)
Taxaceae				
Taxus canadensis	2×	24	1	Dark (1932)
Taxodiaceae				
Sequoia sempervirens	6×	66	0–1	Saylor *et al.* (1970)
Angiosperms—Dicotyledons				
Boraginaceae				
Anchusa azurea	4×	32	1–4	Britton (1951)
Onosma albo-roseum	3×	21	1	Teppner (1971)
Onosma albo-roseum	6×	42	1	Teppner (1971)
Onosma javorkae	2×	14	3–4	Teppner (1971)
Campanulaceae				
Campanula cenisia	2×	34	3	Podlech and Damboldt (1963)
Campanula cochleariifolia	2×	34	0–3	Gadella (1964); Gadella and Kliphius (1970)
Campanula ficarioides	6×	102	2–3	Geslot and Medus (1971)
Campanula hispanica	?	72	0–1	Geslot and Medus (1971)
Campanula latifolia	2×	34	3–5	Gadella (1964)
Campanula oblongifolia	?	90	2	Podlech and Damboldt (1963)
Campanula persicifolia	2×	16	4	Böcher (1964)
Campanula rotundifolia	4×	68	0–4	Böcher (1960, 1963); Gadella (1964)
Campanula rotundifolia	6×	102	0, 3, 5	Geslot and Medus (1971)
Campanula serrata	2×	34	2	Podlech and Damboldt (1963)
Phyteuma spicatum	2×	22	1–4	Ochlewska (1965)
Caprifoliaceae				

TABLE I (*Continued*)

Family, genus, and species	Ploidy	2*n*	Number of Bs	Reference
Flowering plants				
Sambucus adnata	2×	38	0–1	Mehra and Bawa (1968)
Sambucus canadensis	2×	38	0–2	Mehra and Bawa (1968)
Chenopodiaceae				
Salicornia herbacea	4×	36	2	Maude (1940)
Compositae				
Achillea asplenifolia	2×	18	+B	Ehrendorfer (1957, 1959, 1961)
Achillea collina	4×	36	+B	Ehrendorfer (1959, 1961)
Achillea crithmifolia	2×	18	+B	Ehrendorfer (1959, 1961)
Achillea millefolium	6×	54	+B	Ehrendorfer (1960, 1961)
Achillea roseo-alba	2×	18	+B	Ehrendorfer (1959, 1961)
Achillea setacea	2×	18	1–6	Ehrendorfer (1960, 1961)
Ageratum paleaceum[c]	2×	22	1	Turner *et al.* (1962)
Anthemis ruthenica	2×	18	0–3	Mitsuoka and Ehrendorfer (1972)
Aspilia africana[c]	?	68	2	Turner and Lewis (1965)
Aster ageratoides	2×	18	0–6	Matsuda (1970a,b)
Aster ageratoides	4×	36	0–4	Matsuda (1970a,b)
Aster ageratoides	6×	54	0–5	Matsuda (1970a,b)
Aster scaba	2×	18	0–14	Matsuda (1970a,c)
Aster semianplexicaulis	2×	18	1–4	Matsuda (1964)
Astranthium purpurascens	4×	16	1–2	De Jong (1965)
Bahia xylopoda	?	22	1	Powell and Turner (1963)
Bidens taitensis	4× − 2	46	1	Turner and Lewis (1965)
Brachycome curvicarpa	2×	8	4–5	Smith-White *et al.* (1970)
Brachycome lineariloba	2×	4	0–22	Smith-White (1968); Smith-White and Carter (1970); Smith-White *et al.* (1970)
Brachycome marginata	2×?	16	0–4	Smith-White *et al.* (1970)

(*Continued*)

TABLE I (*Continued*)

Flowering plants				
Family, genus, and species	Ploidy	2*n*	Number of Bs	Reference
Brachycome marginata	3×?	24	0–3	Smith-White *et al.* (1970)
Brachycome nivalis	2×?	18	1–2	Smith-White *et al.* (1970)
Brachycome papillosa	2×	8	2	Smith-White *et al.* (1970)
Carduus crispus	2×	16	1–4	Górecka (1956)
Carduus defloratus	2×	20	0–3	Favarger and Küpfer (1970)
Centaurea arguta	?	30	0–1	Bramwell *et al.* (1971)
Centaurea behen	?	36	3	Federov (1969)
Centaurea carduiformis	2×	20	+B	Federov (1969)
Centaurea huettii	?	40	2	Federov (1969)
Centaurea phaeopappoides	?	26	3	Federov (1969)
Centaurea rhenana	2×	18	0–2	Skalińska *et al.* (1959)
Centaurea scabiosa	2×	20	0–22	Fröst (1948, 1956, 1957, 1958a,b)
Centaurea subciliaris	2×	18	2	Phitos and Damboldt (1971)
Centaurea subciliaris	4×	36	0, 4	Phitos and Damboldt (1971)
Centaurea taochia	?	30	+B	Federov (1969)
Chrysanthemum corymbosum	2×	18	0, 2	Dowrick (1952)
Chrysanthemum heterophyllum	8×	72	0, 3	Favarger (1963)
Chrysanthemum ircutianum	4×	36	0–1	Favarger (1963)
Chrysanthemum millefolianum	2×	18	0–1	Dowrick (1952)
Chrysanthemum montanum	6×	54	0–3	Favarger (1963)
Cirsium acaule	2×	34	6	Moore and Frankton (1962)
Cirsium fontinale	2×	34	1	Moore and Frankton (1963)
Cirsium hookerianum	2×	34	2	Moore and Frankton (1965)

TABLE I (*Continued*)

Flowering plants				
Family, genus, and species	Ploidy	2*n*	Number of Bs	Reference
Cirsium scopulorum	2×	34	2	Moore and Frankton (1965)
Crepis capillaris	2×	6	0–4	Abraham *et al.* (1968); Barthelmes and Bauchinger (1962); Röthlisberger (1970); Rutishauser (1960a, 1963); Rutishauser and Röthlisberger (1966); Schweizer (1973)
Crepis conyzaefolia	2×	8	0–8	Fröst and Östergren (1959); Fröst (1962, 1964)
Crepis pannonica	2×	8	0–3	Fröst and Östergren (1959); Fröst (1960)
Crepis sancta	?	?	+B	Nazarova (1969)
Crepis sibirica	2×	10	0–2	Swezy (1935)
Crepis syriaca	2×	10	0–8	Cameron (1934); Babcock (1947)
Dimorphotheca pluvialis	?	17	+B	Gupta (1969)
Franseria chenopodiifolia	4×	72	2	Payne *et al.* (1964)
Gnaphalium luteo-album	2×	14	0–1	Larsen (1960a)
Grindelia arenicola	4×	24	1	Raven *et al.* (1960)
Grindelia rubricaulis	2×	12	1	Raven *et al.* (1960)
Haplopappus gracilis	2×	4	0–6	Jackson (1960); Jackson and Newmark, (1960); Östergren and Fröst (1962); Pritchard (1968)
Haplopappus spinulosus	2×	8	0–4	Raven *et al.* (1960); Li and Jackson (1961)
Haplopappus validus	2×	14	0–6	Smith (1968)
Hedypnois rhagadioloides	?	12	1	Larsen (1956)
Inula graveoleus	?	20	0–2	Nilsson and Lassen (1971)

(*Continued*)

TABLE I (*Continued*)

Flowering plants				
Family, genus, and species	Ploidy	$2n$	Number of Bs	Reference
Kalimeris dentata-incisa[c]	8×	72	2	Shindo (1965)
Kalimeris pseudo-yomena[c]	7×	63	2	Shindo (1965)
Leontodon hispidus	2×	14	1	Bergman (1935)
Leucanthemum croaticum	6×	54	0–2	Papeš (1971)
Leucanthemum croaticum	8×	71, 72	1–2	Papeš (1971)
Leucanthemum leucolepis	4×	36	1	Papeš (1971)
Leucanthemum liburnicum	8×	72	1–2	Papeš (1971)
Leucanthemum montanum	5×	45	1–2	Papeš (1971)
Leucanthemum montanum	8×	70–73	0–2	Papeš (1971)
Leucanthemum species hybrid	7×	63	2–10	Papeš (1971)
Leucanthemum vulgare	4×	36	1–8	Papeš (1971)
Leucanthemum vulgare	6×	54	1–2	Papeš (1971)
Liatris pycknostachya	2×	20	1–3	Gaiser (1949)
Machaeranthera canescens	2×	8	1	Solbrig *et al.* (1964)
Matricaria maritima	2×	18	0–2	Mulligan (1959)
Melampodium flaccidum	?	50	1	Turner and King (1964)
Olearia argophylla	12×	108	1	Solbrig *et al.* (1964)
Parthenium argentatum	2×	36	0–5	Bergner (1946); Catcheside (1950)
Parthenium argentatum	3×	54	+B	Bergner (1946); Catcheside (1950)
Parthenium argentatum	4×	72	+B	Bergner (1946); Catcheside (1950)
Perityle hofmeisteria[c]	?	32	1	Turner and Flyr (1966)
Perityle microglossa[c]	?	34, 36	1	Turner and Flyr (1966)
Perityle microglossa[c]	?	92	8	Turner *et al.* (1962)
Polymnia maculata	?	66	1	Wells (1965)
Pseudoclappia arenaria[c]	?	36	1	Powell and Turner (1963)
Rudbeckia serotina	2×	38	1	Mulligan (1959)
Schkuhria pinnata[c]	4×	40	1	Turner *et al.* (1962)
Senecio adonidifolius	8×	40	2	Palmblad (1965)
Senecio capitatus	?	96	2	Federov (1969)

TABLE I (*Continued*)

Flowering plants				
Family, genus, and species	Ploidy	2*n*	Number of Bs	Reference
Senecio confusus[c]	?	90	5	Turner *et al.* (1962)
Senecio foetidus	?	40	1	Palmblad (1965)
Senecio fremontii	?	40	1	Federov (1969)
Senecio hieronymi	2×	20	0–4	Afzelius (1959)
Senecio pancicalyailatus[c]	?	42	1	Turner and Lewis (1965)
Senecio petasites	?	60	6	Ornduff *et al.* (1963)
Silphium trifoliatum	?	14	1–4	Fisher and Cruden (1962)
Solidago arguta	2×	18	0–4, 6	Beaudry (1963)
Solidago altissima	6×	54	0–2, 5	Beaudry (1963); Beaudry and Chabot (1959)
Solidago bicolor	2×	18	0–1	Beaudry (1963)
Solidago canadensis	2×	18	0–3	Beaudry and Chabot (1959)
Solidago curtisii	2×	18	0–6, 10	Beaudry (1963)
Solidago fistulosa	2×	18	0–10	Beaudry (1963)
Solidago gigantea	2×	18	0–4	Beaudry (1963); Beaudry and Chabot (1959)
Solidago gigantea	4×	36	1	Beaudry and Chabot (1959)
Solidago graminifolia	2×	18	2	Beaudry (1963)
Solidago hispida	2×	18	2	Kapoor and Beaudry (1966)
Solidago lepida	2×	18	0, 1	Beaudry and Chabot (1959)
Solidago leptocephala	6×	54	1	Kapoor and Beaudry (1966)
Solidago macrophylla	2×	18	0–2	Beaudry and Chabot (1959)
Solidago nemoralis	2×	18	0–3	Beaudry and Chabot (1959)
Solidago puberula	2×	18	0–3	Beaudry (1963)
Solidago purshii	2×	18	0–2, 4	Beaudry and Chabot (1959)
Solidago racemosa	6×	54	0–2	Beaudry (1963)
Solidago randii	4×	36	0–1	Beaudry (1963)
Solidago roanensis	2×	18	0–1	Beaudry (1963)
Solidago sempervirens	2×	18	0–1	Beaudry and Chabot (1959)

(*Continued*)

TABLE I (*Continued*)

Flowering plants				
Family, genus, and species	Ploidy	2*n*	Number of Bs	Reference
Taraxacum alpicola	3×	24	2	Takemoto (1954)
Taraxacum japonicum	2×	16	2	Takemoto (1954)
Taraxacum officinale	3×	24	2	Takemoto (1954)
Taraxacum platycarpum	2×	16	2	Takemoto (1954)
Vernonia karwinskiana[c]	?	34	1	Turner *et al.* (1962)
Wulffia baccata[c]	?	60	2	Turner and Irwin (1960)
Xanthisma texanum	2×	8	0–4	Berger and Witkus (1954); Berger *et al.* (1955, 1956); Semple (1972)
Zaluzania montagnifolia[c]	?	34	+B	Powell and Turner (1963)
Zinnia acerosa[c]	?	20, 38	2	Torres (1962)
Zinnia haageana[c]	?	22	2	Powell and Turner (1963)
Zinnia tenella	?	22	1–2	Powell and Turner (1963)
Convolvulaceae				
Ipomoea aquatica	2×	30	1	Krishnappa (1971)
Cornaceae				
Aucuba japonica	4×	32	2	Yamamoto (1937); Viinikka (1970)
Cornus rugosa	2×	22	0–1	Clay and Nath (1971)
Crassulaceae				
Aichryson pachycaulon[c]	4×	60	2	Uhl (1961)
Kalanchoe calycinum	2×	34	4	Warden (1958, 1959)
Kalanchoe gastonis-bonnieri	2×	34	1–4	Uhl (1948); Warden (1959)
Kalanchoe pinnata[c]	2×	40	+B	Warden (1959)
Sempervivum montanum[c]	?	44	+B	Uhl (1961)
Cruciferae				
Arabis alpina[c]	2×	16	1	Titz (1967)
Arabis divarica	?	13, 20	2	Mulligan (1964)
Arabis holboellii	2×	14	+B	Mulligan (1964); Packer (1964)
Arabis holboellii	3×	21	2	Böcher and Larsen (1950); Mulligan (1964)
Arabis procurrens	2×	16	1–2	Burdet (1967)
Arabis procurrens	3×	24	2	Burdet (1967)

TABLE I (*Continued*)

Flowering plants				
Family, genus, and species	Ploidy	2*n*	Number of Bs	Reference
Arabis vochinensis	2×	16	1–2	Burdet (1967)
Cochlearia anglica	8×	48	+B	Saunte (1955)
Cochlearia officinalis	4×	24	+B	Saunte (1955)
Cochlearia pyrenaica	2×	12	0–4	J. J. B. Gill (1971a)
Cochlearia scotia	2×	24	0–4	J. J. B. Gill (1971b)
Diplotaxis muralis	2×	18	2	Baez-Major (1934)
Diplotaxis tenuifolia	?	20	2	Baez-Major (1934)
Draba norvegica	6×	48	4	Böcher (1966)
Iberis saxatile	2×	22	2	Manton (1932)
Iberis semperflorens	2×	22	1	Manton (1932)
Lunaria annua	4×	28	2	Manton (1932)
Lunaria rediva	4×	28	2	Manton (1932)
Matthiola incana	2×	14	1–2	Lesly and Frost (1928)
Sisybrium officinale	2×	14	4	Baez-Major (1934)
Cucurbitaceae				
Melothria maderaspatana	2×	22	1–2	Kumar and Vishveshwaraiah (1951)
Epacridaceae				
Brachyloma preissii	2×	14	0–1	Darlington and Wylie (1955)
Leucopogon oldfieldii	2×	22	0–2	Darlington and Wylie (1955)
Leucopogon revolutus	2×	22	0–3	Darlington and Wylie (1955)
Geraniaceae				
Geranium erianthum	2×	28	1–2	Shimizu (1971)
Geranium tripartitum	2×	28	2	Shimizu (1971)
Geranium wilfordi	2×	28	2	Shimizu (1971)
Labiatae				
Lamium album	2×	18	1	Gill (1970)
Mentha spicata[c]	8×	48	2	Ruttle (1931)
Pycnanthemum flexuosum	?	36	1	Chambers (1961)
Pycnanthemum pilosum	?	78	1	Chambers (1961)
Salvia plebeia	2×	16	1	L. S. Gill (1971a)
Scutellaria repens	2×	20	1	Gill (1970)
Lauraceae				
Beilschmiedia gammieana	2×	24	0–2	Mehra and Bawa (1968)
Neolitsea zeylanica	4×	48	0–5	Mehra and Bawa (1968)

(*Continued*)

TABLE I (*Continued*)

Flowering plants				
Family, genus, and species	Ploidy	2*n*	Number of Bs	Reference
Leguminosae				
Vicia faba	2×	12	0–2	Singh and Singh (1966)
Lobeliaceae				
Lobelia appendiculata	2×	14	0–1	Bowden (1959)
Lobelia brevifolia	2×	14	0–2	Bowden (1960a)
Lobelia cardinalis	2×	14	0–3	Bowden (1960b)
Lobelia elongata	4×	28	0–1	Bowden (1960a)
Lobelia feayana	2×	14	0–1	Bowden (1959)
Lobelia flaccidifolia	2×	14	0–1	Bowden (1960a)
Lobelia glandulosa	4×	28	0–2	Bowden (1960a)
Lobelia puberula	2×	14	0–2	Bowden (1960a)
Lobelia siphilitica	2×	14	0–1	Bowden (1960b)
Lobelia spicata	2×	14	0–3	Bowden (1959)
Loranthaceae				
Phoradendron californicum	2×	28	2–8	Wiens (1964a)
Phoradendron lanatum	2×	28	3–4	Wiens (1964a)
Phoradendron puberulum	2×	28	2	Wiens (1964a,b)
Phoradendron robinsonii	2×	28	2–3	Wiens (1964a,b)
Phoradendron velutinum	2×	28	3	Wiens (1964a)
Phoradendron bolleanum	2×	28	2	Wiens (1964a)
Malvaceae				
Hibiscus vitifolius[c]	2×	34	1	Skovsted (1941)
Napaea dioica	4×	28	1	Skovsted (1935)
Sida rhombifolia	4×	28	0–8	Hazra and Sharma (1971)
Sidalcea candida	?	20	1	Skovsted (1935)
Moraceae				
Ficus krishnae	2×	26	0–2	Joshi and Raghuvanshi (1970)
Onagraceae				
Clarkia amoena	2×	14	+B	Snow (1963)
Clarkia dudleyana	2×	18	0–2	Snow (1960)
Clarkia elegans	2×	18	0–8	Lewis (1951); Lewis and Lewis (1955); Mooring (1960)
Clarkia gracilis	4×	28	0–6	Håkansson (1945, 1950); Lewis and Lewis (1955)
Clarkia lassenensis	2×	14	+B	Lewis and Lewis (1955)

TABLE I (*Continued*)

Flowering plants				
Family, genus, and species	Ploidy	2*n*	Number of Bs	Reference
Clarkia purpurea	?	52	+B	Lewis and Lewis (1955)
Clarkia rhomboidea	?	24	1	Lewis and Lewis (1955)
Clarkia virgata	2×	10	3	Lewis and Lewis (1955); Small (1971)
Clarkia williamsonii	2×	18	0–7	Håkansson (1949); Lewis and Lewis (1955); Wedberg *et al.* (1968)
Oenothera hookeri	2×	14	0–3	Cleland (1951, 1967, 1972); Cleland and Hyde (1963)
Oenothera scintillans	2×	14	+B	Hance (1918a,b)
Oenothera serrulata	2×	14	2, 4	Kurabayashi *et al.* (1962)
Papilionaceae				
Gliricidia sepium	2×	22	0–1	Rao (1972)
Medicago granadensis	2×	16	1	Heyn (1963)
Medicago intertexia	2×	16	1	Heyn (1963)
Medicago rotata	2×	16	1	Heyn (1963)
Medicago sativa	4×	28	0–6	Murray and Craig (1964)
Onobrychis arenaria[c]	2×	14	2	Sacristán (1966)
Tephrosia conzattii	2×	22	1–2	Wood (1949)
Trifolium berytheum	2×	16	0–7	Putiievsky and Katznelson (1970)
Trifolium pratense	2×	14	1	Skovsted (1939)
Trifolium salmoneum	2×	16	0–7	Putiievsky and Katznelson (1970)
Plantaginaceae				
Plantago bigelovii	?	20	1	Bassett (1966)
Plantago coronopus	2×	10	1	Paliwal and Hyde (1958, 1959)
Plantago lanceolata	2×	12	1	Soyano (1959)
Plantago mohnikei	4×	24	2	Matsuura and Suto (1935)
Plantago montana	4×	24	1	Czapska (1959)
Plantago serraria	2×	10	0–3	Böcher *et al.* (1955); Fröst (1959)
Plantago serraria	4×	20	2	Böcher *et al.* (1955)

(*Continued*)

TABLE I (*Continued*)

Flowering plants				
Family, genus, and species	Ploidy	2*n*	Number of Bs	Reference
Polemoniaceae				
Phlox amoena	2×	14	0–1	Smith and Levin (1967)
Phlox bifida	2×	14	0–3	Smith and Levin (1967)
Phlox carolina-heterophylla	2×	14	0, 2	Meyer (1944)
Phlox divarica	2×	14	0–6	Meyer (1944); Levin (1967); Smith and Levin (1967)
Phlox glaberrima	2×	14	0–3	Meyer (1944); Smith and Levin (1967)
Phlox nivalis	2×	14	0, 2–4	Meyer (1944)
Phlox paniculata	2×	14	0–8, 10	Meyer (1944)
Phlox pilosa	2×	14	0–3	Smith and Levin (1967)
Phlox subulata	2×	14	0–13	Meyer (1944)
Phlox suffruticosa	2×	14	0–1	Meyer (1944)
Polygonaceae				
Rumex acetosa	2×	14	0–10	Haga (1961)
Portulacaceae				
Claytonia virginica	?	28	0–5	Lewis (1970); Lewis *et al.* (1971)
Primulaceae				
Primula atricapilla	2×	20	1–3	Bruun (1930, 1931, 1932)
Primula cernua	2×	20	3	Bruun (1931, 1932)
Primula chionantha	2×	22	2–7	Bruun (1930, 1931, 1932)
Primula chionantha	4×	44	6–12	Bruun (1931)
Primula crispa	4×	44	+B	Bruun (1930, 1931, 1932)
Primula denticulata	2×	22	5	Bruun (1930, 1931)
Primula erythrocarpa	2×	22	5	Bruun (1930, 1931, 1932)
Primula jesoana	2×	12	1	Matsuura and Sutô (1935)
Ranunculaceae				
Anemone hepatica	2×	14	2	Suda (1962)
Caltha palustris	4×	32	0–6	Kootin-Sanwu and Woodell (1970)
Caltha palustris	7×	56	0–6	Kootin-Sanwu

TABLE I (*Continued*)

Flowering plants				
Family, genus, and species	Ploidy	$2n$	Number of Bs	Reference
				(1966); Kootin-Sanwu and Woodell (1971)
Caltha radicans	6×	48	2	Löve and Löve (1948)
Hepatica nobilis	2×	14	2	Kurita (1961)
Lycoctonum gigas	2×	16	1–6	Kurita (1961)
Ranunculus acris	2×	14	0–10	Langlet (1927); Walker and Gregson (1966); Fröst (1969a)
Ranunculus ficaria	2×	16	0–7	Larter (1932); McLeish (1954); Walker and Gregson (1966); Marchant and Brighton (1971); Gill *et al.* (1972)
Ranunculus neapolitanus	?	?	+B	Walker and Gregson (1966)
Ranunculus polyanthemus	2×	16	0, 4, 6, 9	Andersson (1958); Böcher (1958)
Ranunculus repens	?	?	+B	Walker and Gregson (1966)
Thalictrum aquilegiefolium	2×	14	+B	Langlet (1927)
Rhamnaceae				
Pomaderris kumeraho	?	12	1	Hair (1963)
Pomaderris phylicifolia	?	24	1	Hair (1963)
Rosaceae				
Rosa spinosissima	4×	28	1	Darlington and Wylie (1955)
Salicaceae				
Salix seringeana[c]	2×	38	1–4	Almeida (1946)
Saxifragaceae				
Heuchera americana	2×	14	4	Skovsted (1934)
Heuchera sanguinea	2×	14	1	Skovsted (1934)
Heuchera villosa	2×	14	1	Skovsted (1934)
Scrophulariaceae				
Collinsia solitaria	2×	14	0–6	Garber (1958); Dhillon and Garber (1962)

(*Continued*)

TABLE I (*Continued*)

Flowering plants				
Family, genus, and species	Ploidy	2*n*	Number of Bs	Reference
Hebe insularis	?	20	1	Hair (1967)
Linaria panciciic[c]	2×	12	4	Heitz (1927)
Mecardonia acuminata[c]	?	42	2	Lewis *et al.* (1962)
Rhinanthus alectorolophus	2×	14	8	Tschermak-Woess and Hasitschka-Jenschke (1963)
Rhinanthus angustifolius[c]	2×	14	6	Hambler (1954)
Rhinanthus hirsutus[c]	2×	14	6	Hambler (1962)
Rhinanthus major	2×	14	8	Fagerlind (1936); Witsch (1950); Wulff (1939)
Rhinanthus minor	2×	14	8	Hambler (1954, 1955, 1958, 1962); Löve (1954); Löve and Löve (1956, 1961)
Rhinanthus serotinus[c]	2×	14	8	Löve and Löve (1956)
Solanaceae				
Nicandra physaloides	2×	20	1	L. S. Gill (1971b)
Petunia parodii	2×	14	1	Sullivan (1947)
Solanum pseudocapsicum	2×	24	+B	L. S. Gill (1971b)
Symplocaceae				
Symplocos glomerata	2×	22	0–2	Mehra and Bawa (1968)
Umbelliferae				
Heracleum lanatum	2×	22	1	Matsuura and Sutô (1935)
Verbenaceae				
Clerodendrum colebrookianum	2×	26	0–2	Mehra and Bawa (1968)
Violaceae				
Viola montana	4×	40	10	Schmidt (1961)
Viola rupestris[c]	2×	20	4–8	Schmidt (1961)
Viola striata	2×	20	2	Clausen (1929)
Angiosperms—Monocotyledons				
Agavaceae				
Hosta coerlea	?	48	+B	Cave (1948)
Amaryllidaceae				
Cooperia brasiliensis[c]	5–6×	69	+B	Traub (1945)
Crinum graminicola	2×	22	3–4	Jones and Smith (1967)

TABLE I (*Continued*)

Flowering plants				
Family, genus, and species	Ploidy	2*n*	Number of Bs	Reference
Crinum longifolium[c]	2×	22	2	Inariyama (1937)
Crinum pedicellatum	2×	22	1–2	Jones and Smith (1967)
Crinum pedicellatum	4×	44	2	Jones and Smith (1967)
Galanthus angustifolius	2×	24	0–3	Sveshnikova (1971)
Galanthus byzantinus	2×	24	0–3	Sveshnikova (1971)
Galanthus corcyrensis	2×	24	0–3	Sveshnikova (1971)
Galanthus elwesii	2×	24	0–3	Sveshnikova (1971)
Galanthus graecu	2×	24	0–3	Sveshnikova (1971)
Galanthus nivalis	2×	24	0–3	Sveshnikova (1971)
Galanthus plicatus	2×	24	0–3	Sveshnikova (1971)
Galanthus reginae-olgae	2×	24	0–3	Sveshnikova (1971)
Haemanthus albiflorus[c]	2×	16	+B	Sâto (1942)
Haemanthus angolensis	2×	16	2	Bronkers (1961)
Haemanthus albreykeri	2×	18	+B	Sharma and Bal (1956)
Hippeastrum equestre[c]	2×	22	+B	Mookerjea (1955)
Hymenocallis harvisiana	?	22	+B	Sharma and Bal (1956)
Lycoris incarnata	?	29	1	Bose (1958, 1961)
Lycoris radiata	?	32	1	Bose (1963)
Narcissus angustifolius	2×	14	1–2	Federov (1969)
Narcissus bernardi	2×	14	1	Wylie (1952)
Narcissus bulbocodium	2×	14	0–5	Fernandes (1943, 1948, 1949); Fernandes and Mesquita (1963); Wylie (1952)
Narcissus calcida	2×	14	1–2	Wylie (1952)
Narcissus cyclamineus	2×	14	1	Wylie (1952)
Narcissus juncifolius	2×	14	1	Fernandes (1939)
Narcissus pseudo-narcissus	2×	14	1–2	Wylie (1952)
Narcissus romieuxii	3×	30	2	Fernandes (1959a,b)
Narcissus tazetta	2×	20	0–3	Weitz and Feinbrun (1972)
Zephyranthes braziliensis	?	69	+B	Traub (1963)
Araceae				
Anthurium crystallinum	2×	30	0–2	Pfitzer (1957a,b)
Anthurium forgetii	2×	30	0–2	Pfitzer (1957a,b)
Anthurium magnificum	2×	30	0–2	Pfitzer (1957a,b)

(*Continued*)

TABLE I (*Continued*)

Flowering plants				
Family, genus, and species	Ploidy	2*n*	Number of Bs	Reference
Bromeliaceae				
Tillandsia pulchella	2×	50	1–2	Marchant (1967)
Commelinaceae				
Commelina benghalensis	2×	22	1–2	Malik (1961)
Cyanotis axillaris	2×	20	+B	Islam and Baten (1952)
Cyanotis cristata	2×	24	+B	Islam and Baten (1952)
Tinantia erecta	2×	34	2	Jones and Jopling (1972)
Tradescantia canaliculata	4×	24	+B	Giles (1941)
Tradescantia crassifolia	2×	12	2	Darlington (1929)
Tradescantia crassifolia	4×	24	2–3	Jones and Jopling (1972)
Tradescantia edwardsiana	2×	12	0–9	Brown (1960); Evans (1956)
Tradescantia fluminensis	?	67	1	Jones and Jopling (1972)
Tradescantia paludosa	2×	12	1–12	Anderson and Sax (1936); Whitaker (1936a); Swanson (1943); Giles (1941)
Tradescantia pedicellata	2×	12	0–1	Celarier (1956)
Tradescantia virginiana	4×	24	1–6	Darlington (1929); Koller (1932); Vosa (1962, 1965)
Tripogandra diuretica	?	62	+B	Jones and Jopling (1972)
Cyperaceae				
Carex bootiana	?	62	+B	Tanaka (1939a,b)
Eriophorum schenchzeri	2×	58	2B	Mosquin and Hayley (1966)
Gramineae				
Aegilops columnaris	4×	28	0–2	Chennaveeraiah and Löve (1959)
Aegilops cylindrica	4×	28	0–1	Chennaveeraiah and Löve (1959)
Aegilops mutica	2×	14	0–5	Mochizuki (1957, 1960); Dover and Riley (1972)

TABLE I (*Continued*)

Flowering plants				
Family, genus, and species	Ploidy	$2n$	Number of Bs	Reference
Aegilops speltoides	2×	14	0–6	Simchen *et al.* (1971); Dover and Riley (1972); Mendelson and Zohary (1972); Zarchi *et al.* (1972)
Agropyron cristatum	2×	14	0–8	Baenziger (1962)
Agropyron cristatum	4×	28	?	Knowles (1955)
Agropyron desertorum	4×	28	0–11	Baenziger (1962); Baenziger and Knowles (1962); Baenziger and Carr (1968)
Agropyron imbricatum	4×	28	0–6	Baenziger (1962); Baenziger and Knowles (1962)
Agrostis canina	4×	28	0–13	Björkman (1951); Jones (1956)
Agrostis castellana	4×	28	0–4	Björkman (1954, 1960)
Agrostis gigantea	6×	42	0–4	Björkman (1954)
Agrostis nevadensis	6×	42	0–10	Björkman (1954, 1960)
Agrostis pourretii	2×	14	0–2	Björkman (1960)
Agrostis rupestris	2×	14	0–2	Björkman (1951, 1960)
Agrostis rupestris	4×	28	0–1	Björkman (1954, 1960)
Agrostis stolonifera	6×	42	0–2	Björkman (1954)
Agrostis tenuis	4×	28	+B	Björkman (1954)
Agrostis trinii	2×	14	2	Federov (1969)
Alopecurus alpinus	?	112+	1–3	Flovik (1938, 1940)
Alopecurus pratensis	4×	28	0–9	Johnsson (1941); Bosemark (1957b); Rapp (1972)
Andropogan gryllus[c]	?	20	1	Mehra *et al.* (1962)
Andropogan gayanus	4×	40	2–4	Singh (1965)
Anthoxanthum alpinum	2×	10	1–2	Rozmus (1958, 1963)
Anthoxanthum aristatum	2×	10	0–4	Östergren (1947); Jones (1964)
Anthoxanthum nivale	4×	20	2	Hedberg (1952)
Avena versicolor	2×	14	1–2	Skalińska (1956)

(*Continued*)

TABLE I (*Continued*)

Flowering plants				
Family, genus, and species	Ploidy	2*n*	Number of Bs	Reference
Bouteloua curtipendula	?	40	0–1	Gould (1959)
Brachiaria plantaginea	4×	36	2	Pohl and Davidse (1971)
Brachiaria ramosa	4×	36	10	Singh (1965)
Briza elatior	2×	14	1–3	Federov (1969)
Briza media	2×	14	0–4	Bosemark (1957b)
Bromus cappadocicus	8×	56	4	Schulz-Schaeffer (1956)
Bromus erectus	6×	42	4	Schulz-Schaeffer (1956)
Bromus inermis	6×	42	20	Schulz-Schaeffer (1956)
Bromus inermis	8×	56	0–11	Hill and Myers (1948); Nielsen (1955)
Bromus scabratus	4×	28	8	Schulz-Schaeffer (1956)
Calamagrostis breweri	4×	28	0, 2	Nygren (1954)
Calamagrostis hakonensis	6×	42	1	Tateoka (1972b)
Calamagrostis hakonensis	8×	56	0–1	Tateoka (1972b)
Calamagrostis koeleri-oides	4×	28	2	Nygren (1954)
Calamagrostis longiseta	4×	28	0–1	Tateoka (1972a)
Catapodium rigidum	?	15–20	?	Singh (1965)
Chionachne koengii	2×	20	0–3	Venkateswarlu *et al.* (1965)
Coix aquatica	2×	10	0–2	Venkateswarlu *et al.* (1965, 1968)
Cymbogon flexuosus	?	20	2	Gupta (1965)
Cymbopogon caesius	?	20	2	Babu (1936)
Cynodon dactylon	4×	36	1–2	Burton (1947); Gould (1966)
Dactylis glomerata	2×	14	0–6	Zohary and Ashkenazi (1958); Jones (1962); Jones and Borrill (1962); Shah (1963, 1964a,b, 1965, 1967); Carroll and Borrill (1965); Borrill and Carroll (1969); Puteyevsky and Zohary (1970)

TABLE I (*Continued*)

Flowering plants				
Family, genus, and species	Ploidy	$2n$	Number of Bs	Reference
Dactylis glomerata	4×	28	0–3	Zohary and Nur (1959); Jones (1962); Borrill and Carroll (1969)
Dendrobium brandisii	6×	72	2	Darlington and Wylie (1955)
Deschampsia alpina	7×	49	2	Jörgensen *et al.* (1958)
Deschampsia bottnica	2×	26	0–2	Albers (1972)
Deschampsia caespitosa	2×	26	0–7	Tateoka (1955); Kawano (1966); Albers (1972)
Deschampsia wibeliana	2×	26	0–2	Albers (1972)
Digitaria decumbens	4×	36	1	Shambulingappa (1968)
Digitaria pentzii	4×	36	3	Shambulingappa (1968, 1970)
Digitaria valida	4×	36	1	Shambulingappa (1968)
Dupontia fisheri	?	44	+B	Flovik (1938, 1940)
Dupontia fisheri	?	88	+B	Flovik (1938, 1940)
Erianthus ravennae[c]	2×	20	1	Janaki-Ammal (1941)
Festuca arundinacea	6×	42	0–3	Crowder (1953); Bosemark (1957b); Malik and Thomas (1966); Borrill *et al.* (1971)
Festuca arundinacea	8×	56	0–4	Borrill *et al.* (1971)
Festuca mairei	4×	28	0–6	Malik and Thomas (1966); Malik and Tripathi (1970)
Festuca polesica	2×	14	0–4	Malik and Thomas (1966)
Festuca pratensis	2×	14	0–21	Bosemark (1950, 1954a,b, 1956a,b, 1957a); Borrill *et al.* (1971)
Festuca rubra	6×	72	+B	Flovik (1938, 1940)
Helictotrichon schellianum	2×	14	1	Sadanaga (1962)
Hemarthria subulata	2×	18	2	Larsen (1963)
Holcus lanatus	2×	14	0–3	Bosemark (1957b); Böcher and Larsen (1958)

(*Continued*)

TABLE I (*Continued*)

Flowering plants				
Family, genus, and species	Ploidy	$2n$	Number of Bs	Reference
Iseilema laxum	2×	28	0–6	Murty (1972); Murty and Satyavathi (1972)
Koeleria alpicola	6×	42	1	Singh and Godward (1963); Singh (1965)
Koeleria cristata	2×	14	1	Tateoka (1955)
Koeleria pubescens	2×	12	0–1	Larsen (1960b)
Lolium perenne	2×	14	0–3	Cameron and Rees (1967); Evans and Macefield (1972, 1973)
Lolium persicum	2×	14	+B	Hovin and Hill (1966)
Lolium remotum	2×	14	+B	Hovin and Hill (1966)
Lolium rigidum	2×	14	0–2	Hovin and Hill (1966); Dahlgren *et al.* (1971)
Lolium strictum	2×	14	+B	Hovin and Hill (1966)
Miscanthus floridulus	2×	38	0–11	Price (1963)
Miscanthus japonicus	2×	38	0, 3	Li and Ma (1950)
Oryzopsis hymenoides	4×	48	0–8	Johnson (1963)
Panicum coloralum	4×	36	0–3	Swaminathan and Nath (1956); Hutchison and Bashaw (1963)
Panicum maximum	4×	32	0–5	Jauhar (1967); Jauhar and Joshi (1969)
Panicum nehruense	6×	54	1–8	Jauhar and Joshi (1968)
Paspalum stoloniferum	2×	20	1–3	Avdulov and Titova (1933)
Pennisetum orientale	2×	18	0–3	Jauhar and Singh (1970)
Pennisetum typhoides	2×	14	0–8	Pantulu (1960); Powell and Burton (1966); Venkateswarlu and Pantulu (1970); Pantulu and Manga (1972)
Phleum montanum	2×	14	+B	Nath and Nielson (1963)

TABLE I (*Continued*)

Flowering plants				
Family, genus, and species	Ploidy	2*n*	Number of Bs	Reference
Phleum montanum	4×	28	+B	Nath and Nielson (1963)
Phleum nodosum	2×	14	0–4	Bosemark (1957b); Fröst (1969a)
Phleum phleoides	2×	14	0–8	Böcher (1950); Bosemark (1956c, 1967)
Phleum phleoides	4×	28	1–2	Böcher (1950)
Phleum pratense	6×	42	1	Nath and Nielson (1961)
Poa alpigena	6×	42	4–5	Flovik (1938, 1940)
Poa alpina	2×	14	0–8	Håkansson (1948b, 1954); Müntzing (1946b, 1948a, 1966); Müntzing and Nygren, (1955); Milinkovic (1957)
Poa alpina	3×	21	?	Håkansson (1948a)
Poa alpina	6×	42	4	Flovik (1938, 1940)
Poa badensis	2×	14	0–1	Nygren (1962)
Poa bulbosa	6×	42	+B	Hartung (1946)
Poa glaucifolia	?	50	0–1	Hartung (1946)
Poa pratensis	5×	35	3	Löve and Löve (1948)
Poa pratensis	9×+	68	1	Hartung (1946)
Poa pratensis	12×	84	1	Skovsted (1939)
Poa scabrella	?	44	0–1	Hartung (1946)
Poa scabrella	?	84	+B	Hartung (1946)
Poa subfastigiata	?	91–92	+B	Hartung (1946)
Poa timoleontis	2×	14	+B	Nygren (1962)
Poa trivialis	2×	14	0–5	Bosemark (1957b)
Poa xerophila	2×	14	0–4	Nygren (1962)
Secale africanum	2×	14	0–2	Emme (1928)
Secale cereale	2×	14	0–8	Nakao (1911); Belling (1925); Gotoh (1924, 1932); Lewitsky (1931); Darlington (1933); Hasegawa (1934); Müntzing and Prakken (1941); Müntzing (1943, 1944, 1945, 1946a, 1948b,c, 1949,

(*Continued*)

TABLE I (*Continued*)

Flowering plants				
Family, genus, and species	Ploidy	2*n*	Number of Bs	Reference
				1950, 1951, 1954a, 1963, 1966, 1970); Müntzing and Akdik (1948); Håkansson (1948a, 1957, 1959); Kishikawa (1962, 1963, 1965, 1968, 1970); Lima-de-Faria (1948, 1949, 1955, 1962); Moss (1966); Fröst (1963); Lee (1963, 1965, 1968)
Secale cereale	3×	21	4	Kishikawa (1966)
Secale cereale	4×	28	0–12	Sarvella (1959); Müntzing (1963, 1966)
Secale fragile	2×	14	0, 2	Emme (1928)
Secale montanum	2×	14	0, 2	Emme (1928)
Secale vavilovii	2×	14	2	Sun (1963); Kranz (1968, 1971)
Sohnsia filifolia	?	20	0–1	Singh (1972)
Sorghum purpureo-sericeum	2×	10	0–6	Janaki-Ammal (1939, 1941); Darlington and Thomas (1941); Garber (1950)
Sorghum verticilliflorum[c]	4×	20	+B	Huskins and Smith (1934)
Spartina pectinata	4×	40	1	Marchant (1968)
Tripsacum dactyloides	?	36	0–4	Chandravadana *et al.* (1970)
Triticum aestivum	6×	42	0–10	Lindström (1965); Müntzing (1970); Müntzing *et al.* (1968, 1969)
Zea mays	2×	20	0–34	Kuwada (1925); Longley (1927, 1938, 1956); Randolph (1928, 1941); McClintock (1933);

TABLE I (*Continued*)

Flowering plants				
Family, genus, and species	Ploidy	2*n*	Number of Bs	Reference
				Humphrey (1935); Darlington and Upcott (1941); Roman (1947, 1948a,b); Catcheside (1956); Blackwood (1956); Rhoades *et al.* (1967); Rhoades and Dempsey (1972, 1973); Himes (1967); Hanson (1969); Carlson (1970, 1973a,b); Ayonoadu and Rees (1968a, 1971); Ward (1973a,b); Nel (1973)
Iridaceae				
Babiana stricta	2×	14	0–1	Zucconi (1957)
Crocus hymenalis	2×	6	0–4	Mather (1932); Karasawa (1940); Feinbrun (1958)
Crocus hymenalis	4×	12	8	Karasawa (1950)
Cypella herbretii	2×	14	1	Covas and Schnack (1947)
Iris pumila	?	30, 31	1	Randolph and Mitra (1961)
Watsonia ardernei	2×	16	2	Riley (1962)
Juncaceae				
Luzula campestris	2×	12	0–2	Câmara *et al.* (1958, 1959); Noronha-Wagner and Castro (1952)
Liliaceae				
Agapanthus orientalis	2×	32	2	Mukerjee and Riley (1961); Riley and Mukerjee (1962)
Allium allegheniense	2×	14	0–4	Levan (1932)
Allium angulosum	2×	16	0–1	Shopova (1966)
Allium angulosum	4×	32	0–1	Shopova (1966)

(*Continued*)

TABLE I (*Continued*)

Flowering plants				
Family, genus, and species	Ploidy	2*n*	Number of Bs	Reference
Allium bimetrale	2×	16	0–4	Bothmer (1970)
Allium carinatum	3×	24	0–3	Shopova (1966)
Allium cepa	2×	16	1	Noda (1953)
Allium cernum	2×	14	0–11	Grun (1959)
Allium flavum	2×	16	2	Cheshmedzhirev (1971)
Allium nutans	2×	16	0–1	Shopova (1966)
Allium nutans	3×	24	0–1	Shopova (1966)
Allium nutans	4×	32	0–1	Shopova (1966)
Allium nutans	5×	40	0–1	Shopova (1966)
Allium paniculatum	2×	16	+B	Ved Brat (1965)
Allium porrum	2×	16	6	Nybom (1947)
Allium porrum	4×	32	0–3	Vosa (1966)
Allium pulchellum	2×	16	0–3	Tschermak-Woess and Schiman (1960)
Allium senescens	4×	32	0–1	Shopova (1966)
Allium sphaerocephalum	2×	16	0–1	Bothmer (1970)
Allium stracheyi	2×	14	2–10	Sharma and Aiyanger (1961)
Allium thunbergii	2×	16	0–4	Noda and Watanabe (1968)
Allium thunbergii	4×	32	0–6	Noda and Watanabe (1961, 1968)
Allium thunbergii	6×	48	0–4	Noda and Watanabe (1968)
Calochortus luteus	2×	14	+B	Beal and Ownbey (1943)
Dipadi serotinum	2×	8	2–16	Resendre and Da Franca (1946); Fernandes *et al.* (1948)
Fritillaria amabilis	2×	22	+B	Noda (1964)
Fritillaria biflora	2×	24	1–8	Snow (1959)
Fritillaria falcata	2×	26	0–1	La Cour (1951)
Fritillaria imperialis	2×	24	0–12	La Cour (1951)
Fritillaria japonica	2×	22	0–2	Noda (1968)
Fritillaria lanceolata	2×	26	0–8	Beetle (1944); La Cour (1951)
Fritillaria nigra	2×	18	0–3	La Cour (1951)
Fritillaria obliqua	2×	24	2	La Cour (1951)
Fritillaria pudica	2×	26	0–1	Beetle (1944)
Fritillaria recurva	2×	26	0–1	Beetle (1944); La Cour (1951)

TABLE I (*Continued*)

Flowering plants				
Family, genus, and species	Ploidy	2*n*	Number of Bs	Reference
Lilium auratum	2×	24	0–2	Stewart (1947); Ogihara (1960, 1962)
Lilium batemanniae	2×	24	1B	Stewart (1943)
Lilium callosum	2×	24	0–5	Kayano (1956a,b, 1957, 1962a,b); Kimura and Kayano (1961)
Lilium canadense	2×	24	1B	Stewart (1943)
Lilium caucasicum	2×	24	+B	Federov (1969)
Lilium formosanum[c]	2×	24	2	Hall (1934)
Lilium henryi	2×	24	1–2	Hall (1934); Mather (1934, 1935); Stewart (1947)
Lilium japonicum	2×	24	1–2	Hall (1934); Mather (1934, 1935); Stewart (1943)
Lilium martagon	2×	24	1–3	Fernandes (1950)
Lilium maximowiczii	2×	24	0–2	Noda (1955, 1956, 1967)
Lilium medeoloides	2×	24	0–11	Matsuura and Sutô (1935); Samejima (1958)
Lilium pumilum	2×	24	1–2	Stewart (1943)
Lilium sargentiae	2×	24	1B	Stewart (1947)
Lilium tsingtauense	2×	24	1B	Stewart (1943)
Lilium willmottiae	2×	24	0–1	Beal (1942); Stewart (1943)
Muscari latifolium	2×	18	1–2	Federov (1969)
Muscari polyanthum	2×	18	1–2	Federov (1969)
Ornithogalum caudatum	?	54	+B	Riley (1962)
Ornithogalum concinnum	2×	36	0–6	Neves (1952)
Ornithogalum ecklonii	?	16	4	Neves (1962)
Ornithogalum flavissimum	2×	12	5	Piennaar (1963)
Ornithogalum gussonei	?	20	+B	Czapik (1965)
Ornithogalum narbonense	6×	54	0–11	Neves (1952)
Ornithogalum pyrenaicum	2×	16	0–3	Neves (1952)
Ornithogalum umbellatum	2×	18	0–6	Neves (1952); Giménez Martin (1958)
Ornithogalum umbellatum	3×	27	0–1	Neves (1952)
Ornithogalum unifolium	2×	34	0–1	Neves (1952)
Puschkinia libanotica	2×	10	0–7	Vosa (1969a); Barlow and Vosa (1969a,b, 1970)

(*Continued*)

TABLE I (*Continued*)

Flowering plants				
Family, genus, and species	Ploidy	$2n$	Number of Bs	Reference
Scilla autumnalis	2×	14	0–8	Battaglia (1963, 1964a)
Scilla scilloides	2×	16	1	Haga (1961)
Scilla scilloides	3×	27	2	Haga and Noda (1956); Haga (1961)
Scilla scilloides	4×	36	5	Haga and Noda (1956); Haga (1961)
Scilla scilloides	5×	44	1	Haga and Noda (1956)
Scutellaria repens	2×	20	2	Gill (1970)
Streptopus japonicus	2×	16	2	Matsuura and Sutô (1935)
Tulipa borszczonii	2×	24	2–7	Hall (1937); Woods and Bamford (1937)
Tulipa galacta	2×	24	1–19	Hall (1937)
Tulipa montana	2×	24	1–2	Woods and Bamford (1937)
Tulipa mucronata	2×	24	1	Federov (1969)
Triteleia ixioides	2×	10	1	Burbanck (1941)
Urginea aurantiaca	2×	10	1	Battaglia (1958)
Urginea fugax	2×	20	0–8	Battaglia (1957, 1964c); Battaglia and Guanti (1966, 1968)
Urginea indica	2×	20	0–4	Raghavan and Venkatasubban (1940)
Urginea rubella	8×	40	+B	De Wet (1957)
Veratrum stamineum	4×	32	2	Matsuura and Sutô (1935)
Musaceae				
Musa acuminata	2×	22	0–1	Govindaswami (1965)
Musa cavendishii	4×	32	2	Matsuura and Sutô (1935)
Orchidaceae				
Dendrobium cretaceum	2×	38	1	Jones (1963)
Dendrobium crumenatum	2×	38	1	Jones (1963)
Dendrobium densiforme	?	40	2	Kosaki (1958)
Dendrobium hildebrandii	2×	38	1	Kosaki (1958); Jones (1963)

TABLE I (*Continued*)

Flowering plants				
Family, genus, and species	Ploidy	2*n*	Number of Bs	Reference
Dendrobium monile	2×	38	1–3	Jones (1963)
Dendrobium moschatum	2×	38	3	Kamemoto and Sagarik (1967)
Epipactis atropurpurea	4×	40	3–11	Meili-Frei (1965)
Epipactis atropurpurea	6×	60	2	Meili-Frei (1965)
Listera borealis	3×	51	+B	Simon (1968)
Listera convallarioides	2×	34	2	Simon (1968)
Listera cordata	2×	34	2	Simon (1968)
Listera ovata	2×	34	0–2	McMahon (1936); Meili-Frei (1965); Vosa (1969b); Vosa and Barlow (1972)
Phalaenopsis amabilis	?	69	3	Sagawa (1962)
Pheione pricei	?	20	1	La Cour (1952)
Tainia laxiflora	2×	36	0–9	Tanaka (1965); Tanaka and Matsuda (1972)
Trilliaceae				
Paris formosana	2×	10	2	Gotoh and Kikkawa (1937)
Paris polyphylla	2×	10	0–2	Darlington (1941); Darlington and Shaw (1959); Larsen (1963)
Paris tetraphylla	2×	10	0–9	Haga (1961); Kayano (1961)
Trillium cernum	2×	10	0–9	Darlington and Shaw (1959); Dyer (1964)
Trillium chloropetalum	2×	10	0–9	Dyer (1964)
Trillium declinatum	2×	10	3	Gotoh (1937)
Trillium erectum	2×	10	0–9	Sparrow *et al.* (1952); Darlington and Shaw (1959); Dyer (1964)
Trillium grandiflorum	2×	10	0–9	Rutishauser (1956a,b, 1960b); Darlington and Shaw (1959); Dyer (1964)
Trillium luteum	2×	10	0–9	Darlington and Shaw (1959); Dyer (1964)

(*Continued*)

TABLE I (*Continued*)

Flowering plants				
Family, genus, and species	Ploidy	2*n*	Number of Bs	Reference
Trillium ovatum	2×	10	0–1	Darlington and Shaw (1959)
Trillium rivale	2×	10	0–9	Dyer (1964)
Trillium sessile	2×	10	0–9	Darlington and Shaw (1959); Dyer (1964)
Zingiberaceae				
Zingiber macrostachyum	2×	22	2	Ramachandran (1969)
Zingiber officinale	2×	22	0–2	Darlington and Wylie (1955)

Animals			
Group, genus, and species	2×	Number of Bs	Reference
Platyhelminths			
Polycelis tenuis	12	0–4	Melander (1950)
Rhynchodemus terrestris	?	?	Melander (1950)
Molluscs			
Helix pomatia	54	0–6	Evans (1960)
Insects			
Aphaniptera			
Nosopsyllus fasciatus	20	0–7	Bayreuther (1969)
Coleopterans			
Acalymma blandulum	18 + X♂	+B	Smith (1972)
Acalymma innubum	18 + X♂	+B	Smith (1972)
Acalymma trivittatum	20 + X♂	+B	Smith (1972)
Amplelasma cavum vicinum	16 + X♂	+B	Smith (1972)
Calligrapha alni	22 + X♂	0–2	Robertson (1966)
Calligrapha bidenticola	22 + X♂	0–1	Robertson (1966)
Calligrapha californica coropsivora	22 + X♂	0–4	Robertson (1966)
Calligrapha multipunctata bigsbyana	22 + X♂	0–3	Robertson (1966)
Calligrapha philadelphica	22 + X♂	0–10	Robertson (1966)
Calligrapha rowena	22 + X♂	12	Robertson (1966)
Cerotoma atrofasciata	32	+B	Smith (1972)
Chilocorus anglolensis	16 + XY♂	+B	Smith (1960)
Chilocorus rubidus	16 + XY♂	+B	Smith (1960)

TABLE I (*Continued*)

Animals			
Group, genus, and species	2×	Number of Bs	Reference
Chilocorus stigma	22 + XXY♂	+B	Smith (1960)
Cleis hudsonica	10 + XY♂	0, 2	Smith (1953)
Diabrotica lemniscta	18 + X♂	0, 2, 3	Ennis (1972a); Smith (1972)
Diabrotica longicornis barberi	18 + X♂	0, 2–4	Ennis (1972a); Smith (1972)
Diabrotica longicornis nigricornis	18 + X♂	+B	Smith (1972)
Diabrotica longicornis longicornis	18 + X♂	+B	Smith (1972)
Diabrotica 12-punctata	18 + X♂	0–4	Stevens (1908)
Diabrotica scutellata	18 + X♂	+B	Smith (1972)
Diabrotica soror	18 + X♂	0–4	Stevens (1908)
Diabrotica tibialis	18 + X♂	+B	Smith (1972)
Diabrotica undecimpunctata howardi	18 + X♂	0–6	Ennis (1972a); Smith (1956, 1972)
Diabrotica undecimpunctata 11-punctata	18 + X♂	+B	Smith (1972)
Diabrotica undecimpunctata tenella	18 + X♂	+B	Smith (1972)
Elaphidion parallelus	16 + XY♂	+B	Smith (1960)
Epicometis hirta	18 + XY♂	+B	Virkki (1954)
Exochomus lituratus	16 + XY♂	0–4	Smith (1965, 1966)
Exochomus quadripustulatus	12 + XY♂	+B	Smith (1960)
Exochomus uropygialis	16 + XY♂	0–6	Smith (1965, 1966)
Gelus californicus	12 + X♂	0–4	Ennis (1972b)
Gynandrobrotica nigrofasciata	16 + X♂	+B	Smith (1972)
Oryctes nasicornis	18	+B	Virkki (1954)
Pyractomena angulata	18 + X♂	0–4	White (1973)
Tribolium madus	18 + XY	+B	Smith (1956)
Diptera			
Anopheles maculipennis	6	0–1	Belcheva and Mihailova (1971)
Anopheles messeae	6	0–1	Belcheva and Mihailova (1971)
Chironomus melanotus	8	1	Keyl and Hägele (1971)
Chironomus plumosus	8	1	Keyl and Hägele (1971)
Hylemya cana	12	0–1	Boyes and Van Brink (1965)
Hylemya cilicrura	12	0–3	Boyes (1954); Boyes and Van Brink (1965)

(*Continued*)

TABLE I (*Continued*)

Animals			
Group, genus, and species	2×	Number of Bs	Reference
Odagmia ornata	?	0–8	Shcherbakov (1966)
Phryne cincta	8 + XY♂	1–7	Wolf (1954, 1961)
Tipula paludosa	8	0–4	Bauer (1931)
Hemiptera			
Antonia pertiosa	10	+B	Ferris, cited in Nur (1962a)
Cimex lectularius	26 + XXY♂	2–14	Ueshima (1967); Darlington (1940)
Orthocephalus fumestus	30 + XY♂	+B	Takenouchi and Muramoto (1972)
Metapodius femoratus	22	0–4	Wilson (1907)
Metapodius granulosus	22	0–5	Wilson (1907)
Metapodius terminalis	22	0–3	Wilson (1907)
Pseudococcus obscurus	10	0–6	Nur (1962a,b, 1966a,b, 1968, 1969b)
Heteroptera			
Anisops niveus	26	2	Jande (1961)
Lepidoptera			
Elbella lamprus	78	1	De Lesse and Brown (1971)
Orthoptera			
Acrida lata	22 + X♂	0–4	Kayano and Sannomiya (1964); Kayano *et al.* (1960, 1970); Sannomiya (1963); Sannomiya and Kayano (1968, 1969)
Aerochoreutes carlinianus	?	+B	White (1973)
Atractomorpha australis	?	+B	White (1973)
Atractomorpha bedeli	18 + X♂	0–7	Sannomiya (1964); Sannomiya and Kayano (1968, 1969)
Atractomorpha crenaticeps	?	+B	White (1957)
Calliptamus palaestinensis	22 + X♂	0–4	Nur (1963)
Camnula pellucida	22 + X♂	0–4	Carroll (1920); Nur (1969a)
Chortoicetes terminifera	22 + X♂	0–1	Hewitt and John (1971)

TABLE I (*Continued*)

Animals			
Group, genus, and species	2×	Number of Bs	Reference
Chrotogonus incertus	18 + X♂	0–2	Srivastava (1954)
Circotettix lobatus	22 + X♂	0–2	Carothers (1917)
Circotettix rabula	?	0–2	White (1951b)
Circotettix thalassinus	?	+B	White (1973)
Circotettix undulatus	?	+B	White (1973, 1951b)
Crytobothrus chrysophorus	?	+B	White (1973)
Euprepocnennis alacris	22 + X♂	+B	Chatterjee *et al.* (1971)
Euprepocnennis roseus	22 + X♂	0–1	Chatterjee *et al.* (1971)
Gelastorrhinus bicolor	22 + X♂	0–13	Sannomiya (1964)
Gonista bicolor	?	+B	Kayano (1971)
Hesperotettix viridis	22 + X♂	1	McClung (1917)
Locusta danica	22 + X♂	0–4	Itoh (1934)
Locusta migratoria	22 + X♂	0–5	Kayano (1971); Nur (1969a); Rees and Jamieson (1954)
Mecostethus grossus	22 + X♂	1	Callan (1941)
Melanoplus borealis stupefactus	?	+B	White (1973)
Melanoplus differentialis	?	+B	Abdel-Hameed *et al.* (1970)
Melanoplus femur-rubrum	22 + X♂	0–5	Stephens and Bregman (1972)
Melanoplus sanguinipes	?	+B	White (1973)
Moraba carissima	?	+B	White (1973)
Moraba viatica	16	+B	White *et al.* (1964)
Myrmeleotettix maculatus	16 + X♂	0–3	Barker (1960, 1966); Gibson and Hewitt (1970, 1972); Hewitt (1972, 1973); Hewitt and Brown (1970); Hewitt and John (1967); Hewitt and Ruscoe (1971); John and Hewitt (1965a,b); Ramel (1969); Southern (1967)
Neopodismopsis abdominalis	18 + X♂	0–3	Rothfels (1950)
Netrosoma fusiforme	?	+B	White (1973)
Oedaleonotus enigma	20 + XY♂	0–2	Hewitt and Schroeter (1968)

(*Continued*)

TABLE I (*Continued*)

Animals			
Group, genus, and species	2X	Number of Bs	Reference
Pantanga japonica	22 + X♂	0–3	Sannomiya (1962, 1964)
Pezotettix giorni	?	+B	Hewitt and John (1972)
Phaulacridium vittatum	22 + X♂	0–2	Jackson and Cheung (1967)
Phaulotettix eurycerуis	?	+B	White (1973)
Phlaeoba infumata	?	+B	White (1973)
Podisma pedestris	22 + X♂	0–2	Hewitt and John (1972)
Stenobothrus lineatus	?	+B	Hewitt and John (1972)
Tettigidea lateralis	12 + X♂	0–2	Fontana and Vickery (1973)
Tetrix ceperoi	12 + X♂	0–2	Henderson (1961)
Trimerotropis cyancipennis	?	+B	White (1973)
Trimerotropis diversellus	?	+B	White (1951b, 1973)
Trimerotropis gracilis	?	+B	White (1973)
Trimerotropis inconspicua	?	+B	White (1973)
Trimerotropis latifasciata	?	1	White (1951b)
Trimerotropis sparsa	?	0–4	White (1951a,b)
Trimerotropis suffusa	22 + X♂	0–3	Carothers (1917); Wenrich (1917); White (1973)
Amphibians			
Acris crepitans	22	0–5	Nur and Nevo (1969)
Leiopelma hochstetteri	24	0, 2, 4	Stephenson *et al.* (1972)
Rana temporaria	26	0–4	Ullerich (1967)
Reptiles			
Tropidurus torquatus	36	0–1	Beçak *et al.* (1972)
Mammals			
Echymipera kalabu	12 + XY/X0	0–5	Hayman *et al.* (1969)
Perognathus baileyi	46–56	0–10	Patton (1972)
Rattus rattus diardii	42	0–4	Yong and Dhaliwal (1972)
Reithrodontomys megalotis	42	0–7	Blanks and Shellhammer (1968); Shellhammer (1969)

TABLE I (*Continued*)

Animals			
Group, genus, and species	2×	Number of Bs	Reference
Schoinobates volans	22	2–6	Hayman and Martin (1965, 1969)
Vulpes vulpes	36	0–4	Buckton and Cunningham (1971); Lin *et al.* (1972)

[a] Classification of the flowering plants into families, and some of the information on ploidy level and basic number, is based on Darlington and Wylie (1955).
[b] Where more than one reference is quoted, the data have been pooled.
[c] Data taken, with references, from Federov (1969). The remainder of the list has been compiled by reference to original papers.

fact, a similar line of reasoning led Darlington (1956a) to estimate that B chromosomes are likely to occur in as many as 10% of angiosperms.

The distribution of B-containing species among families of angiosperms is interesting. The situation is summarized in Table II together with data on their occurrence in relation to polyploidy. Evidently, Bs are much more common in some of the 55 families listed than they are in others. In the dicotyledons, for instance, the Compositae account for almost half (124) of the 306 species. The other half being shared fairly regularly among the other 35 families. Fourteen families of monocotyledons have B-carrying species, and they are concentrated largely in the Gramineae (120), Liliaceae (81), and Amaryllidaceae (31). One interesting question that arises from this situation is that of a possible relationship between B-chromosome occurrence and the degree of evolutionary advancement of angiosperm families. As it happens, the Gramineae in the monocotyledons and the Compositae in the dicotyledons are both highly evolved families. They contrast markedly with other more primitive families like the Malvaceae, Rosaceae, and Juncaceae, which are very low in Bs. But we also have to remember that the Liliaceae, Amaryllidaceae, and Compositae have all been extensively studied on the grounds of cytological suitability, while the Gramineae are of considerable economic importance. Then again, there is the problem of disparity among families, in terms of the numbers of cytologically known species. This can be taken account of though, if we express the figures for numbers of B-containing species per family (from Table II) as a percentage of the number of cytologically known species within a family, which can be roughly estimated from Darlington and Wylie (1955). For what the

TABLE II

CLASSIFICATION OF B-CONTAINING SPECIES ACCORDING TO FAMILY AND PLOIDY LEVEL FOR ANGIOSPERMS, AND ACCORDING TO MAJOR GROUPINGS FOR ANIMALS

Flowering plants									
	Number of species and ploidy level								
Family	2×	3×	4×	5×	6×	7×	8×	?	Σ
Gymnosperms									
Cupressaceae	1	—	—	—	—	—	—	—	1
Ephedraceae	1	—	—	—	—	—	—	—	1
Pinaceae	2	—	—	—	—	—	—	—	2
Taxaceae	1	—	—	—	—	—	—	—	1
Taxodiaceae	—	—	—	—	1	—	—	—	1
Total	5	—	—	—	1	—	—	—	6
Angiosperms—dicotyledons									
Boraginaceae	1	1	1	—	1	—	—	—	4
Campanulaceae	6	—	1	—	2	—	—	2	11
Caprifoliaceae	2	—	—	—	—	—	—	—	2
Chenopodiaceae	—	—	1	—	—	—	—	—	1
Compositae	62	3	13	1	7	2	5	31	124
Convolvulaceae	1	—	—	—	—	—	—	—	1
Cornaceae	1	—	1	—	—	—	—	—	2
Crassulaceae	3	—	1	—	—	—	—	1	5
Cruciferae	11	2	3	—	1	—	1	2	20
Cucurbitaceae	1	—	—	—	—	—	—	—	1
Epacridaceae	3	—	—	—	—	—	—	—	3
Geraniaceae	3	—	—	—	—	—	—	—	3
Labiatae	3	—	—	—	—	—	1	2	6
Lauraceae	1	—	1	—	—	—	—	—	2
Lobeliaceae	8	—	2	—	—	—	—	—	10
Loranthaceae	—	—	—	—	—	—	—	6	6
Malvaceae	1	—	2	—	—	—	—	1	4
Moraceae	1	—	—	—	—	—	—	—	1
Onagraceae	9	—	1	—	—	—	—	2	12
Papilionaceae	10	—	4	—	—	—	—	—	14
Plantaginaceae	3	—	3	—	—	—	—	1	7
Polemoniaceae	10	—	—	—	—	—	—	—	10
Polygonaceae	1	—	—	—	—	—	—	—	1
Portulacaceae	—	—	—	—	—	—	—	1	1
Primulaceae	6	—	2	—	—	—	—	—	8
Ranunculaceae	7	—	1	—	1	1	—	2	12
Rhamnaceae	—	—	—	—	—	—	—	2	2
Rosaceae	—	—	1	—	—	—	—	—	1
Salicaceae	1	—	—	—	—	—	—	—	1
Saxifragaceae	3	—	—	—	—	—	—	—	3

TABLE II (*Continued*)

Flowering Plants

Family	Number of species and ploidy level								
	2×	3×	4×	5×	6×	7×	8×	?	Σ
Scrophulariaceae	8	—	—	—	—	—	—	2	10
Solanaceae	3	—	—	—	—	—	—	—	3
Symplocaceae	1	—	—	—	—	—	—	—	1
Umbelliferae	1	—	—	—	—	—	—	—	1
Verbenaceae	1	—	—	—	—	—	—	—	1
Violaceae	2	—	1	—	—	—	—	—	3
Total	174	6	39	1	12	3	7	66	297
Angiosperms—monocotyledons									
Agavaceae	—	—	—	—	—	—	—	1	1
Amaryllidaceae	24	1	1	—	1	—	—	4	31
Araceae	3	—	—	—	—	—	—	—	3
Bromeliaceae	1	—	—	—	—	—	—	—	1
Commelinaceae	8	—	3	—	—	—	—	2	13
Cyperaceae	1	—	—	—	—	—	—	1	2
Gramineae	52	2	32	—	15	1	6	12	120
Iridaceae	4	—	1	—	—	—	—	1	6
Juncaceae	1	—	—	—	—	—	—	—	1
Liliaceae	62	4	7	2	2	—	1	3	81
Musaceae	1	—	1	—	—	—	—	—	2
Orchidaceae	9	1	1	6	—	—	—	2	19
Trilliaceae	12	—	—	—	—	—	—	—	12
Zingiberaceae	2	—	—	—	—	—	—	—	2
Total	180	8	46	8	18	1	7	26	294
Angiosperm total	354	14	85	9	30	4	14	92	591

Animals

Group	Number of species (all diploid)
Platyhelminths	2
Molluscs	1
Insects	
Aphaniptera	1
Coleoptera	36
Diptera	9
Hemiptera	7
Heteroptera	1
Lepidoptera	1
Orthoptera	48
Amphibians	3
Reptiles	1
Mammals	6
Total	116

comparison is worth, it shows that for those families that have B-containing species the mean percentage of species with Bs per family is 8.9 for the dicotyledons and 9.3 for the monocotyledons. On this basis the Gramineae are about average with 8.8% B-carrying species, while the Liliaceae have 11.6% and the Amaryllidaceae 15.8%. Families with the highest percentage of +B species in the monocotyledons are the Trilliaceae (38.7%) and the Commelinaceae (20.3%). In the dicotyledons the Compositae are also above average with about 14.5% +B species. On this percentage basis, families with the least number of +B species are the Juncaceae and Cyperaceae in the monocotyledons, and the Rosaceae, Umbelliferae, and Polygonaceae in the dicotyledons.

Evidently then, the picture changes according to the way in which the data are presented. When the figures are given on a percentage basis, the Gramineae, for example, which have always been regarded as exceptional in having a very high number of B-containing species, can be seen to have, proportionately, no more than the average monocotyledon family.

Another well-entrenched tenet of B-chromosome philosophy is that Bs are rare in polyploids and favored more by the genetic system of short-lived herbaceous diploids. While it remains true that they are scarce in the woody plants, except for their recent discovery in six woody species from the Himalayas (Mehra and Bawa, 1968) and in the tree *Ficus krishnae* (Joshi and Raghuvanshi, 1970), the extent to which they occur in polyploids has been seriously underestimated. From the data given in Table II and the histogram in Fig. 1, it is evident that about one-third (31%) of all B-containing angiosperms are polyploid. Bearing in mind Stebbin's (1963) estimate that polyploids account for 30–35% of angiosperms, then proportionately Bs are as frequent in polyploids as they are in diploids. Indeed in some cases, as in *Leucanthemum* populations in

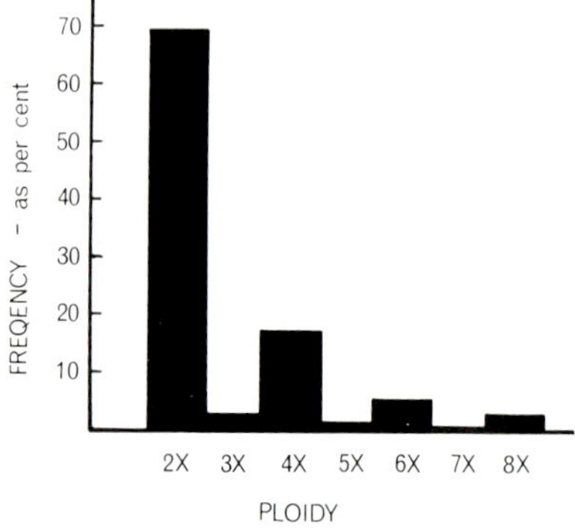

FIG. 1. Distribution of B-containing species according to ploidy level.

Yugoslavia, Bs are entirely restricted to polyploids. Papeš (1971) found B chromosomes in 11 out of 91 populations studied. The populations belonged to six varieties and five ploidy levels. In the diploids, which accounted for 20 of the 91 populations, Bs were not found at all. In other species the opposite can be found, for example, *Ranunculus ficaria* in Britain, where B chromosomes are restricted to diploids and not found in polyploids (Gill *et al.*, 1972).

B Chromosomes are rarely found in inbreeding species. Information assembled by Moss (1969) suggests that there is a strong correlation between the occurrence of B chromosomes and a tendency to outbreeding. The reasons for this have never been adequately explained, but what is known is that forced inbreeding of a naturally outcrossing species, with Bs, rapidly leads to a decline in their frequency. In rye inbred by repeated selection and intercrossing of 2B-containing plants over nine generations, the frequency of B chromosomes was reduced from an average of 2.2 to an average of 1.4 per plant (Müntzing, 1954b).

One of the more striking aspects of B-chromosome systems in both plant and animal species is the way in which Bs vary in frequency among individuals of a population as well as among different populations of a species. The situation is discussed at length in Section VI. For the present, suffice it to say that an idea of the extent of the variation in B numbers among individuals of a species can be obtained from the examples listed in Table III. Under experimental conditions, even higher numbers can be attained, up to 34 in maize (Randolph, 1941) and 22 in *Centauria scabiosa* (Fröst, 1958b).

The main reason for the extensive numerical variation in B-chromosome distribution within species is to be found in their irregular and non-Mendelian mechanism of inheritance. They are unstable at mitosis and/or meiosis, in addition to which they are often capable of undergoing directed segregation either at meiosis or in postmeiotic divisions in plants, thereby boosting their numbers in the functional microspore and megaspore cells. Preferential fertilization by B-containing gametes is also possible in plants. Various mechanisms of meiotic drive occur in some animal species. A detailed consideration of B-chromosome behavior during somatic cell division and during the meiotic cycle is the subject of the following section.

III. Inheritance

A. Somatic Cell Division

In many plants and animals the B chromosomes are completely stable during the cell cycle and are inherited in a constant and unchanging form

TABLE III

THE RANGE OF B CHROMOSOME NUMBERS IN SOME PLANT AND ANIMAL SPECIES

Genus and species	Number of B chromosomes																
	0	1	2	3	4	5	6	7	8	9	10	11	12	13	14	15	16
Centauria scabiosa in Scandinavia and Finland (Fröst, 1958a)	5688	611	564	333	218	128	63	64	39	8	15	1	1	1	2	1	1
Festuca pratensis in Sweden (Bosemark, 1956a)	854	97	68	8	11	3	3	2	—	—	—	—	—	—	—	—	—
Crepis conyzaefolia in Italy (Fröst, 1962)	133	31	11	6	1	—	—	—	—	—	—	—	—	—	—	—	—
Tainia laxiflora in Japan (Tanaka and Matsuda, 1972)	5	10	49	17	24	5	2	1	4	2	—	—	—	—	—	—	—
Trillium erectum from Hump Mountain (Sparrow *et al.*, 1952)	193	34	6	2	2	—	—	—	—	—	—	—	—	—	—	—	—
Ranunculus ficaria in Britain (Gill *et al.*, 1972)	45	8	12	15	5	2	9	3	—	—	—	—	—	—	—	—	—
Reithrodontomys megalotis, the harvest mouse (Shellhammer, 1969)	39	89	94	45	15	2	0	1	—	—	—	—	—	—	—	—	—
Myrmeleotettix maculatus, the mottled grass-hopper (Hewitt, 1972)	90	41	9	—	—	—	—	—	—	—	—	—	—	—	—	—	—

TABLE IV

Stable and Unstable B Chromosomes

Plants	Animals
Stable	
Alopecurus pratensis	*Acrida lata*
Anthoxanthum odoratum	*Chironomus plumosus*
Briza media	*Circotettix undulatus*
Catapodium rigidum	*Diabrotica undecimpunctata*
Clarkia elegans	*Myrmeleotettix maculatus*
Festuca pratensis	*Phaulacridium vittatum*
Holcus lanatus	*Podisma pedestris*
Luzula campestris	*Pyrgomorpha kraussi*
Phleum nodosum	*Pseudococcus obscururs*
Plantago serraria	*Trimerotropis sparsa*
Ranunculus polyanthemus	*Tettigidea lateralis*
Scilla scilloides	
Secale cereale	
Urginea fugax	
Unstable	
Aegilops mutica	*Acris crepitans*
Aegilops speltoides	*Atractomorpha bedeli*
Agropyron cristatum	*Calliptamus palaestinensis*
Agrostis canina	*Camnula pellucida*
Allium angulosum, A. cernum, A. nutans, A. senescens	*Echymipera kalabu*
Aster scaba	*Helix pomatia*
Brachiaria ramosa	*Locusta migratoria*
Claytonia virginica	*Neopodismopsis abdominalis*
Crepis capillaris, C. conyzaefolia, C. pannonica	*Pantanga japonica*
Dactylis glomerata	*Polycelis tenuis*
Ficus krishnae	
Haplopappus gracilis, H. spinulosus	
Iseilema laxum	
Lillium callosum	
Miscanthus floridulus	
Panicum coloratum, P. nehruense	
Paris teraphylla	
Pennisetum orientale, P. typhoides	
Poa alpina, P. timoleontis, P. trivialis	
Ranunculus acris, R. ficaria	
Rumex acetosa	
Sohnsia filifolia	
Sorghum purpureo-sericeum	
Tradescantia virginiana	
Urginea indica	
Xanthisma texanum	

TABLE V

DISTRIBUTION PATTERNS OF UNSTABLE B-CHROMOSOME SYSTEMS IN PLANTS[a]

Genus and species	B Chromosomes			Reference
	Present			
	Constant	Variable	Absent	
Aegilops mutica	Stem, integuments and anthers	—	Roots	Mochizuki (1957)
Aegilops speltoides	Tillers and spikes	—	Roots	Mendelson and Zohary (1972)
Agropyron cristatum	Pmcs, primary roots, and stem meristem	—	Adventitious roots	Baenziger (1962)
Agrostis canina	—	Tissues and organs throughout plant	—	Björkman (1951)
Claytonia virginica	—	Tissues and organs throughout plant	—	Lewis *et al.* (1971)
Allium angulosum	—	Within and between roots	—	Shopova (1966)
Allium nutans	—	Within and between roots	—	Shopova (1966)
Allium senescens	—	Within and between roots	—	Shopova (1966)
Aster scaba	—	Within and between roots	—	Matsuda (1970c)
Ranunculus acris	—	Within and between roots	—	Fröst (1969b)
Ranunculus ficaria	—	Within and between roots	—	Gill *et al.* (1972)
Rumex acetosa	—	Within and between roots	—	Haga (1961)
Allium cernum	—	Within roots, between pmcs	—	Grun (1959)
Ficus krishnae	—	Within roots, between pmcs	—	Joshi and Raghuvanshi (1970)
Brachiaria ramosa	Roots	—	Pmcs	Singh (1965)

Crepis capillaris	Rosette plant	Stem, bracts, young inflorescences, receptacles, and florets	—	Rutishauser and Röthlisberger (1966)
Crepis conyzaefolia	Somatic cells, pmcs	Between somatic cells and pmcs	—	Fröst (1962)
Crepis pannonica	Somatic cells, pmcs	Between somatic cells and pmcs	—	Fröst (1962)
Dactylis glomerata	Within tillers	Between tillers, between spikelets of a panicle	—	Puteyevsky and Zohary (1970)
Haplopappus gracilis	—	Shoot	Roots	Östergren and Fröst (1962)
Haplopappus spinulosus	Roots	Pmcs	—	Li and Jackson (1961)
Poa trivialis	Roots	Pmcs	—	Bosemark (1957b)
Iseilema laxum	—	Pmcs	—	Murty (1972)
Panicum coloratum	—	Pmcs	—	Swaminathan and Nath (1956)
Lillium callosum	—	Seedling	—	Kayano (1962a)
Miscanthus floridulus	—	Somatic cells and pmcs	—	Price (1963)
Panicum nehruense	—	Ears, anthers, and pmcs	—	Jauhar and Joshi (1968)
Pennisetum orientale	—	Ears, anthers, and pmcs	—	Jauhar and Singh (1970)
Pennisetum typhoides	—	Ears, anthers, and pmcs	—	Powell and Burton (1966)
Paris tetraphylla	—	Root tips and ovules	—	Haga (1961)
Poa alpina	—	Central part of primary roots and in germ cells	Leaves and adventitious roots	Müntzing and Nygren (1955)
Poa timoleontis	Pmcs	—	Roots	Nygren (1957)
Sohnsia filifolia	Pmcs	—	Roots	Singh (1972)
Sorghum puroureo-sericeum	Pmcs	Anther wall and ovaries	Roots and sterile flowers	Darlington and Thomas (1941)
Tradescantia virginiana	—	Leaves and stems	—	Vosa (1962)
Urginia indica	—	Within the plant	—	Raghaven and Venkatasubban (1940)
Xanthisma texanum	Shoots	—	Roots	Berger and Witkus (1954)

[a] Modified and enlarged after John and Lewis (1968).

along with the A chromosomes (Table IV). In many other species (Table IV) this is not the case, and they are less stable than the A chromosomes. They are mechanically less efficient, succumbing to loss through anaphase lagging and mitotic nondisjunction. The latter event leads to numerical variation within the individual, or part of the individual, and may be associated with a net gain or loss of Bs, depending on the stage of development at which it occurs and the way in which the dividing cells are subsequently distributed.

Some plant species have an apparent order in their instability, which results in exclusion and/or accumulation of B chromosomes in specific tissues and organs. *Crepis capillaris,* for example, regularly has a directed nondisjunction which is coincident with flower initiation, and this serves to boost the number of Bs destined for the germ line (Röthlisberger, 1970). In *Xanthisma texanum* the Bs are present in the cells of the young embryo, and then are eliminated in the cell lines destined to become roots by anaphase lagging at the late spherical stage in the embryo. Their number is constant in the shoots (Berger and Witkus, 1954). *Poa alpina* has Bs in primary roots and pollen mother cells (pmcs), but they are excluded from adventitious roots; they are absent from young primordia of adventitious roots in the stem interior (Milinkovic, 1957). The various kinds of distribution patterns of unstable B chromosomes in plants are summarized in Table V. There is a precedent of course for chromosome numerical variation in the somatic cells of plants, and this is the common occurrence of endopolyploidy in root meristems. Just as the advantage of polyploidy can be gained by individual cells in otherwise diploid organisms (Lewis, 1967), so it may be that the disadvantages of B chromosomes (see Section V,A) can be overcome in individual cells or even whole root systems by their selective elimination.

Information on unstable B-chromosome systems in animals appears to Table VI. Most of the variation that occurs is found among follicles within testes and is often associated with mechanisms of accumulation of Bs. In *Echymipera kalabu* (Hayman *et al.,* 1969), the instability of the Bs is interesting in that it parallels an instability in the A chromosomes themselves. This organism does not have the full complement of A chromosomes in all its adult somatic tissues. The chromosomes missing are the Y in the male and an X in the female. The full complement is present in the corneal epithelium and the reproductive tissue. The Bs are subject to the same control system as that responsible for the elimination of the sex chromosomes.

B. Meiosis

The B chromosomes are structurally homologous with one another and polysomic. They pair strictly *inter se* and have the capacity to form

TABLE VI

Distribution Pattern of Unstable B Chromosomes in Animals

Genus and species	Type of numerical variation	Reference
Atractomorpha bedeli	Inter- but not intrafollicular	Sannomiya and Kayano (1969)
Calliptamus palaestinensis	Inter- but not intrafollicular	Nur (1963)
Camnula pellucida	Inter- but not intrafollicular	Carroll (1920); Nur (1969a)
Neopodsimopsis abdominalis	Inter- but not intrafollicular	Rothfels (1950)
Melanoplus femur-rubrum	Inter- but not intrafollicular for the two telocentric Bs. The large metacentric B is mitotically stable	Stephens and Bregman (1972)
Locusta migratoria	Inter- and intrafollicular	Kayano (1971); Nur (1969a)
Pantanga japonnica	Inter- and intrafollicular	Sannomiya (1962)
Acris crepitans	Between cells of the testis	Nur and Nevo (1969)
Helix pomatia	Between primary spermatocytes	Evans (1960)
Rana temporaria	Intraindividual variation	Ullerich (1967)
Echymipera kalaba	Found in corneal epithelium and reproductive tissues; absent from adult somatic tissues. The mosaicism parallels that for the sex chromosomes	Hayman *et al.* (1969)
Polycelis tenuis	Elimination from somatic cells in adult animals. Multiplication in ovarial meiosis and in the testis where there is cell-to-cell variation	Melander (1950)

multivalents when more than two are present, although they seldom pair with the same efficiency and regularity as do A chromosomes.

The anaphase behavior of unpaired Bs is variable. They may divide at AI (*Zea mays, Festuca pratensis, Festuca arundinacea, Poa trivialis, Secale cereale, Luzula campestris*), AII (*Pyrgomorpha kraussi, Calliptamus palaestinensis, Holcus lanatus, Anthoxanthum aristatum, Helix pomatia, Myrmeleotettix maculatus, Lilium medeoloides, Cochlearia pyrenaica*), or at either AI or AII (*Locusta migratoria, Camnula pellucida, Pennisetum typhoides*). In most cases there is very little meiotic loss of univalent and unpaired Bs, but the question of loss is an important one since it is one of the means by which the frequency of B chromosomes in the population can be adjusted. There is also reason to believe that the loss rate is genetically determined. Östergren (1947) examined 219 tetrads in *Anthoxanthum aristatum* plants carrying a single B and found no cases of elimination. John and Hewitt (1965a) found no loss at all in *Myrmeleotettix maculatus,* and Nur (1963) likewise detected no lagging or elimination in hundreds of cells examined in *Calliptamus palaestinensis.* There is a very low rate of loss in *Allium cernum* (Grun, 1959), *Godetia nutans* (Håkansson, 1945), *Poa trivialis* (Bosemark, 1957b), *Festuca pratensis* (Bosemark, 1954b), and *Festuca arundinacea* (Bosemark, 1957b). In *Locusta migratoria,* however, Rees and Jamieson (1954) found that the single B may be lost in as many as 20% of meioses. Similarly in *Aegilops speltoides.* When the univalent B is outside the equatorial plate it lags, undergoes precocious division, and fails to be included in the daughter nuclei. It appears at the end of meiosis as micronuclei in 80–85% of pmcs. Thus the loss is appreciable (Mendelson and Zohary, 1972).

In some animal species, notably grasshoppers, the univalent B is often associated with the unpaired X in meiosis in males. *Tetrix ceperoi* (Henderson, 1961) is such a case, in which the univalent B was found to associate with the X in 23 out of 66 first metaphases, as a "quasi bivalent," perfectly coorientated. This X/B quasi bivalent ensures regular segregation of the B at AI, avoiding lagging and elimination. It also results in the B passing to the pole opposite the X, with a consequent preferential distribution of the Bs to the male line. Likewise, in *Phaulacridium vittatum* (Jackson and Cheung, 1967) the B and the X pass to opposite poles at AI in 70% of cases. In *Tettigidea lateralis,* however, there is a persistent heterochromatic association between the B and the X, which leads to preferential migration of the B with the X to the same pole in males with a single B (Fontana and Vickery, 1973). Other species in which the heterochromatic Bs are associated with the X during male meiotic prophase include *Neopodismopsis abdominalis* (Rothfels, 1950) and *Calliptamus palaestinensis* (Nur, 1963).

Univalent B chromosomes in *Lilium callosum* behave in an opportunistic way. They undergo a directed distribution which leads to their accumulation in functional egg cells. This case is discussed in Section III,C.

In many plant species the B chromosomes appear in a univalent condition at MI, irrespective of the number present, and divide as univalents at either AI or AII (Table VII). This applies particularly to those species with very small heterochromatic Bs. Some of these species do show associations between the Bs, but these associations are of a "sticky" nature rather than chiasmatic, for example, *Centauria scabiosa* (Fröst, 1956). In *Parthenium argentatum* the Bs are associated in early meiotic

TABLE VII

DIVISION STAGE OF NONPAIRING B CHROMOSOMES IN PLANTS

Stage AI		Stage AII	
Genus and species	Reference	Genus and species	Reference
Centauria scabiosa	Fröst (1956)	*Allium cernum*	Grun (1959)
Panicum coloratum	Swaminathan and Nath (1956)	*Godetia nutans*	Håkansson (1945)
		Godetia viminea	Håkansson (1949)
Panicum maximum	Jauhar (1967)	*Panicum coloratum*	Swaminathan and Nath (1956)
		Parthenium argentatum	Catcheside (1950)
		Plantago serraria	Fröst (1959)
		Ranunculus acris	Fröst (1969b)

prophase as bivalents and multivalents, but they uncoil, the pairing lapses, and they are all univalent by MI (Catcheside, 1950). Surprisingly, very few of these small Bs are eliminated, even though as in the case of *Allium cernum* (Grun, 1959) they "drift" undivided to the poles. In most cases the segregation of these unpaired Bs is quite at random. In *Plantago serraria,* however, their behavior is special (Fröst, 1959). The univalents localize themselves close to or at one of the poles, and then all become included in only one of the polar groups. At AII they then divide normally. In animal species such unpaired Bs are found in *Diabrotica soror, Diabrotica 12-punctata* (Stevens, 1908), and in the harvest mouse *Reithrodontomys megalotis* (Shellhammer, 1969).

When B chromosomes are present in pairs, their behavior as bivalents is very regular; they show normal orientation, congression, and disjunction. In many instances, however, their pairing efficiency, as judged from the percentage of bivalents at MI, is not as good as in the As (Table VIII). As mentioned earlier, there is an element of genetic control in the degree of pairing and of elimination. Kishikawa (1965) found, for example, that pairing between the Bs in a Japanese variety of rye is very high. Apart from the univalent class there is very little elimination. In the 2B class, 92.4% of cells have bivalents. This contrasts markedly with the situation in other rye strains such as Östgöta Gråråg, in which bivalent frequency is as low as 19.0% (Müntzing and Prakken, 1941). In the variety Vasa II bivalent frequency is 88.9% (Müntzing, 1945). In the Japanese variety there is also a high frequency of occurrence of Bs in the population, namely, in over 90.0% of plants.

Not all bivalent behavior is regular in 2B-containing individuals. The acrocentric bivalents of *Trimerotropsis sparsa* and *Circottettix thalissimus*

TABLE VIII

Pairing Frequencies of B Chromosomes at MI in Some Plant and Animal Species

Genus and species	Two Bs		Three Bs			Four Bs					Reference
	II	I + I	III	II + I	I + I + I	IV	III + I	II + II	II + I + I	I + I + I + I	
Aegilops speltoides	50.0	50.0	—	—	—	—	—	—	—	—	Mendelson and Zohary (1972)
Haplopappus gracilis	82.2	11.8	6.0	83.0	11.0	—	—	—	—	—	Pritchard (1968)
Iseilema laxum	84.9	15.1	—	—	—	—	—	—	—	—	Murty (1972)
Pennesetum typhoides	71.0	29.0	67.5	25.8	6.7	5.3	29.8	47.4	15.8	1.7	Venkateswarlu and Pantulu (1970)
Phleum phleoides	98.0	2.0	2.1	80.9	17.0	30.0	?	60.0	?	?	Bosemark (1956c)
Secale cereale	92.4	7.6	67.0	32.5	0.5	46.2	5.8	38.2	8.2	1.6	Kishikawa (1965)
Camnula pellucida	99.0	1.0	9.6	74.8	15.6	10.3	13.8	51.7	24.2	—	Nur (1969a)
Calliptamus palaestinensis	99.0	1.0	—	—	—	—	—	50.0	50.0	—	Nur (1963)
Helix pomatia	83.0	17.0	—	—	—	—	—	—	—	—	Evans (1960)
Locusta migratoria	99.3	0.7	4.3	87.8	7.9	2.0	8.7	68.9	19.4	1.0	Kayano (1971)
Myrmeleotettix maculatus	50.0	50.0	—	—	—	—	—	—	—	—	John and Hewitt (1965a)
Tetrix ceperoi	96.5	3.5	—	—	—	—	—	—	—	—	Henderson (1961)

frequently orientate themselves on the spindle at MI with both centromeres close to the same pole. This leads to nondisjunction, and one secondary spermatocyte receiving both members of the bivalent (White, 1973).

When more than 2Bs are present, multivalent associations are frequent, although there is still a preference for bivalents and many univalents also occur (Table VIII). In some species there are no associations higher than bivalents, regardless of the number of Bs. This situation exists in *Helix pomatia* (Evans, 1960), *Iseilema laxum* (Murty, 1972), *Poa alpina* (Müntzing, 1948a), and *Tradescantia paludosa* (Whitaker, 1936).

Complications in the pairing relationships of Bs also occur on account of their polymorphism. What frequently happens is that a standard subterminal B type undergoes centromere misdivision, giving rise to telocentrics, or small and large isochromosomes of the two arms, or a combination of both. Telocentrics of different arms are not homologous with one another of course, only with the arms of the standard B from which they derive. Isochromosomes have arms which are mirror images of one another, and these sometimes contrive to pair together and form a ring chromosome, as in *Lilium callosum, Myrmeleottettix maculatus,* and *Secale cereale.* In addition to this, the isochromosomes are homologous with the arm of the standard B from which they originate, and can accommodate two pairing partners, one on each arm. Interarm pairing can also occur in univalent Bs which are not isochromosomes. It has been observed, for example, in *Zea mays* (McClintock, 1933), *Poa trivialis, Holcus lanatus* (Bosemark, 1957b), and several grasshopper species (White, 1973).

In the mealy bug *Pseudococcus obscurus* (Nur, 1962a), the meiotic behavior of the B chromosomes is quite unique. The A chromosomes themselves are specialized. In males the paternal set of As becomes heterochromatic at the blastula stage and remains so throughout development. Spermatogenesis consists of two highly modified divisions. During the first division both heterochromatic and euchromatic sets simply divide mitotically. In the second division the heterochromatic and euchromatic sets are segregated to opposite poles. Of the four products of meiosis, the heterochromatic derivatives degenerate, leaving two functional sperms each with a haploid euchromatic set of chromosomes. The B chromosomes are heterochromatic, like the paternal set of As, up until the late prophase I of meiosis. They then undergo a change in pycnosis and become negatively heteropycnotic. They divide at AI along with the A chromosomes. At the second division they segregate with the euchromatic maternal set of As into the two functional products of meiosis. Thus their activities constitute an accumulation mechanism, since an unreduced number is transmitted by the males.

C. Accumulation Mechanisms

It is evident from direct cytological analysis and from breeding experiments that B chromosomes are inherited in a non-Mendelian manner. In rye, the cross 2B♀ × 2B♂, gives a majority of 4B progeny, and it can be shown by further appropriate crosses than an unreduced number of B chromosomes is passed on through both male and female gametes. In maize only the male gametes carry an unreduced number of Bs, while transmission through the female is normal. In most plants, however, in which breeding tests have confirmed some kind of a boosting system to be in operation, meiosis is normal—except for a certain degree of elimination. Direct cytological observation of the mitotic divisions of the male gametophytes in rye, by Hasegawa (1934), gave the first clue to the existence of *directed postmeiotic segregation.* At the first pollen grain mitosis the Bs undergo directed nondisjunction. Chromatids fail to separate, and they pass together into the generative nucleus (see Fig. 2). Håkansson (1948a) subsequently found that a similar situation exists on the female side in rye, nondisjunction occurring at the first mitosis of the egg cell in such a direction that the doubled-up number of Bs eventually passes into the embryo. In maize, Roman (1947) translocated an A-chromosome segment carrying marker genes onto a B-chromosome centromere, and showed that nondisjunction occurred at the second pollen grain mitosis, followed by preferential fertilization of the egg cell by the B-chromosome-containing gametes. The situation in *Sorghum purpureo-sericeum* is rather extraordinary. Here a large number of Bs leads to extra mitoses of the vegetative nucleus in the pollen grain. At the first of such extra divisions, the Bs pass undivided to the generative pole. When only two generative nuclei are thus formed, one or both may produce a sperm. If more than two are formed, the pollen grain is killed (Darlington and Thomas, 1941). *Lilium callosum* differs from the three cases just cited in that the preferential distribution of the Bs occurs in the meiotic division and not in the mitoses of the gametophyte. Transmission through the pollen is normal, with a 1:1 segregation of 0B and 1B progeny from the cross 0B♀ × 1B♂. The reciprocal of this, however, 1B♀ × 0B♂ gave, in one experiment by Kayano (1956b), 16 0B and 83 1B individuals. The cytological basis of this distorted segregation is a preferential distribution of the unpaired Bs in egg mother cells toward that half of the spindle nearest the micropylar end of the embryo sac. The B chromosome was found to be lying on the micropylar side of the spindle in 63.7% of cells, on the chalazal side in 13.0%, and on the MI plate in 23.2%. If the B outside the MI plate passes undivided to the closer pole, and the B on the plate goes to either pole with equal chance,

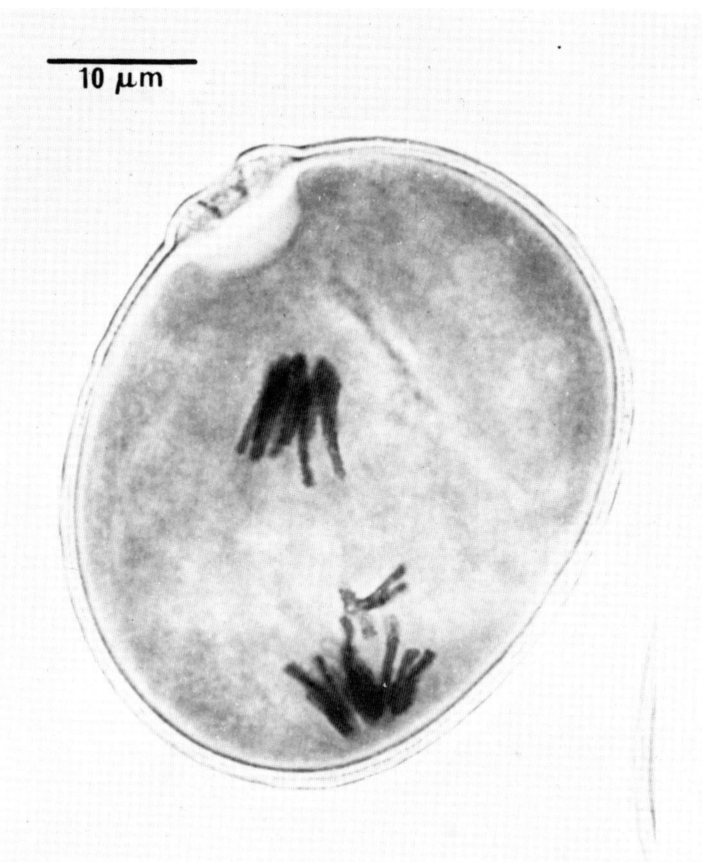

FIG. 2. Chromatid nondisjunction of a rye B chromosome at first pollen grain mitosis.

then at least 75.4% of micropylar nuclei will include the B. The Bs divide normally at the second division, but in *Lilium* the functional megaspore is proximal to the micropyle, so the Bs accumulate in the egg cells. A summary of the various types of accumulation mechanisms operating in flowering plants is shown in Table IX.

In both rye and maize the regulation of nondisjunction in gametophytes has been shown to be autonomously controlled by the B chromosome itself. The Bs of rye still behave in this way even when they are transferred to a strain of hexaploid wheat, and they do it at the corresponding division of the microspore and megaspore (Müntzing, 1970). The stan-

TABLE IX

MECHANISMS OF B CHROMOSOME ACCUMULATION IN PLANTS

Mechanism	Species	Reference
Preferential meiotic segregation in emcs	*Cochlearia pyrenaica*	J. J. B. Gill (1971a)
	Lillium callosum	Kayano (1957)
	Phleum nodosum	Fröst (1969a)
	Plantago serraria	Fröst (1959)
	Tradescantia virginiana	Vosa (1962)
	Trillium grandiflorum	Rutishauser (1956a)
Preferential fertilization by B-carrying male gametes	*Zea mays*	Roman (1948a,b)
Directed nondisjunction		
First pollen grain mitosis	*Aegilops speltoides*	Mendelson and Zohary (1972)
	Alopecurus pratensis	Bosemark (1957b)
	Anthoxanthum aristatum	Östergren (1947)
	Briza media	Bosemark (1957b)
	Dactylis glomerata	Puteyevsky and Zohary (1970)
	Deschampsia bottnica	Albers (1972)
	Deschampsia caespitosa	Albers (1972)
	Deschampsia wibeliana	Albers (1972)
	Festuca arundinacea	Bosemark (1957b)
	Festuca pratensis	Bosemark (1950)
	Haplopappus gracilis	Pritchard (1968)
	Holcus lanatus	Bosemark (1957b)
	Phleum phleoides	Bosemark (1956c)
Second pollen grain mitosis	*Zea mays*	Roman (1947)
First pollen grain mitosis	*Secale cereale*	Hasegawa (1934)
First egg cell mitosis	*Secale cereale*	Håkansson (1948a)
First pollen grain mitosis of extra divisions	*Sorghum purpureo-sericeum*	Darlington and Thomas (1941)
Somatic nondisjunction coincident with flower initiation	*Crepis capillaris*	Rutishauser and Röthlisberger (1966); Röthlisberger (1970)
Endomitotic reduplication in early meiotic prophase?	*Crepis conyzaefolia*	Fröst (1964)
	Crepis pannonica	Fröst (1960)
Nondisjunction at premeiotic mitosis in ♂ and ♀, with polarized distribution	*Achillea* spp.	Ehrendorfer (1961)
Increase during transmission through the ♂, mechanism unknown	*Haplopappus validus*	Smith (1968)
	Clarkia elegans	Mooring (1960)
	Iseilema laxum	Murty (1972)
No apparent mechanism	*Centaurea scabiosa*	Fröst (1957)
	Poa alpina	Håkansson (1954)
	Ranunculus acris	Fröst (1969a)
	Ranunculus ficaria	McLeish (1954)
	Xanthisma texanum	Berger *et al.* (1956)

dard B chromosome in rye is about half the size of the As and has a subterminal centromere, an arm ratio of 5:1, and a characteristic block of heterochromatin near the end of the long arm. Three derived forms can arise from this standard B chromosome. These are an isochromosome for the long arm, an isochromosome of the short arm, and a deleted chromosome lacking the terminal heterochromatic knob. These differing forms may be present individually in a plant, or occur together in various combinations. Both the standard B and the large iso-B are capable of undergoing nondisjunction by themselves. In the case of the standard B the centromere divides normally, along with the rest of the A-chromosome centromeres at anaphase, but the chromatids remain held together at a point close to and on either side of the centromere. Thus the chromatids pass in an unseparated form into the generative nucleus (Fig. 2). The small iso-B and the deleted B are not capable of undergoing nondisjunction when present by themselves, although they both possess the segments responsible for holding the chromatids together. We know this because both will show nondisjunction if they are present together with a standard B. It appears from this that the heterochromatic block at the end of the long arm in the standard B is the controlling element. The situation is summarized in Fig. 3. A similar state of affairs also exists in

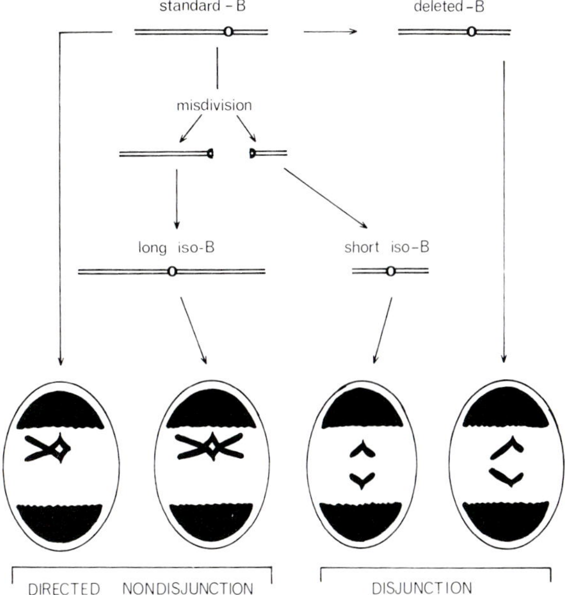

FIG. 3. Behavior of different B types at first pollen grain mitosis in rye.

Festuca pratensis (Bosemark, 1956b). In both species the activity of the controlling heterochromatic segment is interchromosomal.

In *Zea mays,* Ward (1973a) recently obtained a translocation between the B chromosome and chromosome 8 of the normal complement. The breakpoint in the B was found to be near the tip of the long arm. The B^8 chromosome has all the chromatin of the intact B except for the relatively short euchromatic segment at the end of the long arm. Ward found that in the $8B^8$ microspores nondisjunction at the second pollen grain mitosis does not occur, while it is regularly found in microspores of the constitution $8^B B^8$ (the 8^B chromosome having a chromosome-8 centromere and carrying the long arm of the B including the distal euchromatic segment). Apparently, the controlling "gene(s)" for nondisjunction in maize lie in this short euchromatic segment distal to the major heterochromatic region of the long arm, and not in the heterochromatin as was previously thought. Ward favors the hypothesis advanced by Lima-de-Faria (1962) that the controlling element in maize, as in rye, operates by inducing late replication and stickiness in the sensitive regions holding the chromatids together.

One of the most interesting mechanisms of B-chromosome accumulation in animals is that already described (meiosis) for the mealy bug *Pseudococcus obscurus.* Most of the systems in animals though are premeiotic, as in the snail *Helix pomatia.* In *Helix pomatia* the number of Bs varies considerably from one spermatocyte to another in a way which suggests that nondisjunction occurs in spermatogonia. At A1 of meiosis, however, there is a low frequency or absence of normal (0B) cells, some of which are expected to occur by nondisjunction in the low-numbered B classes. This paucity of 0B cells led Evans (1960) to speculate that one or a few Bs may confer an advantage on the cell, resulting in a decrease in the duration of the mitotic cycle and giving a rate of proliferation greater than that of normal diploid cells. Bs will thus be maintained and possibly accumulated in the population by virtue of their advantage over normal cells in the development of gonads. Melander (1950) has claimed that in *Polycelis tenuis* the Bs in the female undergo endomitotic reduplication in the premeiotic resting stage in oogenesis, thereby boosting their numbers. More generally in animals accumulation mechanisms are based on mitotic nondisjunction leading to numerical variation in Bs among different follicles of the testes in males. Consider for example, the data of Kayano for B-chromosome distribution in primary spermatocytes of *Locusta migratoria* (Table X). The number of Bs in cells of the gastric ceca is constant and taken as the standard. In males with one B per cell in the gastric ceca, the mean number of Bs per spermatocyte is greater than 1.0 and ranges in fact from 0 to 4. In males with two Bs per cell

TABLE X

Number of Bs in Primary Spermatocytes of *Locusta migratoria* Males[a]

Male		Percent cells with							Mean number of Bs per cell	Number of cells observed
		0B	1B	2B	3B	4B	5B	6B		
1B[b]	1	1.5	96.2	2.3	—	—	—	—	1.01	1346
	2	1.8	78.3	19.8	0.1	—	—	—	1.18	1384
	3	6.0	72.0	20.0	2.0	—	—	—	1.18	1413
	4	2.4	79.2	12.2	6.2	—	—	—	1.22	1855
	5	5.1	68.6	24.3	—	2.0	—	—	1.25	1561
	6	0.6	61.1	30.3	4.0	4.0	—	—	1.50	1459
2B[b]	1	—	8.0	78.5	9.3	4.1	0.1	—	2.10	1385
	2	—	2.1	86.5	8.3	3.1	—	—	2.12	1772
	3	—	8.0	36.6	34.7	18.3	2.4	—	2.70	924
	4	—	12.4	27.1	29.9	23.5	6.8	0.3	2.86	1327

[a] Kayano (1971).

[b] The number of Bs in cells of the gastric ceca, which is mitotically constant and taken as a standard.

in the gastric ceca the mean is greater than 2.0 in spermatocytes and ranges from 0 to 6. According to Kayano, the pattern of Bs in primary spermatocytes is determined by nondisjunction at the mitoses associated with differentiation of the follicles, and nondisjunction is preferential in that cells with an increased number of Bs are more frequently contributed to the germ line. Nur (1963) has previously argued strongly and convincingly for a similar means of accumulation in *Calliptamus palaestinensis.* He suggested preferential nondisjunction of the Bs at embryonic divisions in which future germ line cells are segregated from future somatic cells. Neither worker has actually witnessed chromatids undergoing mitotic nondisjunction, but rest their case on the scarcity of 0B cells in relation to 2B cells, in individuals having a single B as the basic number. Simple mitotic nondisjunction alone, without some preferential selection for 2B cells, would be expected to result in about equal numbers of 2B and 0B classes in the follicles. In *Calliptamus palaestinensis* follicles with two Bs outnumbered zero Bs by a ratio of 15:1. Similar mechanisms of accumulation are thought to operate in *Camnula pellucida* (Nur, 1969a) and *Neopodismopsis abdominalis* (Rothfels, 1950). Karyotype comparisons of both parents and progeny from single pair crosses in *Myrmeleotettix maculatus* have shown an accumulation of the large

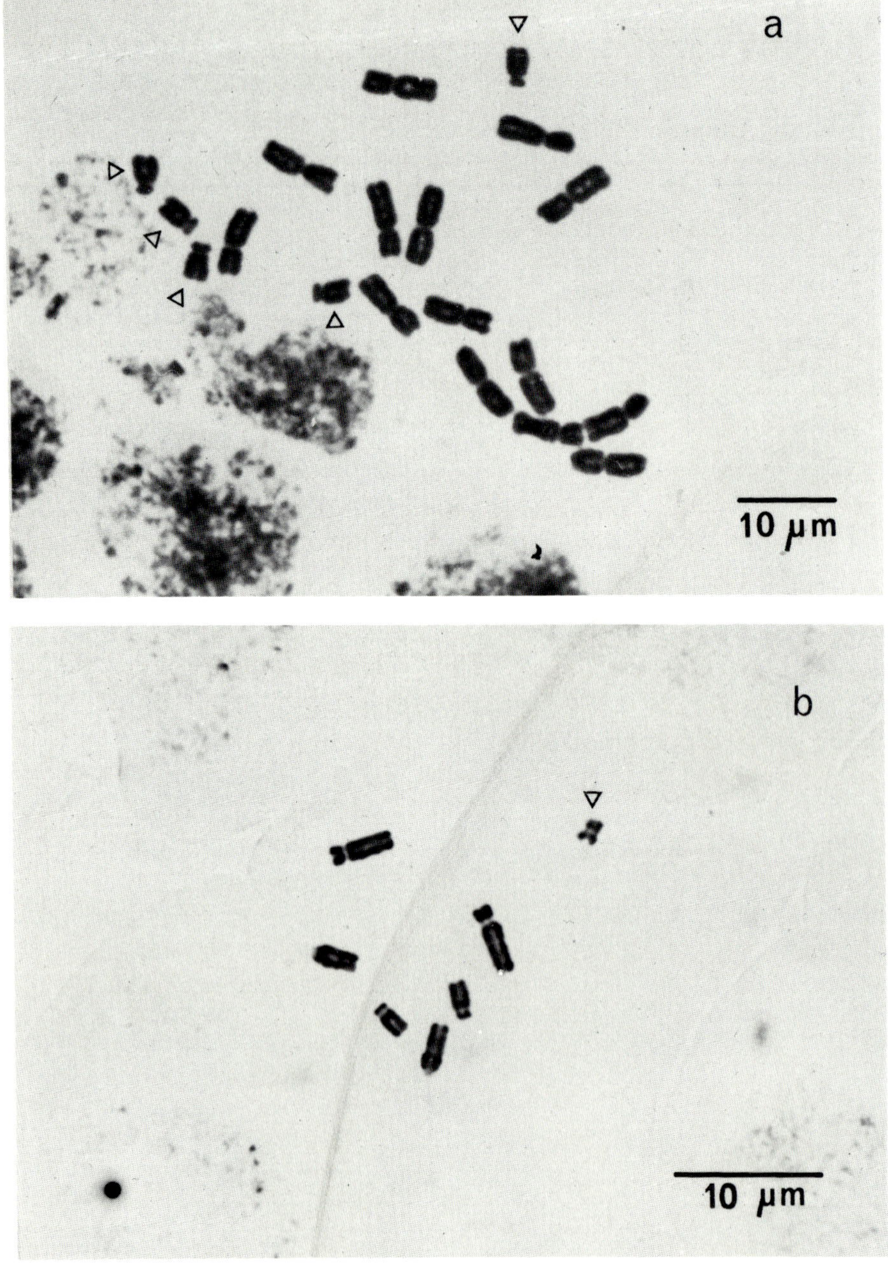

FIG. 4. B chromosomes at C mitosis in root meristem cells of (a) rye and (b) *Crepis capillaris*.

mitotically stable B when transmitted through the female (Hewitt, 1973). Accumulation in this case is presumed to result from a preferential segregation of univalent Bs at the first division of meiosis.

IV. Structure

When B chromosomes are first discovered in a species, they are generally recognized by their small size and lack of homology with the normal set at meiosis. In addition to this of course they are variable with respect to their presence or absence within individuals of a population. Figure 4 shows the B chromosomes in somatic cells of two plant species, rye and *Crepis capillaris.* Apart from their usually smaller size, and sometimes heterochromatic nature, another characteristic feature of B chromosomes is their polymorphism.

As described earlier, the Bs are less stable than the As in their inheritance, and one of the consequences of this is that they frequently undergo misdivision of the centromeres to produce isochromosomes of their two arms. Because Bs are nonessential chromosomes, these derived isochromosomes can persist in the population and even undergo further structural rearrangements to give second-order derivatives. Some such derived types have already been mentioned in connection with pairing properties and with the control mechanism for nondisjunction in pollen grains. The most extreme case of B-chromosome polymorphism occurs in *Aster ageratoides,* in which a standard morphological type and up to 24 derivatives may occur. In *Aster ageratoides leiophyllus* var. *tennuifolium* for example, Matsuda (1970a) has described a standard B, the longest one, and 10 types presumed to be derivatives by deletion, misdivision, or translocation. Five of the ten are classified as primary derivatives, and the other five as secondary derivatives derived from the primary one (Fig. 5). The primary derivatives are named according to length, in descending order, a_p, b_p, d_p, f_p, and iso_p. Their short arms are the same length as the short arm of the standard B. Secondary derivatives are designated l_s, m_s, g_s, h_s, and iso_{s-1}, according to their shape. Various combinations of these 11 types occur within plants, and they show homology with one another at meiosis. Polymorphic Bs are common in animal species, too. In *Melanoplus femur-rubrum* there is a large metacentric B, called B^m, and two telocentrics, B^T and B^t. *Podisma pedestris* has two quite distinct B types. One is large, comparable in size to the X, while the other is smaller than any member of the standard set.

Another important structural difference between As and Bs is the complete absence of any known major genes in the B chromosomes. Randolph (1941) tested several possible B-chromosome loci for allelism

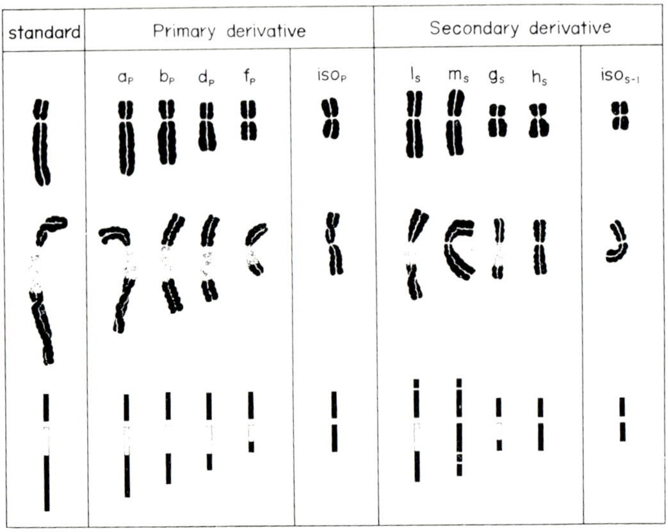

Fig. 5. B-chromosome polymorphism (11 types) in *Aster ageratoides*. Upper row, metaphase. Middle row, prophase. Bottom row, idiograms showing heterochromatic regions in black and euchromatic regions in white. [From Matsuda, 1970a, with permission.]

with a representative group of known mutant genes located in the A chromosomes of maize. Recessive mutants to be tested were crossed with B-chromosome plants carrying the dominant allele, and from the backcross and F_2 data it was possible to see if the presence of Bs in the segregating populations disturbed the expected ratios. Forty-six linked genes distributed among 17 of the 20 arms of the A chromosomes were tested in this way. None of them gave disturbed ratios in combination with the Bs.

Nucleolus organizers are also conspicuously absent in B chromosomes. Powell and Burton (1966) described a B chromosome in *Pennisetum typhoides* that associated with the A-chromosome nucleolus or sometimes organized a separate one. This claim has been disputed, however, by other workers. Bosemark (1957b) reported finding nucleoli organized by the B chromosome of *Alopecurus pratensis*. Keyl and Hägele (1971) also described a nucleolus-organizing B in *Chironomus plumosus*, but in this case they showed from banding pattern analysis in the salivary gland chromosomes that the B was structurally homologous with the centromere region of chromosome 4 of the normal complement. The question of homology is an important one.

In rye, Lima-de-Faria (1952) carried out a detailed pachytene analysis of the chromomere pattern of the standard B chromosome. While it resembles certain of the As more closely than others, there is no doubt that the B chromosome is unique, and in no sense homologous with any of the As. This is confirmed by the absolute lack of pairing between A and B chromosomes at meiosis. Critical surveys of pachytene morphology in well over 50 varieties of maize, including flour, flint, dent, popcorn, and sweet corn also show the B chromosomes to have a very distinctive pachytene structure which is unlike that of any region of comparable length in any of the A chromosomes (McClintock, 1933; Longley, 1938; Randolph, 1941). The situation in rye and maize is probably typical of the B chromosomes in most species. There are cases though, in which extra chromosomes are present that satisfy all the criteria of true Bs, yet show some homology with members of the regular complement. The small centric fragments commonly found in *Phlox* (Meyer, 1944) fall into this category. So do the Bs in *Tradescantia* (Darlington, 1929; Swanson, 1943) and *Lilium* (Mather, 1935). In many animal species the dividing line between sex chromosomes, particularly the X, and the Bs is a narrow one. In fact, the Bs are thought to be derived from the X in some instances (Hewitt and John, 1972; Jackson and Cheung, 1967; Rothfels, 1950).

B Chromosomes are often considered heterochromatic. That is, they appear more densely coiled and deeply staining at interphase, prophase of mitosis, and/or meiosis. In most animal species in which Bs are found this is certainly the case. One notable exception among animals is the snail *Helix pomatia* which has euchromatic Bs (Evans, 1960). Plants are about equally divided between those with heterochromatic and those with euchromatic Bs, as Table XI shows.

Autoradiographic analysis of the DNA replication patterns of heterochromatic Bs in *Crepis capillaris* (Abraham *et al.*, 1968), *Puschkinia libanotica* (Barlow and Vosa, 1969b), and *Secale cereale* (Darlington and Haque, 1966) suggests that they are later replicating relative to the As. Ayonoadu and Rees (1973), however, have recently offered a new interpretation of the situation in rye. They propose that the disproportionately heavy labeling in Bs is attributable to their higher DNA density at metaphase and not to a late completion of their DNA synthesis. In rye the Bs have 1.5 times as much DNA per unit volume as the As (Jones and Rees, 1968).

In terms of gene content and activity (Section V), it is possible that the B chromosomes are in a state of repression (Rees, 1972), or alternatively that they contain a high proportion of their DNA in the form of short reiterated sequences which are not organized into proper functional cistrons. Information on this important question is scant.

TABLE XI

HETEROCHROMATIC AND EUCHROMATIC B CHROMOSOMES IN PLANTS

Heterochromatic	Euchromatic
Alopecurus pratensis	*Allium angulosum*
Anthoxanthum aristatum	*Allium cernum*
Aster ageratoides	*Allium nutans*
Aster scaba	*Allium porrum*
Brachiaria ramosa	*Allium senescens*
Briza media	*Allium thunbergii*
Crepis capillaris	*Andropogan gayanus*
Dactylis glomerata	*Caltha palustris*
Digitaria decumbens	*Catapodium rigidum*
Digitaria pentzii	*Clarkia elegans*
Digitaria valida	*Cochlearia pyrenaica*
Festuca pratensis	*Ficus krishnae*
Holcus lanatus	*Haplopappus gracilis*
Iseilema laxum	*Haplopappus validus*
Luzula campestris	*Koeleria alpicola*
Narcissus bulbocodium	*Listera ovata*
Narcissus juncifolius	*Narcissus bernadi*
Ornithogalum concinnum	*Narcissus calcicola*
Ornithogalum ecklonii	*Narcissus cyclamineus*
Ornithogalum narbonense	*Narcissus pseudonarcissus*
Ornithogalum umbellatum	*Oenothera hookeri*
Parthenium argentatum	*Ornithogalum pyrenaicum*
Phleum nodosum	*Panicum coloratum*
Phleum phleoides	*Paris polyphylla*
Plantago coronopus	*Scilla autumnalis*
Plantago serraria	*Trifolium berytheum*
Poa trivialis	*Trifolium salmoneum*
Puschkinia libanotica	*Trillium cernum*
Secale cereale	*Trillium erectum*
Sorghum purpureo-sericeum	*Trillium grandiflorum*
Trillium ovatatum	*Trillium luteum*
Zea mays	*Trillium sessile*

Hewitt (1972) reports that the B chromosomes of *Myrmeleotettix maculatus* contain a high proportion (25%) of repetitive sequences which could be separated as a satellite by cesium chloride density centrifugation. The heterochromatic Bs of maize (Chilton and McCarthy, 1973) and rye, by contrast, have been found to have a DNA composition very closely related to that of the As. The buoyant densities and renaturation kinetics of DNA from individuals with and without B chromosomes are virtually identical. It seems likely therefore that these Bs must possess a large amount of unique sequence DNA, possibly in the form of functional genes.

V. Effects

A. General

In species of plants and animals that carry B chromosomes, those individuals of a population with and those without Bs cannot generally be distinguished from one another phenotypically. The variation produced by Bs is continuous, like that due to polygenes (Mather, 1945), and direct cytological observation is the only sure way of determining whether B chromosomes are present or absent. In fact, they are most frequently discovered by chance, during cytological studies, in individuals that betray their presence only when the chromosomes are properly counted and classified. There are two exceptional cases, both in plants. In *Haplopappus gracilis* (Jackson and Newmark, 1960), the color of the achenes is changed from brownish-red to dark purple by the presence of Bs, while in *Plantago coronopus* the single B found by Paliwal and Hyde (1959) induced complete male sterility in plants in which it was carried. But the more usual situation, at least in plants, is to find that with low numbers of Bs the effects may be apparently neutral or even stimulatory, while higher numbers almost invariably have an adverse effect upon fertility and vigor. Rye is a good example, as shown in Fig. 6. The data are taken from earlier and extensive work by Müntzing (1963) and are presented here in the form of a graph. Three points must be emphasized: first, the overall negative correlation between vigor and B-chromosome number, which is general in plants; second, the presence of low numbers of Bs is not unduly serious and may even be "beneficial" for some characters; in the case of diploid rye this is particularly true for B chromosomes in the disomic state; third, there is a disproportionately severe effect due to odd-numbered combinations of Bs; this point is taken up

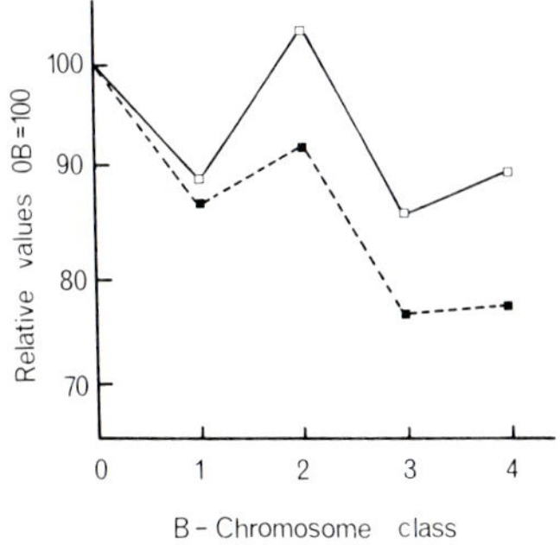

Fig. 6. Straw weight (open squares) and tiller number (solid squares) (mean values over three varieties of rye) plotted against B-chromosome class. (Data from Müntzing, 1963.)

TABLE XII

Effect of Rye B Chromosomes on Fertility

	B class								
	0B	1B	2B	3B	4B	5B	6B	7B	8B
Percent seed set	49.5	31.4	34.2	21.5	5.1	7.1	1.7	0.1	—

more fully later. The effect upon fertility is most severe, as the data in Table XII show. Rye plants with four or more B chromosomes are virtually sterile, and even those without Bs (but within the B-chromosome population) have their fertility reduced to about 50%. Such effects of Bs upon fertility in plants are very common, as the information summarized in Table XIII shows. Much of the earlier work on B-chromosome effects in plants and animals (Table XIII) has already been adequately reviewed (Müntzing, 1958, 1959; Melander, 1950; Battaglia, 1964b; White, 1973), but there are a few cases that warrant further mention. In *Clarkia williamsonii,* for example, Wedberg *et al.* (1968) studied a large number of phenotypic traits, including growth habit, pubescence on leaves and calyx, leaf shape and size, shape of petals, color of flower parts, number of internodes and branches, flowering time, and stigma maturation, without finding any effects at all due to B chromosomes. As far as fertility was concerned, however, they did report that in plants with as many as five Bs the production of seeds was reduced by a half. Likewise, in the mottled grasshopper (*Myrmeleotettix maculatus*), Hewitt and John (1970) did not find any obvious effect of Bs on the "exophenotype." Their study was a detailed one involving five populations with from 10 to 40% B chromosomes, and including measurements of five morphological characters, namely, number of stridulatory pegs per hind leg, average length of tibiae, average length of wings, and number of segments per antenna. They also considered the variation in these characters both within and among populations and again found no significant differences. There are, however, important endophenotypic differences due to the Bs in *Myrmeleotettix,* which are considered in Section V,B. Fertility in animals is also adversely affected by B chromosomes, as the summary in Table XIII makes clear. Briefly then, and notwithstanding the exceptional cases just mentioned, the consequences of B chromosomes at the morphological or exophenotypic level involve a progressive reduction in vigor and especially in fertility. Table XIII tells its own story. As far as outward appearances are concerned, B-chromosome effects are either neutral or, more commonly, negative.

At the level of the endophenotype the story has a positive side as well. Important results have recently been obtained concerning the effects of Bs on A-chromosome behavior at meiosis, and on some fundamental cellular processes governing growth and development. These results, which are leading to new interpretations of the role of B chromosomes in the genetic system, are presented in greater detail below.

B. Recombination

Chiasma frequency and distribution are genetically controlled. They are also subject to environmental modification. With the wisdom of hindsight it is not surprising that B chromosomes should have been found to influence the pattern of A-chromosome recombination at meiosis. First, as chromosomes, they change the genotype. Second, aside from their elusive genetic properties, they modify the immediate environment of the As themselves, that is, the nucleus. Third, they have long been known to "interfere" with many other gene-controlled aspects of growth and development (see Table XIII) in both plant and animal species.

Barker (1960) was the first to discover the effects of Bs on A-chromosome chiasma frequency, in the grasshopper *Myrmeleotettix maculatus*. Catcheside (1950) had earlier studied meiosis in 0B and B-containing plants of *Parthenium argentatum* and found no effect of the Bs on the mean chiasma frequency of the As. In both cases the analyses were limited to mean cell chiasma frequencies and took no account of possible changes in variance. John and Hewitt (1965a,b) subsequently showed that the Bs in *Myrmeleotettix* could affect the distribution of chiasmata among cells as well as increase the mean. In an experimental B-chromosome rye population, Jones and Rees (1967) reported that over the range of zero to eight Bs there was no significant effect of Bs upon the mean A-chromosome chiasma frequency, but they did find an increased "between-cells variance" (i.e., between pollen mother cells within plants), as well as an increase in the variance between bivalents within cells. Several such analyses were subsequently carried out (Table XIV). Some of the cytological data on means and variances are also set out in the form of graphs in Fig. 7.

There is no way of predicting precisely how B chromosomes are likely to modify the regulation of recombination in As. The mean chiasma frequency can be raised (*Festuca mairei*) or lowered (*Lolium perenne*), and this effect upon the mean may or may not be accompanied by changes in the cell and bivalent variances. In the case of variances, which are a measure of the regularity of chiasma distribution among and within cells, the effect is usually one of increasing the variation. Furthermore, the dosage effects of Bs on means and variances are not always additive. In some cases, for example, rye and *Listera* (Fig. 7),

TABLE XIII

GENETIC EFFECTS OF B CHROMOSOMES

Character	Effect	Species	Reference
Plants			
Plant			
Germination	Delayed	*Secale cereale*	Moss (1966)
	Speeded up	*Allium porrum*	Vosa (1966)
	No effect	*Centauria scabiosa*	Fröst (1958b)
Growth and vigor	Reduced	*Aegilops speltoides*	Mendelson and Zohary (1972)
	Reduced	*Anthoxanthum aristatum*	Ögihara (1962)
	Reduced	*Secale cereale*	Müntzing (1943, 1966)
	Reduced	*Zea mays*	Randolph (1941)
	Increased (low number of Bs)	*Centauria scabiosa*	Fröst (1954, 1958b)
	Increased variation	*Lilium auratum*	Ostergren (1947)
	Increased variation	*Secale cereale*	Moss (1966)
	No effect	*Clarkia williamsonii*	Wedberg *et al.* (1968)
Flowering time	Delayed	*Secale cereale*	Kishikawa (1965)
	Delayed	*Zea mays*	Kato (1970)
Fertility	Reduced	*Aegilops speltoides*	Mendelson and Zohary (1972)
	Reduced	*Agropyron desertorum*	Baenziger and Knowles (1962)
	Reduced	*Agropyron imbricatum*	Baenziger and Knowles (1962)
	Reduced	*Anthoxanthum aristatum*	Östergren (1947)
	Reduced	*Centauria scabiosa*	Fröst (1958b)
	Reduced	*Clarkia williamsonii*	Wedberg *et al.* (1968)
	Reduced	*Deschampsia caespitosa*	Albers (1972)
	Reduced	*Deschampsia wibeliana*	Albers (1972)
	Reduced	*Festuca pratensis*	Bosemark (1957a)
	Reduced	*Haplopappus gracilis*	Pritchard (1968)
	Reduced	*Lilium callosum*	Kimura and Kayano (1961)
	Reduced	*Secale cereale*	Müntzing (1943)
	Reduced	*Sorghum purpureo-sericeum*	Darlington and Thomas (1941)
	Reduced	*Trifolium berytheum*	Putiievsky and Katznelson (1970)
	Reduced	*Trifolium solmoneum*	Putiievsky and Katznelson (1970)

	Reduced	*Zea mays*	Randolph (1941)
	Complete male sterility in 1B plants	*Plantago coronopus*	Paliwal and Hyde (1959)
Color of achenes	Change from brown to purple	*Haplopappus gracilis*	Jackson and Newmark (1960)
Seed weight	Increased	*Secale cereale*	Moss (1966)
Cell			
Cell size	Increased	*Secale cereale*	Müntzing and Akdik (1948)
	Increased	*Zea mays*	Randolph (1941)
Pollen size	Increased	*Zea mays*	Randolph (1941)
Pollen grain development	Retarded	*Anthoxanthum aristatum*	Östergren (1947)
	Retarded	*Secale cereale*	Müntzing (1949)
	Retarded	*Sorghum purpureo-sericeum*	Darlington and Thomas (1941)
Mitotic cycle	Increased	*Puschkinia libanotica*	Barlow (1973)
	Increased	*Secale cereale*	Ayonoadu and Rees (1968b)
Spindle	Abnormalities	*Secale cereale*	Håkansson (1957)
Nucleus			
Metabolic activity	Increase in protein and RNA	*Secale cereale*	Kirk and Jones (1970)
	Increase in protein and RNA	*Zea mays*	Ayonoadu and Rees (1971)
	Increase in histone to DNA ratio	*Secale cereale*	Kirk and Jones (1970)
A-chromosome stability	Increased breakage in As	*Crepis capillaris*	Rutishauser (1963)
	Increased breakage in As	*Trillium grandiflorum*	Rutishauser (1956b)
	Loss of knobbed segments of As at the second microspore division	*Zea mays*	Rhoades and Dempsey (1972, 1973)
A-chromosome recombination	Change in chiasma frequency and distribution	See Table XIV	
Animals			
Phenotype	No obvious effect	*Myrmeleotettix maculatus*	Hewitt and John (1970)
	Reduced femur length in males	*Camnula pellucida*	Nur (1969a)
	Increased tibia length	*Pseudococcus obscurus*	Nur (1962b)
Vigor	Reduced	*Polycelis tenuis*	Melander (1950)
	Reduced	*Pseudococcus obscurus*	Nur (1968)
	Increase in disease	*Polycelis tenuis*	Melander (1950)
Hatching	Reduced and retarded	*Polycelis tenuis*	Melander (1950)
Fertility	Reduced	*Camnula pellucida*	Nur (1969a)
	Reduced	*Pseudococcus obscurus*	Nur (1968)
A-Chromosome recombination	Changes in chiasma frequency and distribution	See Table XIV	

TABLE XIV

INFLUENCE OF Bs ON A CHROMOSOME RECOMBINATION

Genus and species	Effect	Reference
Plants		
Aegilops speltoides	Decrease in mean pmc chiasma frequency	Zarchi *et al.* (1972)
Brachycome lineariloba	No effect on mean pmc chiasma frequency or the variance of pmc chiasma frequencies	Carter and Smith-White (1972)
Festuca mairei	Increase in mean pmc chiasma frequency	Malik and Tripathi (1970)
Listera ovata	Increase in mean pmc and emc chiasma frequency	Vosa and Barlow (1972)
Lolium perenne	Decrease in mean pmc chiasma frequency; increased variance between bivalents within pmcs	Cameron and Rees (1967)
Lolium temulentum × *L. perenne* (+Bs)	Suppression of homoeologous pairing	Evans and Macefield (1972, 1973)
Lolium perenne × *Festuca arundinacea*	Suppression of homoeologous pairing	Bowman and Thomas (1973)
Parthenium argentatum	No effect on means pmc chiasma frequency	Catcheside (1950)
Pennisetum typhoides	Increase in mean pmc chiasma frequency	Manga, 1970, cited in Pantulu and Manga (1972)
Puschkinia libanotica	Increase in mean pmc chiasma frequency; change in chiasma distribution	Barlow and Vosa (1970)
Secale cereale	Increase in mean pmc chiasma frequency	Zečević and Paunović (1969)
Secale hybrid (an experimental rye population)	No effect on mean pmc chiasma frequency increased variance between pmcs and between bivalents within pmcs	Jones and Rees (1967)
Triticum aestivum × *Aegilops mutica* (+Bs)	No alteration in the level of pairing at normal temperatures; drop in pmc chiasma frequency at 12°C	Vardi and Dover (1972)
	Suppression of homoeologous pairing in the absence of chromosome 5B of wheat	Mochizuki (1964); Dover and Riley (1972); Vardi and Dover (1972)

TABLE XIV (*Continued*)

Genus and species	Effect	Reference
Triticum aestivum × *Aegilops speltoides* (+Bs)	No alteration in the level of pairing	Vardi and Dover (1972)
	Suppression of homoeologous pairing in the absence of chromosome 5B of wheat	Dover and Riley (1972); Vardi and Dover (1972)
Zea mays	Increase in mean pmc chiasma frequency; increased variance between pmcs within individuals	Ayonoadu and Rees (1968a)
	Increased recombination in chromosome 3	Hanson (1969)
	Increased recombination in chromosome 5	Nel (1973)
	Increased recombination in chromosome 9	Hanson (1969); Nel (1973); Chang and Kikudome (1971); Ward (1973b)
	Chromosome 9 homozygous for a transposed segment of chromosome 3. Increased recombination by as much as 110% between C-Wx.	Rhoades (1968)
	Reduced recombination between Yg-C.	
	Increased intragenic recombination at the Wx locus	Melnyczenko (1970)
Animals		
Melanoplus differentialis differentialis	Increase in mean chiasma frequency, ♂	Abdel-Hameed *et al.* (1970)
Melanoplus femur-rubrum	No effect on mean chiasma frequency. Increase in cell variance within individuals, ♂	Stephens and Bregman (1972)
Myrmeleotettix maculatus	Increase in mean chiasma frequency, ♂	Barker (1960)
	Increase in mean chiasma frequency, increased variance between cells, ♂	John and Hewitt (1965a,b); Hewitt and John (1967); Hewitt and Brown (1970)
Podisma pedestris	No effect of the large B on chiasma frequency, ♂	Hewitt and John (1972)
Tettigidea lateralis	Change in chiasma distribution, ♂	Fontana and Vickery (1973)

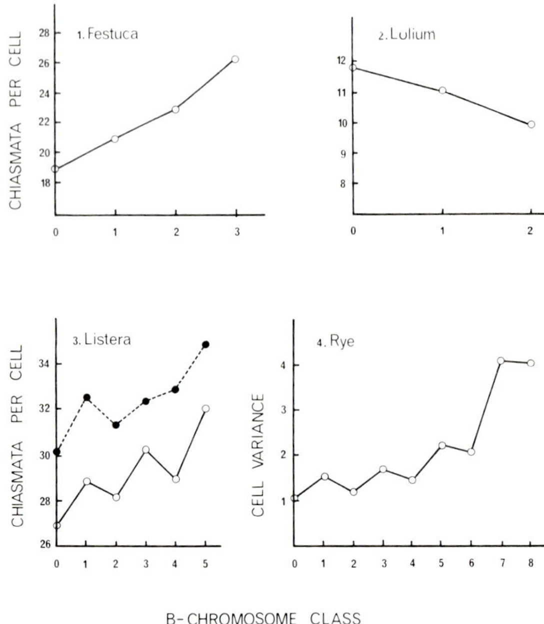

FIG. 7. Effect of B chromosomes on recombination in four plant species. 1, *Festuca mairei* (drawn from data of Malik and Tipathi, 1970); 2, *Lolium perenne* (redrawn from Cameron and Rees, 1967); 3, *Listera ovata* broken line, chiasmata per egg mother cell; solid line, chiasmata per pollen mother cell; (redrawn from Vosa and Barlow, 1972); 4, *rye* (redrawn from Jones and Rees, 1967).

there are differential effects related to odd and even-numbered combinations of Bs. This point is taken up again in Section V,E.

Genetically detected crossing-over between marker genes on homologous chromosome partners arises as a direct consequence of nonsister chromatid exchange in meiotic prophase. Cytologically, these exchanges are expressed in the form of chiasmata which can be quantified at diplotene or metaphase I. Variations in chiasma frequency and distribution should therefore lead to corresponding fluctuations in recombination frequencies. It is particularly satisfying, then, to find that in maize B chromosome effects lead both to an increase in chiasma frequency (Ayonoadu and Rees, 1968a) and to an increase in genetically determined crossing-over (see Table XIV). Changes in the frequency of genetically determined crossing-over in maize due to B chromosomes can even be detected at the intragenic level (Melnyczenko, 1970). In the case of rye it has also been possible to link changes in chiasma frequency and distribution (Jones and Rees, 1967) with increased variation for certain vegetative characters in the progenies of B-chromosome-containing plants (Moss, 1966).

The analysis of B-chromosome effects in some species hybrids has been especially rewarding. Evans and Macefield (1973) found that in the diploid interspecific cross *Lolium temulentum* × *L. perenne* +Bs, the Bs suppress pairing and chiasma formation between homoeologous chromosomes. In the absence of Bs pairing and chiasma formation are very regular (Fig. 8). At the tetraploid level pairing is confined to homologous pairs of chromosomes only, when Bs are present, resulting in exclusive bivalent formation at metaphase I. Thus the tetraploid behaves like an allopolyploid. Without the Bs, multivalent associations are found at metaphase as a consequence of pairing between both homoeologous and homologous chromosomes. The situation is analogous to that pertaining in *Triticum aestivum,* the allohexaploid bread wheat, in which there is a major gene limiting pairing to strict homologs within the three component genomes. Interestingly enough, B chromosomes from diploid *Aegilops mutica* and also from *Aegilops speltoides* have now been introduced into the hexaploid *Triticum aestivum,* and have been found to act as a substitute for the major gene in situations in which it is absent (Mochizuki, 1964; Dover and Riley, 1972).

B Chromosomes have now been found in numerous other species

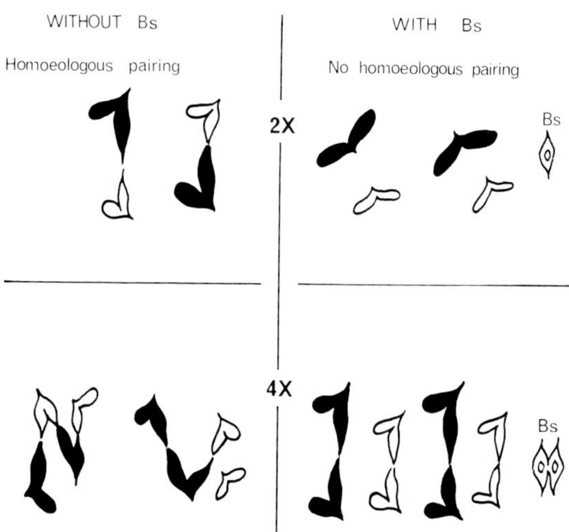

FIG. 8. Diagram summarizing the effect of B chromosomes on A-chromosome pairing in diploid and tetraploid *Lolium temulentum* × *L. perenne* hybrids. Two pairs of chromosomes only are shown, *L. temulentum* in black and *L. perenne* in outline. The Bs are also in outline and labeled. (Redrawn, with permission, from Evans and Macefield, 1973.)

hybrids as well but, with the exception of a case just reported by Bowman and Thomas (1973), no special effects at meiosis have been noted. Bowman and Thomas found results in the hybrid *Lolium perenne* × *Festuca arundinacea* similar to those already described by Evans and Macefield. As a matter of fact, Peto, as long ago as 1933, made a similar cross between *Lolium perenne* ($2n = 14$) and *Festuca arundinaceae* ($2n = 42$) and also found quite low pairing in the F_1 hybrid. Oddly enough, he also reported the occurrence of two extra fragment chromosomes in the hybrid plant, which he accounted for by fragmentation. It seems likely that they were in fact B chromosomes, since both had subterminal centromeres. Peto did not attach much importance to these extra chromosomes, nor did he have an opportunity to compare them with control plants lacking such fragments.

Forcible arguments can now be advanced for an important and adaptive role for B chromosomes in the genetic systems of outbreeding plant and animal species, namely, that of boosting and regulating the release of variability. The control mechanism operates at three levels of organization, namely, (1) within populations through adjustment in mean chiasma frequencies of individuals, (2) within individuals at the cell level, and (3) within cells at the level of individual chromosomes.

C. A-Chromosome Instability

A rather novel B-chromosome effect has lately been discovered in *Zea mays* (Rhoades *et al.*, 1967; Rhoades and Dempsey, 1972, 1973). Certain members of the normal A-chromosome complement possess large heterochromatic knobs at specific sites in the chromosome arms. When two or more Bs are present in a plant, the arms, or parts of arms carrying these knobs, are frequently eliminated at the second microspore division. Little or no loss occurs in microspores with one B, and the rate is not increased with more than two Bs. A Chromosomes that do not possess heterochromatic knobs are stable in their inheritance. This finding is all the more interesting in view of the earlier assertion by Longley (1938) of negative correlations between the number of heterochromatic knobs on the A chromosomes and the presence of Bs in different races of maize. Rhoades and Dempsey have suggested that the control mechanism for this induced chromatin loss is the same as that postulated for the nondisjunction of the Bs themselves, namely, a failure of the heterochromatic knobs to replicate normally during the second pollen grain mitosis. Replication is delayed at the knobbed region, and this results in a failure of the two chromatids to separate at anaphase. As the centromeres move to opposite poles, the chromatids of the chromosome arm carrying the knob are formed into a bridge by conjoining of the knobs. The bridge breaks (in various places) and the knobbed parts of the chromatids ultimately separate, producing one normal chromatid and one cell with a chromatid missing part of a knobbed arm. It is proposed that the distal euchromatic tip of the B, which controls non-disjunction in the B itself, is also responsible for suppressing the replication of these A-chromosome knobs. A-Chromosome instability has also been reported in two other species, *Trillium grandiflorum* (Rutishauser, 1956b) and *Crepis capillaris* (Rutishauser, 1963), in which the presence of Bs is claimed to increase the frequency of spontaneous chromosome breakage.

D. Cell and Nuclear Metabolism

Appropriate methods of quantitative analysis reveal that B chromosomes have wide-ranging effects on the phenotype, especially in plants. It could be that they have a wide spectrum of influence over many different gene-controlled processes or, alternatively, that they operate on some fundamental physiological process with pleiotropic effects. Recent work suggests that the latter situation is the more likely. In rye, for instance, there is a normal mitotic cycle time of 12.75 hours. But in plants containing four B chromosomes this is increased to a mean of 16.71 hours (Ayonoadu and Rees, 1968b). The difference is a large one, of the order of 25%. The number of cells per milligram fresh weight of root meristem is 326×10^3 and 247×10^3 for 0B and 4B plants, respectively—again a difference of 25% (John and Jones, 1970). These two factors namely, extension of the mitotic cycle time and reduction in cell number, coupled together, along with other known effects on cell size, provide a sound physiological basis to account for many B-chromosome effects, such as reduction in vigor, delay in germination, delay in flowering, and so on. The physiological analysis in fact has been taken further than this, and we are now able to trace these B-chromosome effects back to the metabolism of the nucleus itself. Again most of the relevant work has been carried out using plant material of rye and maize.

B Chromosomes contribute extra DNA to the nucleus. In rye, each additional B increases the 0B DNA amount by approximately 5%, so that over the range 0–8B there is a considerable diversity of nuclear DNA values (Jones and Rees, 1968). This quantitative nuclear DNA variation has repercussions on the metabolic activity of the nucleus. One effect, as we have already seen, is extension of the mitotic cycle. In terms of A-chromosome gene activity, some of the consequences of the extra B-chromosome DNA are quite fundamental and regular. Cytochemical estimations on root meristem interphase nuclei of rye have provided information about specific nuclear components, namely, total protein,

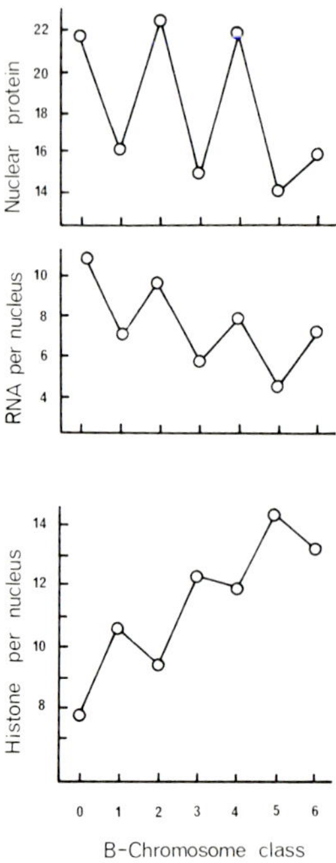

FIG. 9. Relative amounts of total nuclear protein, RNA, and histone per nucleus in interphase root meristem nuclei of rye with a range of B-chromosome numbers. (Data from Kirk and Jones, 1970.)

RNA, and basic histone protein. The results of some of these experiments are summarized in Fig. 9. The average amounts of total nuclear protein and RNA per nucleus decrease with increasing number of B chromosomes (Kirk and Jones, 1970), although not in a strictly linear fashion. There is a differential effect depending on whether the Bs are present in even- or odd-numbered combinations. Quite clearly, the B chromosomes in rye, and also in maize where comparable results have been obtained (Ayonoadu and Rees, 1971), have a repressive effect on the genetic activity of the nucleus. The data on histones are very interesting. Histones are involved in the regulation of gene action in eukaryotes. They serve to mask the DNA templates and thus control, via intermediary molecules (RNA or acidic chromosomal protein), transcription of genes (Clever, 1968). It looks as if the B chromosomes, at least in rye, produce their effects on growth and development by interfering with the normal processes of gene activity in the A chromosomes. The extent to which the influence of Bs on recombination is related to these changes in nuclear metabolism is not yet clear. There is no reason to suppose, however, that the genes that regulate recombination in the As are any less susceptible to repression than genes that govern other processes of growth and development.

E. Odds and Evens Effect

One of the most intriguing aspects of B-chromosome behavior is their differential activity in relation to odd- and even-numbered combinations. The effect is most frequently encountered for characters of the nuclear phenotype (Figs. 7 and 9), but it has also been observed for the whole-plant phenotype in rye (Fig. 6). One instance has been recorded in animals (Shcherbakov, 1966). Here it is found that flies (*Odagmia ornata*) with even numbers of Bs have a higher "adaptive value" than those with odd numbers. In the plant material also, it is the even-numbered combinations that have the least effect upon vigor and aspects of nuclear metabolism, including recombination. It was proposed, when this phenomenon was first demonstrated by Jones and Rees (1969), that the effect may result from a "contiguity effect" between B chromosomes, such that paired combinations in the nucleus act more favorably than unpaired ones. After all, the whole system of B-chromosome inheritance, especially in rye, is adapted, through the nondisjunction mechanism, toward maintaining a preponderance of even-numbered combinations in the population. Furthermore, there is some recent and convincing evidence to show that the A chromosomes themselves have a close homologous association during interphase of the cell cycle. The evidence comes originally from the common wheat *Triticum aestivum* (Feldman *et al.*, 1966). The centromere is found to be primarily responsible for positioning homologs near one another as a result of the attachment of centromeres of these homologs to the same, or closely adjacent, sites on the nuclear membrane (Avivi *et al.*, 1969). Such a mechanism of somatic association could well apply to B chromosomes and even have some functional significance which is expressed as the odds and evens effect.

F. Significance of B-Chromosome Effects

In view of the above account of B-chromosome effects in flowering plants and animals, one could easily be excused for believing that B chromosomes are of no adaptive significance whatsoever. More than this, the case could, and indeed is argued, that they are harmful and are maintained in populations only on account of their deviceful accumulation mechanisms. The claim that they are also genetically inert is, however, no longer tenable. Their genetic effects are manifold and unassailably established. The point at issue now is the significance of these effects to the genetic system, rather than the question whether or not they have any effects.

It was Östergren (1945) who first proposed that B chromosomes be considered parasites, serving no adaptive purpose for the host organism at all but being obliged only to be useful to themselves. This claim has been echoed again from time to time but has never attracted a large following. Of late, however, it has been strongly advocated by Nur (1966a, 1969b), who found that the B chromosomes of *Pseudococcus obscurus* apparently have no effect on the fitness of female mealy bugs, and that they are positively harmful to the males. Their maintenance in the population, he argues, must be due exclusively to the fact that they possess an accumulation mechanism through their preferential segregation at spermatogenesis. Nur extends this argument to include several other animal and plant species. Rhoades and Dempsey (1972) also acquiesce in a similar interpretation, concerning in particular the B chromosomes of maize. According to their scheme, many of the activities of the maize B chromosome can be interpreted as a consequence of its replication without, or with only little, transcription and translation of its DNA. In this way the Bs would have negative effects upon cell activity by competing for nucleic acid precursors and amino acids needed for their replication. This in turn would lead to reduced nuclear genetic activity and extension of the mitotic cycle time, and ultimately give rise to an impairment of growth and vigor. Lengthening of the cell cycle time they argue, could also provide a basis for explaining the effects of B chromosomes on recombination.

The negative effects of B chromosomes must of course be conceded, but so do the positive ones. Darlington (1956a,b) has suggested, and evidence has now been found, that B chromosomes have a constructive role to play in short-lived, outbreeding plants (and the same reasoning can be applied to animals) in regulating the release of genetic variability. The evidence on recombination is extensive, as was seen earlier, and has now been used to establish the case that B chromosomes, far from being parasitic or passive, are in fact a device for controlling recombination, through adjustment of chiasma frequency and distribution at meiosis. This particular point of view has been advanced quite forcibly by Hewitt and John (1967). According to their thesis, B chromosomes add to the range of variability of the population in a more rapid and efficient way than could otherwise be achieved, for example, through selection for polygenes governing chiasma frequency and distribution. B Chromosomes can be gained or lost far more readily than favorable or unfavorable gene combinations. It is on precisely this point that Nur has rejected any suggestion that Bs have a constructive and regulatory function at meiosis. According to Nur, the system is too clumsy and the ratio of advantage to inconvenience too small. What has to be remembered is that Hewitt and John are dealing with an organism that displays no outward morphological sign that it even carries B chromosomes, while Nur's arguments revolve around a species in which the Bs are manifestly harmful to both the vigor and the fitness of the individuals that carry them.

As it happens, the adaptive significance of B chromosomes does not hinge entirely on their effects upon chiasma frequency and distribution. Apart from the increased variability they undoubtedly generate, there are several other important physiological effects, such as enhanced germination in leeks (Vosa, 1966), changes in cell size and metabolism, and so on, which may be of consequence to B-containing individuals of a population, or to the population itself at critical stages of development or under varying environmental circumstances. With this in mind some of the detailed information available on both natural and experimental B-chromosome populations must now be examined.

VI. Populations

In populations that possess B chromosomes the Bs may be thought of as a property of the population rather than of its individual members. In a proportion of individuals they are dispensed with altogether, while other individuals are rendered completely sterile by virtue of their presence in high numbers. Their usefulness seems to consist therefore in their being present in low numbers in a fraction of the population. The general assumption, often regarded as an axiom, is that B chromosomes have a selective value because individuals with and without them compete and coexist in equilibrium. This assumption is an appealing one, but the premises upon which it is based still need to be carefully examined. The first question to be asked is: To what extent do Bs occur among populations of B-containing species, and what are the relative frequencies of the 0B and B-containing individuals among the different populations?

A. Distribution of B Chromosomes in Natural Populations

Population data are not all that extensive in relation to the number of plant and animal species known to carry B chromosomes. Table XV is a summary of the results of the main investigations. While the number of species involved is not large, collectively the data reveal some interesting trends. Where population studies have been made, it is evident that the Bs have a wide territorial distribution and a high incidence of occurrence as among different populations. In some species, such as *Festuca pratensis* (Bosemark, 1956a) and *Tainia laxiflora* (Tanaka and Matsuda, 1972), they are found throughout the geographical range of the species. In

TABLE XV

B-CHROMOSOME DISTRIBUTION IN NATURAL POPULATIONS

Genus and species	Number of populations studied	Number of populations with Bs	B-containing individuals in populations with Bs (%)		Notes	Reference
			Range	Mean		
Plants						
Allium cernum	14	11	10–48	25.4	—	Grun (1959)
Caltha palustris	67	31	4–80	21.8	—	Kootin-Sanwu and Woodell (1969)
Centauria scabiosa	222	121	4–64	27.0	Occurrence of Bs in Sweden correlated with humidity	Fröst (1958a)
Clarkia elegans	32	14	7–100	57.4	—	Mooring (1960)
Clarkia williamsonii	30	11	—	—	No Bs found in populations from highest sites	Wedberg *et al.* (1968)
Crepis conyzaefolia	7	4	3–27	15.5	—	Fröst (1962)
Dactylis glomerata	3	3	21–51	35.1	Different B frequency in populations from different habitats	Zohary and Ashkenazi (1958)
Leucanthemum species	91	71	—	—	Bs restricted to polyploids	Papes (1971)
Lilium auratum	11	7	—	—	—	Ogihara (1962)
Lilium maximowiczii	6	6	3–100	27.0	—	Noda (1956)
Phleum phleoides	96	73	<1–65	15.6	Bs more frequent in soils low in organic matter	Bosemark (1967)
Ranunculus ficaria	148	23	20–100	55.0	Bs restricted to diploids in south England Absent in populations from north England and Scotland	Gill *et al.* (1972)

Secale cereale	11	8	3–9	6.6	Primitive rye strain from Iran	Kranz (1963)
Secale cereale	32	32	2–73	32.2	Korean rye populations	Lee and Min (1965a,b)
Secale cereale	7	7	19–90	54.0	Korean rye populations	Müntzing (1957)
Secale cereale	7	7	1–13	4.5	Turkish rye populations	Müntzing (1950)
Secale cereale	5	1	—	14.1	Afghan rye	Müntzing (1950)
Secale cereale	25	25	10–55	27.7	Yugoslavian rye populations	Zečević and Paunović (1967)
Tainia laxiflora	9	9	80–100	95.8	Bs found throughout range of species distribution	Tanaka and Matsuda (1972)
Zea mays	33	33	—	—	Bs occur in highest frequency in strains with the least number of knobs in the A chromosomes	Longley (1938)
Animals						
Acrida lata	11	11	8–36	22.1	—	Kayano *et al.* (1970)
Myrmeleotettix maculatus	33	17	7–62	32.2	Bs occur in highest frequency where climatic factors provide optimal ecolological conditions	Barker (1966)
Myrmeleotettix maculatus	10	3	13–54	37.2	All B-containing populations found in south Britain	John and Hewitt (1965a)
Myrmeleotettix maculatus	31	27	15–70	—	Populations from south Britain	John and Hewitt (1956b)
Mymreleotettix maculatus	27	13	7–50	30.7	East Anglia populations. Highest B frequency under optimal ecological conditions	Hewitt and Brown (1970)
Pseudococcus obscurus	25	25	1–100	64.5	—	Nur (1962b)
Reithrodontomys megalotis	14	14	60–100	83.0	—	Shellhammer (1969)
Trimerotropis sparsa	23	10	2–18	10.5	—	White (1951a)

many cases Bs can be found in every population studied, and in only a few instances do they occur in less than half the populations. Individually, some of these population studies are extensive and provide valuable information about the distribution pattern of Bs in relation to environmental conditions, for example, *Centauria scabiosa, Myrmeleotettix maculatus* (see below). What is most remarkable about the survey is the degree to which B-chromosome frequency varies among the different populations within a species. In almost every case studied the range is large, from populations with only a few percent of Bs through to those with as high as a 100% in which every individual sampled carried at least one B. In some species, such as *Tainia laxiflora,* in which all the populations sampled carried Bs, the B frequency per population is consistently high at 80–100%. Similarly for the harvest mouse *Reithrodontomys megalotis.*

The question now arises as to what degree these populations frequencies represent situations of equilibrium; that is, situations in which the forces of accumulation and of elimination outlined earlier are in balance. A variation in B frequency among different populations is frequently taken to mean that Bs have some special adaptive significance, and that the relative proportions of 0B and +B individuals are adjusted to the particular environmental circumstances of each population. It is also accepted, and has now been demonstrated, that the means of achieving such adjustments exist, as, for example, by way of changes in the rate of nondisjunction and the rate at which Bs are eliminated at meiosis. What then is the evidence for the existence of populations that are in a state of stable equilibrium? The evidence in fact is meager and comes mostly from studies on a few animal species. Jackson and Cheung (1967) observed stable frequencies of Bs (11.22 ± 0.3% of male carriers) over a period of 9 years in one population of the short-horned grasshopper *Phaulacridium vittatum.* Kayano *et al.* (1970) reported on the variations in frequency of B chromosomes in 11 natural populations of *Acrida lata* in Japan over periods ranging from 3 to 9 years. The frequency of Bs per population ranged from 8 to 36%. In nine of the populations the frequencies of Bs did not differ significantly over the years. In only two of the populations was the frequency shown to vary significantly between years, and here there was evidence of interference by man. In *Pseudococcus obscurus* (Nur, 1966a), the frequency of Bs was found not to be at equilibrium in population 1 from Berkeley, because the frequency was shown to be steadily increasing over a period of 6 years. In other populations the frequencies appeared to be stable, and in equilibrium between the forces of accumulation and loss. In *Myrmeleotettix maculatus* the B chromosomes in natural populations are maintained in relatively constant frequencies from year to year (Hewitt and John, 1967). Equilibrium frequencies can therefore be demonstrated in a few cases, at least on a short-term basis. No cases have been reported of a decline in B frequency. Experimentally, it can be shown that Bs do have the capacity to increase within a population. In an open-pollinating Korean rye population we maintained, the B frequency has increased over three generations from 2.0% in the initial seed sample in 1963 to 8.0% in 1964, 15.0% in 1966, and up to 22.0% in 1972. In cultivated plants the influence of man too has played a part in shaping the distribution pattern of B chromosomes. Many populations of primitive rye (Table XV) are high in Bs, but cultivated and highly bred European varieties have them only in very low frequency. Moss (1966) has attributed this to husbandry practices like selection for uniformity and high fertility, which select against B-containing plants. Other factors that may have an influence on the data for B-chromosome distribution patterns in different populations include errors due to sampling, and certainly some of the samples are suspiciously small, as well as largely unknown influences due to population ancestry and the circumstances of the origin of the Bs. Many populations no doubt, are still in the process of evolving their optimal B frequencies, or even of evolving B chromosomes. These arguments aside, we are obliged to accept that populations of differing habitat and location do invariably differ in the frequency with which they carry B chromosomes, and that in a few cases at least these B frequencies are known to represent states of stable equilibria. It is logical to ask now for the evidence and the mechanism of the adaptation.

B. Adaptation

Fröst (1958a) surveyed 222 populations of *Centauria scabiosa* covering a large part of Scandinavia and Finland. In an attempt to show (among other things) some adaptive significance for the B chromosomes, he analyzed a total of almost 8000 plants, of which over 2000 were found to have Bs. The number of Bs per plant ranged from 0 up to 16, but of the B-containing individuals the majority had either 1 or 2 Bs. The main conclusion was that plants with and without Bs have different selective values under different environmental conditions. Areas with a high frequency of Bs have a more favorable continental climate with lower humidity. Bosemark (1956a, 1967) undertook similar massive surveys with *Festuca pratensis* and *Phleum phleoides*. In *Phleum* Bs were more frequent in soils low in organic matter, while in *Festuca* (in Sweden) there was a positive correlation between B frequency and the clay content of the soil. The main point to emerge from these detailed surveys is that B chromosomes appear to occur in the highest numbers under circumstances most favorable to the existence and growth of the species. On a lesser scale similar claims have been made in respect to the B-chromosome distribution pattern for *Tradescantia edwardsiana* (Brown, 1960) and *Lilium auratum* (Ogihara, 1962). The most convincing evidence to date, however, for an environmentally attuned B-chromosome distribution pattern comes from detailed studies with the grasshopper *Myrmeleotettix maculatus* in Britain. Barker (1966) first mooted the suggestion that B-carrying grasshoppers were found preferentially in areas where climatic factors provide ecological conditions approaching the optimum for the species. Hewitt and John (1967, 1970) subsequently undertook a much more detailed and thorough analysis. They confirmed that Bs are found preferentially in British populations occupying warm, dry habitats, under optimal or near-optimal environmental conditions, and that they are absent in climatically marginal situations. In one analysis they considered a transect of well-isolated natural populations inhabiting the spoil heaps of old lead mines, across the Plynlimon Mountain area of West Wales. The transect, which runs from a warm, dry lowland area near the coast, up to the colder, wetter mountain area, is associated with a cline of decreasing B frequency. In the higher mountain areas the species is not represented at all. There is a strong correlation between rainfall and B frequency. Bearing in mind that these populations have been repeatedly sampled over a period of several years, and are known to have different but stable B frequencies, Hewitt and John have been able to argue a strong case for selection as the causative agent in determining the existing pattern of B-chromosome distribution. Even more striking, in view of the scale of the exercise, is the detailed study by Hewitt and Ruscoe (1971) of the *Myrmeleotettix* populations in one small local area of microhabitat, around an abandoned lead mine at Goginan, near Aberystwyth. The special feature of this area is the steep cline in environmental conditions over a short distance of about 1 mile, from the bottom to the top of the mine area. Using micrometeorological techniques, Hewitt and Ruscoe established the existence of clines of temperature, humidity, and wind speed on the mine, of which they consider temperature to be the most important factor influencing the distribution of grasshopper populations. The steep cline in temperature corresponds well to the stable cline in B frequency, warm, dry conditions at the bottom of the mine favoring the presence of B chromosomes. These detailed studies on *Myrmeleotettix* undoubtedly demonstrate that B chromosomes have some adaptive value and that selection must have played an important role in prescribing the characteristic and stable frequencies of individual populations. But to precisely what end selection is operating remains something of a pregnant question.

Notwithstanding this impasse, one still inevitably enquires into the nature of the selection mechanism. Hewitt and Ruscoe have suggested a kind of "physiological adaptation" in *Myrmeleotettix* involving differential rates of growth between B-containing and 0B individuals. More concrete and informative evidence comes from another favored experimental organism *Secale cereale.* Kishikawa (1970) performed quite a neat experiment involving a study of the B-chromosome frequency in progeny populations derived from rye plants grown under different temperature and soil moisture conditions. The progeny were derived from intercrossing 2B × 2B and 4B × 4B plants, after first cloning the parental material to eliminate the effect of differences due to genotype. The frequency of Bs in the progeny populations derived from the crosses was lower under the high-temperature or dry soil conditions than under the low-temperature or moist soil conditions. Differences in B frequency among the progeny population plants were attributed to observed changes in meiotic pairing behavior among the Bs, and to possible changes in nondisjunction rates at the first pollen grain mitosis in the parent plants. Hence we have at least one clear demonstration that selection can operate through variable rates of transmission of B chromosomes from one generation to the next. A direct selection mechanism of a different kind, involving differential mortality in rye and *Lolium* seedlings, under conditions of severe environmental stress, was experimentally demonstrated by Rees and Hutchinson (1973). The system they used for rye was to grow seeds from a standard B-chromosome population, in which the frequency of different B-chromosome classes was known, at various densities (5, 10, 25, 50, 75, 100, 150) in 6.5-cm, plastic multipots, and then to screen the surviving plants at maturity for their B class (using MI of meiosis). At the high densities mortality was severe, over 80%. The mean B frequency was found to decrease with increasing plant density, and to be closely correlated with mortality. In the very low-density populations, however, the B frequency was significantly higher than that in the standard population. The evidence of this experiment can be used to foster the argument introduced earlier that B chromosomes are harmful and maintained in populations solely on account of their accumulation mechanisms, by suggesting that conditions in the low-density populations were even more favorable than the conditions under which the standard population was maintained.

Alternatively, as Rees and Hutchinson convincingly argue, it can be asserted that "the fitness of plants with B's under low density is superior to that of plants without B's." Further evidence for such a positive and adaptive role for Bs comes from the *Lolium* experiment. It follows a design similar to that already outlined for rye, and the results appear

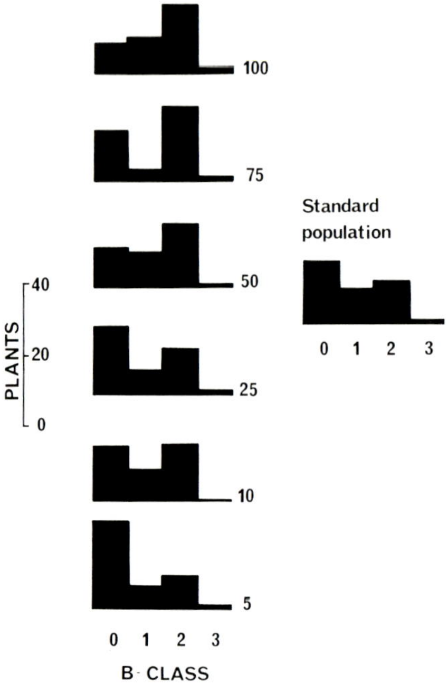

FIG. 10. Distribution of B chromosomes among survivors of *Lolium* plants sown as seed in multipots at densities of 5, 10, 25, 50, 75, and 100 seeds per pot. (Data from Rees and Hutchinson, 1973.)

in Fig. 10. The mean B frequency increases significantly with increasing plant density, showing quite manifestly that, under the enforced conditions of fierce competition, the B-containing individuals have the advantage in terms of survival. The outcome of these competition-type experiments is hardly compatible with the view that Bs are harmful and maintained in populations by potent mechanisms of accumulation.

VII. Origin

The question of the origin of B chromosomes is one that has so persistently nagged cytologists over the years that many have felt obliged to present some scheme or other no matter how fragile and speculative. One particular point of weakness in a chromosome complement has been repeatedly seized upon as a likely site at which breakage could occur and ultimately give rise to Bs, namely, the nucleolus organizer region. Certainly, it succumbs readily to a little extra pressure on the cover slip, but this has been taken as a point in favor of what Battaglia

(1964b) calls the "hypothesis of disarticulation at the secondary constriction." Gotoh put forward such a scheme to account for the origin of the 2 extra chromosomes in 16-chromosome rye, as long ago as 1924. He was supported by Stolze (1925), Emme (1928), and others. Emme (1928) claimed that the total length of the chromosomes in rye with 14 and 16 chromosomes was the same, but careful measurements by Lewitsky (1931) showed the two extra chromosomes to be new additions to the complete set. Schulz-Schaeffer (1966) recently put forward detailed proposals for the possible origin of B chromosomes in *Alopecurus pratensis, Haplopappus spinolosus,* and *Zea mays,* according to the disarticulation hypothesis. The looseness of attachment of the satellite of chromosome 5 in *A. pratensis* to the parent chromosome is a key feature of the mechanism for spontaneous production of "fragment chromosomes." Apparently, the satellite of chromosome 5, after spontaneous disarticulation, must be able "to assume centromere activity." The scheme is not convincing. Kranz (1971) is less specific, and considers that the B chromosome of rye originates by fragmentation of an A chromosome in the ancestral form of the genus *Secale.* This again is assuming a lot for, as we already know, the As and Bs in rye are not at all homologous with one another. Some of the more acceptable schemes for the origin of B chromosomes come from those organisms in which the Bs (if we may still call them that) show some homology with members of the regular set. For *Tradescantia paludosa,* for example, Swanson (1943) has put forward a viable hypothesis for their origin by translocation events. Lewis (1951, 1968) considers that the Bs in *Clarkia* arise from trisomics resulting from nondisjunction in translocation heterozygotes, followed by loss of part or all of the euchromatic segment of the extra chromosome. Another scheme for the origin of Bs in plants, and one that commands a fair measure of credibility, considers the Bs the products of chromosome evolution. They supposedly arise during the process of reduction in basic number as small centric fragments from unequal interchanges. Such a means of species evolution is known in the genus *Crepis* (Tobgy, 1943) and, moreover, the derived species with the lowered basic number (*C. fuliginosa*) possess B chromosomes. If this scheme is a realistic one, then a cross between *C. fuliginosa* with Bs and *C. neglecta,* from which it is supposed to arise, should show some pairing between the Bs of *C. fuliginosa* and the centromere region of one of the A chromosomes in *C. neglecta.* This kind of system for B-chromosome origin has likewise been proposed for *Haplopappus garcilis* (Jackson, 1960) and *Paspalum stoloniferum* (Avdulov and Titova, 1933).

In animal species the sex chromosomes are often looked upon as likely candidates for the ancestors of Bs. The X chromosome in particular has been favored. Jackson and Cheung (1967) find that in *Phaulacridium vitattum* the B chromosome is clearly heterochromatic, like the X, and forms physical connections with it during meiotic prophase. They claim that the association is not of the usual type of stickiness associated with heterochromatin, but rather that it indicates a degree of homology between the B and the X. Hewitt and John (1972) likewise reason that the meiotic associations between the B and the X in *Podisma pedestris* differ from conventional heterochromatic associations in that they take the form of parallel alignment and are persistent in nature, even up to first metaphase. In *Neopodismopsis abdominalis* the B is considered by Rothfels (1950) to represent a distally deficient X chromosome.

Evidently, there are many different ways in which chromosomes might have arisen. This is reflected in the heterogeneity of B types, which range from the true Bs showing a complete lack of homology or structural relationship to any of the As, through those of some insects with affinities for the X chromosome, to the minority of cases like those in *Phlox* and *Chironomus* in which the Bs represent centric fragments of certain of the A chromosomes. The origin of true Bs remains an enigma.

VIII. Conclusions

B chromosomes break most of the laws of classic genetics, including those of Mendelian heredity and apparently those too of natural selection. Yet Darlington (1956a,b) has seen fit to incorporate them into what he calls the *genetic system*—"the organization of the reproduction and heredity of a group of organisms as a unit in evolution"—and even ascribe to them an important role as agents for boosting the variability, hence the adaptability, of species. Östergren (1945) has taken the contrary position and labeled them parasitic, maintained in populations only on account of their devious mechanisms of accumulation and infiltration. The arguments for and against both these points of view have already been discussed, and on balance it seems that the former case is more strongly advocated than the latter.

One of the stronger points in favor of the latter case, as Rhoades and Dempsey (1972) only recently reminded us, is the nature of B-chromosome DNA. It is specialized in so far as it does not appear to have any proper functional cistrons, and may not even be transcribable at all. In this sense its mere existence, with all the necessary requisite demands made upon the resources of the cell's metabolic machinery, can be expected to account for many, if not all, of the physiological consequences of B chromosomes. But, if we are to accept that these arguments have any cogency at all, what are we then to say about the occurrence of repetitious DNA now so firmly established as a component of the genotype in many plant and animal species? The properties of B-chromosome and of repetitious DNA do not seem to be all that different in many respects, and the same may well be said of some of their consequences (Rees, 1972). The one major difference is that within species individuals do not differ with respect to their repetitious DNA, like they do with Bs, so we are not able to study its consequences with the same facility. The argument amounts to saying, then, that B chromosomes and B-chromosome effects should not be considered in isolation but in the wider context of genome organization.

References

Abdel-Hameed, F., Rootham, D. L., and Flinn, R. R. (1970). *Genetics* **64**, Suppl., 1.

Abraham, S., Ames, I. H., and Smith, H. H. (1968). *J. Hered.* **59**, 297.

Afzelius, K. (1959). *Acta Horti Bergiani* **19**, 1.

Albers, F. (1972). *Beitr. Biol. Pflanz.* **48**, 1.

Almeida, J. L. F. (1946). *Bolm. Soc. Broteriana* **20**, 201.

Anderson, E., and Sax, K. (1936). *Bot. Gaz. (Chicago)* **97**, 433.

Andersson, H. (1958). *Bot. Notis.* **111**, 237.

Avdulov, N., and Titova, N. (1933). *Tr. Prikl. Bot., Genet. Selek., Ser. 2* p. 165.

Avivi, L., Feldman, M., and Bushuk, W. (1969). *Genetics* **62**, 745.

Ayonoadu, U., and Rees, H. (1968a). *Genetica* **39**, 75.

Ayonoadu, U., and Rees, H. (1968b). *Exp. Cell Res.* **52**, 284.

Ayonoadu, U., and Rees, H. (1971). *Heredity* **27**, 365.

Ayonoadu, U., and Rees, H. (1973). *Heredity* **30**, 233.

Babcock, E. B. (1947). *Univ. Calif., Berkeley, Publ. Bot.* **21**, 1.

Babu, C. N. (1936). *Curr. Sci.* **4**, 739.

Baenziger, H. (1962). *Can. J. Bot.* **40**, 549.

Baenziger, H., and Carr, R. B. (1968). *Can. J. Genet. Cytol.* **10**, 813.

Baenziger, H., and Knowles, R. P. (1962). *Crop Sci.* **2**, 417.

Baez-Major, A. B. (1934). *Cavanillesia* **6**, 59.

Barker, J. F. (1960). *Heredity* **14**, 211.

Barker, J. F. (1966). *Evolution* **20**, 665.

Barlow, P. W. (1973). *In* "The Cell Cycle in Development and Differentiation" (M. Balls and F. S. Billet, eds.), pp. 133–165. Cambridge Univ. Press, London and New York.

Barlow, P. W., and Vosa, C. G. (1969a). *Chromosoma* **27**, 436.

Barlow, P. W., and Vosa, C. G. (1969b). *Chromosoma* **28**, 457.

Barlow, P. W., and Vosa, C. G. (1970). *Chromosoma* **30**, 344.

Barthelmes, A., and Bauchinger, M. (1962). *Genetica* **33**, 165.

Bassett, I. J. (1966). *Can. J. Bot.* **44**, 467.

Battaglia, E. (1957). *Caryologia* **9**, 234.

Battaglia, E. (1958). *Caryologia* **11**, 79.

Battaglia, E. (1963). *Caryologia* **16**, 609.

Battaglia, E. (1964a). *Caryologia* **17**, 65.

Battaglia, E. (1964b). *Caryologia* **17**, 245.

Battaglia, E. (1964c). *G. Bot. Ital.* **71**, 1.

Battaglia, E., and Guanti, G. (1966). *Caryologia* **19**, 375.
Battaglia, E., and Guanti, G. (1968). *Caryologia* **21**, 283.
Bauer, H. (1931). *Z. Zellforsch. Mikrosk. Anat.* **14**, 139.
Bayreuther, K. (1969). *Chromosoma* **27**, 20.
Beal, J. M. (1942). *Bot. Gaz.* (*Chicago*) **103**, 617.
Beal, J. M., and Ownbey, M. (1943). *Bot. Gaz.* (*Chicago*) **104**, 553.
Beaudry, J. R. (1963). *Can. J. Genet. Cytol.* **5**, 150.
Beaudry, J. R., and Chabot, D. L. (1959). *Can. J. Bot.* **37**, 209.
Beçak, M. L., Beçak, W., and Denaro, L. (1972). *Caryologia* **25**, 313.
Beetle, D. E. (1944). *Madrono* **7**, 133.
Belcheva, R. G., and Mihailova, P. V. (1971). *Dokl. Bolg. Akad. Nauk* **24**, 1255.
Belling, J. (1925). *J. Hered.* **16**, 360.
Berger, C. A., and Witkus, E. R. (1954). *Bull. Torrey Bot. Club* **81**, 489.
Berger, C. A., McMahon, R. M., and Witkus, E. R. (1955). *Bull. Torrey Bot. Club* **82**, 377.
Berger, C. A., Feeley, E. J., and Witkus, E. R. (1956). *Bull. Torrey Bot. Club* **83**, 428.
Bergman, B. (1935). *Sv. Bot. Tidskr.* **29**, 155.
Bergner, A. D. (1946). *U. S. Dept. Agr., Tech. Bull.* No. 918, p. 1.
Björkman, S. O. (1951). *Hereditas* **37**, 465.
Björkman, S. O. (1954). *Hereditas* **40**, 254.
Björkman, S. O. (1960). *Symb. Bot. Upsal.* **17**, 1.
Blackwood, M. (1956). *Heredity* **10**, 353.
Blanks, G. A., and Shellhammer, H. S. (1968). *J. Mammal.* **49**, 726.
Böcher, T. W. (1950). *Bot. Notis.* p. 353.
Böcher, T. W. (1958). *Bot. Tidsskr.* **54**, 160.
Böcher, T. W. (1960). *Biol. Skr. Dan. Vidensk. Selsk.* **11**, 1.
Böcher, T. W. (1963). *Bot. Notis.* **116**, 113.
Böcher, T. W. (1964). *Sv. Bot. Tidskr.* **58**, 1.
Böcher, T. W. (1966). *Biol. Skr. Dan. Vidensk. Selsk.* **14**, 1.
Böcher, T. W., and Larsen, K. (1950). *Medd. Groenland* **147**, 1.
Böcher, T. W., and Larsen, K. (1958). *Bot. Notis.* **111**, 289.
Böcher, T. W., Larsen, K., and Rahn, K. (1955). *Hereditas* **41**, 423.
Borrill, M., and Carroll, C. P. (1969). *Cytologia* **34**, 6.
Borrill, M., Tyler, B., and Lloyd-Jones, M. (1971). *Cytologia* **36**, 1.
Bose, S. (1958). *Plant Life* (*Herbertia*) **14**, 37.
Bose, S. (1961). *Proc. Int. Bot. Congr., 9th, Montreal, 1959* **2**, 41.
Bose, S. (1963). *Nature* (*London*) **197**, 1229.
Bosemark, N. O. (1950). *Hereditas* **36**, 366.
Bosemark, N. O. (1954a). *Hereditas* **40**, 346.
Bosemark, N. O. (1954b). *Hereditas* **40**, 425.
Bosemark, N. O. (1956a). *Hereditas* **42**, 189.
Bosemark, N. O. (1956b). *Hereditas* **42**, 235.
Bosemark, N. O. (1956c). *Hereditas* **42**, 443.
Bosemark, N. O. (1957a). *Hereditas* **43**, 211.
Bosemark, N. O. (1957b). *Hereditas* **43**, 236.
Bosemark, N. O. (1967). *Hereditas* **57**, 239.
Bothmer, R. (1970). *Bot. Notis.* **123**, 519.
Bowden, W. M. (1959). *Can. J. Genet. Cytol.* **1**, 49.
Bowden, W. M. (1960a). *Can. J. Genet. Cytol.* **2**, 11.

Bowden, W. M. (1960b). *Can. J. Genet. Cytol.* **2**, 234.
Bowman, J. G., and Thomas, H. (1973). *Nature (London), New Biol.* **245**, 80.
Boyes, J. W. (1954). *Can. J. Zool.* **32**, 39.
Boyes, J. W., and Van Brink, J. M. (1965). *Can. J. Genet. Cytol.* **7**, 537.
Bramwell, D., Humphries, C. J., Murray, B. G., and Owens, S. J. (1971). *Bot. Notis.* **124**, 376.
Britton, D. M. (1951). *Brittonia* **7**, 233.
Bronkers, F. (1961). *Bull. Jard. Bot. (Brussels)* **31**, 429.
Brown, W. V. (1960). *Southwest. Natur.* **5**, 49.
Bruun, H. G. (1930). *Sv. Bot. Tidskr.* **24**, 468.
Bruun, H. G. (1931). *Proc. Int. Bot. Congr., 5th, Cambridge 1930* p. 243.
Bruun, H. G. (1932). *Symb. Bot. Upsal.* **1**, 1.
Buckton, K. E., and Cunningham, C. (1971). *Chromosoma* **33**, 268.
Burbanck, M. P. (1941). *Bot. Gaz. (Chicago)* **103**, 247.
Burdet, H. M. (1967). *Candollea* **22**, 107.
Burton, G. W. (1947). *J. Amer. Soc. Agron.* **39**, 551.
Callan, H. G. (1941). *J. Hered.* **32**, 296.
Câmara, A., Castro, D., and Noronha-Wagner, M. (1958). *Proc. Int. Congr. Genet., 10th, Montreal* **2**, 41.
Câmara, A., Castro, D., and Noronha-Wagner, M. (1959). *Agron. Lusitana* **21**, 193.
Cameron, D. R. (1934). *Univ. Calif., Berkeley, Publ. Agr. Sci.* **6**, 257.
Cameron, F. M., and Rees, H. (1967). *Heredity* **22**, 446.
Carlson, W. (1970). *Chromosoma* **30**, 356.
Carlson, W. (1973a). *Chromosoma* **42**, 127.
Carlson, W. (1973b). *Theor. Appl. Genet.* **43**, 147.
Carothers, E. E. (1917). *J. Morphol.* **28**, 445.
Carroll, C. P., and Borrill, M. (1965). *Genetica* **36**, 65.
Carroll, M. (1920). *J. Morphol.* **34**, 375.
Carter, C. R., and Smith-White, S. (1972). *Chromosoma* **39**, 361.
Catcheside, D. G. (1950). *Genet. Iber.* **2**, 139.
Catcheside, D. G. (1956). *Heredity* **10**, 345.
Cave, M. S. (1948). *Amer. J. Bot.* **35**, 343.
Celarier, R. P. (1956). *Field Lab.* **24**, 5.
Chambers, H. L. (1961). *Brittonia* **13**, 116.
Chandravadana, P., Rao, B. G. S., and Galinat, W. C. (1970). *Maize Genet. Coop. Newslett.* **44**, 134.
Chang, C. C., and Kikudome, G. Y. (1971). *Maize Genet. Coop. Newslett.* **45**, 134.
Chatterjee, K., Majhi, A., and Barik, S. K. (1971). *Caryologia* **24**, 447.
Chennaveeraiah, M. S., and Löve, A. (1959). *Can. J. Genet. Cytol.* **1**, 26.
Cheshmedzhirev, I. V. (1971). *Bot. Zh. (Leningrad)* **56**, 662.
Chilton, M. D., and McCarthy, B. J. (1973). *Genetics* **74**, 605.
Clausen, J. (1929). *Ann. Bot. (London)* **43**, 741.
Clay, S. N., and Nath, J. (1971). *Cytologia* **36**, 716.
Cleland, R. E. (1951). *Evolution* **5**, 165.
Cleland, R. E. (1967). *Evolution* **21**, 341.
Cleland, R. E. (1972). *In* "Oenothera: Cytogenetics and Evolution," pp. 79–81. Academic Press, New York.
Cleland, R. E., and Hyde, B. B. (1963). *Amer. J. Bot.* **50**, 179.
Clever, U. (1968). *Annu. Rev. Genet.* **2**, 11.
Covas, G., and Schnack, B. (1947). *Rev. Argent. Agron.* **14**, 224.

Crowder, L. V. (1953). *Amer. J. Bot.* **40**, 348.
Czapik, R. (1965). *Acta Biol. Cracov., Ser. Bot.* **8**, 21.
Czapska, D. (1959). *Acta Soc. Bot. Pol.* **28**, 129.
Dahlgren, R., Karlsson, T., and Lassen, P. (1971). *Bot. Notis.* **124**, 249.
Dark, S. O. S. (1932). *Ann. Bot. (London)* **46**, 965.
Darlington, C. D. (1929). *J. Genet.* **21**, 207.
Darlington, C. D. (1933). *Cytologia* **4**, 444.
Darlington, C. D. (1940). *J. Genet.* **39**, 101.
Darlington, C. D. (1941). *Ann. Bot. (London)* **5**, 203.
Darlington, C. D. (1956a). *Proc. Roy. Soc., Ser. B* **146**, 350.
Darlington, C. D. (1956b). "Chromosome Botany," pp. 19–27. Allen & Unwin, London.
Darlington, C. D., and Haque, A. (1966). *Chromosomes Today, Proc. Oxford Chromosome Conf., 1st, 1964* p. 102.
Darlington, C. D., and Shaw, G. W. (1959). *Heredity* **13**, 89.
Darlington, C. D., and Thomas, P. T. (1941). *Proc. Roy. Soc., Ser. B* **130**, 127.
Darlington, C. D., and Upcott, M. B. (1941). *J. Genet.* **41**, 275.
Darlington, C. D., and Wylie, A. P. (1955). "Chromosome Atlas of Flowering Plants." Allen and Unwin, London.
De Jong, D. C. D. (1965). *Publ. Mus. Mich. State Univ., Biol. Ser. 2* **9**, 433.
De Lesse, H., and Brown, K. S. (1971). *Bull. Soc. Entomol. Fr.* **76**, 131.
De Wet, J. M. J. (1957). *Cytologia* **22**, 145.
Dhillon, T. S., and Garber, E. D. (1962). *Amer. J. Bot.* **49**, 168.
Dover, G. A., and Riley, R. (1972). *Nature (London)* **240**, 159.
Dowrick, G. J. (1952). *Heredity* **6**, 365.
Dyer, A. F. (1964). *Cytologia* **29**, 155.
Ehrendorfer, F. (1957). *Naturwissenschaften* **14**, 405.
Ehrendorfer, F. (1959). *Chromosoma* **10**, 365.
Ehrendorfer, F. (1960). *Z. Vererbungsl.* **91**, 400.
Ehrendorfer, F. (1961). *Chromosoma* **11**, 523.
Emme, H. (1928). *Z. Indukt. Abstamm. Vererbungsl.* **47**, 99.
Ennis, T. (1972a). *Can. J. Genet. Cytol.* **14**, 113.
Ennis, T. (1972b). *Can. J. Genet. Cytol.* **14**, 851.
Evans, G. M., and Macefield, A. J. (1972). *Nature (London), New Biol.* **236**, 110.
Evans, G. M., and Macefield, A. J. (1973). *Chromosoma* **41**, 63.
Evans, H. J. (1960). *Heredity* **15**, 129.
Evans, W. L. (1956). *Cytologia* **21**, 417.
Fagerlind, F. (1936). *Hereditas* **22**, 189.
Favarger, C. (1963). *Bull. Soc. Neuchatel. Sci. Natur.* **86**, 101.
Favarger, C., and Küpfer, P. (1970). *Ber. Schweiz. Bot. Ges.* **80**, 269.
Federov, A. A., ed. (1969). "Chromosome Numbers of Flowering Plants." Acad. Sci. USSR, Leningrad.
Feinbrun, N. (1958). *Genetica* **29**, 172.
Feldman, M., Mello-Sampayo, T., and Sears, E. R. (1966). *Proc. Nat. Acad. Sci. U. S.* **56**, 1192.
Fernandes, A. (1939). *Sci. Genet.* **1**, 141.
Fernandes, A. (1943). *Bolm. Soc. Broteriana* **17**, 251.
Fernandes, A. (1948). *Bolm. Soc. Broteriana* **22**, 119.
Fernandes, A. (1949). *Bolm. Soc. Broteriana* **23**, 1.
Fernandes, A. (1950). *Agron. Lusitana* **12**, 551.

Fernandes, A. (1959a). *C. R. Acad. Sci.* **248,** 3672.
Fernandes, A. (1959b). *Bolm. Soc. Broteriana* **33,** 103.
Fernandes, A., and Mesquita, J. F. (1963). *Port. Acta Biol., Ser. A* **7,** 139.
Fernandes, A., Garcia, J., and Fernandes, R. (1948). *Mem. Soc. Broteriana* **4,** 1.
Fisher, T. R., and Cruden, R. C. (1962). *Ohio J. Sci.* **62,** 258.
Fisk, E. L. (1925). *Proc. Nat. Acad. Sci. U. S.* **11,** 352.
Flovik, K. (1938). *Hereditas* **24,** 265.
Flovik, K. (1940). *Hereditas* **26,** 430.
Fontana, P. G., and Vickery, V. R. (1973). *Chromosoma* **43,** 75.
Fröst, S. (1948). *Hereditas* **34,** 255.
Fröst, S. (1954). *Hereditas* **40,** 529.
Fröst, S. (1956). *Hereditas* **42,** 415.
Fröst, S. (1957). *Hereditas* **43,** 403.
Fröst, S. (1958a). *Hereditas* **44,** 75.
Fröst, S. (1958b). *Hereditas* **44,** 112.
Fröst, S. (1959). *Hereditas* **45,** 191.
Fröst, S. (1960). *Hereditas* **46,** 497.
Fröst, S. (1962). *Hereditas* **48,** 667.
Fröst, S. (1963). *Hereditas* **50,** 150.
Fröst, S. (1964). *Hereditas* **52,** 237.
Fröst, S. (1969a). *Hereditas* **61,** 317.
Fröst, S. (1969b). *Hereditas* **62,** 421.
Fröst, S., and Östergren, G. (1959). *Hereditas* **45,** 211.
Frost, H. B. (1931). *Proc. Nat. Acad. Sci. U. S.* **17,** 499.
Gadella, T. W. J. (1964). *Wentia* **11,** 1.
Gadella, T. W. J., and Kliphius, E. (1970). *Rev. Gen. Bot.* **77,** 487.
Gaiser, L. O. (1949). *Amer. J. Bot.* **36,** 122.
Garber, E. D. (1950). *Univ. Calif., Berkeley, Publ. Bot.* **23,** 283.
Garber, E. D. (1958). *Bot. Gaz. (Chicago)* **120,** 55.
Geslot, A., and Medus, J. (1971). *Can. J. Genet. Cytol.* **13,** 888.
Gibson, I., and Hewitt, G. M. (1970). *Nature (London)* **225,** 67.
Gibson, I., and Hewitt, G. M. (1972). *Chromosoma* **38,** 121.
Giles, N. (1941). *Bull. Torrey Bot. Club* **68,** 207.
Gill, J. J. B. (1971a). *Caryologia* **24,** 173.
Gill, J. J. B. (1971b). *Ann. Bot. (London)* **35,** 947.
Gill, J. J. B., Jones, B. M. G., Marchant, C. J., McLeish, J., and Ockendon, D. J. (1972). *Ann. Bot. (London)* **36,** 31.
Gill, L. S. (1970). *Phyton* **17,** 177.
Gill, L. S. (1971a). *Experientia* **27,** 596.
Gill, L. S. (1971b). *Bull. Torrey Bot. Club* **98,** 281.
Giménez Martin, G. (1958). *Phyton* **10,** 51.
Górecka, A. (1956). *Acta Soc. Bot. Pol.* **25,** 719.
Gotoh, K. (1924). *Shokubutsugaku Zasshi* **38,** 135.
Gotoh, K. (1932). *Nippon Idengaku Zasshi* **7,** 172.
Gotoh, K. (1937). *Nippon Idengaku Zasshi* **13,** 209.
Gotoh, K., and Kikkawa, R. (1937). *Nippon Idengaku Zasshi* **13,** 241.
Gould, F. W. (1959). *J. Range Manage.* **12,** 25.
Gould, F. W. (1966). *Can. J. Bot.* **44,** 1683.
Govindaswami, S. (1965). *Cytologia* **30,** 42.
Grun, P. (1959). *Amer. J. Bot.* **46,** 218.

Gupta, P. K. (1965). *Proc. Indian Acad. Sci., Sect. B* **62,** 155.
Gupta, P. K. (1969). *Cytologia* **34,** 429.
Haga, T. (1961). *Proc. Jap. Acad.* **37,** 627.
Haga, T., and Noda, S. (1956). *Senshokutai* **27/28,** 948.
Hair, J. B. (1963). *N. Z. J. Bot.* **1,** 243.
Hair, J. B. (1967). *N. Z. J. Bot.* **5,** 322.
Håkansson, A. (1945). *Bot. Notis.* p. 1.
Håkansson, A. (1948a). *Hereditas* **34,** 35.
Håkansson, A. (1948b). *Hereditas* **34,** 233.
Håkansson, A. (1949). *Hereditas* **35,** 375.
Håkansson, A. (1950). *Hereditas* **36,** 39.
Håkansson, A. (1954). *Hereditas* **40,** 523.
Håkansson, A. (1957). *Hereditas* **43,** 603.
Håkansson, A. (1959). *Hereditas* **45,** 623.
Hall, A. D. (1934). *Lily Yearb.* p. 35.
Hall, A. D. (1937). *J. Linn. Soc. London, Bot.* **50,** 481.
Hambler, D. J. (1954). *Nature (London)* **174,** 838.
Hambler, D. J. (1955). *Proc. Bot. Soc. Brit. Isles* **1,** 384.
Hambler, D. J. (1958). *Watsonia* **4,** 101.
Hambler, D. J. (1962). *Cytologia* **27,** 343.
Hance, R. T. (1918a). *Genetics* **3,** 225.
Hance, R. T. (1918b). *Biol. Bull.* **35,** 33.
Hanson, G. P. (1969). *Genetics* **63,** 601.
Hartung, M. E. (1946). *Amer. J. Bot.* **33,** 516.
Hasegawa, N. (1934). *Cytologia* **6,** 68.
Hayman, D. L., and Martin, P. G. (1965). *Aust. J. Biol. Sci.* **18,** 1081.
Hayman, D. L., and Martin, P. G. (1969). *In* "Comparative Mammalian Genetics" (K. Benirschke, ed.), pp. 191–217. Springer-Verlag, Berlin and New York.
Hayman, D. L., Martin, P. G., and Waller, P. F. (1969). *Chromosoma* **27,** 371.
Hazra, R., and Sharma, A. (1971). *Cytologia* **36,** 285.
Hedberg, O. (1952). *Hereditas* **38,** 256.
Heitz, E. (1927). *Abh. Naturwiss. Vereins Hamburg* **21,** 47.
Henderson, S. A. (1961). *Chromosoma* **12,** 553.
Hewitt, G. M. (1972). *Chromosomes Today* 3, 208.
Hewitt, G. M. (1973). *Chromosoma* **40,** 83.
Hewitt, G. M., and Brown, F. M. (1970). *Heredity* **25,** 363.
Hewitt, G. M., and John, B. (1967). *Chromosoma* **21,** 140.
Hewitt, G. M., and John, B. (1970). *Evolution* **24,** 169.
Hewitt, G. M., and John, B. (1971). *Chromosoma* **34,** 302.
Hewitt, G. M., and John, B. (1972). *Chromosoma* **37,** 23.
Hewitt, G. M., and Ruscoe, C. (1971). *J. Anim. Ecol.* **40,** 753.
Hewitt, G. M., and Schroeter, G. (1968). *Chromosoma* **25,** 121.
Heyn, C. (1963). *Scr. Hierosolymitana, Publ. Heb. Univ., Jerusalem* **12,** 1.
Hill, H. D., and Myers, W. M. (1948). *J. Amer. Soc. Agron.* **40,** 466.
Himes, M. (1967). *J. Cell Biol.* **35,** 175.
Hovin, A. W., and Hill, H. D. (1966). *Amer. J. Bot.* **53,** 702.
Humphrey, L. M. (1935). *Iowa State Coll. J. Sci.* **9,** 549.
Hunziker, J. H. (1961). *Rev. Invest. Agr.* **15,** 169.
Huskins, C. L., and Smith, S. G. (1934). *J. Genet.* **28,** 387.
Hutchison, D. J., and Bashaw, E. C. (1963). *Can. J. Genet. Cytol.* **5,** 281.

Inariyama, S. (1937). *Sci. Rep. Tokyo Bunrika Daigaku, Sect. B* 3, 95.
Islam, A. S., and Baten, A. (1952). *Nature (London)* **169**, 457.
Itoh, H. (1934). *Nippon Idengaku Zasshi* **10**, 115.
Jackson, R. C. (1960). *Evolution* **14**, 135.
Jackson, R. C., and Newmark, P. (1960). *Science* **132**, 1316.
Jackson, W. D., and Cheung, D. S. M. (1967). *Chromosoma* **23**, 24.
Janaki-Ammal, E. K. (1939). *Curr. Sci.* **8**, 210.
Janaki-Ammal, E. K. (1941). *J. Genet.* **41**, 217.
Jande, S. S. (1961). *Chromosoma* **12**, 318.
Jauhar, P. P. (1967). *Curr. Sci.* **36**, 244.
Jauhar, P. P., and Joshi, A. B. (1968). *Caryologia* **21**, 105.
Jauhar, P. P., and Joshi, A. B. (1969). *Cytologia* **34**, 222.
Jauhar, P. P., and Singh, U. (1970). *Genet. Iber.* **22**, 53.
Jörgensen, C. A., Sörensen, T. H., and Westergaard, M. (1958). *Biol. Skr. Dan. Vidensk. Selsk.* **9**, 1.
John, B., and Hewitt, G. M. (1965a). *Chromosoma* **16**, 548.
John, B., and Hewitt, G. M. (1965b). *Chromosoma* **17**, 121.
John, B., and Lewis, K. R. (1968). *Protoplasmatologia* **6A**, 45.
John, P. C. L., and Jones, R. N. (1970). *Exp. Cell Res.* **63**, 271.
Johnson, B. L. (1963). *Amer. J. Bot.* **50**, 228.
Johnsson, H. (1941). *Acta Univ. Lund.* **37**, 1.
Jones, K. (1956). *J. Genet.* **54**, 370.
Jones, K. (1962). *Genetica* **32**, 272.
Jones, K. (1963). *Amer. Orchid Soc., Bull.* **32**, 634.
Jones, K. (1964). *Chromosoma* **15**, 248.
Jones, K., and Borrill, M. (1962). *Genetica* **32**, 296.
Jones, K., and Jopling, C. (1972). *J. Linn. Soc. London, Bot.* **65**, 129.
Jones, K., and Smith, J. B. (1967). *Caryologia* **20**, 163.
Jones, R. N., and Rees, H. (1967). *Heredity* **22**, 333.
Jones, R. N., and Rees, H. (1968). *Chromosoma* **24**, 158.
Jones, R. N., and Rees, H. (1969). *Heredity* **24**, 265.
Joshi, S., and Raghuvanshi, S. S. (1970). *Ann. Bot. (London)* **34**, 1037.
Kamemoto, H., and Sagarik, R. (1967). *Amer. Orchid Soc., Bull.* **36**, 889.
Kapoor, B. M., and Beaudry, J. R. (1966). *Can. J. Genet. Cytol.* **8**, 422.
Karasawa, K. (1940). *Jap. J. Bot.* **11**, 129.
Karasawa, K. (1950). *Genetica* **25**, 188.
Kato, T. A. (1970). *Maize Genet. Coop. Newslett.* **44**, 18.
Kawano, S. (1966). *Shokubutsugaku Zasshi* **79**, 293.
Kayano, H. (1956a). *Mem. Fac. Sci., Kyushu Univ., Ser. E* **2**, 45.
Kayano, H. (1956b). *Mem. Fac. Sci., Kyushu Univ., Ser. E* **2**, 53.
Kayano, H. (1957). *Proc. Jap. Acad.* **33**, 553.
Kayano, H. (1962a). *Evolution* **16**, 86.
Kayano, H. (1962b). *Evolution* **16**, 246.
Kayano, H. (1971). *Heredity* **27**, 119.
Kayano, H., and Sannomiya, M. (1964). *Nippon Idengaku Zasshi* **39**, 352.
Kayano, H., Sannomiya, M., and Nakamura, K. (1960). *Nippon Idengaku Zasshi* **35**, 95.
Kayano, H., Sannomiya, M., and Nakamura, K. (1970). *Heredity* **25**, 113.
Keyl, H. G., and Hägele, K. (1971). *Chromosoma* **35**, 403.
Kimura, M., and Kayano, H. (1961). *Genetics* **46**, 1699.

Kirk, D., and Jones, R. N. (1970). *Chromosoma* **31**, 241.
Kishikawa, H. (1962). *Chromosome Inform. Serv.* **3**, 25.
Kishikawa, H. (1963). *Chromosome Inform. Serv.* **4**, 10.
Kishikawa, H. (1965). *Saga Daigaku Nogaku Iho* **21**, 1.
Kishikawa, H. (1966). *Nippon Idengaku Zasshi* **41**, 427.
Kishikawa, H. (1968). *Nippon Idengaku Zasshi* **43**, 33.
Kishikawa, H. (1970). *Jap. J. Breeding* **20**, 269.
Knowles, R. P. (1955). *Can. J. Bot.* **33**, 534.
Koller, P. C. (1932). *J. Genet.* **26**, 81.
Kootin-Sanwu, M. (1966). *Chromosomes Today, Proc. Oxford Chromosome Conf., 1st, 1964* p. 266.
Kootin-Sanwu, M., and Woodell, S. R. J. (1969). *Chromosomes Today* **2**, 192.
Kootin-Sanwu, M., and Woodell, S. R. J. (1970). *Caryologia* **23**, 225.
Kootin-Sanwu, M., and Woodell, S. R. J. (1971). *Heredity* **26**, 121.
Kosaki, K. (1958). *Proc. World Orchid Congr., 2nd, Cambridge, Mass.* p. 25.
Kranz, A. R. (1963). *Z. Pflanzenzuecht.* **50**, 44.
Kranz, A. R. (1968). *Z. Pflanzenzuecht.* **60**, 144.
Kranz, A. R. (1971). *Z. Pflanzenzuecht.* **66**, 317.
Krishnappa, D. G. (1971). *Proc. Indian Acad. Sci., Sect. B* **73**, 179.
Kruklis, M. V. (1971). *Dokl. Akad. Nauk SSSR* **196**, 1213.
Kumar, L. S., and Vishveshwaraiah, S. (1951). *Curr. Sci.* **20**, 211.
Kurabayashi, M., Lewis, H., and Raven, P. H. (1962). *Amer. J. Bot.* **49**, 1003.
Kurita, M. (1961). *Ehime Daigaku Kiyo, Dai-2-Bu* **4**, 251.
Kuwada, Y. (1915). *Shokubutsugaku Zasshi* **29**, 83.
Kuwada, Y. (1925). *Shokubutsugaku Zasshi* **39**, 227.
La Cour, L. F. (1951). *Heredity* **5**, 37.
La Cour, L. F. (1952). *Annu. Rep. John Innes Hort. Inst.* **42**, 47.
Langlet, O. F. I. (1927). *Sv. Bot. Tidskr.* **21**, 1.
Larsen, K. (1956). *Bot. Notis.* **109**, 293.
Larsen, K. (1960a). *Biol. Skr. Dan. Vidensk. Selsk.* **11**, 1.
Larsen, K. (1960b). *Hereditas* **46**, 312.
Larsen, K. (1963). *Dan. Bot. Ark.* **20**, 211.
Larter, L. N. H. (1932). *J. Genet.* **26**, 255.
Lee, W. J. (1963). *Korean J. Bot.* **6**, 15.
Lee, W. J. (1965). *Korean J. Bot.* **9**, 33.
Lee, W. J. (1968). *Proc. Int. Congr. Genet., 12th, Tokyo* **2**, 118.
Lee, W. J., and Min, B. R. (1965a). *Wheat Inform. Serv.* **21**, 27.
Lee, W. J., and Min, B. R. (1965b). *Korean J. Bot.* **8**, 1.
Lesly, M. M., and Frost, H. (1928). *Amer. Natur.* **62**, 22.
Levan, A. (1932). *Hereditas* **16**, 257.
Levin, D. A. (1967). *Evolution* **21**, 92.
Lewis, H. (1951). *Evolution* **5**, 142.
Lewis, H. (1968). *Proc. Int. Congr. Genet., 12th, Tokyo* **2**, 121.
Lewis, H., and Lewis, M. E. (1955). *Univ. Calif., Berkeley, Publ. Bot.* **20**, 241.
Lewis, K. R. (1967). *Nucleus* **10**, 99.
Lewis, W. H. (1970). *Science* **168**, 115.
Lewis, W. H., Stripling, H. L., and Ross, R. G. (1962). *Rhodora* **64**, 147.
Lewis, W. H., Oliver, R. L., and Luikart, T. J. (1971). *Science* **172**, 564.
Lewitsky, G. A. (1931). *Tr. Prikl. Bot., Genet. Selek.* **27**, 19.

Li, H. W., and Ma, T. H. (1950). *Proc. Congr. Int. Soc. Sugar-Cane Technol., Brisbane* p. 277.
Li, N., and Jackson, R. C. (1961). *Amer. J. Bot.* **48,** 419.
Lima-de-Faria, A. (1948). *Port. Acta Biol., Ser. A* **2,** 167.
Lima-de-Faria, A. (1949). *Hereditas* **35,** 77.
Lima-de-Faria, A. (1952). *Chromosoma* **5,** 1.
Lima-de-Faria, A. (1955). *Chromosoma* **7,** 51.
Lima-de-Faria, A. (1962). *Genetics* **47,** 1455.
Lin, C. C., Johnston, D. H., and Ramsden, R. O. (1972). *Can. J. Genet. Cytol.* **14,** 573.
Lindström, J. (1965). *Hereditas* **54,** 149.
Löve, A. (1954). *Vegetatio* **5/6,** 212.
Löve, A., and Löve, D., (1948). *Univ. Inst. Appl. Sci. Dep. Agr. Rep. (Reykjavik), Ser. B* No. 3, p. 1.
Löve, A., and Löve, D. (1956). *Acta Horti Gotoburg.* **20,** 65.
Löve, A., and Löve, D. (1961). *Opera Bot.* **5,** 1.
Longley, A. E. (1927). *J. Agr. Res.* **35,** 769.
Longley, A. E. (1938). *J. Agr. Res.* **56,** 177.
Longley, A. E. (1956). *Amer. J. Bot.* **43,** 18.
McClintock, B. (1933). *Z. Zellforsch. Mikrosk. Anat.* **19,** 191.
McClung, C. E. (1917). *J. Morphol.* **29,** 519.
McLeish, J. (1954). *Annu. Rep. John Innes Hort. Inst.* **45,** 21.
McMahon, B. (1936). *Cellule* **45,** 209.
Makino, S. (1951). "An Atlas of Chromosome Numbers in Animals," 2nd. Ed. Iowa State Coll. Press, Ames.
Malik, C. P. (1961). *Sci. Cult.* **27,** 197.
Malik, C. P., and Thomas, P. T. (1966). *Caryologia* **19,** 167.
Malik, C. P., and Tripathi, R. C. (1970). *Z. Biol. (Munich)* **116,** 321.
Manton, I. (1932). *Ann. Bot. (London)* **46,** 509.
Marchant, C. J. (1967). *Kew Bull.* **21,** 161.
Marchant, C. J. (1968). *J. Linn. Soc. London, Bot.* **60,** 411.
Marchant, C. J., and Brighton, C. A. (1971). *Chromosoma* **34,** 1.
Mather, K. (1932). *J. Genet.* **26,** 129.
Mather, K. (1934). *Roy. Hort. Soc. Lily Handb.* **3,** 38.
Mather, K. (1935). *Cytologia* **6,** 354.
Mather, K. (1945). *Proc. Roy. Soc., Ser. B* **132,** 308.
Matsuda, T. (1964). *Shokubutsugaku Zasshi* **77,** 139.
Matsuda, T. (1970a). *J. Sci. Hiroshima Univ., Ser. B, Div. 2 (Bot.)* **13,** 1.
Matsuda, T. (1970b). *J. Sci. Hiroshima Univ., Ser. B, Div. 2 (Bot.)* **13,** 65.
Matsuda, T. (1970c). *J. Sci. Hiroshima Univ., Ser. B, Div. 2 (Bot.)* **13,** 81.
Matsuura, H., and Suto, T. (1935). *J. Fac. Sci. Hokkaido Imp. Univ., Ser. 5* **5,** 33.
Maude, P. F. (1940). *New Phytol.* **39,** 17.
Mehra, K. L., Subramanyam, K. M., and Swaminathan, A. S. (1962). *J. Indian Bot. Soc.* **41,** 491.
Mehra, P. N., and Bawa, K. S. (1968). *Chromosoma* **25,** 90.
Meili-Frei, E. (1965). *Ber. Schweiz. Bot. Ges.* **75,** 219.
Melander, Y. (1950). *Hereditas* **36,** 19.
Melnyczenko, W. I. (1970). *Maize Genet. Coop. Newslett.* **44,** 203.
Mendelson, D., and Zohary, D. (1972). *Heredity* **29,** 329.

Meyer, J. R. (1944). *Genetics* **29**, 199.
Milinkovic, V. (1957). *Hereditas* **43**, 583.
Mitsuoka, S., and Ehrendorfer, F. (1972). *Oesterr. Bot. Z.* **120**, 155
Mochizuki, A. (1957). *Wheat Inform. Serv.* **5**, 9.
Mochizuki, A. (1960). *Wheat Inform. Serv.* **11**, 31.
Mochizuki, A. (1964). *Nippon Idengaku Zasshi* **39**, 356.
Moir, R. B., and Fox, D. P. (1972). *Silvae Genet.* **21**, 182.
Mookerjea, A. (1955). *Caryologia* **7**, 1.
Moore, R. J., and Frankton, C. (1962). *Can. J. Bot.* **40**, 281.
Moore, R. J., and Frankton, C. (1963). *Can. J. Bot.* **41**, 1553.
Moore, R. J., and Frankton, C. (1965). *Can. J. Bot.* **43**, 597.
Mooring, J. S. (1960). *Amer. J. Bot.* **47**, 847.
Mosquin, T., and Hayley, D. E. (1966). *Can. J. Bot.* **44**, 1209.
Moss, J. P. (1966). *Chromosomes Today, Proc. Oxford Chromosome Conf., 1st, 1964* p. 15.
Moss, J. P. (1969). *Chromosomes Today* **2**, 208.
Müntzing, A. (1943). *Hereditas* **29**, 91.
Müntzing, A. (1944). *Hereditas* **30**, 231.
Müntzing, A. (1945). *Hereditas* **31**, 457.
Müntzing, A. (1946a). *Hereditas* **32**, 97.
Müntzing, A. (1946b). *Hereditas* **32**, 127.
Müntzing, A. (1948a). *Heredity* **2**, 49.
Müntzing, A. (1948b). *Hereditas* **34**, 161.
Müntzing, A. (1948c). *Hereditas* **34**, 435.
Müntzing, A. (1949). *Proc. Int. Congr. Genet., 8th, Stockholm, 1948* p. 402.
Müntzing, A. (1950). *Hereditas* **36**, 507.
Müntzing, A. (1951). *Port. Acta Biol., Ser. A* Goldschmidt Vol., p. 831.
Müntzing, A. (1954a). *Hereditas* **40**, 459.
Müntzing, A. (1954b). *Caryologia* **6**, Suppl., 282.
Müntzing, A. (1957). *Hereditas* **43**, 682.
Müntzing, A. (1958). *Trans. Bose Res. Inst., Calcutta* **22**, 1.
Müntzing, A. (1959). *Proc. Int. Congr. Genet., 10th, Montreal, 1958* **1**, 453.
Müntzing, A. (1963). *Hereditas* **49**, 371.
Müntzing, A. (1966). *Chromosomes Today, Proc. Oxford Chromosome Conf., 1st, 1964* p. 7.
Müntzing, A. (1970). *Hereditas* **66**, 279.
Müntzing, A., and Akdik, S. (1948). *Hereditas* **34**, 248.
Müntzing, A., and Nygren, A. (1955). *Hereditas* **41**, 405.
Müntzing, A., and Prakken, R. (1941). *Hereditas* **27**, 273.
Müntzing, A., Jaworska, H., and Carlibom, C. (1968). *Wheat Inform. Serv.* **26**, 21.
Müntzing, A., Jaworska, H., and Carlibom, C. (1969). *Hereditas* **61**, 179.
Mukerjee, D., and Riley, H. P. (1961). *Bull. Ass. Southeast. Biologists* **8**, 22.
Mulligan, G. A. (1959). *Can. J. Bot.* **37**, 81.
Mulligan, G. A. (1964). *Can. J. Bot.* **42**, 1509.
Murray, B. E., and Craig, I. L. (1964). *Can. J. Genet. Cytol.* **6**, 170.
Murty, U. R. (1972). *Genetica* **43**, 84.
Murty, U. R., and Satyavathi, D. (1972). *Genetica* **43**, 575.
Nakao, M. (1911). *J. Coll. Agr., Tohoku Imp. Univ.* **4**, 173.
Nath, J., and Nielsen, E. L. (1961). *Amer. J. Bot.* **48**, 772.
Nath, J., and Nielsen, E. L. (1963). *Euphytica* **12**, 161.

Nazarova, A. (1969). *Bot. Zh. (Leningrad)* **54,** 929.
Nel, P. (1973). *Theor. Appl. Genet.* **43,** 196.
Neves, J. B. (1952). *Bolm. Soc. Broteriana* **26,** 5.
Neves, J. B. (1962). *Bolm. Soc. Broteriana* **34,** 151.
Nielsen, E. L. (1955). *Bot. Gaz. (Chicago)* **116,** 293.
Nilsson, O., and Lassen, P. (1971). *Bot. Notis.* **124,** 270.
Noda, S. (1953). *Mem. Fac. Sci., Kyushu Univ., Ser. E* **1,** 139.
Noda, S. (1955). *Senshokutai* **22/24,** 779.
Noda, S. (1956). *Mem. Fac. Sci., Kyushu Univ., Ser. E* **2,** 95.
Noda, S. (1964). *Bull. Osaka Gakuin Univ.* **2,** 125.
Noda, S. (1967). *Jap. J. Breeding* **17,** 173.
Noda, S. (1968). *Bull. Osaka Gakuin Univ.* **10,** 127.
Noda, S., and Watanabe, H. (1961). *Nippon Idengaku Zasshi* **36,** 391.
Noda, S., and Watanabe, H. (1968). *Bull. Osaka Gakuin Univ.* **11,** 105.
Noronha-Wagner, M., and Castro, D. (1952). *Sci. Genet.* **4,** 149.
Nur, U. (1962a). *Chromosoma* **13,** 249.
Nur, U. (1962b). *Genetics* **47,** 1679.
Nur, U. (1963). *Chromosoma* **14,** 407.
Nur, U. (1966a). *Genetics* **54,** 1225.
Nur, U. (1966b). *Genetics* **54,** 1239.
Nur, U. (1968). *Proc. Int. Congr. Genet., 12th, Tokyo* **2,** 119.
Nur, U. (1969a). *Chromosoma* **27,** 1.
Nur, U. (1969b). *Chromosoma* **28,** 280.
Nur, U., and Nevo, E. (1969). *Caryologia* **22,** 97.
Nybom, N. (1947). *Hereditas* **33,** 571.
Nygren, A. (1954). *Hereditas* **40,** 377.
Nygren, A. (1957). *Kgl. Lantbruks Hoegsk. Ann.* **23,** 489.
Nygren, A. (1962). *Ann. Acad. Regiae Sci. Upsal.* **6,** 5.
Ochlewska, M. (1965). *Acta Biol. Cracov., Ser. Bot.* **8,** 135.
Östergren, G. (1945). *Bot. Notis.* **2,** 157.
Östergren, G. (1947). *Hereditas* **33,** 261.
Östergren, G., and Fröst, S. (1962). *Hereditas* **48,** 363.
Ogihara, R. (1960). *Senshokutai* **44/45,** 1500.
Ogihara, R. (1962). *Senshokutai* **53/54,** 1785.
Ornduff, R., Raven, P. H., Kyhos, D. W., and Kruckeberg, A. R. (1963). *Amer. J. Bot.* **50,** 131.
Packer, J. G. (1964). *Can. J. Bot.* **42,** 473.
Paliwal, R. L., and Hyde, B. B. (1958). *Proc. Int. Congr. Genet., 10th, Montreal* **2,** 212.
Paliwal, R. L., and Hyde, B. B. (1959). *Amer. J. Bot.* **46,** 460.
Palmblad, I. G. (1965). *Can. J. Bot.* **43,** 715.
Pantulu, J. V. (1960). *Curr. Sci.* **29,** 28.
Pantulu, J. V., and Manga, V. (1972). *Cytologia* **37,** 389.
Papeš, D. (1971). *Acta Biol. Iugoslav., Genet.* **3,** 261.
Patton, J. L. (1972). *Chromosoma* **36,** 241.
Payne, W. W., Raven, P. H., and Kyhos, D. W. (1964). *Amer. J. Bot.* **51,** 419.
Peto, F. H. (1933). *J. Genet.* **28,** 113.
Pfitzer, P. (1957a). *Chromosoma* **8,** 436.
Pfitzer, P. (1957b). *Chromosoma* **8,** 545.
Phitos, D., and Damboldt, P. J. (1971). *Ann. Naturhistor. Mus. Wien* **75,** 157.

Piennaar, R. V. (1963). *J. S. Afr. Bot.* **29**, 111.
Podlech, D., and Damboldt, J. (1963). *Ber. Deut. Bot. Ges.* **76**, 360.
Pohl, R. W., and Davidse, G. (1971). *Brittonia* **23**, 293.
Powell, A. M., and Turner, B. L. (1963). *Madrono* **17**, 128.
Powell, J. B., and Burton, G. W. (1966). *Crop Sci.* **6**, 131.
Price, S. (1963). *J. Hered.* **54**, 13.
Pritchard, E. (1968). *Can. J. Genet. Cytol.* **10**, 928.
Puteyevsky, E., and Zohary, D. (1970). *Chromosoma* **32**, 135.
Putiievsky, E., and Katznelson, J. (1970). *Chromosoma* **30**, 476.
Raghavan, T. S., and Venkatasubban, K. R. (1940). *Cytologia* **11**, 55.
Ramachandran, K. (1969). *Cytologia* **34**, 213.
Ramel, C. (1969). *Hereditas* **63**, 464.
Randolph, L. F. (1928). *Anat. Rec.* **41**, 102.
Randolph, L. F. (1941). *Genetics* **26**, 608.
Randolph, L. F., and Mitra, J. (1961). *Amer. J. Bot.* **48**, 862.
Rao, C. K. (1968). *Curr. Sci.* **37**, 591.
Rao, C. K. (1972). *Proc. Indian Acad. Sci., Sect. B* **75**, 117.
Rapp, K. (1972). *Blyttia* **30**, 101.
Raven, P. H., Solbrig, O. T., Kyhos, D. W., and Snow, R. (1960). *Amer. J. Bot.* **47**, 124.
Rees, H. (1972). *Brookhaven Symp. Biol.* **23**, 394.
Rees, H., and Hitchinson, J. (1973). *Cold Spring Harbor Symp. Quant. Biol.* **38**, 175.
Rees, H., and Jamieson, A. (1954). *Nature (London)* **173**, 43.
Reeves, R. G. (1925). *Proc. Iowa Acad. Sci.* **32**, 171.
Resende, F., and Da Franca, P. (1946). *Port. Acta Biol., Ser. A* **1**, 289.
Rhoades, M. M. (1968). *In* "Replication and Recombination of Genetic Material" (W. J. Peacock and R. D. Brock, eds.), p. 229. Australian Acad. Sci. Canberra.
Rhoades, M. M., and Dempsey, E. (1972). *Genetics* **71**, 73.
Rhoades, M. M., and Dempsey, E. (1973). *J. Hered.* **64**, 13.
Rhoades, M. M., Dempsey, E., and Ghidoni, A. (1967). *Proc. Nat. Acad. Sci. U. S.* **57**, 1626.
Riley, H. P. (1962). *Can. J. Genet. Cytol.* **4**, 50.
Riley, H. P., and Mukerjee, D. (1962). *Cytologia* **27**, 325.
Robertson, J. G. (1966). *Can. J. Genet. Cytol.* **8**, 695.
Röthlisberger, E. (1970). *Ber. Schweiz. Bot. Ges.* **80**, 195.
Roman, H. (1947). *Genetics* **32**, 391.
Roman, H. (1948a). *Genetics* **33**, 122.
Roman, H. (1948b). *Proc. Nat. Acad. Sci. U. S.* **34**, 36.
Rothfels, K. (1950). *J. Morphol.* **87**, 287.
Rozmus, M. (1958). *Acta Biol. Cracov., Ser. Bot.* **1**, 171.
Rozmus, M. (1963). *Acta Biol. Cracov., Ser. Bot.* **6**, 115.
Rutishauser, A. (1956a). *Heredity* **10**, 195.
Rutishauser, A. (1956b). *Heredity* **10**, 367.
Rutishauser, A. (1960a). *Beih. Z. Schweiz. Forstwereins* **30**, 93.
Rutishauser, A. (1960b). *Heredity* **15**, 241.
Rutishauser, A. (1963). *Genet. Today, Proc. Int. Congr., 11th, The Hague* **1**, 119.
Rutishauser, A., and Röthlisberger, E. (1966). *Chromosomes Today, Proc.* Oxford Chromosome Conf., *1st, 1964* p. 28.
Ruttle, M. L. (1931). *Gartenbauwissenschaft* **4**, 428.

Sacristán, M. D. (1966). *Ann. Estac. Exp. Aula Dei* **8**, 1.
Sadanga, K. (1962). *Can. J. Genet. Cytol.* **4**, 302.
Sagawa, Y. (1962). *Amer. Orchid Soc., Bull.* **31**, 459.
Samejima, J. (1958). *Cytologia* **23**, 159.
Sannomiya, M. (1962). *Chromosome Inform. Serv.* **3**, 30.
Sannomiya, M. (1963). *Chromosome Inform. Serv.* **4**, 5.
Sannomiya, M. (1964). *Nippon Idengaku Zasshi* **39**, 366.
Sannomiya, M., and Kayano, H. (1968). *Proc. Int. Congr. Genet., 12th, Tokyo* **2**, 116.
Sannomiya, M., and Kayano, H. (1969). *Nippon Idengaku Zasshi* **44**, Suppl. 1, 84.
Sarvella, P. (1959). *Hereditas* **45**, 505.
Sâto, D. (1942). *Jap. J. Bot.* **12**, 57.
Saunte, H. L. (1955). *Hereditas* **41**, 499.
Saylor, L. C., and Simons, H. A. (1970). *Cytologia* **35**, 294.
Schmidt, A. (1961). *Oesterr. Bot. Z.* **108**, 20.
Schulz-Schaeffer, J. (1956). *Z. Pflanzenzuecht.* **35**, 297.
Schulz-Schaeffer, J. (1966). *Z. Pflanzenzuecht.* **55**, 21.
Schweizer, D. (1973). *Chromosoma* **40**, 307.
Semple, J. C. (1972). *Science* **175**, 666.
Shah, S. S. (1963). *Chromosoma* **14**, 162.
Shah, S. S. (1964a). *Heredity* **19**, 736.
Shah, S. S. (1964b). *Cytologia* **29**, 387.
Shah, S. S. (1965). *Heredity* **20**, 470.
Shah, S. S. (1967). *Heredity* **22**, 443.
Shambulingappa, K. G. (1968). *Cytologia* **33**, 539.
Shambulingappa, K. G. (1970). *J. Agr. Univ. P. R.* **54**, 401.
Sharma, A. K., and Aiyanger, H. R. (1961). *Chromosoma* **12**, 310.
Sharma, A. K., and Bal, A. K. (1956). *Cytologia* **21**, 329.
Shcherbakov, E. S. (1966). *Genetika* **4**, 30.
Shellhammer, H. S. (1969). *Chromosoma* **27**, 102.
Shimizu, M. (1971). *Shokubutsu Kenkyu Zasshi* **46**, 60.
Shindo, K. (1965). *Shokubutsugaku Zasshi* **78**, 374.
Shopova, M. (1966). *Chromosoma* **19**, 149.
Simchen, G., Zarchi, Y., and Hillel, J. (1971). *Chromosoma* **33**, 63.
Simon, W. (1968). *Caryologia* **21**, 181.
Singh, D. N. (1965). *Caryologia* **18**, 547.
Singh, D. N. (1972). *Can. J. Genet. Cytol.* **14**, 175.
Singh, D. N., and Godward, M. B. E. (1963). *Heredity* **18**, 538.
Singh, D. N., and Singh, U. (1966). *Genet. Iber.* **18**, 205.
Skalińska, M. (1956). *Acta Soc. Bot. Pol.* **25**, 713.
Skalińska, M., Czapik, R., and Piotrowicz, M. *et al.* (1959). *Acta Soc. Bot. Pol.* **28**, 487.
Skovsted, A. (1934). *Dan. Bot. Ark.* **8**, 1.
Skovsted, A. (1935). *J. Genet.* **31**, 263.
Skovsted, A. (1939). *Ct. R. Trav. Lab. Carlsberg, Ser. Physiol.* **22**, 427.
Skovsted, A. (1941). *Ct. R. Trav. Lab. Carlsberg, Ser. Physiol.* **23**, 195.
Small, E. (1971). *Evolution* **25**, 330.
Smith, E. B. (1968). *Evolution* **22**, 748.
Smith, D. M., and Levin, D. A. (1967). *Amer. J. Bot.* **54**, 324.
Smith, S. G. (1953). *Heredity* **7**, 31.

Smith, S. G. (1956). *J. Hered.* **47**, 157.
Smith, S. G. (1960). *Can. J. Genet. Cytol.* **2**, 66.
Smith, S. G. (1965). *Can. J. Genet. Cytol.* **7**, 363.
Smith, S. G. (1966). *Can. J. Genet. Cytol.* **8**, 744.
Smith, S. G. (1972). *Chromosomes Today* **3**, 197.
Smith-White, S. (1968). *Chromosoma* **23**, 359.
Smith-White, S., and Carter, C. R. (1970). *Chromosoma* **30**, 129.
Smith-White, S., Carter, C. R., and Stace, H. M. (1970). *Aust. J. Bot.* **18**, 99.
Snow, R. (1959). *Madrono* **15**, 81.
Snow, R. (1960). *Amer. J. Bot.* **47**, 302.
Snow, R. (1963). *Amer. J. Bot.* **50**, 337.
Solbrig, O. T., Anderson, L. C., Kyhos, D. W., Raven, P. H., and Rüdenberg, L. (1964). *Amer. J. Bot.* **51**, 513.
Southern, D. I. (1967). *Chromosoma* **22**, 227.
Soyano, S. (1959). *Senshokutai* **40**, 1362.
Sparrow, A. H., Pond, V., and Sparrow, R. C. (1952). *Amer. Natur.* **86**, 277.
Srivastava, M. D. L. (1954). *J. Genet.* **52**, 480.
Stebbins, G. L. (1963). "Variation and Evolution in Plants." Columbia Univ. Press, New York.
Stephens, R. T., and Bregman, A. A. (1972). *Chromosoma* **38**, 297.
Stephenson, E. M., Robinson, E. S., and Stephenson, N. G. (1972). *Can. J. Genet. Cytol.* **14**, 691.
Stevens, N. M. (1908). *J. Exp. Zool.* **5**, 453.
Stewart, R. N. (1943). *Bot. Gaz. (Chicago)* **104**, 620.
Stewart, R. N. (1947). *Amer. J. Bot.* **34**, 9.
Stolze, K. (1925). *Bibl. Genet.* **8**, 1.
Suda, Y. (1962). *Sci. Rep. Tohoku Univ., Ser. 4* **28**, 97.
Sullivan, T. D. (1947). *Bull. Torrey Bot. Club* **74**, 453.
Sun, S. (1963). Unpublished data.
Sveshnikova, L. I. (1971). *Bot. Zh. (Leningrad)* **56**, 118.
Swaminathan, M. S., and Nath, J. (1956). *Curr. Sci.* **25**, 123.
Swanson, C. P. (1943). *Bot. Gaz. (Chicago)* **105**, 108.
Swezy, O. (1935). *Amer. Natur.* **69**, 383.
Takemoto, T. (1954). *Nippon Idengaku Zasshi* **29**, 177.
Takenouchi, Y., and Muramoto, N. (1972). *Nippon Idengaku Zasshi* **47**, 291.
Tanaka, N. (1939a). *Cytologia* **10**, 51.
Tanaka, N. (1939b). *Nippon Idengaku Zasshi* **15**, 96.
Tanaka, R. (1965). *Shokubutsu Kenkyu Zasshi* **40**, 65.
Tanaka, R., and Matsuda, T. (1972). *Shokubutsugaku Zasshi* **85**, 43.
Tateoka, T. (1955). *Cytologia* **20**, 296.
Tateoka, T. (1972a). *Bull. Nat. Sci. Mus., Tokyo* **15**, 385.
Tateoka, T. (1972b). *Bull. Nat. Sci. Mus., Tokyo* **15**, 461.
Teppner, H. (1971). *Oesterr. Bot. Z.* **119**, 196.
Titz, W. (1967). *Ber. Deut. Bot. Ges.* **79**, 474.
Tobgy, H. (1943). *J. Genet.* **45**, 67.
Torres, A. M. (1962). *Amer. J. Bot.* **49**, 1033.
Traub, H. P. (1945). *Herbertia* **12**, 38.
Traub, H. P. (1963). *In* "The Genera of Amaryllidaceae." Amer. Plant Life Soc., La Jolla, California.

Tschermak-Woess, E., and Hasitschka-Jenschke, G. (1963). *Oesterr. Bot. Z.* **110,** 468.
Tschermak-Woess, E., and Schiman, H. (1960). *Oesterr. Bot. Z.* **107,** 212.
Turner, B. L., and Flyr, D. (1966). *Amer. J. Bot.* **53,** 24.
Turner, B. L., and Irwin, H. S. (1960). *Rhodora* **62,** 122.
Turner, B. L., and King, R. M. (1964). *Southwest. Natur.* **9,** 27.
Turner, B. L., and Lewis, W. H. (1965). *J. S. Afr. Bot.* **31,** 207.
Turner, B. L., Powell, M., and King, R. M. (1962). *Rhodora* **64,** 251.
Ueshima, N. (1967). *Chromosoma* **20,** 311.
Uhl, C. H. (1948). *Amer. J. Bot.* **35,** 695.
Uhl, C. H. (1961). *Amer. J. Bot.* **48,** 114.
Ullerich, F. H. (1967). *Chromosoma* **21,** 345.
Vardi, A., and Dover, G. A. (1972). *Chromosoma* **38,** 367.
Ved Brat, S. (1965). *Chromosoma* **16,** 486.
Venkateswarlu, J., and Pantulu, J. V. (1970). *Cytologia* **35,** 444.
Venkateswarlu, J., Rao, P. N., and Chaganti, R. S. K. (1965). *Maize Genet. Coop. Newslett.* **39,** 182.
Venkateswarlu, J., Rao, P. N., and Chaganti, R. S. K. (1968). *Maize Genet. Coop. Newslett.* **42,** 2.
Viinikka, Y. (1970). *Ann. Bot. Fenn.* **7,** 203.
Virkki, N. (1954). *Ann. Acad. Sci. Fenn., Ser. A4* **26,** 1.
Vosa, C. G. (1962). *Chromosome Inform. Serv.* **3,** 26.
Vosa, C. G. (1965). *Heredity* **20,** 467.
Vosa, C. G. (1966). *Chromosomes Today, Proc. Oxford Chromosome Conf., 1st, 1964* p. 24.
Vosa, C. G. (1969a). *Chromosomes Today* **2,** 189.
Vosa, C. G. (1969b). *Chromosomes Today* **2,** 209.
Vosa, C. G., and Barlow, P. W. (1972). *Caryologia* **25,** 1.
Walker, S. W., and Gregson, N. (1966). *Chromosomes Today, Proc. Oxford Chromosome Conf., 1st, 1964* p. 266.
Ward, E. J. (1973a). *Genetics* **73,** 387.
Ward, E. J. (1973b). *Chromosoma* **43,** 177.
Warden, J. W. (1958). *Port. Acta Biol., Ser. A* **5,** 186.
Warden, J. W. (1959). *Bol. Soc. Argent. Bot.* **7,** 209.
Wedberg, H. L., Lewis, H., and Venkatesh, C. S. (1968). *Evolution* **22,** 93.
Weitz, S., and Feinbrun, N. (1972). *Isr. J. Bot.* **21,** 9.
Wells, J. R. (1965). *Brittonia* **17,** 144.
Wenrich, D. H. (1917). *J. Morphol.* **29,** 471.
Whitaker, T. W. (1936). *Amer. J. Bot.* **23,** 517.
White, M. J. D. (1951a). *Evolution* **5,** 376.
White, M. J. D. (1951b). *Advan. Genet.* **4,** 267.
White, M. J. D. (1957). *Annu. Rev. Entomol.* **2,** 71.
White, M. J. D. (1973). "Animal Cytology and Evolution," 3rd Ed. Cambridge Univ. Press, London and New York.
White, M. J. D., Carson, H. L., and Cheney, J. (1964). *Evolution* **18,** 417.
Wiens, D. (1964a). *Amer. J. Bot.* **51,** 1.
Wiens, D. (1964b). *Brittonia* **16,** 11.
Wilson, E. B. (1907). *Science* **26,** 870.
Witsch, H. (1950). *Nachr. Akad. Wiss. Goettingen, Math.-Phys. Kl., 2A* **4,** 21.
Wolf, B. E. (1954). *Verh. Deut. Zool. Ges.* p. 282.

Wolf, B. E. (1961). *Verh. Deut. Zool. Ges.* p. 110.
Wood, C. E. (1949). *Rhodora* **51**, 193.
Woods, M. W., and Bamford, R. (1937). *Amer. J. Bot.* **24**, 175.
Wulff, H. D. (1939). *Ber. Deut. Bot. Ges.* **57**, 84.
Wylie, A. P. (1952). *Heredity* **6**, 137.
Yamamoto, Y. (1937). *Cytologia Fujii Vol.*, p. 181.
Yong, H. S., and Dhaliwal, S. S. (1972). *Chromosoma* **36**, 256.
Zarchi, Y., Simchen, G., Hillel, J., and Schapp, T. (1972). *Chromosoma* **38**, 77.
Zečević, L., and Paunović, D. (1967). *Biol. Plant.* **9**, 205.
Zečević, L., and Paunović, D. (1969). *Chromosoma* **27**, 198.
Zohary, D., and Ashkenazi, I. (1958). *Nature (London)* **182**, 477.
Zohary, D., and Nur, U. (1959). *Evolution* **13**, 311.
Zucconi, L. (1957). *Caryologia* **9**, 226.

The Intracellular Neutral SH-Dependent Protease Associated with Inflammatory Reactions

HIDEO HAYASHI

Department of Pathology, Kumamoto University Medical School, Kumamoto, Japan

I. Introduction

Like most fundamental concepts of pathology, inflammation is difficult to define. In vertebrates it can perhaps be described as the local reaction to injury of the living microcirculation and its associated tissues, which includes blood leukocytes and such features of perivascular tissue as mast cells and histiocytes. Any noxious stimulus can provoke an inflammatory reaction. When such injury is associated with the death of tissue, the reaction is often found at the periphery of the lesion where the microcirculation survives.

The inflammatory response has attracted renewed interest in recent years with the realization that many diseases exist in which a large portion of tissue damage results from the inflammatory response itself. Under these conditions the stimulus provoking the reaction, unlike simple bacterial invasion, would be innocuous were it not for its ability to induce an inflammatory reaction. Allergic diseases fall into this category.

The inflammatory response is a series of events rather than a single event, and follows a course that is essentially uniform. Different types of injury do, however, lead to variation in the relative intensity and duration of particular aspects of the reaction. Burdon-Sanderson (1888, quoted in Florey, 1970) defined the process of inflammation as "the succession of changes which occurs in a living tissue when it is injured, provided that the injury is not of such a degree as at once to destroy its structure and vitality."

The events of inflammation fall naturally into two broad divisions involving the fluid and cellular phases of circulation. The fluid-phase reaction consists of transient vasoconstriction followed by substained dilation of arterioles, capillaries, and venules, during which blood flow is increased and subsequently decreased, and permeability to plasma protein is raised. The cellular response consists of changed function of tissue cells (including local histiocytes and mast cells), migration of various types of leukocytes from blood vessels, and proliferation of migrated cells.

In spite of long-standing recognition of the inflammatory reaction and of its importance as a fundamental pathological process, present knowledge remains rudimentary. Spector and Willoughby (1968) have arrayed and criticized, in their monograph entitled *The Pharmacology of Inflammation,* the substances that have been suspected as chemical mediators of inflammation. The work of Virchow (1858), Krogh (1922), and Lewis (1927) suggested the basic concept of humoral factors in inflammation. Since the advent of techniques for isolating individual substances, considerable confusion has unfortunately reigned in this field as a result of inadequate characterization of hypothetical mediators crudely separated from inflammatory tissues or exudates; and the criteria for determining whether a certain substance is involved in inflammation have been on occasion surprisingly feeble. Accordingly, all the so-called inflammatory chemical substances suggested should be carefully reexamined according to the criteria established for them.

As is well known, Menkin (1940) emphasized in his monograph that all the reactions in inflammation including wound healing are brought about by the release of endogenous substances; and *leukotaxine* was demonstrated as one of his major findings. This material is capable of increasing vascular permeability and of attracting leukocytes. He claimed that this substance is a chemically pure material, but other workers have reported that an increase in vascular permeability induced by leukotaxine is due to its histamine-liberating action (Miles and Miles, 1952; Yoshinaga *et al.*, 1966b), and that it consists of a mixture of peptides. There has been no consideration of the criteria necessary for inflammatory chemical substance. Also, many of Menkin's (1950, 1956) factors are now regarded as highly controversial and certainly many are now no longer in favor. However, there is no doubt that Menkin made important contributions to the concept of biochemical mechanisms of inflammation, and his work has served as a fruitful stimulus to subsequent investigators. The fundamental understanding of such an inflammatory process certainly needs careful study from the viewpoint of both biochemistry and morphology, because a biochemical change in each event always precedes the associated morphological change.

There have been several review articles on the biochemical mechanisms of inflammation, which are well-written and thoughtful (Menkin, 1956; Miles, 1956; Spector and Willoughby, 1963; Wilhelm, 1962; Cochrane, 1967; Movat, 1971). In this article particular attention is paid to the criteria for determining whether chemical substances isolated from inflammatory tissues or exudates, especially intracellular neutral SH-dependent protease are involved in inflammation.

II. Intracellular Neutral SH-Dependent Protease

As is well known, the role of proteases in inflammation has received increased attention. Since the description of leukoprotease by Opie (1906), the involvement of proteases in the mechanism of inflammation has been strongly suggested by several investigators. Among these reports, Menkin's (1940) observations that tryptic digests of protein developed properties similar to those of leukotaxine (capable of increasing vascular permeability and of attracting leukocytes) were perhaps the most stimulating.

Protease systems, which may be concerned with inflammatory reactions, fall naturally into two broad divisions, namely, the plasma protease system and the cellular protease system. The activation of the plasma protease system was first associated with anaphylaxis and in part with allergic inflammation. As seen in a review on the early stage of the problem by Ungar and Hayashi (1958), the significance of this association remained controversial. The current interest in plasma proteases stems from the fact that they induce the formation of plasma kinins, showing a most potent vascular permeability-increasing effect, as discussed in the reviews by Miller and Melmon (1970), Wilhelm (1971), and Donaldson (1970). However, the cellular protease system, which may participate in the production of inflammatory reactions, has been less well studied. The first such enzyme described in some detail was the neutral SH-dependent protease isolated from active Arthus lesions, cutaneous burns, and cultured mononuclear cells or histiocytes (Hayashi, 1955; Hayashi *et al.*, 1958, 1960).

Neutral cellular proteases have been demonstrated in a wide variety of human and animal tissues and in specific types of cells including fibroblasts (Schwabe and Kalnistsky, 1966), mast cells (Pastan and Almquist, 1966a,b), polymorphonuclear (PMN) leukocytes (Janoff, 1972; Janoff and Scherrer, 1968; Janoff and Zeligs, 1968; Lazarus *et al.*, 1967; Ziff *et al.*, 1960; Evanson *et al.*, 1967; Weissman and Spilberg, 1968; Davies *et al.*, 1971; Wasi *et al.*, 1966; Kouno, 1971; Kouno *et al.*, 1972), mononuclear cells or histiocytes (Hayashi *et al.*, 1960; Hayashi, 1967; Udaka and Udaka, 1969; Cohn and Wiener, 1963a,b; Kato *et al.*, 1972), and lymphocytes (Kambara, 1973). To date, five different cathepsins, A through E, have been described with pH optima, in the main, well below neutrality. Cathepsins D and E appear to be responsible for the major part of the acid protease activity of lysosomes from rabbit, bovine, and human spleen. The widespread distribution of neutral tissue proteases suggested that these enzymes are directed at multiple physiological substrates; and the degradation of numerous proteins by such proteases from different tissues or cells has naturally received particular attention.

A. Criteria for Inflammatory Protease

With the advance of techniques to isolate and purify enzymes, several investigations have been performed to determine the protease systems responsible for inflammatory reactions, but present knowledge of these systems is still incomplete. Although several types of proteases have been proposed as agents for inflammatory reactions, satisfactory criteria for the proteases responsible for the reactions have not yet been established. It is therefore important that the following criteria are applied to the suspected proteases, and that they fulfill some if not all of the conditions listed (Hayashi, 1967; Hayashi *et al.*, 1969).

1. Proteases should be locally available to induce inflammatory reactions.
2. Protease activities should parallel the time course of the reactions.
3. The optimum pH of proteases should be similar to the pH value at the inflamed site.
4. Proteases should produce morphological changes similar to the reactions when injected in concentrations reasonably comparable to those detected at the inflamed site.
5. Proteases, when injected locally in reasonable amounts, should produce the chemical mediators associated with individual inflammatory

manifestations, such as vascular permeability changes or leukocyte migration, in the reactions.

6. Proteases should be inhibited by a specific antagonistic substance locally available.

7. Inflammatory reactions should be suppressed by a specific antagonistic substance.

8. Depletion of the proteases or their precursors should cause a decrease in inflammatory reactions.

B. Time Course of Proteolytic Activity in the Inflammatory Process

As is well-known, the activity of many types of proteases has been demonstrated in inflammatory tissue. However, in spite of the special importance of criteria 1 and 2 described above, studies of the proteolytic mechanism in parallel with the inflammatory process have been surprisingly few. A protease, which hydrolyzes casein or hemoglobin at neutral pH, was extracted in the euglobulin fraction of inflammatory tissue; its activity increased with the development of the inflammatory reaction, reaching a peak at the maximum stage of the reaction and decreasing as the reaction subsided. For instance, the proteolytic activity in an active Arthus reaction in the skin of rabbit increased shortly after intradermal injection of antigen (2.5 mg bovine serum albumin), reached its peak in about 12 hours, and almost disappeared in about 72 hours (Hayashi *et al.*, 1958, 1965a; Udaka, 1963a) (criteria 1 and 2) (Fig. 1). Similar

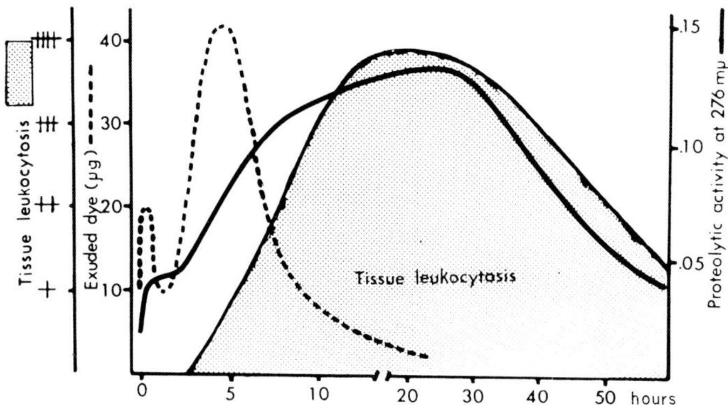

Fig. 1. Time course of increased vascular permeability, PMN migration, and neutral SH-dependent protease activity in an active Arthus reaction (induced by 2.5 mg BSA) in rabbit. Vascular permeability change (broken line) showing two phases—immediate and delayed—was followed by PMN migration. In connection with these inflammatory events, the activity of neutral SH-dependent protease (solid line) was demonstrated.

proteolytic activity in a moderate thermal injury in the skin of rabbit also increased shortly after heating, reached its maximum in 3–4 hours, and then declined in intensity (Sakabe, 1962; Tokaji, 1971) (criteria 1 and 2).

A protease, which hydrolyzes tosyl-L-arginine methyl ester, was demonstrated in the extract of skin homograft in rat; and its activity was correlated with the process of the rejection (Ungar and Ungar, 1966) (criteria 1 and 2). Some proteases, like cathepsins B and D or collagenolytic cathepsin, were extracted from inflammatory granulomas induced by turpentine or polyvinyl sponges in the skin of rat; and the activity of these proteases roughly paralleled the inflammatory process (Bazin and Delaunay, 1969) (criteria 1 and 2). The activity of cathepsins has been also detected by other investigators (Woessner and Boucek, 1961; Luscombe, 1963; Smith *et al.*, 1965; Freedman *et al.*, 1967). However, the optimum pH of cathepsins suggested was in an acid range far from the pH of body fluid or tissues. Establishment of the time course of proteolytic activity in inflammatory tissue is of great importance in determining the presence of an inflammatory protease.

C. Physicochemical Properties of SH-Dependent Protease

As summarized in Table I, a neutral SH-dependent protease was extracted in the euglobulin fraction of an inflammatory lesion such as an active Arthus reaction (Hayashi *et al.*, 1958, 1965a; Udaka, 1963a) or a thermal injury (Tokaji, 1971) (criterion 1). The enzyme sample was further fractionated on DEAE-Sephadex and GE-cellulose, in that order; and it was confirmed that the enzyme was a mixture of two neutral SH-dependent proteases (Koono *et al.*, 1968; Tokaji, 1971). Protease I was eluted in 0.02 *M* phosphate buffer (pH 7.4) on GE-cellulose, and protease II eluted in 0.1 *M* phosphate buffer (pH 7.4) on GE-cellulose. These protease samples are both homogeneous in electrophoresis.

The optimum pH of both these proteases was 7.1 when tested against casein, but was in the range of 6.0–7.0 when tested against hemoglobin. Both enzyme preparations contain no acid or alkaline protease. It is of significance to note that the optimum pH of these enzymes is adequate for action in inflammatory tissue (criterion 3), because the pH of inflamed tissue (Yasuhira, 1955) or inflammatory exudates (Edlow and Sheldon, 1971; Hutchins and Sheldon, 1972) has been shown to be only slightly acidic.

These proteases are both nondiffusible and thermolabile. The activity of these proteases is SH-dependent, because the enzymes can be inhibited by *p*-chloromercuribenzoate (PCMB) or molecular oxygen, but reactivated by thiol compounds (cysteine and reduced glutathione). The

TABLE I

PROCEDURES OF ISOLATION AND PURIFICATION OF NEUTRAL SH-DEPENDENT PROTEASE FROM INFLAMMATORY TISSUE[a]

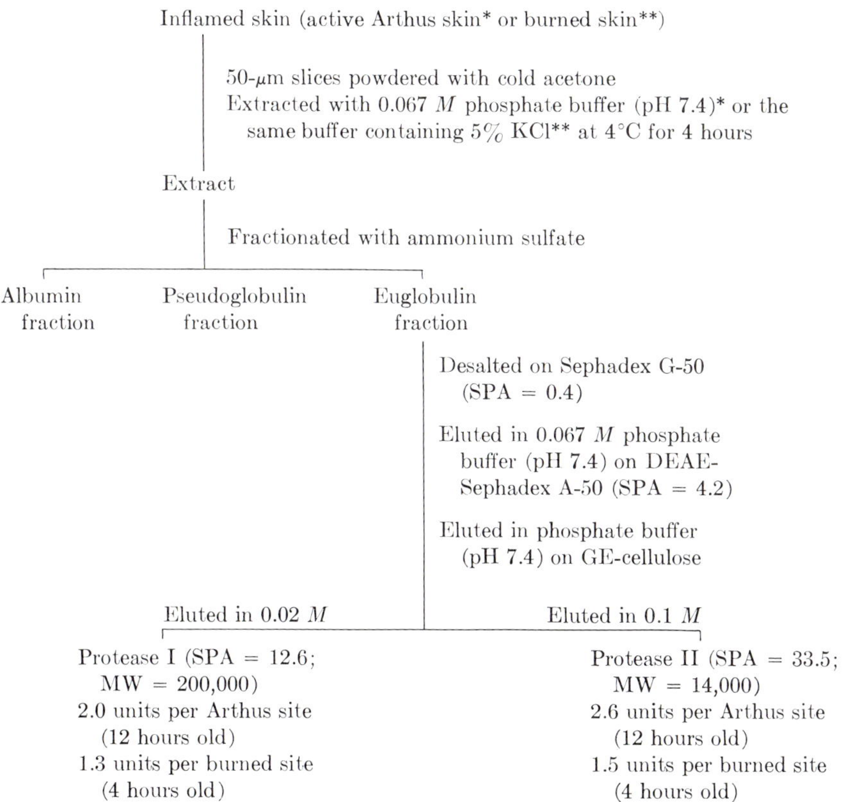

[a] SPA, Proteolytic unit per milligram of protein (Hayashi *et al.*, 1974). Caseinolytic activity was measured at pH 7.1 and ionic strength 0.25 according to a modification (Hayashi *et al.*, 1962a) of Kunitz' (1947) method. The arbitrary unit of proteolytic activity was defined as the amount of enzyme that caused an increase of 0.001 unit of extinction per minute of digestion (Koono and Hayashi, 1969); 1 proteolytic unit represents 0.06 absorbance units at 276 nm. Neutral SH-independent protease was eluted in 1.0 *M* sodium chloride on DEAE-Sephadex (SPA = 1.8). Molecular weight estimated by gel filtration.

molecular weight of protease I was about 200,000 when estimated by gel filtration, and that of protease II was about 14,000 (Koono and Hayashi, 1969).

These proteases did not hydrolyze some known substrates, such as benzoyl-L-arginine ethyl ester (for trypsin or cathepsin B), glycyl-L-phenylalanine amide (for cathepsin C), acetyl-L-tyrosine ethyl ester (for C'1 esterase), and tosyl-L-arginine methyl ester (for trypsin or plasmin). Some known esters such as L-tyrosine ethyl ester, L-glycine ethyl ester, L-leucine ethyl ester, L-valine methyl ester, and L-histidine methyl ester also were not digested by these proteases. Thus it is suggested that the neutral SH-dependent protease has no esterase activity, and is different from such proteases as plasmin, kallikrein, C'1 esterase, and cathepsins B and C. Cathepsin A has little action on protein itself, but acts synergistically with endopeptidases such as cathepsin D; it is unaffected by cysteine and iodoacetamide. Cathepsin B, active in the presence of thiol and chelating reagents, acts as an endopeptidase with broad specificity similar to that of papain, but with hemoglobin its pH optimum is near 4; it has a molecular weight of about 24,000 (Snellman, 1971). Cathepsin C, requiring thiol reagent and Cl^- for maximal activity, is an exopeptidase with broad specificity for the splitting of dipeptide naphthylamides and the removal of dipeptides sequentially from the amino terminus of a polypeptide chain (McDonald *et al.*, 1971). Cathepsin D is an endopeptidase with pepsinlike specificity for polypeptides, but having narrower specificity and slower reaction rates with low-molecular-weight substrates; it is unaffected by thiol reagents and by thiol-blocking reagents. The optimum pH of cathepsin D is 3–4, with most substrates, and its molecular weight is between 40,000 and 60,000 (Keilova, 1971; Barrett, 1971; Lapresle, 1971). Cathepsin E is an endopeptidase of much more limited tissue distribution than cathepsin D, but similar specificity; it has a molecular weight of about 100,000 and a lower pH optimum with human serum albumin as substrate (Lebez *et al.*, 1971). Thus it is suggested that physicochemical properties of SH-dependent protease differ from those of known cathepsins, neutral collagenase which is inhibited by 10^{-2} *M* EDTA (Lazarus *et al.*, 1967), and neutral elastase which is inhibited by SBTI (Janoff and Scherer, 1968).

When SH-dependent protease I was rechromatographed on GE-cellulose, the enzyme was found to convert to SH-dependent proteases II and III. Furthermore, protease II converted to protease III under the same conditions of rechromatography. The molecular weight of protease III was about 7000 when measured by gel filtration (Koono and Hayashi, 1974). It was thus assumed that protease I may convert gradually to protease II in inflammatory tissue. Similar observations have been made on the conversion of cathepsin D; the production of different forms (more than 10) of the enzyme was confirmed during chromatography (Press *et al.*, 1960). The mechanism of SH-dependent protease conversion is obscure, but the mechanism of production of π, δ, γ, and α forms of chymotrysin from chymotrypsinogen is of interest (Bettelheim and Neurath, 1955; Rovery *et al.*, 1955).

D. Inflammation-Inducing Potency of SH-Dependent Protease

The activity of protease I at a *single* burned site (3-hours old) showing maximal reaction was calculated to be about 1.3 units, and that of protease II calculated to be 1.5 units (Tokaji, 1971). Similarly, the activity of protease I at a *single* active Arthus site showing maximal reaction (12-hours old) was calculated to be about 2.0 units, and that of protease II calculated to be 2.6 units (Hayashi *et al.*, 1969, 1973). As shown in Table II, it is of extreme importance that the enzymes can produce rapid development of local cutaneous lesions strikingly similar to those of thermal injury or active Arthus reactions in both gross and microscopic features when injected intradermally in concentrations comparable to those detected at the inflamed sites (criterion 4). Diffuse edema and erythema occurred in lesions 30 minutes old, becoming maximal in 3–6 hours and remaining considerable for several hours. In contrast, other known proteases at the same level of activity produced little or no inflammatory reaction, as summarized in Table II.

Furthermore, since such inflammatory reactions could be similarly induced by intradermal SH-dependent protease in rabbits whose serum complement was apparently depleted by anticomplementary factor from cobra venom (Ballow and Cochrane, 1969; Cochrane *et al.*, 1970), it

TABLE II

Comparison of Inflammation-Inducing Potency of Various Proteases

Proteases injected	Proteolytic units	Macroscopic inflammatory reactions[a]		
		1 hour	3 hours	6 hours
Neutral SH-dependent	5.0	±	++	+++
protease I	2.0[b]	±	+ − ++	++
	1.2	−	+	+
Neutral SH-dependent	5.0	+	+++	++++
protease II	2.5[b]	+	++	+++
	1.2	±	++	++
Neutral SH-independent	5.0	−	±	−
protease	1.7	−	−	−
α-Chymotrypsin	5.0	−	−	−
Trypsin	5.0	−	±	−
Cathepsin D	5.0	−	−	−
Cathepsin E	5.0	−	−	−
Papain	5.0	±	+	±

[a] The intensity of inflammatory reactions provoked was recorded according to the method of Hayashi *et al.* (1962a).

[b] Proteolytic activity was comparable to that detected at a single active Arthus site (12 hours old) whose inflammatory reactions were recorded as ++++.

seemed reasonable to assume that the inflammation-inducing effect of the enzyme was not influenced by the depleted condition of serum complement (Hayashi, 1974; Nishiura *et al.*, 1974b). The production of leucoegresin by the enzyme also was not influenced by the complement-depleted condition as described in Section J-6.

The skin lesions at 6 hours were characterized by (1) diffuse and local (perivascular) infiltration by inflammatory cells, chiefly consisting of PMN leukocytes; (2) fibrinoid exudation characterized by eosinophilic swelling of collagenous fibers and by interfibrillary deposition of amorphous eosinophilic materials; (3) hemorrhage from, and the formation of hyaline or fibrinous thrombi in, the small blood vessels; (4) vesiculation between the epidermis and dermis; and (5) vascular inflammation of a peri- and panarteritis nodosa type in varying degrees (Figs. 2 and 3). The microscopic changes produced were clearly associated with the activity of the SH-dependent proteases injected. Thus it seemed reasonable that the induction of such inflammatory reactions was associated with the combined action of SH-dependent proteases I and II. Whether or not the SH-dependent protease from thermal lesions is identical with that from active Arthus lesions remains to be determined, because the microscopic changes induced by the former appeared to differ in some respects from those induced by the latter. When the intradermal effects of these proteases were compared at a similar activity, the following were observed: (1) more intense vesiculation between the epidermis and dermis, (2) less marked infiltration of PMN leukocytes, and (3) less marked deposition of fibrinoid materials. In fact, as mentioned in Section J-6, the amount of chemotactic factor (*leucoegresin*) for PMN leukocytes separated from burned skin lesions or from protease-induced skin lesions was smaller. Perhaps they closely resemble but seem to represent different types of neutral SH-dependent protease.

E. Inhibitors of SH-Dependent Protease

The inhibitor of SH-dependent protease was first found in the euglobulin fraction of burned and active Arthus skin sites with decreased or healed inflammation (showing no proteolytic activity) (Matsuba, 1960a; Hayashi *et al.*, 1965a), and from a culture of peritoneal mononuclear cells (Tokuda *et al.*, 1960); and it was shown that this inhibitory euglobulin clearly suppressed the inflammatory reaction when injected locally (Matsuba, 1960b) (criteria 6 and 7). The inhibitor was purified by gel filtration with Sephadex G-50 and chromatography on DEAE-cellulose and CM-cellulose, and then by precipitation at its isoelectric point (pH 6.6) (Udaka and Hayashi, 1965a). This inhibitor was shown to be homogeneous on electrophoresis and ultracentrifugation, and had a sedimentation coefficient of 1.21S. It seemed to be a peptide (Udaka

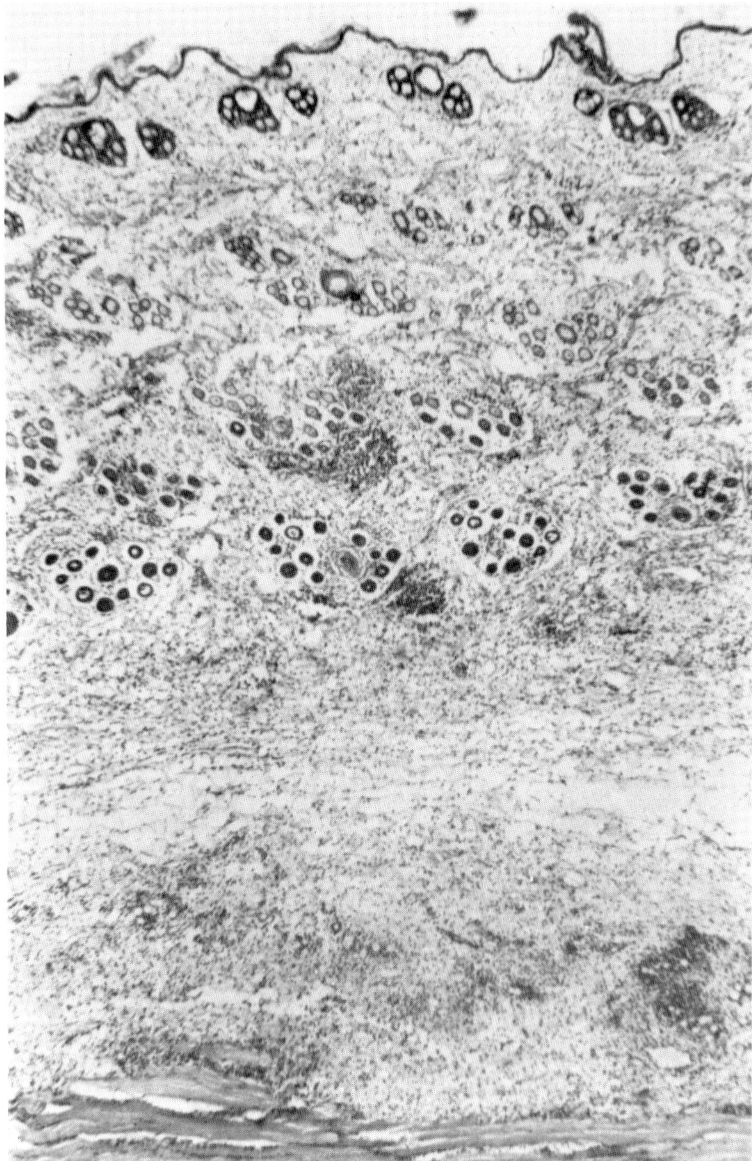

FIG. 2. Marked inflammatory reactions induced by intradermal SH-dependent protease II (5.0 units), closely resembling those in an active Arthus reaction (12-hours old). Rabbit, 6 hours after injection. Hematoxylin–eosin. ×50.

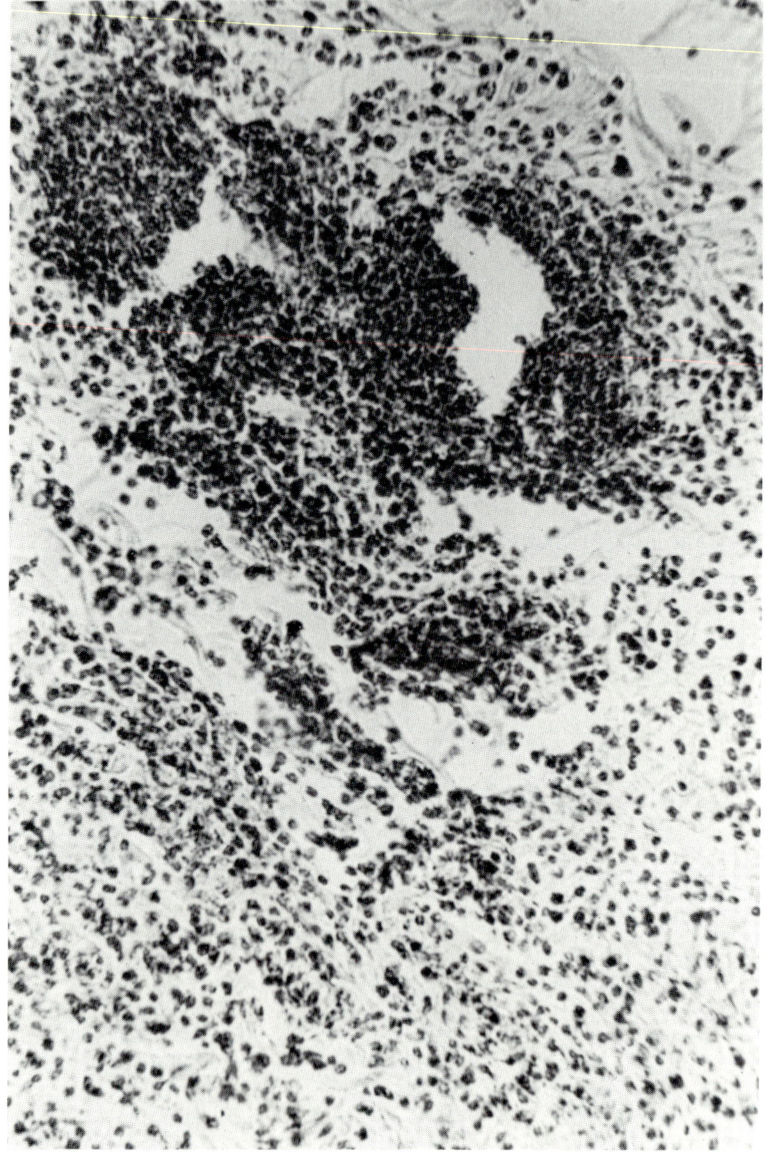

FIG. 3. Vascular inflammation of peri- and panarteritis nodosa type induced by intradermal SH-dependent protease II (5.0 units). Rabbit, 6 hours after injection Hematoxylin–eosin. ×250.

and Hayashi, 1965b). In addition to SH-dependent protease, this inhibitor also inactivated papain but not trypsin or α-chymotrypsin. This inhibitor was nondialyzable, although thermostable, nonprecipitable with 3% trichloroacetic acid, and soluble in salt solutions but insoluble in water.

It was of importance to note that in skin lesions concentration of the inhibitor increased as the inflammation subsided (Udaka, 1963b; Hayashi *et al.*, 1965a); the observations seemed to explain the decrease in protease activity, which became noticeable as the reactions subsided (criterion 6). In this connection it was interesting that Janoff *et al.* (1965) reported that extracts from granules (lysosomes) of peritoneal mononuclear cells inhibited the inflammatory response produced with a PMN leukocyte granule extract containing SH-dependent protease. As is widely accepted, in a later stage of inflammation, for example, an active Arthus reaction or a thermal injury, PMN leukocytes decreased in number, and the inflammation was characterized by a mononuclear cell response (Movat, 1956). It was quite natural that this inhibitor was released from tissue mononuclear cells, and that it was isolated from a late stage of an Arthus reaction or thermal injury. However, local synthesis of the inhibitor in a later stage of inflammation was also suggested (Kambara *et al.*, 1968).

Another inhibitor of SH-dependent protease was isolated from the sera of rabbit, guinea pig, and bovine (Tokaji, 1971). This inhibitor was precipitated from acidified sera by repeated fractionations with ammonium sulfate, and then eluted on Sephadex G-50 and CM-cellulose; it was contained in the α_1-acid glycoprotein fraction. This substance was nondialyzable and thermostable, and its molecular weight was about 40,000–50,000 when estimated by gel filtration. The inhibitor also inactivated papain, but not trypsin or α-chymotrypsin. It was probably a glycoprotein, and its intradermal injection suppressed considerably the inflammatory response due to moderate heat (criterion 7). It was thus suggested that SH-dependent protease at the inflamed sites was inactivated first by a *serum* inhibitor which may be exuded by the mechanism of increased vascular permeability, and successively by a *tissue* inhibitor which may be released from mononuclear cells or synthesized locally. Accordingly, it was assumed that the extent, intensity, and duration of the inflammatory reaction were probably correlated with the balance between SH-dependent protease and these inhibitors.

F. Activation of SH-Dependent Protease in and Release by Cells

SH-dependent protease has been shown to be present in several types of cells. For instance, the enzyme was first demonstrated in peritoneal mononuclear cells of rabbit (Hayashi *et al.*, 1960; Kato *et al.*, 1972), perivascular histiocytes in the skin of rabbit (Udaka and Udaka, 1969), and PMN leukocytes of rabbit (Wasi *et al.*, 1966; Kouno, 1971; Kouno *et al.*, 1974); the enzyme seemed to be largely located in the granules or lysosomes of the cells. The release of the enzyme by antigen was detectable in cultures of mononuclear cells or histiocytes (Hayashi *et al.*, 1960; Hayashi, 1967), simultaneous with morphological changes in the cultured cells, consisting of retraction of pseudopods, swelling of the cells, formation of a fibrous reticulum in the ground cytoplasm, and decreased movement of mitochondria (Hayashi *et al.*, 1955b; Ono, 1958). The enzyme, which was detected at the earliest stage of an active Arthus reaction, was presumed to be released from perivascular sensitized histiocytes (Hisaka, 1966; Kinuwaki *et al.*, 1967; Sonoda, 1967), because antigen given intradermally first affected the sensitized histiocytes. In the use of labeled antigen (Congo red–azo-bovine serum albumin), it became clear that the earliest morphological change in an active Arthus reaction was the swelling of the sensitized histiocytes which were characterized by very fine granules of labeled antigen distributed in a diffuse or fibrous manner on the cell surface or in the ground cytoplasm (Hayashi *et al.*, 1958). By contrast, nonsensitized histiocytes showed no edematous change, but incorporated the labeled antigen as solitary and larger granules. Such characteristic swelling and distribution of labeled antigen was confirmed in cultures of mononuclear cells (Hayashi *et al.*, 1955a) or histiocytes (Kuze, 1960); it became clear that the characteristic swelling of the cells was the earliest morphological indication of the reaction, and that SH-dependent protease was activated in and released by such swollen cells. In view of the cellular reaction, since an active Arthus reaction is characterized by the swelling of histiocytes followed by an intense accumulation of PMN leukocytes and, in a later stage of the reaction, by mononuclear cells, it is possible that the SH-dependent protease in the reaction may be derived mainly from the lysosomes of these cells. In thermal injury SH-dependent protease seems to be first released from the epidermal cells by the direct effect of heating, because the morphological change at the earliest stage of the inflammatory response was characterized by marked vesiculation between the epidermis and dermis (Tokaji, 1971). SH-dependent protease from PMN leukocytes and mononuclear cells may be subsequently involved in the proteolytic mechanism in the thermal injury.

SH-dependent protease of PMN leukocytes has been shown to be released by some stimuli, for example, antigen–antibody complexes (Kouno, 1971); it was purified by chromatography with DEAE-Sephadex and CM-Sephadex, in that order (Table III). As is well-known, PMN leukocytes and mononuclear cells have been shown to phagocytose immune complexes under a variety of conditions in experimental animals (Cochrane *et al.*, 1965; Cochrane, 1963; Movat *et al.*, 1963) and in man (Cohn and Morse, 1960; Hollander *et al.*, 1965; Movat *et al.*, 1964), thus exposing the complexes to lysosomal proteases. As a result, immune complexes or free immunoglobulins could be exposed to proteolytic enzymes both intracellularly and extracellularly.

Menkin's (1940) necrosin, described in inflammatory exudates, was probably a protease. It was found in the euglobulin fraction of exudates and shown to have proteolytic activity. Since it was not encountered in

TABLE III

PROCEDURES OF ISOLATION OF NEUTRAL SH-DEPENDENT PROTEASE FROM RABBIT PMN LEUKOCYTES[a]

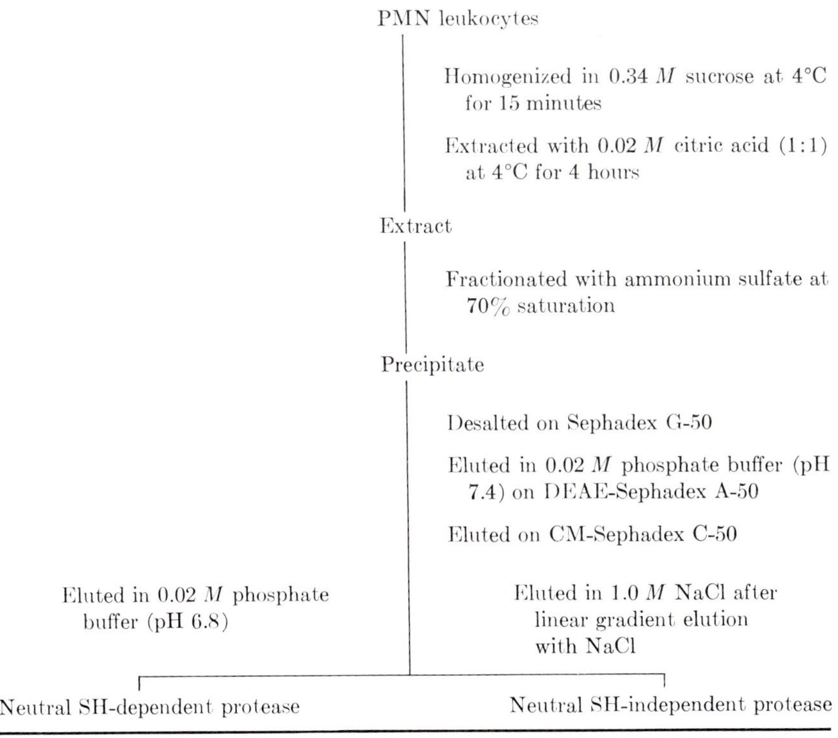

[a] Rabbit PMN leukocytes were collected from glycogen-induced peritoneal exudates.

all inflammatory exudates, he suggested that it may be readily neutralized by antiprotease (Menkin, 1956). The source of necrosin in exudates has not yet been determined, but it is tempting to speculate that the necrosin demonstrable in exudates stems from the lysosomes of PMN leukocytes, because the cells are abundantly found in the exudates.

However, a cellular protease active in the range of neutral pH can be extracted from human spleen lysosomes and fractionated on agarose; its molecular weight is 15,000–30,000 (LoSpalluto *et al.*, 1960, 1971). The enzyme is stimulated by cysteine, strikingly inhibited by diisopropyl fluorophosphate (DFP), and not significantly affected by iodoacetamide. As for other known intracellular proteases, cathepsin A is not enhanced by cysteine when assayed with a synthetic substrate, and has a pH optimum of 5.7 (Lichtenstein and Fruton, 1960). Cathepsins B and C are stimulated by cysteine but, in contrast to this protease, are markedly inhibited by iodoacetamide (Greenbaum and Fruton, 1957; Planta and Gruber, 1964). Cathepsins D and E, described by Lapresle and Webb (1962), have clearly different pH profiles. Although the enzyme has not yet been characterized sufficiently to compare with the neutral SH-dependent protease from inflammatory cells or tissues, it is speculated that the enzyme may be similar to the SH-dependent protease described above. For this purpose the separation of SH-dependent protease from human spleen following the method summarized in Table I is needed. The spleen tissue contains a large number of mononuclear cells even under physiological conditions.

It is of interest to speculate that the release of intracellular proteases concerned with the production of inflammatory reactions might occur selectively according to the nature of the stimuli. Chemical substances capable of inducing a selective release of proteases might be produced in the tissue cells stimulated. There is an experiment to suggest such selective release of intracellular proteases. Although rat ascites hepatoma cells contain at least three different proteases, that is, two neutral SH-independent proteases with different electrophoretic mobility and an alkaline SH-dependent protease (Koono *et al.*, 1974b), the release of these proteases is induced by chemical factors specific for each protease, which are separated from tumor tissues (Koono *et al.*, 1974a). The chemical factor capable of releasing a neutral protease with fast mobility was found to be a heat-stable peptide; it was demonstrated that the neutral protease released produced extracellularly a factor chemotactic for cancer cells (Hayashi *et al.*, 1970; Yoshida *et al.*, 1970). This chemotactic factor locally induced extravascular migration of circulating cancer cells and formation of metastatic secondary tumors (Ozaki *et al.*, 1971).

G. Reactivation of SH-Dependent Protease and of Inflammatory Reactions

In tissue showing healing inflammation, SH-dependent protease existed in an inactive state by forming a complex with the *tissue* inhibitor. Protease from healed skin sites was activated *in vitro* by adding reduced glutathione (GSH) or cysteine far more significantly than that from normal untreated skin sites; and intradermal injection of activated protease induced the rapid development of local inflammatory reactions (Hayashi *et al.*, 1962b). In contrast to virgin sites, intradermal injection of GSH at the healing sites of an active Arthus reaction (Hayashi and Nitta, 1961; Hayashi *et al.*, 1962a), thermal injury (Nitta, 1965; Miyoshi, 1965), or Shwartzman reaction (Nitta, 1965) produced a recurrence of the inflammatory process having the same appearance as the original reactions. It was thus thought that such recurrence of inflammatory reactions was due to reactivation of SH-dependent protease by GSH, because the protease, which had the same properties at the pH optimum, SH dependency, inflammation-inducing potency, or inactivation by the *tissue* inhibitor as those of SH-dependent protease from the original lesions, was isolated from the sites of recurrent inflammation (Hayashi *et al.*, 1962a). It seemed possible that GSH acted on the *tissue* inhibitor, since it was shown that this inhibitor lost its activity *in vitro* under the influence of reducing agents (Harada, 1960). It was therefore assumed that higher susceptibility to GSH of the protease at the healing sites may be related to the ease of inactivation of the peptide inhibitor by GSH.

The susceptibility to GSH of the protease from healing sites decreased with increasing age of the lesion; for instance, 7 weeks after complete healing of an active Arthus reaction, it was completely abolished. This was correlated with the fact that intradermal injection of GSH into 7-week-old healing sites induced little inflammatory reaction (Hayashi *et al.*, 1964c). Although cysteine was less active *in vivo*, no difference in the *in vitro* effects of GSH and cysteine was observed. It was of interest that such reactivation of inflammation was not provoked by other reducing agents, such as L-ascorbic acid, 6,8-disulfhydryloctanoic acid, and thiosulfate (Hayashi *et al.*, 1962a). Negative results with ascorbic acid appeared to be ascribed to its higher potential; at pH 7 the redox potential (E^0) of ascorbic acid is +0.060 V, while that of GSH is about −0.40 V (Barron, 1951). Such a strong reducing power of GSH, almost close to that of hydrogen, was thought to be one of the important conditions for the recurrence of inflammation.

These observations were interpreted to indicate again that SH-dependent protease is responsible for the induction of inflammatory reactions. As is well-known, GSH is an important hydrogen carrier, and its amount may vary under different physiological or pathological conditions (Lazarow, 1954; Jocelyn, 1959); the observations with GSH or cysteine might explain the recurrence frequently observed in human inflammatory diseases. Clinical occurrences such as polycyclic rheumatic attacks and repeated episodes of the recurrence of various diseases such as hepatitis, pancreatitis (Paxton and Payne, 1948), glomerulonephritis (Kusunoki, 1963), and bronchitis accompanied by asthma onset (Ostwald, 1957; Mise, 1963) have been reported. These episodes seemed to be associated with recurrent inflammatory processes at the primarily affected sites (Klinge, 1931); and explanation has been made only in terms of repeated attacks of antigens or inflammatory irritants themselves (Cochrane *et al.*, 1959). For instance, an injection of antigen into healing Arthus sites induced reactions similar to those at the original lesion, which were ascribed to the effects of a concentration of antigen–antibody complexes in the wall of blood vessels (Cochrane *et al.*, 1959). However, the mechanism of reactivation of inflammatory reactions mentioned above was clearly different and not related to the antigen–antibody reaction. Since the recurrence of human allergic diseases frequently takes place in the absence of definite signs of antigen–antibody reaction, the observations seem to be important in regard to the mechanism of recurrence of these diseases.

In contrast to the healing sites of an active Arthus reaction, thermal injury, or Shwartzman reaction, after croton oil- and ultraviolet-induced injury healing sites were not reactivated by intradermal injection of GSH or cysteine (Nitta, 1965). It was therefore suggested that tissue injuries may be theoretically divided into two groups: (1) lesions characterized by reactivation of healing site by GSH, and (2) lesions unaffected by GSH, which might be due to the activation of an SH-independent protease as revealed in radiation-induced cell injury (Hayashi *et al.*, 1963). The activation of SH-dependent protease was also demonstrated in croton oil-induced skin lesions in rabbit (Miyoshi, 1965) and in specimens of human skin under the influence of croton oil *in vitro* (Ungar *et al.*, 1961), but the increase in protease activity was not significant. Why inflammatory lesions are unaffected by GSH remains to be determined.

H. Release of Mucopolysaccharide by SH-Dependent Protease from the Protein Complex in Inflammation

As is well-known, the sites of experimental injury or inflammatory diseases, for example, nodules of rheumatic arthritis, myocardial lesions of acute rheumatic fever and systemic lupus erythematosus, or early granulation tissue in the skin of animals (Altschuler and Angevine, 1949; Klemperer, 1948; Bensley, 1934; Sylvén, 1941), show varying degrees of metachromatic reaction with toluidine blue, but the mechanism has been simply considered to be due to an increased amount of acid mucopolysaccharides in the ground substance of the tissues (Altschuler and Angevine, 1949; Angevine, 1951).

However, it was demonstrated that purified mucopolysaccharides (free of protein) themselves, in dried smears, exhibited intense metachromasia even in a very low concentration, but the reaction became negative when the mucopolysaccharides (such as hyaluronic acid and chondroitin sulfate) were previously "masked" by forming a complex with protein (serum protein or egg albumin); the masked mucopolysaccharides became metachromatic again after mild treatment with proteases such as pepsin (Hayashi *et al.*, 1955c). In spite of the presence of a satisfactory amount of mucopolysaccharides, the ground substance of normal skin of rabbit showed little or no metachromasia, whereas it became intensely metachromatic soon after an injury such as an active Arthus reaction or a moderate burn (Kasai, 1959; Hayashi, 1967); and the intensity of the metachromasia roughly paralleled the local increase in SH-dependent protease activity. A similar degree of metachromasia was found in skin lesions induced by SH-dependent protease; and the inflammatory reactions were suppressed by the *tissue* inhibitor of the protease (Udaka, 1963b). Furthermore, when normal skin sections were incubated with SH-dependent protease, immediate and intense metachromasia of the skin was produced (Kasai, 1959). Therefore these observations suggested that mucopolysaccharides in the ground substance were unmasked by SH-dependent protease activated after injury, which resulted in the production of metachromasia. Although the mechanism of increased metachromasia in inflammation may be more complex, such unmasking of mucopolysaccharides seemed to explain the metachromatic reaction at least at an early stage of inflammation.

Also, it was of interest that excess vitamin A acted to cause degradation of cartilage matrix both *in vitro* and *in vivo* by the release of proteases from cartilage lysosomes (Lucy *et al.*, 1961; Thomas *et al.*, 1960; Weissman and Dingle, 1961; Weissman and Fell, 1962); and it was hypothesized that lysosomal enzymes released in various forms of acute tissue injury could account for hydrolysis of connective tissue macromolecules (Fell and Dingle, 1962). Thus lysosomal proteases present in cartilage and in leukocytes have been shown to be capable of degrading cartilage matrix, isolated proteoglycans, and simple chemical substrates (Weissman and Spilberg, 1968; Davies *et al.*, 1971). Proteolytic activity was attributed to cathepsin D and the neutral protease of leukocytes (Weissman *et al.*, 1972; Janoff, 1972).

In an advanced stage of inflammation, metachromasia was found to decrease, probably because of the depolymerization of mucopolysaccharides by the action of hyaluronidase or hyaluronidaselike enzymes. The metachromatic reaction of mucopolysaccharides decreased *in vitro* in proportion to depolymerization by treatment with testicular hyaluronidase, but the reactivity to periodic acid–Schiff (PAS) reagents of mucopolysaccharides increased in proportion to depolymerization (Hayashi *et al.*, 1955d); thus advanced inflammatory lesions showed decreased metachromasia, but the PAS reaction was increased. The increased coloration of depolymerized hyaluronic acid with Schiff reagent was due to the liberation of reactive units by cleavage of the polysaccharide molecule, since the PAS reaction was considered to be a common reaction of all potential aldehydes unmasked by the action of periodic acid. The faint reaction of purified hyaluronic acid with PAS reagent was in agreement with the observations of Meyer and Felling (1950). Since the mucopolysaccharide level in the serum was increased in patients with arthritis or rheumatic fever (Shetlar *et al.*, 1950; Kelley and Good, 1949), and in dogs with inflammatory reactions induced by turpentine or talc (Shetlar *et al.*, 1949), the role of mucopolysaccharides released or depolymerized at the inflamed sites was of interest.

I. Production of Inflammatory Permeability Factor (Vasoexin) by SH-Dependent Protease

1. *Estimation of Increased Vascular Permeability*

The endothelial wall of capillaries and venules is a barrier freely permeable to water and electrolytes but only very slightly permeable to plasma proteins. The term increased vascular permeability therefore refers to an alteration in the vascular wall leading to an accelerated rate of passage of plasma proteins into the extravascular tissues. This phenomenon is undoubtedly one of the cardinal features of inflammation.

Permeability changes in inflammation have been particularly investigated in the skin of experimental animals into which some vital dye, such as pontamine sky blue, Evans blue, or trypan blue, has been injected (Menkin, 1938; Rawson, 1943). It has been accepted that when given intravenously these dyes become attached to circulating plasma proteins, particularly to albumins, and that their accumulation at inflamed sites indicates an exudation of plasma proteins. However, evaluation of experimental results in such tests often lacks precision

(Wilhelm, 1962), although there have been attempts to estimate quantitatively the amount of accumulated dye at inflamed skin sites, for instance, in terms of mean lesion diameter (Miles and Wilhelm, 1955) or by comparing the color intensity with a series of standard colors (Lockett and Jarman, 1958).

Several attempts have been made to extract the dye from the skin and estimate it quantitatively. The technique of Judah and Willoughby (1962) involves disintegration of frozen rat skin in a percussion mortar, followed by extraction with ether and ethanol to remove unwanted lipids and then by extraction of trypan blue with hot aqueous pyridine. The method of Nitta *et al.* (1963) consists of extraction of pieces of rabbit skin with an ethanol–dioxan mixture, dioxan, and a chloroform–methanol mixture, in that order, to remove the lipids, and extraction of pontamine sky blue in hot aqueous pyridine and then reading the resultant color at 620 nm. The technique of Carr and Wilhelm (1964) depends on homogenization in a concentrated solution of urea, followed by splitting of Evans blue in protein complex and precipitation of the protein with acetone and then Somogyi reagents.

Another procedure (Wasserman *et al.*, 1955) involves the intravenous injection of serum albumin labeled with ^{131}I and determination of its subsequent concentration in the thoracic duct lymph and blood plasma. This technique has the advantage of being quantitative and, by varying the choice of lymphatic vessel cannulized, its usefulness can be extended, for example, to changes confined to all or part of one limb. Alternatively, the radioactivity of an inflammatory exudate is measured and, by comparison with the radioactivity of the blood, it is possible to calculate the rate at which the vessels are leaking proteins (Spector, 1956). The method of Ascheim (1965) consists of injecting radiolabeled serum albumin and counting localized radioactivity at the selected skin site, making proper corrections from normal symmetric sites and taking into account the blood concentrations. A modified convenient technique using radiolabeled serum albumin has been developed by Jullien-Vitoux *et al.* (1973).

A less frequently used method for the detection of exudation is estimation of edema in the inflamed tissue (Sevitt, 1958; 1964; Spector and Willoughby, 1959). This is done by comparing the wet weight with the dry weight of the skin excised within a certain radius. However, the method is less sensitive than the dye technique.

For demonstrating morphological change in vascular permeability, several investigations have been performed following the method of intravenous injection of shellac-free colloidal carbon during the development of inflammation to demonstrate when the blood vessels are involved in the leaking of plasma proteins (Majno *et al.*, 1961; Majno, 1964; Cotran and Majno, 1964a,b). The particles of carbon have an average diameter of 200 A. The dyes described above are useful for estimating increased vascular permeability in inflammation, but not for demonstrating the individual blood vessels involved, because the dyes escape from the tissue during the preparation of histological sections. The cremaster muscle of rat (Majno *et al.*, 1961) and panniculus carnosus muscle of rabbit (Hayashi *et al.*, 1965b; Tasaki, 1968b) are readily removed from the tissues and provide favorable preparations for light microscopic and electron microscopic observation on the blood vessels involved; the vascular trees in these tissues can be easily traced in unstained preparations.

However, there are two criticisms of this method. It is generally believed that carbon is cleared from the body by the reticuloendothelial system. When injected intravenously, carbon particles, like other macromolecules, are phagocytosed also by PMN leukocytes whose number decreases temporarily as a result of aggregation (Movat *et al.*, 1965). Thus some of the vascular labeling could be due to accumulation of phagocytosing aggregates of leukocytes and platelets. The second criticism is that colloidal carbon has been shown to be pyrogenic (Hanna and Watson, 1965). These criticisms also apply to the use of colloidal carbon as an electron-dense tracer.

It has not yet been established whether or not the accumulation of carbon in the venules and capillaries reflects leakage of plasma proteins from these blood vessels. This problem can be solved by utilizing the fluorescent antibody technique (Hayashi *et al.*, 1965b). At various intervals after injury, rabbits are given bovine serum albumin intravenously. The panniculus carnosus muscle of the animals is removed and stained by direct or indirect fluorescent antibody technique; and the results are compared with the features of carbon leakage from the venules and capillaries. It has been confirmed that carbon leaking from blood vessels imitates the leaking of bovine serum albumin. Experimentation using antibody against rabbit serum albumin is needed.

2. *Criteria for Inflammatory Permeability Factor*

With the advent of techniques allowing investigators to estimate or demonstrate vascular permeability changes, different types of permeability factors have been reported from many laboratories. However, present knowledge of the natural mediators is still incomplete to draw a conclusion; there is considerable confusion due to inadequate characterization of isolated chemical mediators. It is therefore emphasized that the following criteria should be applied to suspected mediators and that they fulfill some if not all of the conditions listed (Hayashi, 1967; Tasaki *et al.*, 1969).

1. Permeability factor should be locally available to initiate and maintain the characteristic vascular response—immediate or delayed.
2. Amount (or activity) of the permeability factor should parallel the time course of the vascular response.
3. Action of the permeability factor should be specific, being active in enhancing vascular permeability but not in causing leukocyte migration and hemorrhagic change.
4. Permeability factor, when given in concentrations reasonably comparable to those detected at the inflamed site during the response, should produce a vascular response (including morphological change in the small blood vessels) similar to the reaction.
5. Permeability factor for the delayed response should elicit a long-acting effect which could account for the prolonged duration of the delayed response.
6. Effect of the permeability factor should be inhibited by the specific antagonistic substance locally available.
7. Vascular response should be suppressed by a specific antagonistic substance.
8. Depletion of permeability factor or of its precursor should cause a decrease in the vascular response.
9. Precursor of permeability factor, or the enzyme associated with its production, should be detected at the inflamed site during the vascular response.

3. *Time Course of Increased Vascular Permeability in Inflammatory Process*

The general pattern of increased vascular permeability in injury in experimental animals consists of two different phases—immediate and delayed. For instance, when a polished steel disc (1.0 cm in diameter) closing one end of a steel tube and maintained by circulating water at a stated temperature (56°C for 20 seconds for rabbit and 54°C for 20 seconds for rat) was applied to the clipped skin of animals and locally accumulated pontamine sky blue extracted, the permeability changes appeared diphasic in rabbit and rat: immediate–less intense and transient (lasting less than 10 minutes); delayed–more intense and prolonged (beginning in about 1 hour, reaching a maximum in 2 hours in rabbit and in 3 hours in rat, and disappearing in about 6 hours) (Hayashi *et al.*, 1964b). The permeability changes in an active Arthus reaction (induced by 2.5 mg bovine serum albumin) in the skin of rabbit also consist of two different phases: immediate—transient and mild (lasting less than 10 minutes); delayed—more intensive and prolonged (beginning in about 1 hour, reaching a maximum in 4–5 hours, and disappearing in 12 hours) (Hayashi *et al.*, 1964b) (Fig. 1). Although there are some variations in the pattern and duration according to the nature of injury, the delayed response, the major permeability event, is essentially a characteristic feature, common to experimental injuries. In contrast, the immediate response is the minor permeability change, probably nonspecific, in inflammation.

Such diphasic vascular response has been confirmed by many investigators in thermal injury (Sevitt, 1958; Wilhelm and Mason, 1960; Spector and Willoughby, 1959; Allison and Lancaster, 1959; Hayashi *et al.*, 1964b), bacterial infections (Burke and Miles, 1958; Petri *et al.*, 1952; Elder and Miles, 1957; MacFarlane and Knight, 1941), active Arthus reaction (Hayashi *et al.*, 1964b), tuberculin reaction (Voisin and Toullet, 1963), ultraviolet injury (Logan and Wilhelm, 1963), and ultrasonic injury (Burke, see Wilhelm, 1962). Confirmation of the time course of increased vascular permeability in each type of inflammation is highly important for verifying the chemical substance responsible for the vascular response.

4. *Physicochemical and Biological Properties of Vasoexin*

As described above, the delayed response was elicited more intensely and regularly than the immediate response. In contrast to the immediate response, the delayed response was almost unaffected by local antihistamines (triprolidine or mepyramine) given simultaneously with irritation, or repeatedly during maturation of the response; the unsusceptibility of the response to antihistamines was useful in determining the permeability factor responsible for the response.

Permeability factor was extracted in the pseudoglobulin fraction of inflammatory tissue such as thermal injury or an active Arthus reaction in the skin of rabbit (Hayashi *et al.*, 1964b) (criterion 1). This substance was not susceptible to antihistamines, soybean trypsin inhibitor (SBTI), or DFP; it failed to induce contraction of guinea pig ileum or extravascular migration of leukocytes, and it had no proteolytic activity. α-Chymotrypsin, which inactivates plasma kinins such as bradykinin and kallidin, showed no effect on this permeability factor (Yoshinaga, 1964). A proteolytic enzyme, extracted in the euglobulin fraction of inflammatory tissue showing decreased vascular response, suppressed the effects of this permeability factor, but had no effect on leukotaxine, bradykinin, kallidin, globulin permeability factor, and kallikrein (Hayashi *et al.*, 1964b; Koono, 1966). These observations suggested that this permeability factor may be different from the various mediators that have been proposed as responsible for the increased vascular permeability in injured tissue (criterion 6). Antipermeability factor abolished its activity after heating to 60°C for 30 minutes, or to 75°C for 15 minutes. However, the tissue inhibitor of SH-dependent protease (Udaka and Hayashi, 1965a), described above, had no effect on the permeability factor. Observations indicated that the antipermeability factor of the euglobulin fraction was different from the tissue inhibitor also present in the euglobulin fraction. It is equally evident that this permeability factor is distinct from SH-dependent protease.

Since its activity clearly paralleled the time course of the delayed permeability change in these types of inflammation, it seemed reasonable that this permeability factor may be a chemical mediator of the delayed response (criterion 2). As summarized in Table IV, the permeability factor of the pseudoglobulin fraction was further fractionated through a column of DEAE-cellulose by stepwise changes in concentration of eluting buffers; and it was demonstrated that this permeability factor was a complex of at least three peptides, suggesting that the permeability-increasing effects of the pseudoglobulin fraction were due to the combined action of these peptides. Peptide b, eluted in 0.067 *M* phosphate buffer (pH 7.4) on DEAE-cellulose, was further purified by chromatography on hydroxylapatite and DEAE-Sephadex, in that order (Yoshinaga *et al.*, 1966a); peptide c, eluted in 0.067 *M* phosphate plus 0.2 *M* sodium chloride (pH 7.4) on DEAE-cellulose, was purified by chromatography on DEAE-Sephadex and hydroxylapatite, in that order (Tasaki *et al.*, 1966; Hayashi, 1968a; Hayashi *et al.*, 1967; Tasaki, 1968a). The thermostable peptides b and c behaved as homogeneous substances on electrophoresis and ultracentrifugation, and their molecular weights were both about 7000 (Hayashi, 1967; Tasaki, 1968a). Purification of peptide a, eluted in 0.0067 *M* (pH 6.8) on DEAE-cellulose, is in progress.

The action of these permeability factors was apparently long-lasting, in contrast to the brief action of histamine, bradykinin, kallidin, and leukotaxine (Yoshinaga *et al.*, 1966b). The leaking of circulating pontamine sky blue as an indicator occurred shortly after intradermal injection of these permeability factors, reached its peak in about 1.5 hours, and became almost negative in 3–4 hours. It was thus assumed that the combined action of these peptides could account for the prolonged duration of the delayed vascular response because of their characteristic long-lasting action (criterion 5) (Table V).

The amount of purified permeability factor b from a *single* active

TABLE IV

PROCEDURES OF ISOLATION AND PURIFICATION OF VASOEXIN[a]

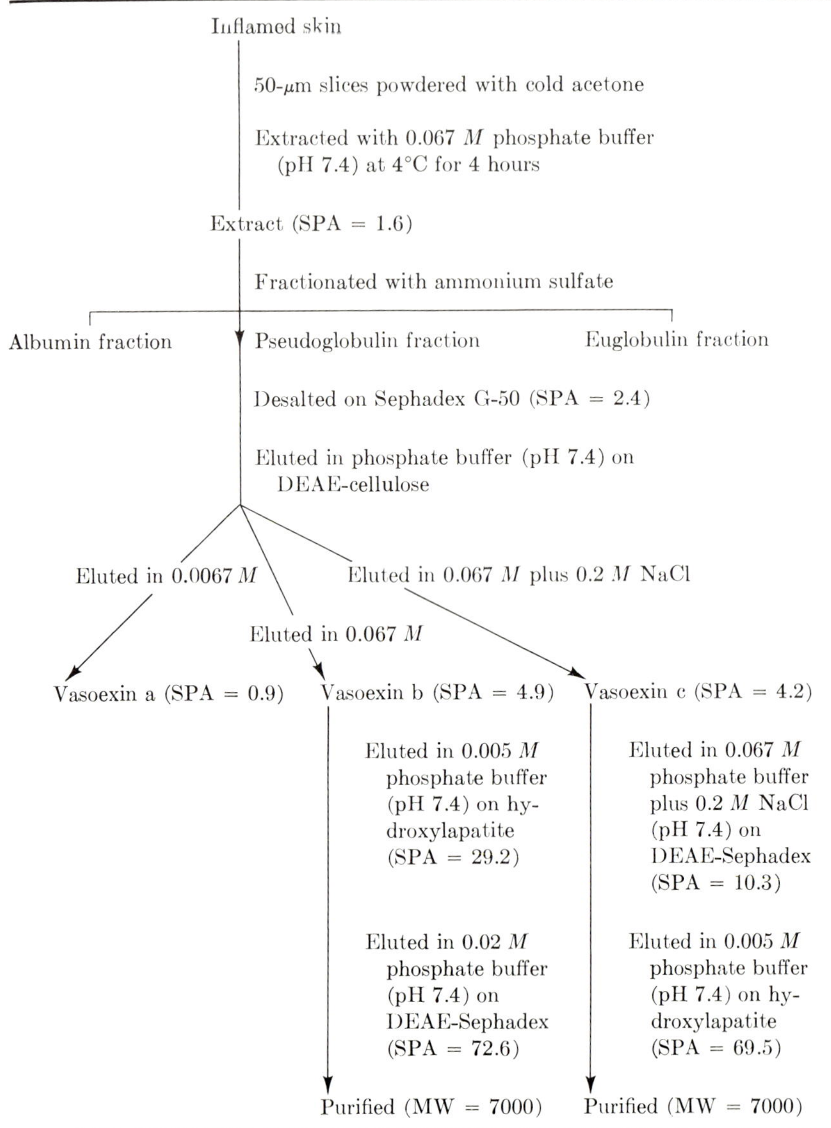

[a] SPA, Specific activity, in micrograms of extravascular dye/$E_{280\ \mathrm{nm}}$. Immediately after intradermal injection of vasoexin (0.1 ml) into clipped flanks, rabbits were injected with pontamine sky blue intravenously (60 mg/kg as a 5% solution in 0.425% saline), and 60 minutes later the locally accumulated dye was extracted and estimated according to the method of Nitta *et al.* (1963).

TABLE V

COMPARISON OF PEPTIDE PERMEABILITY FACTORS

	Molecular weight	Antihistamine	Contraction of guinea pig ileum	PMN migration	α-Chymotrypsin	Antivasoexin[a]	Action
Bradykinin	1060	−	+	+	+	−	Short
Kallidin	1200	−	+	+	+	−	Short
Leukotaxine	—	+	−	+	Not tested	−	Short
Vasoexin	7000	−	−	−	−	+	Long

[a] A SH-independent protease separated from inflamed tissue showing decreased permeability change (Koono, 1966).

Arthus skin site showing maximal permeability change was about 13.4 μg, and that of purified permeability factor c was about 11.9 μg; and such amounts of these factors could reproduce to a large extent the original Arthus delayed permeability change at the maximal stage, when given intradermally (Hayashi, 1967; Yoshinaga and Hayashi, 1969) (criterion 4). Similarly, the amount of purified factor b from a *single* burned skin site was about 13.7 μg, and that of purified factor c about 12.4 μg; and such amounts of these factors reproduced the original thermal delayed response at the maximal stage on experimental injection (Tasaki, 1968a) (criterion 4). It was of significance that these peptides induced marked edema which was accompanied by strong dye leakage, but did not cause leukocyte migration or hemorrhagic change (criterion 3), indicating successful separation of permeability factors from a chemotactic factor (leucoegresin) also present in the pseudoglobulin fraction of inflammatory tissue. The extravascular leaking of plasma proteins by these permeability factors was characterized by a structural change in the intercellular junction in the form of a wide open gap (Ogata, 1971; Tasaki, 1968b), but in the migration of PMN leukocytes such an open gap was not observed at all stages of migration (Ogata, 1971).

Since permeability factors similar to factors a, b, and c were also isolated from human inflammatory exudates from various sources (including rheumatoid arthritis, tuberculous pleuritis, and staphylococcal peritonitis) (Yoshinaga *et al.*, 1968), and from human primary tumors or experimentally transplanted tumors in rat (Nishimura, 1970), these factors have been named, respectively, vasoexin a, b, and c because of their generation in various types of tissue injury (Hayashi *et al.*, 1969). Thus it became clear that vasoexin a, b, and c may be associated with the delayed permeability change common in experimental injuries. The activity of vasoexin was not revealed at the site of the immediate permeability change.

5. *Production of Vasoexin*

Intradermal injection of SH-dependent protease in a reasonable amount (0.35 unit), comparable to that of the enzyme detected in an early stage of inflammation, clearly produced biphasic vascular responses closely resembling those seen in an active Arthus reaction (criterion 4) (Muto, 1969; Hayashi *et al.*, 1969); the immediate response provoked was inhibited by local antihistamines, but the delayed response induced was not influenced by the antihistamines. From protease-induced skin lesions during the period of the delayed response, three permeability factors were separated and then purified according to the above methods. Three permeability factors obtained were indistinguishable from vasoexin a, b, and c in physicochemical and biological properties, as mentioned above. It was thus presumed that vasoexin may be in part produced by the action of SH-dependent protease from substrate protein in skin tissue.

However, a similar delayed permeability factor was shown in culture to be released from sensitized histiocytic cells which were characteristically swollen by the specific antigen, after the immediate permeability factor susceptible to the antihistamines (probably histamine) (Hayashi *et al.*, 1964a); its action was not influenced by the antihistamines but inactivated by the antivasoexin described above (Hayashi *et al.*, 1964b; Kinuwaki, 1966). Release of the immediate and delayed permeability factors by the antigen was greatly diminished when the tissue inhibitor (Udaka and Hayashi, 1965a) was previously incorporated by the cells (Hayashi *et al.*, 1965c; Hayashi, 1967, 1968b). This tissue inhibitor itself showed no toxic action on the cells. When tested by phase-contrast cinemicrophotography, characteristic early morphological changes by the antigen, consisting of retracted pseudopods, decreased undulation of cell membrane, and decreased movement of mitochondria (Takaba, 1965; Hayashi, 1966; Kudo, 1967), were induced independent of the presence of the inhibitor (Hayashi, 1968b). These observations suggested that the protease inhibitor, which had been taken up by the cells, specifically inactivated the intracellular SH-dependent protease activated by the antigen and resulted in prevention of the release of permeability factors. No inhibition was observed with SBTI that had been taken up by the cells. It was thus presumed that vasoexin was in part produced by SH-dependent protease intracellularly.

J. Production of Inflammatory Chemotactic Factor for PMN Leukocytes (Leucoegresin) by SH-Dependent Protease

1. *Estimation of Leukocyte Chemotaxis*

As is well-known, most vessels of the terminal vascular bed open, dilate, and carry blood at a rapid rate (*hyperemia*) soon after injury. When the blood flow slows down, although the vessels remain dilated and filled with blood (*prestasis*), leukocytes appear in the marginal plasmatic zone (*margination*) apart from the central stream. The marginating leukocytes tend to adhere to the endothelium. First, the sticking is of short duration and the cells are carried away by the bloodstream along the vessel wall. Eventually, the cells adhere firmly to the endothelium and can no longer be dislodged by the bloodstream. Cohnheim (1873, 1889) believed that migration of leukocytes did not depend on spontaneous ameboid movement of the cells, but was due mainly to molecular changes in the vessel wall. Metchnikoff (1893) stressed the active part played by the migrating cell. These early observations were subsequently extended by many investigators (Clark and Clark, 1935; Clark *et al.*, 1936; Zweifach, 1953; Kageyama, 1963; Kageyama and Fukuda, 1967). Slowing of the bloodstream, margination and sticking of the leukocytes, and changes in the vascular wall enhance migration, but the passage of leukocytes through the blood–tissue barrier is accomplished by ameboid movements of the cell. These morphological observations were of importance in the evaluation of chemotactic factors.

In spite of the significance of leukocyte migration in inflammation, the problem of its mediation was unsolved for many years; a voluminous amount of literature existed concerning endogenous substances that might be responsible for migration of leukocytes from inflamed vessels, but most of this work is now of historical interest only. The reason for its irrelevance lies to a large extent in the unsatisfactory method used to demonstrate stimulation of leukocyte movement (Harris, 1960). Following the introduction of a new technique developed by Boyden (1962) into this field, numerous investigations were aimed at determining the naturally occurring chemotactic host factor for leukocytes; thus different types of chemotactic factors have been proposed by many laboratories. However, present knowledge of the natural mediators of leukocyte migration in inflammation is still incomplete.

Modifications of Boyden's (1962) method were made by many workers for the design of culture chambers (Hurley, 1963a; Ward *et al.*, 1965;

Cornely, 1966; Yamamoto *et al.*, 1971; Yoshinaga *et al.*, 1972a), filter membranes (Keller and Sorkin, 1967; Snyderman *et al.*, 1968; Horwitz and Garret, 1971), and culture media (Ward *et al.*, 1965; Keller and Sorkin, 1967). It was demonstrated that chemotaxis of PMN leukocytes of various animals including humans was measurable by the use of millipore filters with a pore size of 0.65, 1.2, or 3.0 μm; filters of 3.0-μm pore size gave the greatest number of migrating cells, but the background number was also greater than that for other filters. Since PMN leukocytes that migrated through a 3.0-μm pore did not stick firmly to the lower surface of the filters but tended to drop down into the culture medium of the lower compartment, it was possible to count only the cells that were still sticking to the lower surface. Accordingly, other Millipore filters were found to be useful (Ward *et al.*, 1965; Keller and Sorkin, 1967; Snyderman *et al.*, 1968); and in some experiments (Gallin *et al.*, 1973), ^{51}Cr-labeled cells were utilized for higher sensitivity and accuracy. As to whether or not the observations by Boyden's method indicated the directional movement of leukocytes, it was suggested that, when assayed by both Boyden's method and the cellophane square test, chemotaxis of slime mold amebas and leukocytes was essentially similar (Bonner *et al.*, 1971). Such directional movement of leukocytes was further demonstrated by phase-contrast cinemicroscopy (Ramsey, 1972) or the cover slip technique (Zigmond and Hirsch, 1973). Thus it seems reasonable that Boyden's method can mainly measure directional movement of leukocytes *in vitro* at a considerable quantitative level, but whether or not the observations obtained reflect the *in vivo* movement of leukocytes is still to be determined.

Therefore any investigator should perform both *in vivo* and *in vitro* experiments to elucidate the mechanism underlying chemotaxis of leukocytes or the effect of chemotactic factors (Hayashi, 1967). The cremaster muscle (Majno *et al.*, 1961) and panniculus carnosus muscle (Hayashi *et al.*, 1965b; Tasaki, 1968b) of rat and rabbit gave preparations favorable for searching detailed morphological changes in the blood vessels involved with leukocyte migration. Cinemicrophotographic observation in the ear chamber (Clark *et al.*, 1936) or the mesentery of rat and guinea pig (Kageyama, 1963) has also made a great contribution to the analysis of morphological sequence in leukocyte migration.

2. *Criteria for Inflammatory Chemotactic Factor*

Following the introduction of Boyden's method, different types of chemotactic factors have been proposed from many laboratories. However, present knowledge of the natural mediators of leukocyte migration in inflammation is still incomplete, because of inadequate characterization of suspected mediators. It is thus postulated that the following criteria should be applied to suspected mediators and that they fulfill some if not all of the conditions listed (Hayashi, 1967; Hayashi *et al.*, 1969).

1. Chemotactic factor should be locally available to induce inflammatory leukotaxis.
2. Amount (or activity) of the chemotactic factor should parallel the time course of leukotaxis.
3. Chemotactic factor, when injected in concentrations reasonably comparable to those detected at the inflamed site, should produce morphological changes similar to the reaction.
4. Action of the chemotactic factor should be specific, being active in causing leukocyte migration but not in inducing vascular permeability and hemorrhagic changes.
5. Chemotactic factor should be inhibited by the specific antagonistic substance locally available.
6. Inflammatory leukotaxis should be suppressed by a specific antagonistic substance.
7. Depletion of chemotactic factor or its precursor should cause a decrease in leukotaxis.
8. Precursor of a chemotactic factor or the enzyme associated with production of chemotactic factor, should be locally available.

3. *Time Course of Leukocyte Migration in the Inflammatory Process*

As described above, increased vascular permeability and leukocyte migration are consistent and significant events in inflammation. In general, vascular permeability change precedes migration of PMN leukocytes (Fig. 1); and migrated PMN leukocytes often become replaced by mononuclear cells (i.e., monocytes, macrophages, and their derivatives) as well as by lymphocytes, although there are some modifications according to the nature of inflammatory stimuli. Until quite recently it was assumed that leukocyte migration occurred at the same time and from the same blood vessels as leakage of plasma protein. It has been shown, however, that after injury leukocyte migration does not commence until much later than increased vascular permeability. For instance, in the active Arthus reaction described above, the increase in vascular permeability reaches its peak in about 4.5 hours, but leukocyte migration in about 12 hours. Similarly, in thermal injury, the increase in vascular permeability reaches its peak in about 2 hours, but leukocyte migration in 4–6 hours. The time course of these inflammatory events is distinctive, and suggests involvement of different chemical substances responsible for each inflammatory event. By the mechanism of increased vascular permeability, various types of plasma proteins are exuded into the extravascular tissue before leukocytes migrate. As described in Section J-6, immunoglobulins among plasma proteins exuded are locally converted to chemotactic factors for PMN leukocytes or lymphocytes, or mononuclear cells.

4. *Physicochemical and Immunological Properties of Leucoegresin*

In studies of the mechanism of the inflammatory process, the great effort to identify natural mediators that may be responsible for characteristic leukocyte migration has been continued, and Boyden's (1962) work has particularly stimulated further work in this field.

Chemotactic factor was extracted in the pseudoglobulin fraction of inflammatory tissue from thermal injury or an active Arthus reaction in the skin of rabbit (Hayashi, 1967; Yoshida *et al.*, 1968; Yoshinaga *et al.*, 1971b; Maeda *et al.*, 1970) (criterion 1). Since the chemotactic activity of the protein fraction roughly paralleled the time course in the degree of tissue leukocytosis observed in these injuries (criterion 2), it was suggested that the chemotactic factor of the protein fraction may be a mediator of inflammatory leukotaxis. For purification of this chemotactic factor, the pseudoglobulin fraction was eluted on DEAE-Sephadex, CM Sephadex (Yoshinaga *et al.*, 1971b), and an antirabbit IgG affinity column (Yoshinaga *et al.*, 1972a), in that order, as summarized in Table VI.

This chemotactic factor was further eluted on Sephadex G-200; the elution profile of the protein showed a symmetric homogeneous pattern, and its chemotactic activity paralleled the absorbancy at 280 nm. Its molecular weight was about 140,000 when measured by gel filtration. Its homogeneity was further confirmed by boundary electrophoresis and ultracentrifugation. Its isoelectric point was pH 5.0, and its sedimentation coefficient ($s_{w,20}$) was 6.58S. This substance was a protein, and its chemotactic activity was relatively thermostable. It was examined with antibodies against the chemotactic factor or extracts of Arthus skin or burned skin in agar immunoelectrophoresis and immunodiffusion. In all cases only a single precipitin line was revealed, and it was confirmed that the chemotactic factor obtained did not contain any immunologically detectable impurity. This chemotactic factor has been named *leucoegresin* because of its characteristic properties, as described below (Hayashi, 1967; Yoshida *et al.*, 1968).

Leucoegresin was difficult to relate to complement or its products. Its

TABLE VI

Procedure for Isolation and Purification of Leucoegresin[a]

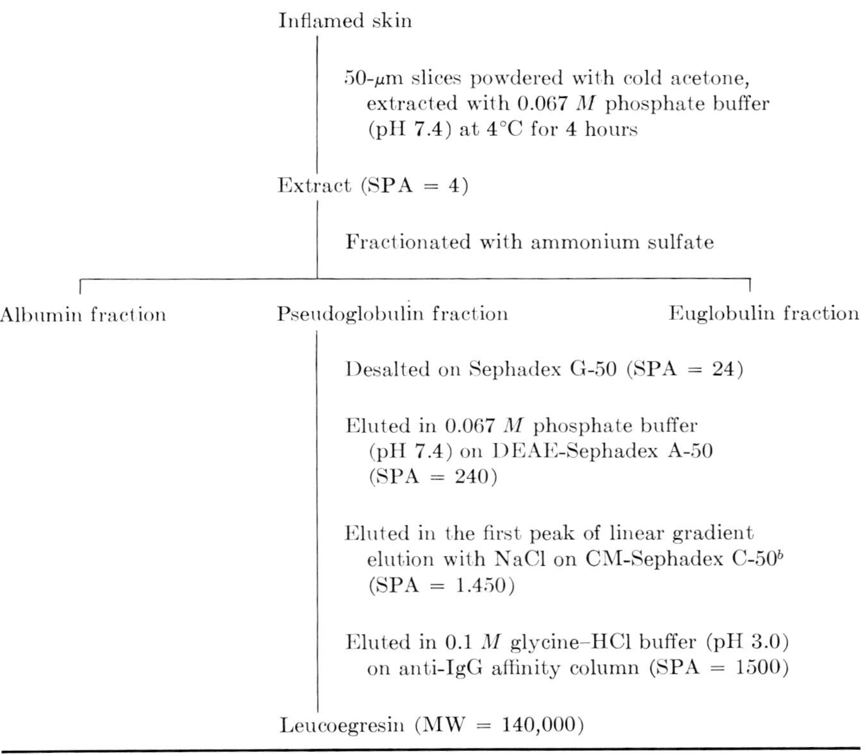

[a] SPA, specific activity, numbers of PMN leukocytes migrated/$E_{280\ nm}$. Molecular weight estimated by gel filtration.

[b] If necessary, this step was repeated until the elution profile became homogeneous.

activity was not influenced by *n*-CBZ-α-glutamyl-L-tyrosine which inactivates the chemotactic complex of C′567 (Ward *et al.*, 1965, 1966); the molecular weight of the trimolecular complex was more than 300,000. The molecular weight of previously reported complement-derived chemotactic factors was clearly less than that of leucoegresin (Hayashi *et al.*, 1974). The molecular weight of chemotactic substances derived from C'_5 was reported to range from 8000 to 23,000 and even up to 70,000 (Ward and Hill, 1970, 1972). It was of interest to note that the chemotactic substance from chick embryos carrying a viral infection had a molecular weight similar to that of leucoegresin (Ward *et al.*, 1972).

Rabbit leucoegresin shared some of the antigenicity with rabbit IgG, because it gave a single precipitin line with antirabbit serum, anti-IgG, anti-Fc, or antilight chain antibody in agar immunoelectrophoresis (Yamamoto *et al.*, 1971). Leucoegresin was also absorbed by antileucoegresin, anti-IgG, anti-Fc, or antilight chain antibody at their optimal conditions; and its activity completely disappeared from the fluid phase, indicating that its activity was associated with leucoegresin itself, but not with any contaminant, if present (Yamamoto *et al.*, 1971).

It was of importance to note that most (about 70%) of the chemotactic activity of the extracts of active Arthus and burned skin lesions at the maximum stage of PMN leukocyte migration was associated with leucoegresin itself, because antileucoegresin gave a single precipitin line in agar immunoelectrophoresis with the skin extracts, and that the chemotactic activity of the extracts was significantly absorbed by adding antileucoegresin in amounts capable of inducing an optimal precipitation (Hayashi *et al.*, 1974) (Table VII). These observations indicate that leucoegresin must play a significant part in inflammatory leukotaxis. At present, it is assumed that the remaining chemotactic activity (about 30%) of the extracts might be concerned with complement-derived chemotactic factor for PMN leukocytes, because such chemotactic factors have been demonstrated in the extracts of skin lesions of immunological vasculitis (Ward and Hill, 1972) and in the synovial fluid of rheumatoid arthritis (Ward and Zvaifler, 1971).

TABLE VII

EFFECT OF ANTI-LEUCOEGRESIN OF CHEMOTACTIC ACTIVITY OF LEUCOEGRESIN AND EXTRACTS OF ACTIVE ARTHUS SKIN LESIONS

Samples in lower compartment	Number of PMNs migrated
Leucoegresin[a] plus buffer	225
Buffer[b]	12
Leucoegresin plus antileucoegresin[c] plus buffer	8
Leucoegresin plus normal goat IgG[d]	228
Antileucoegresin plus buffer	11
Extract[e] plus buffer	215
Extract plus antileucoegresin plus buffer	70
Extract plus normal goat IgG plus buffer	212
Normal goat IgG plus buffer	13

[a] 100 μg (in phosphate buffer, 0.067 *M*, pH 7.4).
[b] Phosphate buffer, 0.067 *M*, pH 7.4.
[c] 500 μg (in phosphate buffer, 0.067 *M*, pH 7.4).
[d] 500 μg (in phosphate buffer, 0.067 *M*, pH 7.4).
[e] From 12-hour-old skin lesion of active Arthus reactions, 10.0 at 280 nm.

5. *Biological Action of Leucoegresin*

Rabbit leucoegresin was active not only for PMN leukocytes of rabbit but also for those of guinea pig, rat, and mouse. In contrast to the inflammatory permeability factor (vasoexin) separated from the same protein fraction of inflammatory tissue, leucoegresin induced pronounced migration of PMN leukocytes but did not cause vascular permeability change or hemorrhagic change on experimental injection (criterion 4) (Hayashi, 1967; Yoshinaga *et al.*, 1971b). PMN leukocytes were found sticking to the endothelium of the venules within 10–30 minutes of intradermal injection, and migrating through the venular wall (not through the capillary wall) after 30–60 minutes; such migration of PMN leukocytes became maximal at 3–4 hours of injection. The cells migrated infiltrate in a diffuse and focal (perivascular) pattern throughout the skin; some cells enter the lymphatic vessels. No migration of cells other than neutrophilic PMN leukocytes is observed, indicating its specific action (criterion 4). Five micrograms of purified leucoegresin were clearly positive for migration of PMN leukocytes. The amount of leucoegresin was about 250 μg per 12-hour-old Arthus site and that of leucoegresin was about 50 μg per 4-hour-old burned site (criterion 3).

The morphological sequence of migration of PMN leukocytes induced by leucoegresin (50–70 μg) was studied in the cremaster muscle of rat and rabbit by means of electron microscopy (Ogata, 1971). Migration of the cells was essentially found at the site of venules 30–50 μm in diameter. As shown in Figs. 4–9, PMN leukocytes characteristically accumulated in the lumina of the venules as the first step of migration, and then adhered to the endothelial surface. The cell frequently became flattened against the vessel wall and increased the contact surface, leaving a narrow space (100–400 A in distance). The sticking cell developed a small pseudopod from the flattened cytoplasm and inserted it into the junction of two adjacent endothelial cells. As the cell started its passage through the intercellular junction by extending cytoplasm, the marginal fold–like processes were formed from the endothelial cytoplasm, and the cell was enveloped partially or completely. The appearance of such a characteristic cytoplasmic process was noted only in endothelial cells with PMN leukocytes adhering to them, suggesting reasonable fixation of the cell to the endothelial surface and successful prevention of dislodging by the bloodstream. While the PMN leukocyte was completely surrounded by such cytoplasmic processes at the luminal side and intercepted from the lumen, the cell itself penetrated by ameboid motion into the intercellular junction and then reached the space lying between the endothelial cells and periendothelial cells. As the PMN leukocyte

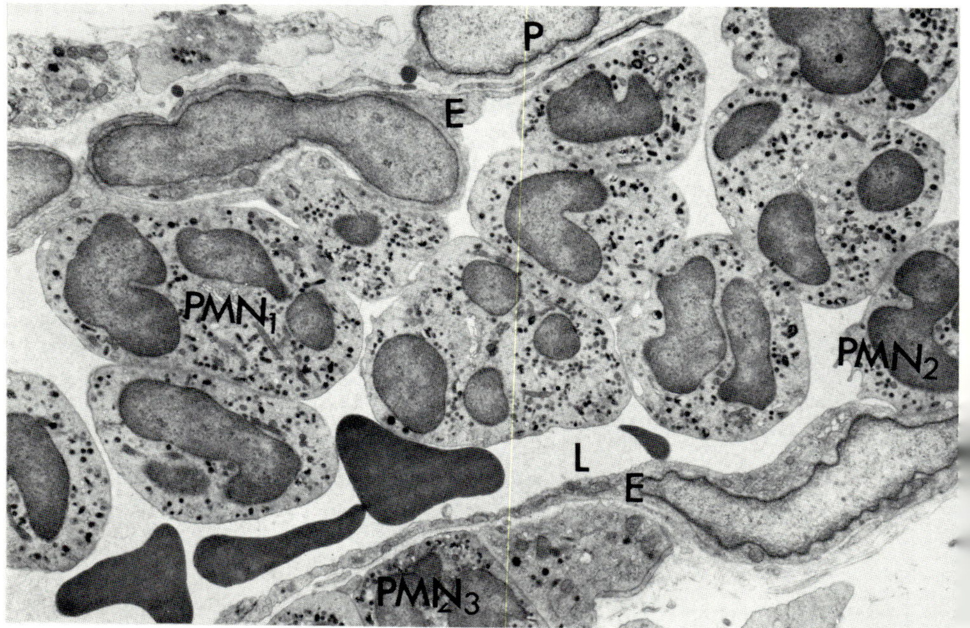

FIG. 4. Characteristic accumulation of several neutrophilic PMN leukocytes in the lumen (L) of the venule in the cremaster muscle of the rat. The cells adhere loosely to each other. PMN_1 and PMN_2 adhere to the endothelial surface (E). PMN_3 has already passed through the venular wall. R, Red blood cell; P, periendothelial cell. The photograph was taken 2 hours after injection of leucoegresin (50 μg). ×4500.

moved, the intercellular junction was promptly re-formed by the cytoplasmic processes described above; and no free endothelial gap communicating from the lumen to the periendothelial sheath throughout the course of migration was observed. Finally, the PMN leukocyte passed through the space between the neighboring periendothelial cells and migrated into the adjacent connective tissue. Migration of other blood cells, such as monocytes, lymphocytes, red blood cells, and platelets, was not found. Electron micrographs of the migration of PMN leukocytes by leucoegresin strikingly resembled those observed in experimentally induced pancreatitis in dogs (Williamson and Grisham, 1961). By contrast, extravascular leaking of plasma protein by histamine, bradykinin, and vasoexin was characterized by a structural change in the intercellular junction in the form of a wide open gap (Tasaki, 1968b; Ogata, 1971), as seen in Fig. 10. The formation of such an open gap was not observed at all stages of PMN migration. Such difference in the electron

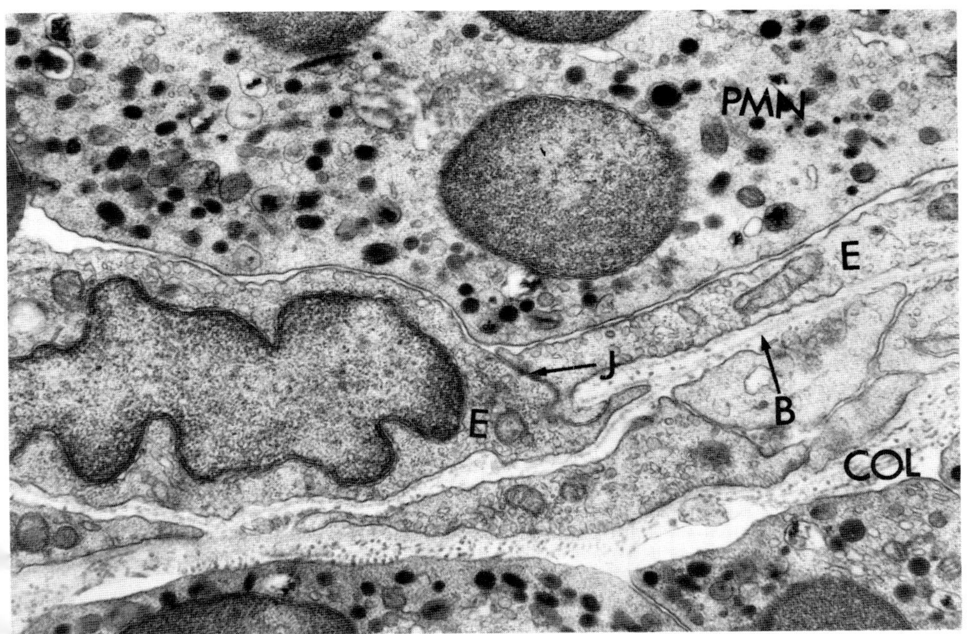

FIG. 5. Adhering of PMN leukocyte to endothelial cells (E). It has become flattened against the endothelial cells and has increased the area of contact, leaving a narrow space. The cell has started to extend a pseudopod into the intercellular junction (J). Demonstrable structural change was not found in the endothelial cells. B, Basement membrane; P, periendothelial cell; COL, collagen fibers. The photograph was taken 2 hours after injection of leucoegresin. ×25,000.

micrographs seemed to explain the biological difference between leucoegresin and vasoexin.

6. *Production of Leucoegresin*

As mentioned above, when SH-dependent protease II was injected intradermally in a reasonable amount (1.65 units), there was observed a pronounced infiltration of PMN leukocytes showing its peak at 6 hours after an increase in vascular permeability. Chemotactic factor for PMN leukocytes was separated from the protease-induced skin lesion after 6 hours, and then purified according to the procedures summarized in Table VI. The chemotactic factor obtained was indistinguishable from leucoegresin in physicochemical, immunological, and biological properties (Hayashi *et al.*, 1969; Muto, 1969). From a *single* protease-induced skin lesion, about 210 μg of leucoegresin was isolated; about 250 μg of leuco-

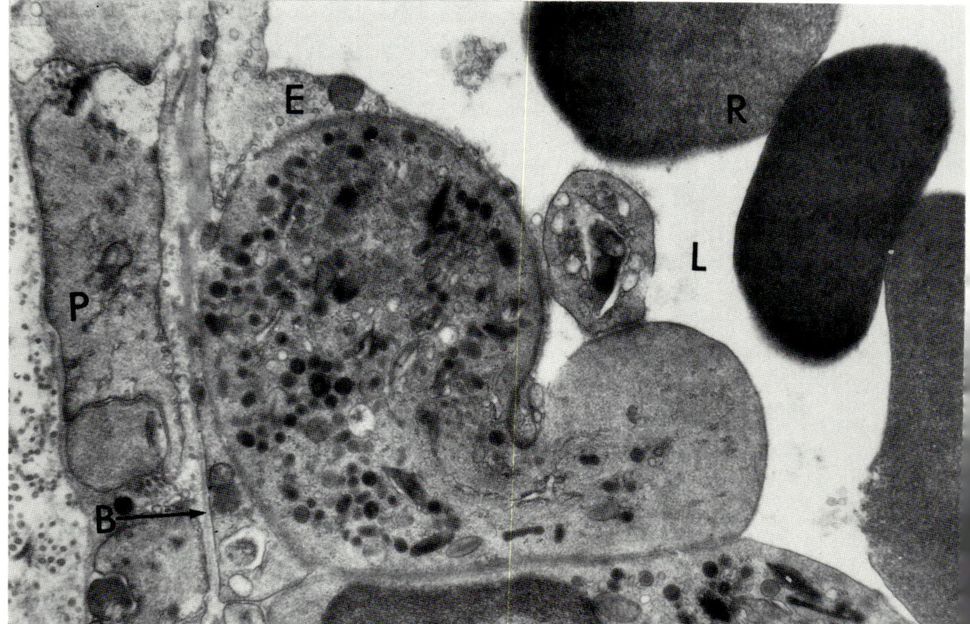

FIG. 6. PMN leukocyte penetrating the intercellular junction is tightly embraced by the marginal fold-like cytoplasmic process from adjacent endothelial cell (E), by which a PMN leukocyte is fixed on the endothelial lining against the bloodstream. L, Vessel lumen; P, periendothelial cell. ×20,000.

egresin was isolated from a *single* Arthus site at 12 hours. These observations postulated that local production of leucoegresin was probably associated with the action of the SH-dependent protease injected, and the precursor of leucoegresin, which is probably among the plasma proteins exuded by the action of permeability factors described above.

In view of observations that leucoegresin has common antigenicity with serum IgG (Yamamoto *et al.,* 1971), it was also assumed that the precursor of leucoegresin could be found among the plasma proteins exuded. By preparative zone electrophoresis on Pevikon C-870, six protein fractions were separated from normal rabbit serum. Only the γ_2-globulin fraction from rabbit serum became strongly chemotactic *in vitro* for PMN leukocytes after mild treatment with a small amount (0.33 unit) of SH-dependent protease II (Yoshinaga *et al.,* 1971a). None of other protein fractions showed chemotactic effects under the same conditions. SH-dependent protease itself was ineffective on PMN leukocytes.

It was further confirmed that the IgG from rabbit and man had no

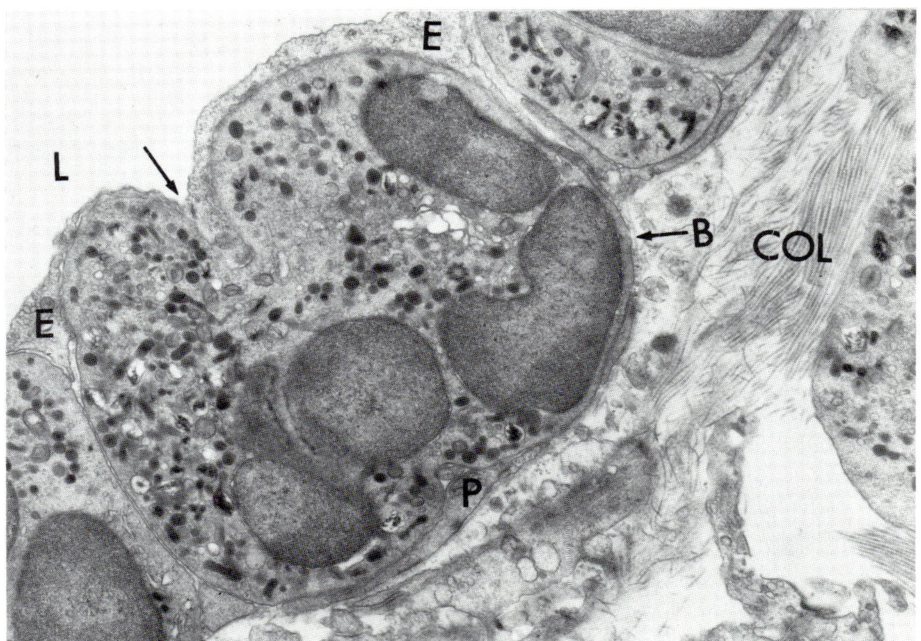

FIG. 7. A PMN leukocyte perfectly enveloped by marginal fold-like cytoplasmic processes. A hollow in the cell at the luminal side (arrow) suggests the existence of the mechanical force of these cytoplasmic processes. PMN migration appeared to be prevented temporarily by the periendothelial cell (P). L, lumen. ×12,500.

chemotactic activity on the cells but became strongly chemotactic when treated with SH-dependent protease II (0.33 unit) (Yoshinaga *et al.*, 1971a) (criterion 5). *In vitro* production of similar chemotactic factor was confirmed with all IgG subclass samples tested, that is, human IgG_1, IgG_2, IgG_3, and IgG_4 (Nishiura *et al.*, 1974a), although some differences in chemotactic effects were observed between the IgG subclasses; IgG_2 and IgG_4 were more active in production of chemotactic factors than IgG_1 and IgG_3, suggesting some relation to a particular structural specificity of the IgG molecule. The molecular weight of the chemotactic factors produced was about 140,000, similar to that of leucoegresin.

Although papain also induced *in vitro* chemotactic generation of IgG, this was associated more clearly with susceptibility to the enzyme, that is, to the specificity of the heavy-chain structure of the IgG molecule (Yoshinaga *et al.*, 1972b); papain-resistant IgG, such as electrophoretically fast rabbit IgG, human IgG_2 and IgG_4, and mouse IgG_1, became strongly chemotactic by mild treatment with a small amount of

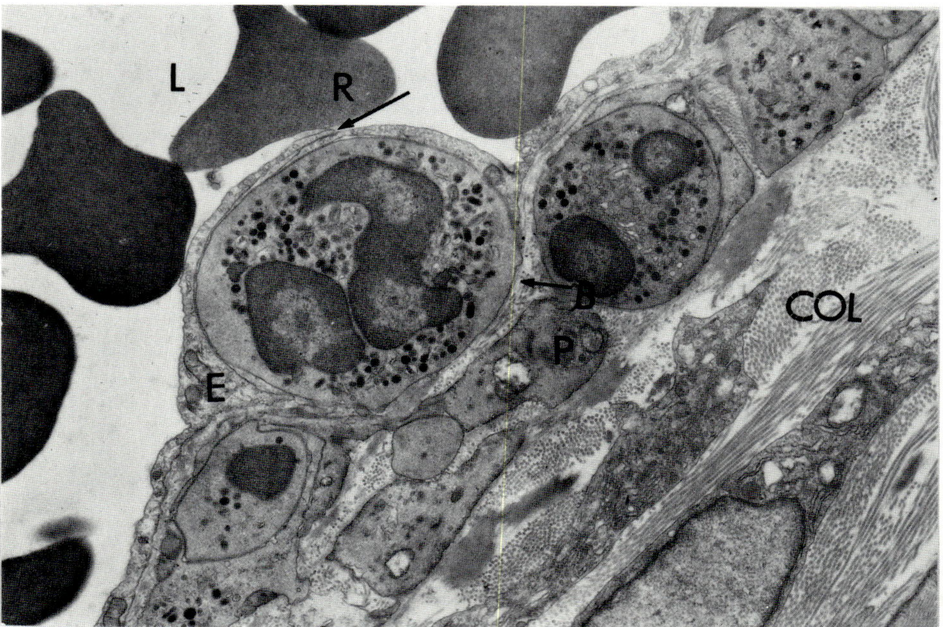

FIG. 8. As the PMN leukocyte passes through the intercellular junction, cytoplasmic processes from the adjacent endothelial cells (E) pile up and form a tight junction (arrow). The basement membrane (B) was extended by a migrating PMN leukocyte and became irregular. L, Vessel lumen; R, red blood cell; P, periendothelial cell; COL, collagen fibers. The photograph was taken 2 hours after injection of leucoegresin. ×7500.

papain (0.33 unit) or water-insoluble papain, but papain-sensitive IgG, that is, electrophoretically slow rabbit IgG, human IgG_1 and IgG_3, and mouse IgG_{2a} and IgG_{2b} showed no chemotactic activity after the same enzymic treatment. The molecular weight of the chemotactic factors was similarly about 140,000.

The chemotactic generation of IgG by SH-dependent protease seemed to be associated with some minor structural change in the IgG molecule, because the molecular weight of chemotactic factor produced was similar to that of leucoegresin or untreated IgG, while the Fab and Fc fragments produced by papain had no chemotactic potency (Hayashi *et al.*, 1974). Such chemotactic generation of IgG occurred only under favorable conditions, that is, mild short-termed digestion of IgG with a small amount of SH-dependent protease.

It was of significance to note that production of chemotactic factor by

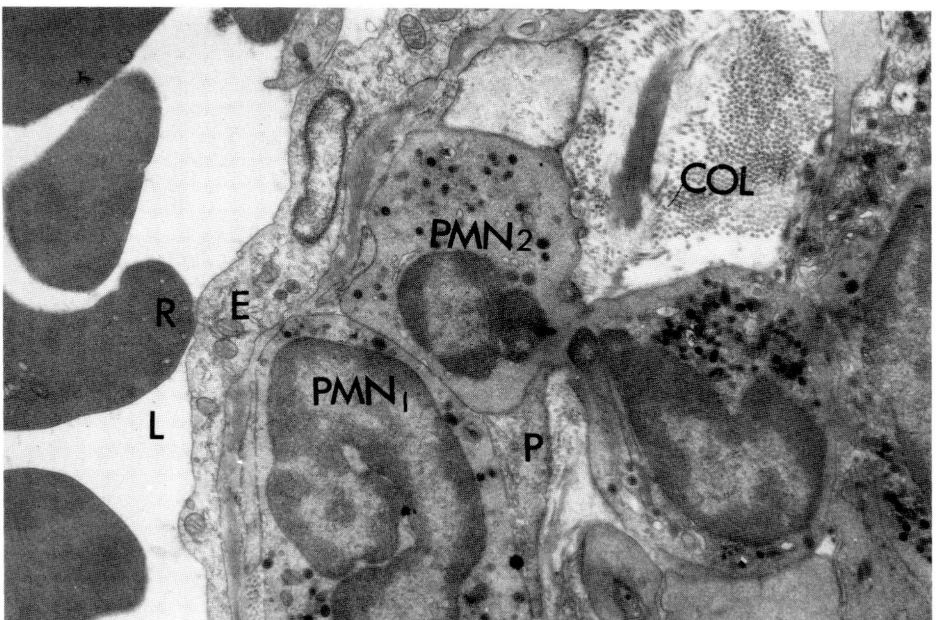

FIG. 9. PMN_1 leukocyte appearing in the periendothelial sheath. PMN_2 leukocyte migrating from the periendothelial sheath into the connective tissue. R, Red blood cell; L, vessel lumen; E, endothelial cell; COL, collagen fibers. ×21,000.

SH-dependent protease was accompanied by a release of inactive dialyzable peptides from IgG molecule; such release of peptides was already apparent at an early stage (1 hour) of the digestion when ^{125}I-labeled IgG was used. In contrast to papain, SH-dependent protease was characterized by the failure to produce fragments like Fab or Fc even after prolonged digestion of IgG (24 hours) (Yamamoto *et al.*, 1973). Chemotactic activity was decreased in parallel with prolongation of the digestion period; such a decrease in chemotactic activity was due to advanced degradation of chemotactic factor, as revealed by increased formation of dialyzable peptides. It was suggested that the structural change in the IgG molecule required for such a chemotactic generation occurred in the Fc portion of the IgG molecule within at least 1 hour under these enzymic conditions (Yamamoto *et al.*, 1973).

In an early work by Hurley and Spector (1961), it was demonstrated that leukocyte migration could be elicited when the saline extract of burned skin was injected into the skin of rats. It could also be induced with serum incubated with extracts of various tissues, for example, liver,

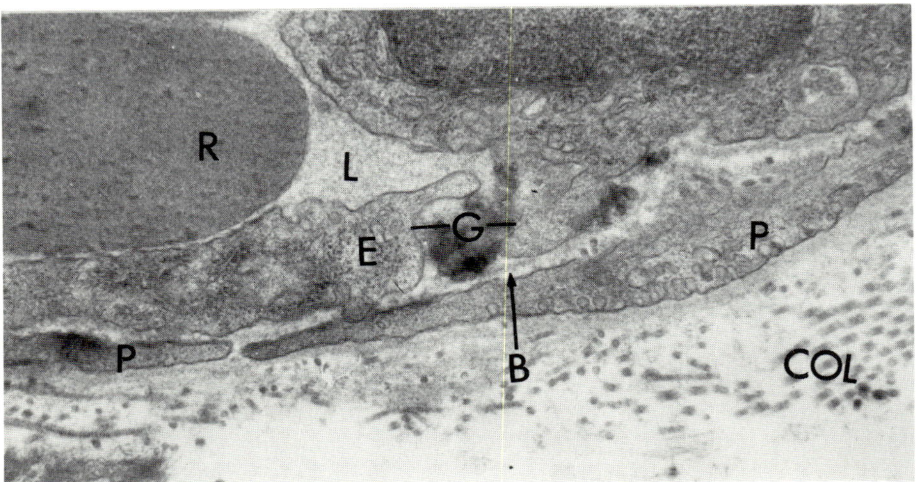

FIG. 10. Characteristic leaking of carbon particles in the wide open gap (G) between two adjacent endothelial cells (E) of the venule in the cremaster muscle of the rat. The endothelial cells did not incorporate carbon particles. The photograph was taken 1 hour after injection of vasoexin (20 μg). $\times$25,000.

heart muscle, spleen, and PMN leukocytes (Hurley, 1963a). These investigators suggested that a specific endogenous system induces leukocyte migration after injury and that it consists of a precursor in serum, an activator in certain tissues, and an activated principal. The activated substance was found to be a nondialyzable protein. It was first concluded that the endogenous factor acted on the vessel wall, but later Hurley (1963b, 1964) demonstrated that the substance capable of causing PMN leukocyte migration upon injection into the dermis was also chemotactic when tested *in vitro*. Their observations were quite reasonable in connection with the production of leucoegresin. The suspected precursor in serum corresponded to IgG; the suspected activator corresponded to SH-dependent protease from inflammatory cells; and the suspected activated principal corresponded to leucoegresin.

It was further demonstrated that production of leucoegresin by SH-dependent protease at the inflamed site was not related to the complement system, because a similar amount of leucoegresin was isolated from active Arthus skin lesions or from protease-induced skin lesions in rabbits whose serum complement was apparently depleted by an anticomplementary factor from cobra venom (Hayashi, 1974; Nishiura *et al.*, 1974b). The *in vivo* effect of leucoegresin was not influenced by such depletion of serum complement. In view of the finding that this anticomplementary factor can deplete serum complement components such as C'_3 and C'_5 (Cochrane *et al.*, 1970), it was clear that these components are not involved in the production of leucoegresin.

K. Production of Chemotactic Factor for Lymphocytes by SH-Dependent Protease

As mentioned above, serum IgG was confirmed to be a precursor of an inflammatory chemotactic factor (leucoegresin). In view of observations that PMN leukocyte migration is followed by lymphocyte migration in inflammation, this type of experiment was extended as follows.

Rabbit IgG or IgM itself had no chemotactic potency for lymphocytes obtained from the thoracic lymph of rat, but became strongly chemotactic after mild treatment (3-hour digestion) with a small amount (0.08–0.33 unit) of SH-dependent protease separated from PMN leukocyte lysosomes; and it was found that IgM was more significant than IgG as a precursor in chemotactic generation (Higuchi and Hayashi, 1974). Production of chemotactic factor was clearly associated with the activity of the enzyme added. Most of the cells collected were small lymphocytes and favorable for chemotactic assay. Subsequently, chemotactic factor was successfully separated from PMN leukocyte chemotactic factor (MW 140,000) by gel filtration on Sephadex G-200; the molecular weight of the chemotactic factor was assumed to be similar to that (14,000) of SH-dependent protease II. Chemotactic activity was decreased in a parallel with prolongation of enzymic digestion (10 and 24 hours). It was thus assumed that SH-dependent protease released from PMN leukocytes converted immunoglobulins, especially IgM, to a chemotactic factor for lymphocytes at the inflamed sites. Unlike PMN leukocytes or mononuclear cells, lymphocytes do not pass through intercellular junctions but cross the vascular wall by entering the endothelial cells and transversing their cytoplasm. Therefore the question whether the chemotactic factor injected can produce such a characteristic morphological change in lymphocyte migration should be answered.

In contrast, no chemotactic generation of IgG and IgM for lymphocytes was revealed after treatment with SH-independent protease from PMN leukocytes at the same level of activity (0.33 unit, 3-hour digestion). However, IgG was not chemotactic for mononuclear cells collected from oil-induced peritoneal exudates of guinea pigs, but it became chemotactic when treated with SH-independent protease (0.33 unit, 6-hour digestion) (Maeda *et al.*, 1970). In contrast, SH-dependent protease did not generate chemotactic activity during incubation with IgG molecule when tested at the same level of activity. These observations suggested that the production of different types of chemotactic factors might be associated with the mode of cleavage of immunoglobulin molecules by different proteases. Recently, a chemotactic factor for mononuclear cells was extracted from a 24-hour-old skin site of a reaction in guinea pigs, which was induced by intradermally injected tuberculin purified protein derivative (PPD) (Kambara *et al.*, 1972). This chemotactic factor was largely concentrated in the pseudoglobulin fraction of the skin extracts. The way in which these two chemotactic factors are related is unknown.

III. Concluding Remarks

Among the intracellular proteases, neutral SH-dependent protease was shown to satisfy many criteria making it acceptable as an *inflammatory protease.* In contrast to other proteases, the enzyme characteristically provoked morphological changes resembling the original inflammatory reactions when injected in concentrations reasonably comparable to those detected at the inflamed site. A long-acting peptidic permeability factor (vasoexin), which was locally produced by SH-dependent protease, also satisfied several criteria for an *inflammatory permeability factor* of delayed vascular permeability change. A chemotactic factor specific for PMN leukocytes (leucoegresin), which was locally produced by SH-dependent protease, also satisfied many criteria making it acceptable as an *inflammatory chemotactic factor* of tissue leukotaxis.

It was thus suggested that SH-dependent protease is activated in mesenchymal connective tissue cells such as histiocytes by inflammatory stimulation, for example, by an antigen–antibody reaction, and releases histamine and then vasoexin (the main source of histamine is the mast cell). Following the increase in vascular permeability induced by these mediators, the SH-dependent protease released first converts the exuded IgG molecules to a chemotactically active molecule, namely, leucoegresin. The latter could cause the migration of PMN leukocytes into the inflammatory lesion. Later, the neutral SH-dependent protease, which was activated in and released from the migrated PMN leukocytes by antigen–antibody complexes, might produce from the exuded immunoglobulin (especially IgM) molecules a chemotactic factor which might induce the accumulation of lymphocytes at the inflamed site. The migration of mononuclear cells might be associated with chemotactic factor produced from IgG by neutral SH-independent protease released from PMN leukocytes. The role of this enzyme in inflammation remains to be studied.

Neutral SH-dependent protease, which was termed an inflammatory protease, was inactivated by two types of inhibitors (serum and tissue).

It was suggested that the balance between the enzyme and these inhibitors may control the intensity, extent, and duration of inflammatory reactions.

It was further postulated that immunoglobulin molecules, besides their function as antibodies, may play an immunologically nonspecific part as precursors of chemotactic factors in inflammation. In relation to these observations, it was of special interest that the amount of various types of IgG products increased in the lymph of severely burned patients (Goldberg and Whitehouse, 1970), and that a phagocytosis-stimulating peptide (tuftsin) was derived from IgG molecules (Nishioka *et al.*, 1972, 1973), suggesting the dynamic metabolism of immunoglobulins under physiological and pathological conditions. The biological action of these products should be ascertained.

Acknowledgments

The author is very grateful to Dr. M. Koono and Dr. M. Yoshinaga for discussion, and to Dr. G. H. Bourne for encouragement in the course of preparation of this review article. Our investigations referred to in this article were in part supported by grants from the Japanese Ministry of Education, the U. S. Army Research and Development Group, Far East, the Mitsubishi Foundation, Tokyo, the Biological Institute of Shionogi Pharmaceutical Company, Osaka, and the Society for Metabolism Research, Tokyo.

The author wishes to acknowledge the permission of the *Kumamoto Medical Journal* to reproduce Figs. 4–10.

References

Allison, F., Jr., and Lancaster, M. G. (1959). *Brit. J. Exp. Pathol.* **40,** 324.

Altschuler, C. H., and Angevine, M. (1949). *Amer. J. Pathol.* **25,** 1061.

Angevine, M. (1951). *Conf. Connect. Tissues, Trans.* **1,** 44.

Ascheim, E. (1965). *Fed. Proc., Fed. Amer. Soc. Exp. Biol.* **24,** 1104.

Ballow, M., and Cochrane, C. G. (1969). *J. Immunol.* **103,** 944

Barrett, A. J. (1971). *In* "Tissue Proteinases" (A. J. Barrett and J. T. Dingle, eds.), p. 109. North-Holland Publ., Amsterdam.

Barron, E. S. G. (1951). *Advan. Enzymol.* **11,** 201.

Bazin, S., and Delaunay, A. (1969). *In* "Inflammation Biochemistry and Drug Interaction" (A. Bertelli and J. C. Houck, eds.), p. 21. Excerpta Med. Found., Amsterdam.

Bensley, S. H. (1934). *Anat. Rec.* **60,** 93.

Bettelheim, F. R., and Neurath, H. (1955). *J. Biol. Chem.* **212,** 241.

Bonner, J. T., Hirshfield, M. F., and Hall, E. M. (1971). *Exp. Cell Res.* **68,** 61.

Boyden, S. (1962). *J. Exp. Med.* **115,** 454.

Burke, J. F., and Miles, A. A. (1958). *J. Pathol. Bacteriol.* **76,** 1.

Carr, J., and Wilhelm, D. L. (1964). *Aust. J. Exp. Biol. Med. Sci.* **42,** 511.

Clark, E. R., and Clark, E. L. (1935). *Amer. J. Anat.* **57,** 385.

Clark, E. R., Clark, E. L., and Rex, R. O. (1936). *Amer. J. Anat.* **59,** 123.

Cochrane, C. G. (1963). *J. Exp. Med.* **118,** 489.

Cochrane, C. G. (1967). *Progr. Allergy* **11,** 1.

Cochrane, C. G., Weigle, W. O., and Dixon, F. J. (1959). *Proc. Soc. Exp. Biol. Med.* **101,** 695.

Cochrane, C. G., Unaue, E. R., and Dixon, F. J. (1965). *J. Exp. Med.* **122,** 99.

Cochrane, C. G., Müller-Eberhard, H. J., and Aikin, B. S. (1970). *J. Immunol.* **105,** 55.

Cohn, Z. A., and Morse, S. I. (1960). *J. Exp. Med.* **111,** 667.

Cohn, Z. A., and Wiener, E. (1963a). *J. Exp. Med.* **118,** 991.

Cohn, Z. A., and Wiener, E. (1963b). *J. Exp. Med.* **118,** 1009.

Cohnheim, J. (1873). "Neuere Untersuchungen über die Entzündung." Hirschwald, Berlin.

Cohnheim, J. (1889). "Lectures on General Pathology." New Sydenham Soc., London.

Cornely, H. P. (1966). *Proc. Soc. Exp. Biol. Med.* **122,** 831.

Cotran, R. S., and Majno, G. (1964a). *Ann. N. Y. Acad. Sci.* **116,** 750.

Cotran, R. S., and Majno, G. (1964b). *Amer. J. Pathol.* **45,** 261.

Davies, P., Krakauer, K., Rita, R. A., and Weissman, G. (1971). *Biochem. J.* **123,** 559.

Donaldson, V. H. (1970). *Ser. Haematol.* 3, 39.

Edlow, D. W., and Sheldon, W. H. (1971). *Proc. Soc. Exp. Biol. Med.* **137,** 1328.

Elder, J. M., and Miles, A. A. (1957). *J. Pathol. Bacteriol.* **76,** 1.

Evanson, J. M., Jeffrey, J. J., and Krane, S. M. (1967). *Science* **158,** 499.

Fell, H. B., and Dingle, J. T. (1962). *Biochem. J.* **87,** 403.

Florey, H. W., ed. (1970). "General Pathology," 4th Ed., p. 22. Lloyd-Luke, London.

Freedman, H. L., Taichman, N. S., and Keystone, J. (1967). *Proc. Soc. Exp. Biol. Med.* **125,** 1209.

Gallin, J. I., Clark, R. A., and Kimball, H. R. (1973). *J. Immunol.* **110,** 233.

Goldberg, C. B., and Whitehouse, E., Jr. (1970). *Nature (London)* **228,** 160.

Greenbaum, L. M., and Fruton, J. S. (1957). *J. Biol. Chem.* **266,** 713.

Hanna, R. E., and Watson, D. W. (1965). *Proc. Soc. Exp. Biol. Med.* **118,** 865.

Harada, Y. (1960). *Mie Med. J.* **10,** 47.

Harris, H. (1960). *Bacteriol. Rev.* **24,** 3.

Hayashi, H. (1955). *Seitai No Kagaku* **7,** 124.

Hayashi, H. (1966). *Tohoku J. Exp. Med.* **89,** 341.

Hayashi, H. (1967). *Trans. Soc. Pathol. Jap.* **56,** 37. (In Jap.)

Hayashi, H. (1968a). *In* "Immunopharmacology" (H. O. Schild, ed.), p. 91. Pergamon, Oxford.

Hayashi, H. (1968b). *In* "Biochemistry of the Acute Allergic Reactions" (K. F. Austen and E. L. Becker, eds.), p. 141. Blackwell, Oxford.

Hayashi, H. (1974). *Proc. Int. Congr. Allergol., 8th, Tokyo.* p. 252.

Hayashi, H., and Nitta, R. (1961). *Proc. Soc. Exp. Biol. Med.* **107,** 1002.

Hayashi, H., Ono, T., and Funaki, T. (1955a). *Mie Med. J., Suppl.* **2,** 85.

Hayashi, H., Ono, T., and Matsumoto, T. (1955b). *Mie Med. J., Suppl.* **2,** 119.

Hayashi, H., Funaki, T., Udaka, K., and Kato, Y. (1955c). *Mie Med. J., Suppl.* **2,** 143.

Hayashi, H., Udaka, K., and Funaki, T. (1955d). *Mie Med. J., Suppl.* **2,** 159.

Hayashi, H., Tokuda, A., Matsuba, K., Inoue, T., Udaka, K., Kuze, T., and Ito, K. (1958). *Mie Med. J.* **8,** 329.

Hayashi, H., Tokuda, A., and Udaka, K. (1960). *J. Exp. Med.* **112,** 237.

Hayashi, H., Miyoshi, H., Nitta, R., and Udaka, K. (1962a). *Brit. J. Exp. Pathol.* **43,** 564.
Hayashi, H., Udaka, K., Koono, M., and Yoshimura, M. (1962b). *Brit. J. Exp. Pathol.* **43,** 575.
Hayashi, H., Tokuda, A., Ono, T., and Takaba, Y. (1963). *Brit. J. Exp. Pathol.* **44,** 1.
Hayashi, H., Kinuwaki, Y., Koono, M., and Takaba, Y. (1964a). *Lab. Invest.* **13,** 1124.
Hayashi, H., Yoshinaga, M., Koono, M., Miyoshi, H., and Matsumura, M. (1964b). *Brit. J. Exp. Pathol.* **45,** 419.
Hayashi, H., Nitta, R., and Miyoshi, H. (1964c). *Proc. Soc. Exp. Biol. Med.* **116,** 417.
Hayashi, H., Udaka, K., Miyoshi, H., and Kudo, S. (1965a). *Lab. Invest.* **14,** 665.
Hayashi, H., Yoshinaga, M., Nitta, R., Yamamoto, S., and Tasaki, I. (1965b). *J. Jap. Coll. Angiol.* **5,** 81. (In Jap.)
Hayashi, H., Kinuwaki, Y., and Yoshinaga, M. (1965c). *Nature (London)* **208,** 1007.
Hayashi, H., Tasaki, I., and Yoshinaga, M. (1967). *Nature (London)* **215,** 759.
Hayashi, H., Koono, M., Yoshinaga, M., and Muto, M. (1969). *In* "Inflammation Biochemistry and Drug Interaction" (A. Bertelli and J. C. Houck, eds.), p. 34. Excerpta Med. Found., Amsterdam.
Hayashi, H., Yoshida, K., Ozaki, T., and Ushijima, K. (1970). *Nature (London)* **226,** 174.
Hayashi, H., Koono, M., Yamamoto, S., and Yoshinaga, M. (1973). *In* "Proteolytic Enzymes in the Regulation of Biological Defense" (T. Murachi, T. Asada, and S. Fujii, eds.), p. 99. Univ. of Tokyo Press, Tokyo. (In Jap.)
Hayashi, H., Yoshinaga, M., and Yamamoto, S. (1974). *In* "Chemotaxis: Its Biology and Biochemistry" (E. Sorkin, ed.), p. 296. Karger, Basel.
Higuchi, Y., and Hayashi, H. (1974). *Cell. Immunol.* In press.
Hisaka, M. (1966). *Kumamoto Med. J.* **19,** 160.
Hollander, J. L., McCarthy, D. J., Astorga, G., and Castro-Murillo, E. (1965). *Ann. Intern. Med.* **62,** 271.
Horwitz, D. A., and Garret, M. A. (1971). *J. Immunol.* **106,** 649.
Hurley, J. V. (1963a). *Nature (London)* **198,** 1212.
Hurley, J. V. (1963b). *Aust. J. Exp. Biol. Med.* **41,** 171.
Hurley, J. V. (1964). *Ann. N. Y. Acad. Sci.* **116** 918.
Hurley, J. V., and Spector, W. G. (1961). *J. Pathol. Bacteriol.* **82,** 403.
Hutchins, G. M., and Sheldon, W. H. (1972). *Proc. Soc. Exp. Biol. Med.* **140,** 623.
Janoff, A. (1972). *Amer. J. Pathol.* **68,** 579.
Janoff, A., and Scherer, J. (1968). *J. Exp. Med.* **128,** 1137.
Janoff, A., and Zeligs, J. G. (1968). *Science* **161,** 702.
Janoff, A., Schaefer, S., Shcerer, J., and Bean, M. A. (1965). *J. Exp. Med.* **122,** 841.
Jocelyn, P. C. (1959). *In* "Glutathione" (E. M. Crook, ed.), p. 43. Cambridge Univ. Press, London and New York.
Judah, J. D., and Willoughby, D. A. (1962). *J. Pathol. Bacteriol.* **83,** 567.
Jullien-Vitoux, D., Voisin, G. A., and Nemirovky, M. (1973). *Brit. J. Exp. Pathol.* **54,** 20.
Kageyama, K. (1963). *Acta Pathol. Jap.* **13,** 209.
Kageyama, K., and Fukuda, J. (1967). *Arerugi* **16,** 144.
Kambara, T. (1973). *Proc. Int. Congr. Allergol., 8th, Tokyo* p. 24. (Abstr.)
Kambara, T., Aimoto, T., and Hayashi, H. (1968). *Tohoku J. Exp. Med.* **94,** 237.
Kambara, T., Katsuya, H., and Maeda, S. (1972). *Acta Pathol. Jap.* **22,** 465.
Kasai, H. (1959). *Mie Med. J.* **9,** 187.
Kato, T., Kojima, K., and Murachi, T. (1972). *Biochim. Biophys. Acta* **289,** 187.

Keilova, H. (1971). *In* "Tissue Proteinases" (A. J. Barrett and J. T. Dingle, eds.), p. 45. North-Holland Publ., Amsterdam.
Keller, H. U., and Sorkin, E. (1967). *Proc. Soc. Exp. Biol. Med.* **126**, 677.
Kelley, V. C., and Good, R. A. (1949). *Fed. Proc., Fed. Amer. Soc. Exp. Biol.* **8**, 359.
Kinuwaki, Y. (1966). *Kumamoto Med. J.* **19**, 90.
Kinuwaki, Y., Sonoda, T., Yamamoto, S., and Nishimura, K. (1967). *Saibo Kagaku Shimpojumu* **18**, 93. (In Jap.)
Klemperer, F. P. (1948). *Ann. Intern. Med.* **28**, 1.
Klinge, F. (1931). *Arch. Pathol. Anat. Physiol.* **279**, 16.
Koono, M. (1966). *Kumamoto Med. J.* **19**, 79.
Koono, M., and Hayashi, H. (1969). *Jap. J. Biochem. Soc.* **41**, 436. (In Jap.)
Koono, M., and Hayashi, H. (1974). In preparation.
Koono, M., Muto, M., and Hayashi, H. (1968). *Tohoku J. Exp. Med.* **94**, 231.
Koono, M., Katsuya, H., and Hayashi, H. (1974a). *Int. J. Cancer* **13**, 334.
Koono, M., Ushijima, K., and Hayashi, H. (1974b). *Int. J. Cancer* **13**, 105.
Kouno, T. (1971). *Kumamoto Med. J.* **24**, 135.
Kouno, T., Katsuya, H., Maeda, S., and Hayashi, H. (1972). *Arerugi* **21**, 156.
Kouno, T., Higuchi, Y., and Hayashi, H. (1974). In preparation.
Krogh, A. (1922). "Anatomy and Physiology of Capillaries." Yale Univ. Press, New Haven, Connecticut.
Kudo, S. (1967). *J. Kumamoto Med. Soc.* **41**, 115. (In Jap.)
Kunitz, M. (1947). *J. Gen. Physiol.* **30**, 291.
Kusunoki, N. (1963). *Soogo-Rynshoo* **12**, 2116.
Kuze, T. (1960). *Mie-Igaku* **4**, 1613.
Lapresle, C. (1971). *In* "Tissue Proteinases" (A. J. Barrett and J. T. Dingle, eds.), p. 135. North-Holland Publ., Amsterdam.
Lapresle, C., and Webb, T. (1962). *Biochem. J.* **84**, 455.
Lazarow, A. (1954). *In* "Glutathione" (S. Colowick, A. Lazarow, E. Racker, D. R. Schwarz, E. Stadtman, and H. Waelsch, eds.), p. 231. Academic Press, New York.
Lazarus, G. S., Brown, R. S., Daniels, J. R., and Fullmer, H. M. (1967). *Science* **159**, 1483.
Lebez, D., Kopitar, M., Turk, V., and Kregar, I. (1971). *In* "Tissue Proteinases" (A. J. Barrett and J. T. Dingle, eds.), p. 167. North-Holland Publ., Amsterdam.
Lewis, T. (1927). "The Blood Vessels of the Human Skin and Their Responses." Shaw, London.
Lichtenstein, N., and Fruton, J. S. (1960). *Proc Nat. Acad. Sci. U. S.* **46**, 787.
Lockett, M. F., and Jarman, D. A. (1958). *Brit. J. Pharmacol.* **13**, 11.
Logan, G., and Wilhelm, D. L. (1963). *Nature (London)* **198**, 968.
LoSpalluto, J., Chegoriansky, J., Lewis, A., and Ziff, M. (1960). *J. Clin. Invest.* **39**, 473.
LoSpalluto, J., Fehr, K., and Ziff, M. (1971). *In* "Tissue Proteinases" (A. J. Barrett and J. T. Dingle, eds.), p. 263. North-Holland Publ., Amsterdam.
Lucy, J. A., Dingle, J. T., and Fell, H. B. (1961). *Biochem. J.* **79**, 500.
Luscombe, M. (1963). *Nature (London)* **197**, 1010.
McDonald, J. K., Callahan, P. X., Ellis, S., and Smith, R. E. (1971). *In* "Tissue Proteinases" (A. F. Barrett and J. T. Dingle, eds.), p. 69. North-Holland Publ., Amsterdam.
MacFarlane, M. G., and Knight, B. C. J. G. (1941). *Biochem. J.* **35**, 884.
Maeda, S., Katsuya, H., and Hayashi, H. (1970). *Arerugi* **19**, 110.

Majno, G. (1964). *In* "Injury, Inflammation, and Immunity" (L. Thomas, J. W. Uhr, and L. Grant, eds.). Williams & Wilkins, Baltimore, Maryland.

Majno, G., Palade, G. E., and Schoefel, G. I. (1961). *J. Biophys. Biochem. Cytol.* **11,** 607.

Matsuba, K. (1960a). *Mie-Igaku* **4,** 1720.

Matsuba, K. (1960b). *Mie Med. J.* **10,** 21.

Menkin, V. (1938). *J. Exp. Med.* **67,** 129.

Menkin, V. (1940). "Dynamics of Inflammation." Macmillan, London.

Menkin, V. (1950). "Newer Concept of Inflammation." Thomas, Springfield, Illinois.

Menkin, V. (1956). "Biochemical Mechanisms in Inflammation." Thomas, Springfield, Illinois.

Metchnikoff, E. (1893). "Lectures on the Comparative Pathology of Inflammation." Kegan Paul, London.

Meyer, K., and Fellig, J. (1950). *Experientia* **6,** 186.

Miles, A. A. (1956). *Ann. N. Y. Acad. Sci.* **66,** 356.

Miles, A. A., and Miles, E. M. (1952). *J. Physiol. (London)* **118,** 228.

Miles, A. A., and Wilhelm, D. L. (1955). *Brit. J. Exp. Pathol.* **36,** 71.

Miller, R. L., and Melmon, K. L. (1970). *Ser. Haematol.* 3, 5.

Mise, J. (1963). *Soogo-Rynshoo* **12,** 2101.

Miyoshi, H. (1965). *Kumamoto Med. J.* **18,** 80.

Movat, H. Z. (1956). *Beitr. Pathol. Anat. Allg. Pathol.* **116,** 238.

Movat, H. Z. (1971). *In* "Inflammation, Immunity and Hypersensitivity" (H. Z. Movat, ed.), p. 2. Harper, New York.

Movat, H. Z., Fernando, N. V. P., Uriuhara, T., and Weiser, W. T. (1963). *J. Exp. Med.* **118,** 557.

Movat, H. Z., Uriuhara, T., MacMorine, D. R. L., and Burke, J. S. (1964). *Life Sci.* **3,** 1025.

Movat, H. Z., Uriuhara, T., Murray, R. K., MacMarine, D. R. L., Wasei, S., Tachiman, N. S., and Franklin, A. E. (1965). "Studies of Rheumatic Disease," Proc., 3rd Can. Conf. Rheumatic Disease. Univ. of Toronto Press, Toronto.

Muto, M. (1969). *J. Kumamoto Med. Soc.* **43,** 690. (In Jap.)

Nishimura, K. (1970). *J. Kumamoto Med. Soc.* **44,** 298. (In Jap.)

Nishioka, K., Constantopoulos, A., Satoh, P. S., and Najjar, V. A. (1972). *Biochem. Biophys. Res. Commun.* **47,** 172.

Nishioka, K., Satoh, P. S., Constantopoulos, A., and Najjar, V. A. (1973). *Biochim. Biophys. Acta* **310,** 230.

Nishiura, M., Yamamoto, S., and Hayashi, H. (1974a). *Immunology.* Submitted.

Nishiura, M., Yamamoto, S., and Hayashi, H. (1974b). *Immunology.* In press.

Nitta, R. (1965). *Kumamoto Med. J.* **18,** 72.

Nitta, R., Hayashi, H., and Norimatsu, K. (1963). *Proc. Soc. Exp. Biol. Med.* **113,** 185.

Ogata, T. (1971). *Kumamoto Med. J.* **24,** 103.

Ono, T. (1958). *Mie Med. J.* **8,** 211.

Opie, E. L. (1906). *J. Exp. Med.* **8,** 410.

Ostwald, N. C. (1957). *Amer. Rev. Tuberc. Pulm. Dis.* **75,** 341.

Ozaki, T., Yoshida, K., Ushijima, K., and Hayashi, H. (1971). *Int. J. Cancer* **7,** 93.

Pastan, I., and Almquist, S. (1966a). *Endocrinology* **78,** 350.

Pastan, I., and Almquist, S. (1966b). *Endocrinology* **78,** 361.

Paxton, J. R., and Payne, J. H. (1948). *Surg. Gynecol. Obstet.* **86,** 69.

Petri, G., Csipak, J., Kovacs, A., and Bentzik, M. (1952). *Acta Med. Acad. Sci. Hung.* **3,** 347.

Planta, R. J., and Gruber, M. (1964). *Biochim. Biophys. Acta* **89**, 403.
Press, E. M., Porter, R. R., and Cebra, G. (1960). *Biochem. J.* **74**, 501.
Ramsey, W. S. (1972). *Exp. Cell Res.* **70**, 129.
Rawson, R. A. (1943). *Amer. J. Physiol.* **138**, 708.
Rovery, M., Poilroux, M., Curnier, A., and Desnuelle, P. (1955). *Biochim. Biophys. Acta* **17**, 565.
Sakabe, S. (1962). *Mie-Igaku* **6**, 138.
Schwabe, C., and Kalnistsky, G. (1966). *Biochemistry* **5**, 158.
Sevitt, S. (1958). *J. Pathol. Bacteriol.* **75**, 27.
Sevitt, S. (1964). *In* "Injury, Inflammation, and Immunity" (L. Thomas, J. W. Uhr, and L. Grant, eds.). Williams & Wilkins, Baltimore, Maryland.
Shetlar, M. R., Bryan, R. S., Foster, J. V., Shetlar, C. L., and Everett, M. R. (1949). *Proc. Soc. Exp. Biol. Med.* **72**, 294.
Shetlar, M. R., Shetlar, C. L., Richmond, V., and Everett, M. R. (1950). *Cancer Res.* **10**, 681.
Smith, J. G., Church, C. F., and Burk, P. G. (1965). *J. Clin. Invest.* **44**, 1098.
Snellman, O. (1971). *In* "Tissue Proteinases" (A. J. Barrett and J. T. Dingle, eds.), p. 29. North-Holland Publ., Amsterdam.
Snyderman, R., Gewurz, H., and Mergenhagen, S. E. (1968). *J. Exp. Med.* **128**, 259.
Sonoda, T. (1967). *Kumamoto Med. J.* **20**, 185.
Spector, W. G. (1956). *J. Pathol. Bacteriol.* **72**, 367.
Spector, W. G., and Willoughby, D. A. (1959). *J. Pathol. Bacteriol.* **78**, 121.
Spector, W. G., and Willoughby, D. A. (1963). *Bacteriol. Rev.* **27**, 117.
Spector, W. G., and Willoughby, D. A. (1968). "The Pharmacology of Inflammation." English Univ. Press, London.
Sylven, B. (1941). *Acta Chir. Scand., Suppl.* **86**, 66.
Takaba, Y. (1965). *Kumamoto Med. J.* **18**, 31.
Tasaki, I. (1968a). *Kumamoto Med. J.* **21**, 13.
Tasaki, I. (1968b). *Kumamoto Med. J.* **21**, 1.
Tasaki, I., Yoshinaga, M., and Hayashi, H. (1966). *Biochim. Biophys. Acta* **130**, 260.
Tasaki, I., Yoshinaga, M., and Hayashi, H. (1969). *Tampakushitsu, Kakusan, Koso* **14**, 1336.
Thomas, L., McClusky, R. F., Potter, J. L., and Weissman, G. (1960). *J. Exp. Med.* **111**, 705.
Tokaji, G. (1971). *Kumamoto Med. J.* **24**, 68.
Tokuda, A., Hayashi, H., and Matsuba, K. (1960). *J. Exp. Med.* **112**, 249.
Udaka, K. (1963a). *Kumamoto Med. J.* **16**, 55.
Udaka, K. (1963b). *Kumamoto Med. J.* **16**, 70.
Udaka, K., and Hayashi, H. (1965a). *Biochim. Biophys. Acta* **97**, 251.
Udaka, K., and Hayashi, H. (1965b). *Biochim. Biophys. Acta* **104**, 600.
Udaka, K., and Udaka, K. (1969). *Fed. Proc., Fed. Amer. Soc. Exp. Biol.* **28**, 1990.
Ungar, G., and Hayashi, H. (1958). *Ann. Allergy* **16**, 542.
Ungar, G., and Ungar, A. L. (1966). *Proc. Soc. Exp. Biol. Med.* **121**, 1098.
Ungar, G., Yamura, T., Isola, J. B., and Kobrins, S. (1961). *J. Exp. Med.* **113**, 359.
Virchow, R. (1858). Die Cellularpathologie in ihre Begründung auf physiologische und pathologische Gewebelehre," Hirschwald, Berlin.
Voisin, G. A., and Toullet, F. (1963). *Ann. Inst. Pasteur, Paris* **104**, 169.
Ward, P. A., and Hill, J. H. (1970). *J. Immunol.* **104**, 535.
Ward, P. A., and Hill, J. H. (1972). *J. Immunol.* **108**, 1137.

Ward, P. A., and Zvaifler, N. J. (1971). *J. Clin. Invest.* **50,** 606.
Ward, P. A., Cochrane, G. G., and Müller-Eberhard, H. J. (1965). *J. Exp. Med.* **122,** 327.
Ward, P. A., Cochrane, C. G., and Müller-Eberhard, H. J. (1966). *Immunology* **11,** 141.
Ward, P. A., Cohen, S., and Flanagan, T. D. (1972). *J. Exp. Med.* **135,** 1095.
Wasi, S., Murray, R. K., Macmorine, D. R. L., and Movat, H. Z. (1966). *Brit. J. Exp. Pathol.* **47,** 411.
Wasserman, K., Loeb, L., and Mayerson, H. S. (1955). *Circ. Res.* **3,** 594.
Weissman, G., and Dingle, J. T. (1961). *Exp. Cell Res.* **25,** 207.
Weissman, G., and Fell, H. B. (1962). *J. Exp. Med.* **116,** 365.
Weissman, G., and Spilberg, I. (1968). *Arthritis Rheum.* **11,** 162.
Weissman, G., Zurier, R. B., and Hoffstein, S. (1972). *Amer. J. Pathol.* **68,** 539.
Wilhelm, D. L. (1962). *Pharmacol. Rev.* **14,** 251.
Wilhelm, D. L. (1971). *Annu. Rev. Med.* **22,** 63.
Wilhelm, D. L., and Mason, B. (1960). *Brit. J. Exp. Pathol.* **41,** 487.
Williamson, J. R., and Grisham, J. W. (1961). *Amer. J. Pathol.* **39,** 239.
Woessner, J. F., Jr., and Boucek, R. J. (1961). *Arch. Biochem. Biophys.* **93,** 95.
Yamamoto, S., Yoshinaga, M., and Hayashi, H. (1971). *Immunology* **20,** 803.
Yamamoto, S., Nishiura, M., and Hayashi, H. (1973). *Immunology* **24,** 791.
Yasuhira, K. (1955). *Nippon Ketsueki Gakkai Zasshi* **18,** 215. (In Jap.)
Yoshida, K., Yoshinaga, M., and Hayashi, H. (1968). *Nature (London)* **218,** 977.
Yoshida, K., Ozaki, T., Ushijima, K., and Hayashi, H. (1970). *Int. J. Cancer.* **6,** 123.
Yoshinaga, M. (1964). *Kumamoto Med. J.* **17,** 31.
Yoshinaga, M., and Hayashi, H. (1969). *Rinsho-Ketsueki* **10,** 294.
Yoshinaga, M., Tasaki, I., and Hayashi, H. (1966a). *Biochim. Biophys. Acta* **127,** 172.
Yoshinaga, M., Tasaki, I., Hisaka, M., and Sonoda, T. (1966b). *Kumamoto Med. J.* **19,** 39.
Yoshinaga, M., Matsumoto, K., and Hayashi, H. (1968). *J. Jap. Coll. Angiol.* **8,** 103. (In Jap.)
Yoshinaga, M., Yamamoto, S., Maeda, S., and Hayashi, H. (1971a). *Immunology* **20,** 809.
Yoshinaga, M., Yoshida, K., Tashiro, A., and Hayashi, H. (1971b). *Immunology* **21,** 281.
Yoshinaga, M., Yamamoto, S., and Hayashi, H. (1972a). *In* "Immunology" (T. Kuroyanagi, Y. Ohtaka, and C. Matsuhashi, eds.), Vol. 7, p. 31. Igaku-Shoin, Tokyo. (In Jap.)
Yoshinaga, M., Yamamoto, S., Kiyota, S., and Hayashi, H. (1972b). *Immunology* **22,** 393.
Ziff, M., Gribetz, H. J., and LoSpalluto, J. (1960). *J. Clin. Invest.* **39,** 405.
Zigmond, S. H., and Hirsch, J. G. (1973). *J. Exp. Med.* **137,** 387.
Zweifach, B. W. (1953). *In* "The Mechanism of Inflammation" (G. Jasmin and A. Robert, eds.). Acta Inc., Montreal.

The Specificity of Pituitary Cells and Regulation of Their Activities[1]

VLADIMIR R. PANTIĆ

Serbian Academy of Sciences and Arts, Belgrade, Yugoslavia

I. Introduction

As a result of recent advances in hormone research, 10 pituitary and most of the hypothalamic hormones have been isolated and their

[1] The following abbreviations are used in this article: ACTH, adrenocorticotropin; AER, agranular (smooth) endoplasmic reticulum; cAMP, cyclic adenosine monophosphate; CRF, ACTH-releasing hormone, factor; ER, endoplasmic reticulum; FRF, FSH-releasing hormone, factor; FSH, follicle-stimulating hormone; GER, granular (rough) endoplasmic reticulum; ICSH, interstitial cell-stimulating hormone (LH); IF, inhibiting factor; LH, luteinizing hormone; LPH, lipotropin hormones; LRF, LH-releasing hormone; ME, eminentia mediana, median eminence; MIF, MSH release inhibiting hormone, factor; MRF, MSH-releasing hormone, factor; MSH, melanocyte-stimulating hormones, melanotropins (MSH-α and MSH-β); NLT, nucleus lateralis tuberi; NPV, nucleus paraventricularis; NSO, nucleus supraopticus; NVM, nucleus ventromedialis; PAS, periodic acid–Schiff; PIF, prolactin, PRL release inhibiting hormone, factor; PRF, prolactin-releasing hormone; PRL, prolactin (luteotropic, lactogenic) hormone (LTH); RF, releasing hormone, factor; SRF, somatotropin-releasing hormone; SRIF, somatotropin release inhibiting hormone, somatostatin; STH, somatotropin (growth hormone); T_3, triiodothyronine; T_4, thyroxine; TRF, thyrotropic releasing hormone, factor; TRIF, TSH release inhibiting factor; TSH, thyroid-stimulating hormone, thyrotropin.

chemical structures determined (for reviews see Li, 1972; Blackwell and Guillemin, 1973).

Based on the fact that the amino acid sequences of pituitary hormones have certain common features and overlapping biological activities, Li (1972) proposed three groups of pituitary hormones: (1) MSH, ACTH, and LPH; (2) LTH and STH; and (3) the glycoprotein hormones (gonadotropins and TSH). Each of these groups of hormones has one common ancestral molecule (Li, 1972).

Pituitary hormones are synthesized and secreted by well-defined specific cells in the pituitary. These cells have been identified by the following criteria: size, shape, location, affinity of specific granules for staining, immunohistochemical techniques, and granule size, shape, and location (Farquhar, 1961; Van Oordt, 1965, 1968; Ganong *et al.*, 1966; Fawcett *et al.*, 1969; Ball and Baker, 1969; Nakane, 1970; Farquhar, 1971; Sage and Bern, 1971; Sage and Bromage, 1971; Mattheij, 1970; Mattheij *et al.*, 1971; Mirecka and Pearse, 1971; Tixier-Vidal *et al.*, 1971, 1972; Herlant *et al.*, 1972).

On the basis of these criteria, the chromophobes and six specific pituitary cells can be clearly distinguished. However, if we assume that FSH and LH cells are two different types of gonadotropic cells, then pituitary hormones are produced by seven specific cell types.

Many of the papers cited in this article deal with studies of the pituitary cells of various species of vertebrates. In summarizing some of the general properties such as common cellular features and the ability of certain cells to synthesize hormones from the same ancestor molecule, the genesis of specific pituitary cells from three main groups of cells can be considered: (1) ACTH and MSH cells, (2) LTH and STH cells, and (3) glycoprotein hormone-producing cells. This classification of cell types can perhaps best be used to understand the mechanisms involved in the following processes: regulation of specific cells during differentiation, specificity during ontogenesis, cell interrelationships and properties during various stages of activity and in the case of tumorigenesis.

In order to better understand the role of ancestral cell types in the genesis of specific pituitary cells, only the following data are mentioned.

Two cell types of the adenohypophysis of the hagfish (*Myxine glutinosa*) producing protein hormones have been identified by Fernholm (1972). Cell type 1 produces a hormone similar to ACTH and MSH, which later gives rise to ACTH and MSH cells which produce the corresponding polypeptides; cell type 2 produces hormones similar to LTH and STH. These cells contain granules with a mean diameter of 176 nm, and they "might have a descendent form of cellular line giving rise to the cell types producing STH and LTH." In the pituitary of this fish, gonadotropic and thyrotropic activity was not observed (Fernholm, 1972).

STH could not be detected in the salamander, and changes in the LTH/STH concentration ratio in reptilian, avian, and mammalian pituitaries have been investigated by Nicoll and Licht (1971).

Ovine prolactin stimulates somatic growth in various species of mammals, birds, and amphibians (Bern and Nicoll, 1968; Nicoll and Bern, 1971).

Only one type of gonadotropic cell in teleost fish pituitary has been distinguished (Sage and Bromage, 1970a,b); FSH in the lizard is at least 1000-fold more active than LH (Licht and Papkoff, 1972).

For pituitary hormone species-specific effects see Fontaine and Fontaine (1962) and Fontaine (1969).

The specificity of pituitary cells is mainly characterized by a close relationship between differentiation of the specific cells and the degree of nervous system development, and the ability to respond to the following hormones: factors produced by the cells of hypothalamic nuclei, hormones of target organs, and the corresponding pituitary hormones. However, during tumorigenesis the pituitary cells lose these properties, and they are not able to store hormones.

It is now accepted that the cells of the hypothalamic nuclei produce releasing factors and inhibitors, which are peptides of low molecular weight. These hormones flow via circulation, mostly by the portal vessels, to the hypophysis and to other organs. Most of them have been isolated in pure form, and their amino acid sequences established and confirmed by synthesis as a result of intensive research by Guillemin and his co-workers (for reviews see Guillemin, 1971; Blackwell and Guillemin, 1973) and by Schally and his collaborators (1968, 1969, 1973).

As an organ of the central nervous system, the hypothalamus provides the basis for integration of information pathways from the extrahypothalamic nervous centers and from target and other organs. This information stimulates or inhibits the activity of the pituitary cells. Hypothalamic control of the anterior pituitary has been considered by Szentagothai *et al.* (1968) and Alešin (1971). The properties of neurosecretory cells were the subject of investigations by Scharrer (1967), Polenov and Garlov (1973), and others.

In order to study the pathways between the extrahypothalamic centers and hypothalamic nuclei, various methods were used. Only some of the data are mentioned. A direct and always bilateral projection of the retinal fibers to the suprachiasmatic nucleus exists (Hendrickson *et al.*, 1972; Villings, 1973). The anterior hypothalamic-preoptic area has been suggested as a positive feedback site for estrogen in the immature rat (Davidson, 1969). The MBH appears to be involved in the negative feedback action of estrogen (Ramirez and McCann, 1964).

The characteristic changes in hypothalamic nuclei and in the pituitary cells, and their interrelationship under various experimental conditions, such as adrenalectomy, thyroidectomy, gonadectomy, and subsequent administration of hormones, were the subject of our investigations (Pantić *et al.*, 1967; Žgurić *et al.*, 1968; Stošić *et al.*, 1969; Žgurić and Pantić, 1969, 1970a,b; Žgurić-Hristić and Pantić, 1972, 1974; Martinović *et al.*, 1969).

There is evidence that ACTH release is regulated by the anterior and medial hypothalamic region (Žgurić and Pantić, 1969, 1970a,b,c; Žgurić-Hristić and Pantić, 1972).

Pituitary gonadotropic activity is regulated in teleost fish by the lateral group of neurons, probably by way of type-B fibers making synaptoid contact with gonadotropic and other pituitary cells (Zambrano, 1971; Zambrano and Hurrixa, 1973). In mammals the regulation of gonadal activity might be indirectly concerned with the arcuate ME system (Zambrano and De Robertis, 1968; Fuxe and Hokfelt, 1969; Dukić, 1972). NPV and arcuate intrahypothalamic loci are responsible for the induction of androgen sterilization in neonatal female rats (Nadler, 1972a,b).

The release of TSH from pituitary TSH cells in chicken is regulated by a region of the anterior hypothalamus extending from the ventrocaudal area to the optic chiasma (Kanematsu and Mikami, 1970). In the rat the area extending from the NPV to the infundibulum is concerned with TSH release (Green and Erwin, 1954).

Nervous system involvement in the "Everett effect" has been suggested (Everett and Sawyer, 1949; Sawyer *et al.*, 1949). Sawyer (1972), in studying the influence of pituitary target organs on brain pituitary function to control pituitary secretion, pointed out that the action of estrogen may involve the amygdala, whereas adrenal steroids seem to exert an important influence on the hippocampus. He noted that the corticomedial amygdala of the female rat appears to contain two functional groups of neurons: (1) cells inhibitory to gonadotropic function in general, and (2) cells that facilitate the ovulatory surge of pituitary LH release. Both appear to project to the hypothalamus via the stria terminalis.

The role of cathecholamine in the regulation of gonadotropin secretion was first observed by Sawyer *et al.* (1947). The catecholaminergic, indolaminergic, and cholinergic pathways and the hypothalamic neurohormones involved in the control of gonadotropin secretion were investigated by Kamberi *et al.* (1970, 1971; Kamberi, 1973).

Feedback control of gonadotropin secretion has also been studied (Hirono *et al.*, 1970; Stetson, 1972; Karsch *et al.*, 1973).

It was established that the cells of the NPV, NVM, and ME are the cells most sensitive to thyroidectomy and to the action of thyroid hormones (Stošić *et al.*, 1965; Stošić and Pantić, 1973).

The focus of present research is on the mechanism of action of pituitary hormones on target cells and hormone feedback action. It is pertinent to mention that a homeostatic interrelationship exists which depends mainly on the concentration of target organ hormones and related hypothalamic factors.

During the last few years the number of articles discussing the mechanism of hormone action rapidly increased. For example, the mechanism of action of releasing factors on pituitary cells has been reviewed by Geschwind (1969, 1972), Blackwell and Guillemin (1973), and Schally *et al.* (1973). Also, the mechanism of steroid hormone action at the molecular level, the initial event being the combination of hormone with receptor protein, was reviewed by Villee (1964, 1968).

Now most of the hormones have been isolated and their structures elucidated. However, the sites of synthesis of hypothalamic hormones, the mechanism of the regulation of hormone synthesis and accumulation, the mode of hormone action at the molecular level, and the interrelationship between brain centers, the hypothalamic nuclei and the endocrine cells are of prime interest to investigators involved in further studies related to neuroendocrinology. This is of interest in biology as well as in medicine.

This article does not pretend to review all articles and monographs dealing with the morphology, biochemistry, physiology, and so on, of pituitary cells and hormones. The main purpose is to select some of the data in the literature contributing to the progress in neuroendocrinology at the present time to facilitate understanding of the main concepts in this field.

II. Chromophobic Cells

Chromophobic cells are known as a cell type usually devoid of specific granules (Fig. 1). (Pituitary cells from *Torpedo ocellata*, sacrificed during a period of high gonadotropic activity, were used for the electron micrographs reproduced in this article.) However, they may contain some granules, so that it is rather difficult to distinguish them as a specific type of pituitary cell.

The smallest of the chromophobic cells, which have mostly polysomes in the cytoplasm, may be the undifferentiated cells. In the larger cells the cytoplasm contains poorly developed ER and some dilated cavities of

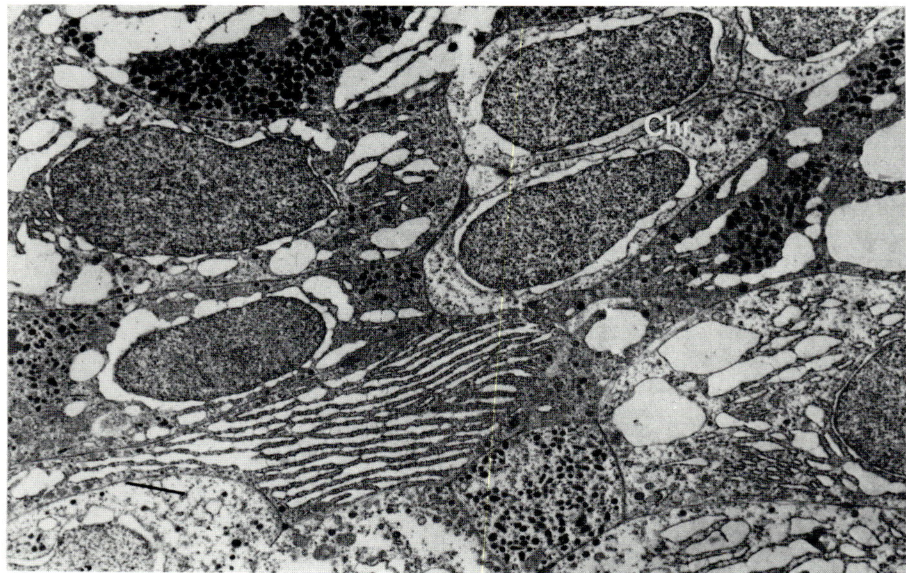

FIG. 1. Chromophobic cells can be clearly distinguished from the differentiated pituitary cells.

GER or vesicles. The Golgi zone usually occupies a small cytoplasmic area. The number of these cells increases under various experimental conditions, for example, after treatment with a single or repeated doses of estrogen (Pantić and Genbačev, 1969, 1970, 1972; Pantić and Škaro, 1974). These cells were considered the cells most sensitive to estrogen (Pantić and Genbačev, 1969).

Chromophobic cells have been investigated in the pituitaries of all vertebrates. They are described as follicular, or folliculostellate, cells. Lagios (1973) investigated the presence of a follicle boundary in the adenohypophysis of some fish, showing many characteristics of organization and ultrastructure similar to follicular cells. These cells elaborate glycocalyx, microvilli, junctional complexes, and colloid droplets. These and other characteristics show that they have a role in active exchange between the cells and the follicular content, as has been demonstrated for their mammalian counterparts (Lagios, 1973).

Folliculostellate cells, which are agranular and represent a large part of the chromophobic cells, are characterized by their multiple cytoplasmic processes and by their organization around cavities (Vila Porcile, 1972). They are involved in the digestion of waste material from other cells (Farquhar, 1971; Dingemans and Feltkamp, 1972).

III. Pituitary ACTH, MSH, and LPH Cells

A. Common General Properties

ACTH and MSH cells may be distinguished in the pituitaries of all vertebrates. It seems that they are a source of LPH-β and LPH-γ as well. Synthesis and biological properties of ACTH peptides have been investigated (Li *et al.*, 1960; Li, 1962).

A primitive molecule in the evolutionary processes for the synthesis of ACTH, MSH, and LPH is a heptapeptide (Li, 1972). Li (1972) suggested that this heptapeptide, Met-Glu-His-Phe-Arg-Trp-Glu, possesses ACTH and MSH activity and has melanotropic and lipolytic activities *in vitro*.

Two different sizes of storage granules with STH or ACTH activity from the pellets of acidophils have been isolated from rat anterior pituitaries (Ishikawa *et al.*, 1972).

B. ACTH Cells

ACTH cells are usually stellate or irregular in shape. In the pituitaries of teleosts, such as *Poecilia reticulata*, ACTH cells lie between the prolactin cells and the neurohypophysis (Fig. 2). They are localized in the cephalic region of the pars distalis of pituitaries of birds. In mammals they are distributed as single cells or are scattered in groups usually at the periphery of the pars anterior (posterior) of the adenohypophysis. They constitute only 0.16% of the pituitary cell population (Siperstein, 1963).

The rostral zone of the pars intermedia of the mouse is almost exclusively composed of typical ACTH cells located around and even within the neural stalk. In the rat they are found among the MSH cells of the rostral zone and in the neurohypophysis as well. These cells are in direct contact with aminergic and neurosecretory nerve fibers (Stoeckel *et al.*, 1973).

The ACTH cells of the pars distalis are smaller than the MSH cells and are characterized by undeveloped GER and Golgi complexes. The diameter of their specific granules is on the average about 200 nm (160–250 nm). These specific granules with ACTH activity have been isolated (Ishikawa *et al.*, 1972). The sequence of changes in the ACTH cells at the ultrastructural level has been described in female rats bearing regenerating adrenals (Nakayama *et al.*, 1969; Siperstein and Miller, 1970; Nakayama and Nickerson, 1973a).

The second cell type of ACTH cells, which are usually scarce, but more numerous in rat than in the mouse, contain smaller granules (100–130 nm) (Stoeckel *et al.*, 1971, 1973). The accumulation of specific

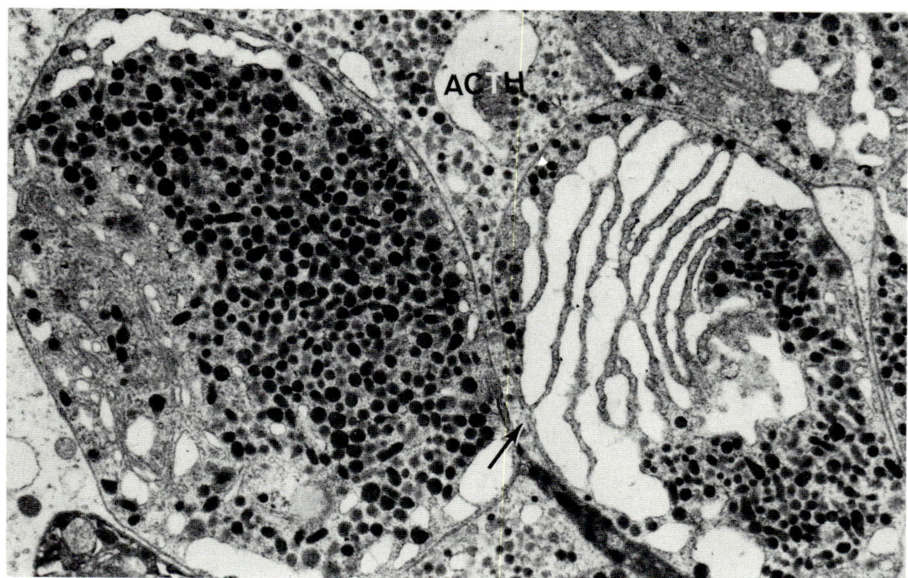

FIG. 2. ACTH cells with the cytoplasmic extension (arrow) located between the LTH cells.

granules in the cytoplasm on the neurohypophysial side (Sage and Bromage, 1970a,b) can be considered a tendency of these cells to be polarized. The properties of immature and mature specific granules have been described (Nakayama *et al.*, 1969; Nakayama and Nickerson, 1973a).

During intensive activity ACTH cells are degranulated. However, specific granules and ER are absent in ACTH-producing MtT tumor cells (Pantić *et al.*, 1971a,b).

The activity of ACTH cells is regulated by hypothalamic hormones, ACTH, and adrenocortical hormones. The last-mentioned involve feedback control of ACTH. Ontogenesis of the pituitary-adrenal axis in the chick embryo has been described (Woods *et al.*, 1971).

1. *ACTH*

ACTH molecules, peptide chains of 39 amino acids, with an N-terminal serine and C-terminal phenylalanine, were first isolated from ovine pituitaries (Li *et al.*, 1954, 1961). Two years later, Li and Dixon (1956) isolated and determined the properties of bovine ACTH. Studies of properties of ACTH showed that the portion of the C-terminal sequence that includes positions 25 to 39 does not lose ACTH potency (Cole *et al.*, 1956). It has been concluded that the entire molecule is not necessary for the stimulation of the adrenal cortex (Li, 1962; Hofmann and Yajima, 1962).

By using Sephadex gel filtration and radioimmunoassay, "big" and "little" forms of ACTH were detected: the minor fraction in the normal pituitary and the "big" fraction in tumor extracts (Gewirtz and Yalow, 1973). However, "big" ACTH has no significant biological activity *in vitro* (Schneider *et al.*, 1973).

2. *CRF*

It has been known for more than 20 years that extracts of hypothalamus or neurohypophysis stimulate ACTH release *in vivo* and *in vitro*. Many attempts have been made to prove that a true CRF of hypothalamic origin controls the release of ACTH. However, vasopressin was proposed as the sole hypothalamic mediator of ACTH release (Guillemin, 1971). Yates *et al.* (1971) suggested that vasopressin may release endogenous CRF from the hypothalamus and act as a CRF or with CRF to release ACTH. The presence of CRF in human hypothalamic extract has been shown (Schilling and Blobel, 1973), and the role of CRF and vasotocin in ACTH release has been examined (Salem *et al.*, 1970a,b).

3. *Feedback Action of Corticosteroids*

Examining the location and specificity of corticosteroid feedback receptors at the hypothalamohypophysial level, Bohus and Strashimirov (1970) suggested that corticosteroid receptors are present in both the hypothalamus and the pituitary.

Studying the changes in the anterior MBH and the ultrastructure of pituitary cells in adrenalectomized rats at various times after the operation, by substitution of mineralocorticoids or/and glycocorticoids, Žgurić and Pantić (1970a,b,c) showed the following: a close relationship between neurosecretory cell structure and plasma concentration of glycocorticoids and mineralocorticoids; a deficiency in corticosteroids after adrenalectomy causes significant hypertrophy of the cells of the NPV and NSO and nuclei of NVM cells. The accumulation of neurosecretory substance in NSO, NPV, ME, and neurohypophysis is clearly expressed in intact or adrenalectomized rats treated with ACTH or Hy.

As a result of the suppressive effect of hydrocortisone on ACTH release, the number and size of specific granules in the ACTH cells of adrenalectomized rats increased.

The role of extrahypothalamic centers, such as the amygdala and the hippocampus, in ACTH control via the basal hypothalamus was suggested

(Weitzman *et al.*, 1971; Allen, 1973). Inhibitory action on basal and/or diurnal aspects of hypothalamohypophysial adrenal activity has been reported (Newman-Taylor *et al.*, 1973). The inhibition of ACTH secretion by norepinephrine-secreting neurons in the hypothalamus has also been mentioned (Ganong *et al.*, 1973).

There is no doubt that advances have been made in understanding the role of the hypothalamic nuclei and extrahypothalamic brain centers in the regulation of ACTH cell activity. However, many problems remain, including the brain mechanism controlling CRF secretion, sites of CRF synthesis, and the nature of these CRF molecules. The characteristic response of ACTH cells to CRF and other factors, and the feedback action of corticosteroids, still remain to be elucidated.

C. MSH Cells

MSH cells are known as hypobasophilic cells of the pars intermedia of the adenohypophysis. In pituitaries of fish these cells can be stained by lead hematoxylin (Baker, 1972). These cells are involved in the biosynthesis and secretion of MSH-α and MSH-β, and their number and activity are closely related to the ability of animals to change color. In animals such as lizards that change color rapidly, the pars intermedia is the largest division of the pituitary and constitutes 70% of its total volume (Saint Girons, 1961).

It should be mentioned that the pars intermedia of deer adenohypophysis is exceptionally well developed (Pantić, 1965). The structure and role of the cells in the pars intermedia has been studied in other species (Forbes, 1972; Gosbee *et al.*, 1970; Roux, 1971; Herlant *et al.*, 1972).

Two types of MSH cells have been described: "light" and "dark" cells. Light cells are the predominant form and contain pale basophilic granules. These granules are often fused, and large irregular bodies are formed. Discontinuities between the granular and the plasma membranes have been observed.

There are fewer dark cells than light cells. Their granules vary in size greatly and have long dimensions of approximately 500–1200 nm (Forbes, 1972). In the pars intermedia a third stellate cell type has been described as well.

The release of MSH from MSH cells is closely related to Ca^{2+} concentration. However, it seems that the role of Ca^{2+} is secondary to an initial action of MIFs on the cell membrane, that is, on the permeability of the cell to K^{+} and/or Na^{+} ions (Bower and Hadley, 1972).

In addition to the secretion of MSH, the pars intermedia may have additional functions, such as synthesis and secretion of ACTH, the hormone with the same ancestral molecule.

FIG. 3. Portion of neurohypophysis composed of nerve terminals which contain granules varying in diameter and density.

Mass action direct feedback affects the secretion of both MSH and ACTH (Kastin *et al.*, 1971). The other feedback mechanisms regulating MSH cell activities are not known at the present time.

Two types of neurons that penetrate the pars intermedia have been distinguished: the nerve terminals containing large electron-dense granules and another type with smaller (monaminergic) granules (Enemar *et al.*, 1967; Jørgensen and Vijayakumar, 1971). The pars intermedia in *Rana esculenta* is innervated with nerves without granules which might be cholinergic (Doerr-Schott and Follenius, 1970). However, terminals contain a variable amount of electron-dense granules in all vertebrates (Fig. 3).

1. *MSH-α and MSH-β*

Two melanotropins causing darkening of the skin have been established in the vertebrates and have been designated MSH-α and MSH-β. The most potent melanocyte-stimulating hormone is MSH-α. The amino acid sequence of this molecule is identical with the first 13 residues in ACTH. MSH-α was first isolated from porcine pituitaries (Harris and Lerner,

1957). Three years later Dixon and Li (1960) isolated and determined the chemical structure of equine MSH-α.

Porcine MSH-β has been isolated, and this polypeptide is composed of 18 amino acids (Geschwind *et al.*, 1956, 1957). Species differences have been reported, and it is interesting that human MSH-β has four additional amino acid residues at the N_2-terminal of the nonprimate hormones (see review in Li, 1972, Fig. 3).

The role of 3′,3′-aminophosphate in melanin dispersion has been discussed (Hallahan and Orsi, 1972).

2. *MSH-Releasing and -Inhibiting Factors*

a. *MRF.* The presence of MRF in hypothalamic extracts has been reported (Teleisnik and Orias, 1965), and the role of this factor in the regulation of MSH release established (Kastin *et al.*, 1970). This pentapeptide has been isolated from the hypothalamus (Celis *et al.*, 1971).

b. *MIF.* MIF has been proposed as the tripeptide C-terminal of oxytocin by Celis *et al.* (1971). However, MIF-II has been isolated and determined as a pentapeptide and is not as potent as MIF-I (Nair *et al.*, 1971).

D. LPH-β and LPH-γ

Two lipotropins, LPH-β and LPH-γ, were isolated in pure form and their structures elucidated (Birk and Li, 1964; Li *et al.*, 1965, 1966; Chretien and Li, 1967). LPH-β is a polypeptide consisting of 90 amino acids, and LPH-γ is a polypeptide with 58 amino acids. Chretien and Li (1967) showed that the entire LPH-γ molecule is represented by sequence 1 to 58 or LPH-β.

ACTH and MSH cells are involved in the biosynthesis and secretion of lipotropins. However, the role of STH and LTH cells in the production of these hormones still has to be determined. The main role of lipotropins is the mobilization and hydrolysis of fat (Li, 1972).

IV. LTH and STH Cells

A. Common General Properties

Both cell types are acidophilic if stained by classic methods. They are characterized by highly developed GER, and synthesized hormones are accumulated in the largest specific granules (Fig. 4).

The degree of development of GER and the amount of specific granules are closely related to cellular activity, that is, they are the most pronounced in the very active cells of normal pituitaries. These cells undergo vacuolation of the cytoplasm and may degenerate (Fig. 5).

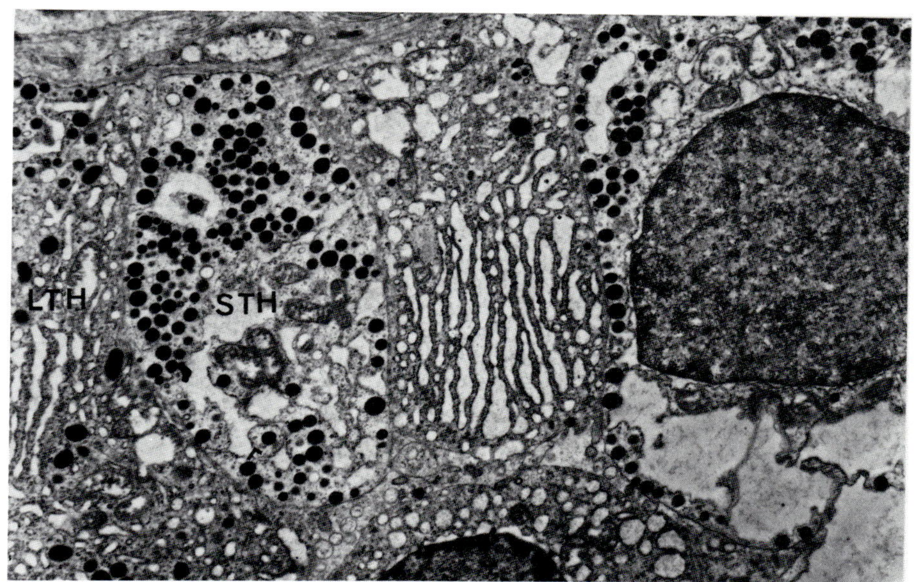

FIG. 4. STH and LTH cells with granular ER and specific granules.

These cells lose the ability to accumulate hormones in the neoplastic state, as for example in transplantable MtT tumor cells (Pantić *et al.*, 1971a,b; Nakayama and Nickerson, 1973b).

By using cytological methods, with both light and electron microscopy, two distinct cell types could be clearly distinguished even in an adenohypophysis transplanted to the allantochorion (Pantić and Djuričić, 1966).

The ultrastructure of LTH and STH cells of rat pituitaries after 1–4 months' treatment with single or repeated doses of estrogen has been investigated. The results were compared with disc electrophoretic patterns of LTH and STH, and the ability of these cells to synthesize and release hormones studied (Pantić, 1968a,b; Pantić and Genbačev, 1969, 1972; Genbačev and Pantić, 1972).

Recently, a significant positive correlation of human STH and human LTH content of fetal pituitary gland was observed, and the ratio between these two hormones was 1:250 (Aubert *et al.*, 1973).

A prolactinlike molecule is the ancestor of LTH and STH in nonprimate species and evolves into a single hormone, h-STH, in man (Li, 1972). Both hormones are single polypeptide chains. The evolutionary biology of LTH and STH has been studied (Nicoll and Licht, 1971).

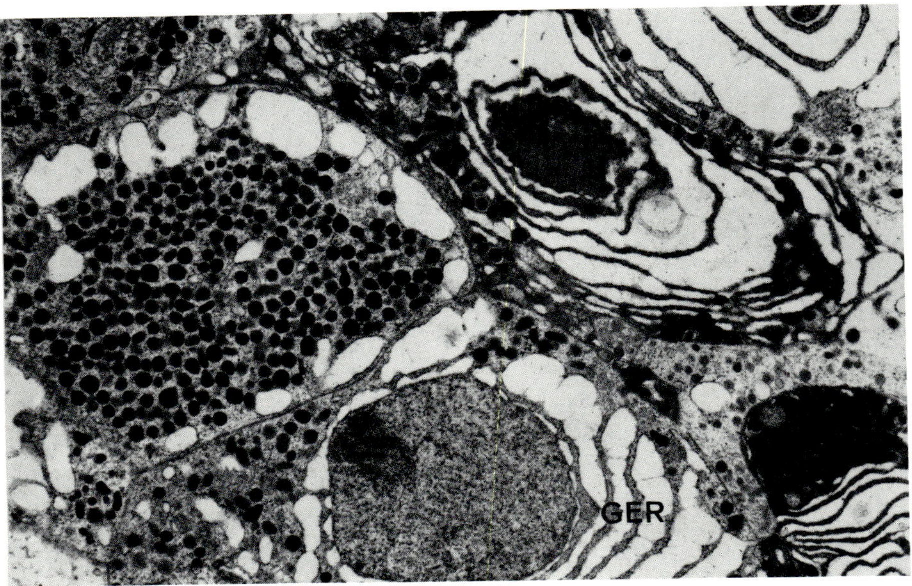

FIG. 5. Markedly dilated cavities of GER. Mature secretory granules located throughout the cytoplasm and large vesicles can be seen. Some cells are undergoing degeneration.

cAMP stimulates both STH and LTH release. The release is maximal 4 minutes after the addition of N^6-monobutyric cAMP into the incubation medium and usually remains linear for 1 hour. In some STH and LTH cells exocytosis is seen at the shortest time (5 minutes). One hour after the addition of cAMP the cells were almost completely depleted of mature secretory granules. The remaining granules were located along the plasma membrane, and immature ones were seen in the Golgi zone (Pelletier *et al.*, 1971).

B. LTH CELLS

LTH cells are identified in the pituitaries of most vertebrate as acidophilic and erythrosinophilic. By means of electron microscopy and immunofluorescence techniques, the properties of LTH cells have been described in various animals under various physiological and experimental conditions, especially during pregnancy, lactation, and after treatment with single or repeated doses of estrogen (Emmart *et al.*, 1963, 1965, 1966; Emmart and Mossakowski, 1967; Bern and Nicol, 1968; Van Oordt, 1969; Emmart, 1969; Farquhar, 1969, 1971; Pantić and Genbačev,

1969, 1972; Nakane, 1970; Pieters and Herlant, 1972; Tixier-Vidal *et al.*, 1971, 1972; Kalra *et al.*, 1973; Pantić and Škaro, 1974; Vellano *et al.*, 1973).

The localization of prolactin cells in the pituitary of fish has been examined (Olivereau, 1969, 1972; Olivereau and Dimovska, 1969) using various histochemical methods and fluorescent antibody techniques. They are present in the rostral pars distalis, and in *Poecilia reticulata* they occupy the greatest part of the rostral pituitary area (Emmart, 1969; Sage and Bromage, 1970a,b; Aler, 1970, 1971).

In avian pituitaries they are localized in both gland regions (Brasch and Betz, 1971). They have been identified in the primate pituitary by immunofluorescence and other methods (Goluboff and Ezrin, 1969; Herbert and Hayashida, 1970).

LTH cells are characterized by the following properties: They are more numerous in female than in male pituitaries. The number of these cells increases during pregnancy and lactation. Their ultrastructure has been investigated as well (Smith and Farquhar, 1966). The size of these cells varies mostly from 6 to 10 μm. In the active LTH cells of some teleosts the diameter of the nuclei is between 3 and 4 μm, and nucleoli are particularly prominent. In comparison to other pituitary cells ER is the most developed in LTH cells, forming a so-called *Nebenkern* (Pantić and Genbačev, 1969, 1972). This concentrically oriented reticulum undergoes transformation into lysosomal bodies.

This proliferation of ER forming the concentrically organized membranous system known as a *Nebenkern* is one of the most pronounced and characteristic properties of LTH cells in their response to estrogen. More than one *Nebenkern* was observed in LTH cells of male rats treated with single or repeated doses of estrogen (Pantić and Genbačev, 1969, 1972).

Regressive changes in the LTH cells of the rat pituitary immediately after the removal of suckling young have been demonstrated (Smith and Farquhar, 1966).

Of particular interest is the genesis of membranous and vacuolar "inclusions" identified within the nucleus of LTH cells in the pituitary of the adult female Mongolian gerbil (Nakayama and Nickerson, 1972). It was observed that after testosterone treatment increased infolding of nuclear membrane and connection between the membrane of vacuolar inclusions and the cytoplasm were much more frequent. A channel system formed by invagination of the nuclear membrane into the karyoplasm has been observed (Abraham, 1971). Mitotic activity of pituitary cells in pregnant and lactating rabbits has also been reported (Allanson *et al.*, 1969).

The specific granules are the largest, and their diameter varies from 100 to 900 nm. They are on the average smaller in pituitaries of fish as compared to mammals (Fig. 6). The size of these granules in fish varies from 100 to 480 nm (Dharmamba *et al.*, 1967; Dharmamba and Nishioka, 1968). Nagahama *et al.* (1973) showed that within 3 hours after transfer to fresh water the prolactin cells of *Gillichtys* exhibited definite functional activity, that is, development of GER, the Golgi system, and mitochondria, and secretion by exocytosis of granules. Seasonal modifications of LTH cells have been described (Pieters and Herlant, 1972).

A rapid discharge of granules occurs by exocytosis during intensified LTH cell activity, for example, suckling and stimulation of release by LRF.

There is a close correlation between the size of LTH cells, degree of granulation, and LTH content. However, pituitary cells of MtT transplantable mammotropic tumors lose their ultrastructural properties, and GER and specific granules are nearly completely absent. The cells are able to synthesize and secrete, but accumulation of hormones in the granules does not occur (Pantić *et al.*, 1971a,b).

LTH secretion in estrogen- and progesterone-pretreated rats is

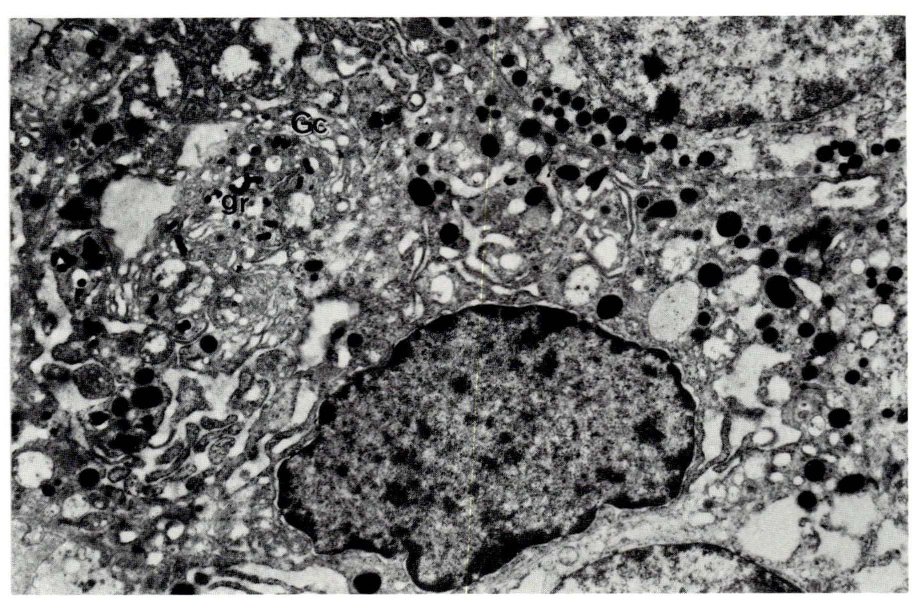

FIG. 6. LTH cell with specific granules (gr) enclosed within Golgi vesicles (Gc) and in the cytoplasmic matrix.

stimulated by TRF (Vale *et al.*, 1973a,b). The pituitary as a possible site of LTH feedback in autoregulation has also been considered (Spies and Clegg, 1971). Target organs of LTH are little understood.

1. *LTH*

Highly purified ovine prolactin has been isolated, and the complete amino acid sequence of this single polypeptide chain has been elucidated (Li *et al.*, 1940, 1942, 1970). It consists of 198 amino acids and has a molecular weight of 23,000. Prolactin from the pituitaries of other species has also been isolated (see review in Li, 1972, Table 7).

It has been noted that LTH exists in different forms in the pituitary and that each method apparently detects different properties of LTH; "depletion" by 180′ of 35, 60, and 70% using disc electrophoresis (DE), radioimmunoassay and bioassay (BA), respectively (Nicoll *et al.*, 1973).

Many of the biological activities of LTH have been described. However, some biological properties have been particularly stressed such as stimulation of growth and regeneration in reptiles and amphibians; stimulation of the thyroid in fish (Olivereau, 1966); stimulation of melanocytic proliferation in some fish (teleosts); stabilizing action on lysosomes in tails from *Xenopus laevis* (Campantico *et al.*, 1972); survival and effect on plasma osmotic pressure in hypophysectomized *Telapia mossambica* (Dharmamba *et al.*, 1967); synergistic action with steroids in the female reproductive tract; feeding of young birds; stimulation of mammary gland and prolonged secretion of corpus luteum in some animals (for LTH molecular properties, see Li, 1972).

The antigonadal properties of LTH and its reciprocal relationship with gonadotropins in birds (Multon and Erickson, 1970; Pantić and Škaro, 1974; Pantić and Kosanović, 1973) and in mammals (Pantić and Genbačev, 1972) have been established.

Most of the 82 different effects of LTH have been classified as follows: 52% of all of the reported actions have a STH-like effect; about 38% are related to reproduction; LTH is involved in water and electrolyte metabolism, integumentary (ectodermal) structure and, finally, actions of LTH that involve synergism with steroid effects on organs which are also influenced by steroids (Nicoll and Bern, 1971).

2. *PIF and PRF*

The concept of some tonic inhibition of LTH release by an inhibitor (PIF) was suggested for more than a decade to explain the role of the hypothalamus in the regulation of LTH release. Recently, the existence of two hypothalamic factors, namely, prolactin inhibitor (PIF) and prolactin-releasing factor (PRF), has been considered. The presence of PRF synthetase has also been suggested (Mitnick *et al.*, 1973). Kamberi *et al.* (1971b) demonstrated a significant decrease in LH and an increase in LH and FSH in plasma of rat after infusion of fragment equivalents of hypothalamus into the hypophysial portal vessel. Based on their results, it is suggested that dopamine influences the release of some hypothalamic factor which decreases LTH release. Pituitary cells in culture with hypothalamic extracts added to the incubation medium have been examined (Vale *et al.*, 1972a,b).

From studies of the action of thyroid hormones at the level of the hypothalamus, it was suggested that T_4 may have a role in the regulation of LTH release (Knigge and Silverman, 1972; and others). The relationship between thyroid hormones and prolactin secretion may be more intimate than previously thought, since it has been realized that TRF provokes a rapid rise in plasma LTH (Bowers *et al.*, 1971; Jacobs *et al.*, 1971).

The chemical nature of PIF or PRF has not been elucidated.

C. STH Cells

STH cells are acidophilic cells with granules that are clearly stained red by Mallory's trichrome and Kresazan (Romeis, 1940) or orange by the PAS–orange G method. STH cells are most numerous in the pituitary of almost all vertebrates (up to 50%) and can be distinguished as specific cells at the end of the first trimester of mammalian embryogenesis (Pantić, 1965; Stošić and Pantić, 1966). In fish, such as the teleost *Poecilia reticulata*, they are intermingled with TSH cells.

STH cells are localized in the pituitaries of birds in the caudal region, and in mammals they are present in the so-called acidophilic region.

GER in STH cells is very well developed, but not to such an extent as in LTH cells (Fig. 7). Newly synthesized STH is transported from GER to the Golgi complex via 50-nm vesicles within 45 minutes of its biosynthesis. Storage granule formation occurs within 75 minutes. Secretion of newly synthesized hormone occurs maximally from 90 minutes on. Transfer of STH to the Golgi system, storage of hormone, formation of specific granules, and secretion by exocytosis are all ATP-dependent, but do not require continuing protein synthesis (Howell, 1972).

Ca^{2+} and Mg^{2+} have important and probably opposite effects on the synthesis of STH and LTH (Ratner and Peake, 1972). STH is stored in specific granules, usually round or oval in shape. They are numerous and densely packed and have a maximal diameter of 400 nm. STH cells are degranulated during rapid growth, as for example in *Poecilia*

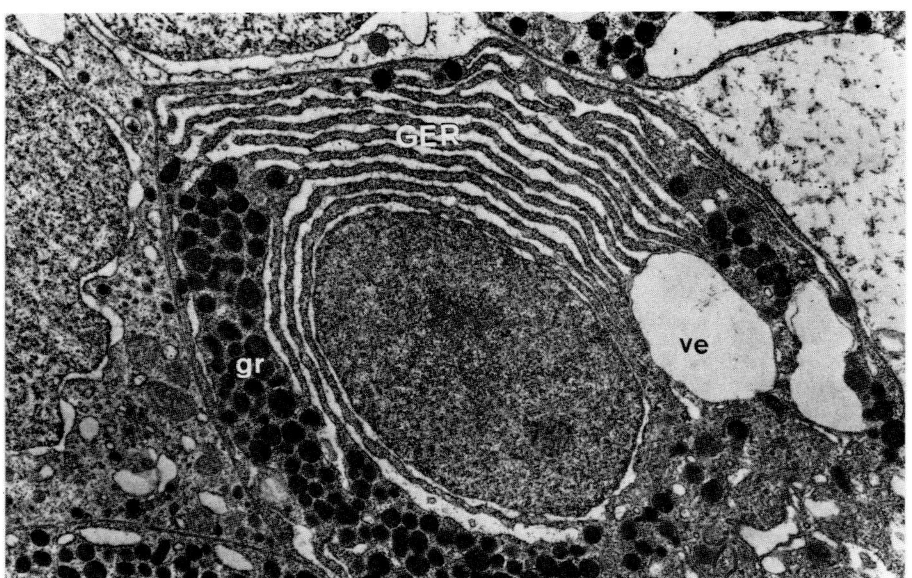

FIG. 7. Electron micrograph of STH cell with concentric GER, specific granules (gr), and some large vesicles (ve).

reticulata (Sage and Bromage, 1970a,b). When growth is arrested, the cytoplasm is again full of granules.

1. *STH*

STH, a single polypeptide chain, consists of 191 amino acids and has a molecular weight of 21,500. This hormone was first isolated and characterized by Li and Evans in 1944. A minor component of "big" STH (MW approx. 40,000) has been demonstrated (Wright and Goodman, 1973; Stachura and Frohman, 1973). It has been shown that synthetic human growth hormone has both STH and LTH activity (Li and Yamashiro, 1970). By examining the major regions of STH molecules, it has been established that "the total of all the alignments between human STH and LTH is 51 identical residues, 46 highly acceptable and 29 unacceptable replacements" (Li, 1972). It was believed that there are sufficient similarities in the sequences of STH from various species to indicate their separate evolution from a common ancestor molecule. STH can be identified in the rudiments of the pharyngeal pituitary. However, this hormone could not be detected in the salamander. Usually, high prolactin levels and relatively low STH concentrations have been observed in several species of amphibian, but low concentrations of both LTH and STH in reptilian and avian pituitaries. Moreover, high STH levels in comparison to LTH have been established in mammals (Bern and Nicoll, 1968; Nicoll and Licht, 1971). However, it is known that ovine LTH stimulates somatic growth of various species of mammals, birds, amphibians, and so on (see Bern and Nicoll, 1968).

2. *SRF and SRIF*

a. *SRF*. STH-releasing factor has been isolated from porcine hypothalamus, and its structure elucidated by Schally *et al.* (1969, 1971). SRF_I releases bioassayable growth hormone in the rat but does not affect the release of immunoreactive STH. The presence of a second hormone, SRF_{II}, acting on the release of STH in porcine hypothalamus, has been suggested (Sandow and Locke, 1973).

b. *SRIF*. SRIF, a tetradecapeptide (H-Ala-Gly-Cys-Lys-Asn-Phe-Phe-Trp-Lys-Thr-Phe-Thr-Ser-Cys-OH), was isolated from ovine hypothalamic extract (Vale *et al.*, 1973b). This peptide inhibits the basal rate of STH secretion and inhibits the TRF-mediated secretion of TSH *in vivo* and *in vitro* (Brazeau *et al.*, 1973; Vale *et al.*, 1973b). It was observed that LTH secretion in estrogen-, and progesterone-pretreated rats is stimulated by TRF.

V. Glycoprotein-Producing (FSH, LH, ICSH, and TSH) Cells

A. Common General Properties

Glycoprotein-producing cells are basophilic when stained with Mallory's trichrome, Kresazan (Romeis, 1940), or aldehyde thionine–PAS–orange G stain. These cells are usually larger than acidophilic cells. ER is moderately developed in the form of variable vesicles. In the absence of corresponding hormones, or target organs, they are hypertrophic; prominence of hypertrophy of Golgi complex and vacuolization of the cytoplasm is prominent, leading to formation of "gonadectomy" or "thyroidectomy" cells. These cells are less heavily granulated than the acidophilic cells. The diameter of their granules is the smallest (100–250 nm).

Glycoprotein hormones (FSH, LH, and TSH) are synthesized in the corresponding cell cytoplasm and may be stored in granules. The diameter and other properties of granules are used as criteria for distinguishing specific types of cells specialized for synthesis of specific glycoproteins, that is, gonadotropins, or TSH.

Each of the glycoprotein hormones has two dissimilar subunits designated α and β. The FSH-α subunit is similar to ICSH-α, and it is therefore possible that they have a common ancestor (Li, 1972). The structures of ovine ICSH-α and bovine TSH-α are nearly identical (Li, 1972). Thyrotropic cells can be easily distinguished from gonadotropic cells.

B. Gonadotropic Cells

Gonadotropic cells are elongated or stellate basophilic cells which are stained blue by Mallory's trichrome, Kresazan, or aldehyde thionine–PAS–orange G; they are not stained with aldehyde fuchsin.

The differentiation and activity of gonadotropic cells are closely related to the development and function of hypothalamic nuclei and gonads (Fig. 8).

In normal teleost *Poecilia reticulata* they start to differentiate 15 days after birth; during the period of gonad maturation the gonadotropic cells gradually assume their characteristic shape and activity; thereafter the cells decrease when the development of gonads is completed. These cells lie in the ventral part of the proximal pars distalis of *Poecilia* (Sage

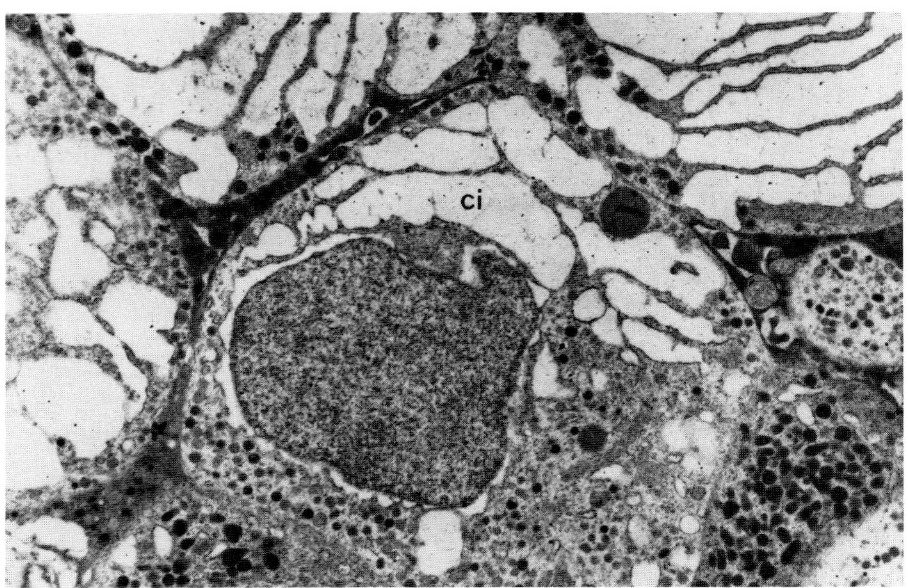

Fig. 8. Gonadotropic cell with cisternoid cavities (ci) of GER, specific granules, and dense bodies.

and Bromage, 1970a,b). They are localized in groups in the pituitaries of other fish as well. Cyclic changes in these cells have been investigated (Knowles and Vollrath, 1966; Pavlović and Pantić, 1971; Pantić, 1972).

In the adenohypophysis of the chick embryo, Guedent (1972) described the Golgi apparatus which has saccules and detached vesicles containing glycoprotein on approximately the fifth day of incubation. Glycoprotein granules were observed on day 8 of incubation (Mikami *et al.*, 1973). TSH cells are localized in the cephalic lobe, and LH cells in the caudal lobe (Brasch and Betz, 1971). These two lobes develop independently and differently from each other during an early stage of ontogenesis (Mikami *et al.*, 1973).

In younger rats typical gonadotropic cells occur in the rostral zone. It is suggested that typical gonadotropic cells in adults are derived from previously differentiated gonadotropes (Allanson *et al.*, 1969; Zambrano, 1971). The production of LH and FSH in clone cultures of rat pituitary cells has been followed (Steinberg and Chowdhury, 1971). An attempt has been made to distinguish two types of gonadotropic (FSH and LH) cells. In the rat anterior pituitary, LH cells differ from FSH cells (Steinberg and Chowdhury, 1971). The identification of these two types of cells has not been confirmed using electron microscopy (Farquhar, 1971) or immunochemical methods (Mirecka and Pearse, 1971). However, Tixier-Vidal *et al.* (1972) identified LH and FSH cells at the level of the electron microscope.

Many other studies have dealt with the identification of these two types of cells, and their main characteristics may be summarized.

Both LH and FSH cells are medium-sized and are mostly localized along blood capillaries. LH cells are usually round or polygonal, and FSH cells are mostly round and larger than LH cells. Secretory granules vary in diameter between 150 and 300 nm. They are electron-dense, usually spherical, and their interior is homogeneous. On the basis of granule size, four different gonadotropic cells have been described (Nakayama *et al.*, 1970).

In these cells, large 700- to 100-nm granules have been reported (Kurosumi, 1968). It seems that granules with a maximal diameter up to 300 nm are specific and characteristic of normal gonadotropic pituitary cells. Larger granules may result from inhibition of hormone release as a consequence of the effects of radiation or other agents (Pantić and Stošić, 1974).

The extrusion of granules from gonadotropic cells was observed during the spawning season in fish, and at 1, 3, and 9 minutes following the administration of synthetic LRF (Mendoza *et al.*, 1973).

The best material for the identification and characterization of FSH and LH cells as separate gonadotropic cells are the pituitaries of bats, in which these two hormones are secreted at different times. However, it is necessary to improve the existing methods and to compare the results obtained with immunocytochemical and other techniques.

GER membranes may develop in specific concentric whorls in the form of *Nebenkerne* only after castration (Farquhar and Rinehart, 1954; Zambrano, 1970a,b; Stetson and Erickson, 1971). This was considered a sign of increased cellular activity. The degree of ER proliferation and formation and the dilatation of cisternae never coalesce in reptiles as in the other vertebrate species. The appearance of gonadectomy and thyroidectomy cells after castration or administration of radioiodine has been reported (Jovanović *et al.*, 1959; Pantić and Stošić, 1966; Halász and Gorski, 1967). An elevation of plasma LH and FSH was observed within 48 hours after castration of adult male mice (Beamer *et al.*, 1972).

There is a close correlation between an increased amount of granules and the development of ER and the concentration of gonadotropic hormones. LH concentration increased and was greater in males than in females prior to day 130 of fetal lamb development (Foster *et al.*, 1972).

1. *Gonadotropic Hormones*

a. *FSH.* FSH is synthesized in gonadotropic cells. Ovine and human FSH were the earliest hormones isolated and purified (Papkoff *et al.*, 1967a,b). The structure of FSH-α is similar to ICSH-α, and this molecule appears to be evolutionarily an ancestral molecule of the glycoproteins (Li, 1972). The molecular weight of FSH is 32,000. However, highly purified quantities of FSH are not as yet available. Serum FSH activity appears by 19 weeks in the female human fetus and by 23 weeks in the male, rising to higher mean levels in females and then falling after 28 weeks of fetal age. Serum activity was higher in females than in males (Levina, 1972).

It was thought that the mammalian FSH molecule possesses the properties of FSH and LH reptilian hormones, and it was proposed that the terms FSH and LH be replaced by the more general term gonadotropin when referring to the pituitary-gonad system of reptiles (Licht, 1972). Licht and Papkoff (1972) observed that the ovine LH molecule possesses instrinsic gonadotropic activity in the lizard, although it is considerably less potent than FSH. The great disparity in relative potencies of these hormones may be explained by differences in their biological half-life. The half-life of FSH in the circulation is longer than that of LH and seems to be about 2 hours.

Site of FSH action. The plasma membrane of spermatogonia is proposed as a site of FSH action. This may be supported by findings that maximal FSH stimulation of adenyl cyclase was detected in the seminiferous tubules of rats older than 20 days when spermatogonial and Sertoli cells were present. Adenyl cyclase activity decreases when spermatocytes appear. Therefore it does not appear that spermatocytes possess FSH-responsive adenyl cyclase. Alternatively, cyclase activity may be repressed (Braun and Sepsenwol, 1973). As cAMP levels decrease, cellular differentiation proceeds to mature spermatozoa (Christiansen and Desantel, 1973).

The effect of FSH on the development of follicles in the ovary seems to be expressed in granulosa cells. In immature, intact rat, FSH activates or induces binding sites for LH (Zeleznik and Midgley, 1973).

b. *LH (ICSH)*. LH is a glycoprotein which consists of two subunits (ICSH-α and ICSH-β) and has a molecular weight of 16,000. ICSH was first isolated and purified from ovine pituitary (Li *et al.*, 1940) and later from other species.

ICSH subunits are not identical, and the amino acid and carbohydrate compositions differ in various species (Li and Starman, 1964). Activity for each individual subunit has not been observed. The amino acid sequences of both subunits of ovine ICSH have been determined (Papkoff *et al.*, 1965, 1971a,b). This hormone consists of 216 amino acid residues and contains 26% carbohydrate (Papkoff *et al.*, 1971a). ICSH-α consists of 96 residues and ICSH-β of 120. Both the immunological and biological activities of ICSH reside in its β subunit (Papkoff *et al.*, 1971b; Yang *et al.*, 1972).

The main effect of ICSH is a pronounced development of the interstitial cells both in testis and in ovaries. The effect of mammalian gonadotropins on the gonads in various species of animals has been investigated (Fontaine and Fontaine, 1962; Fontaine, 1969). The presence of a specific receptor for ICSH in ovarian tissues has been observed (Rajaniemi and Vanha-Perttulla, 1972). It has been suggested that cAMP may be involved in the luteinizing action of ICSH (Ellsworth and Armstrong, 1973).

2. *Gonadotropic Releasing Factors*

The hypothalamic factors that control the release of LH and FSH have been isolated as a nonapeptide from bovine hypothalamus (Schally *et al.*, 1971) and from ovine hypothalamic homogenate (Amoss *et al.*, 1971).

However, the primary structure of porcine LRF, a decapeptide (pGlu-His-Trp-Ser-Tyr-Gly-Leu-Arg-Pro-Gly-NH_2), was determined by Matsuo *et al.* (1971). It was established that the primary structure of ovine LRF is identical to that of porcine LRF (Burgus *et al.*, 1971, 1972).

Synthetic LRF has been produced, and its effect found to be identical to that of the native hormone (Matsuo *et al.*, 1971; Monahan *et al.*, 1971).

On the basis of experimental and clinical data obtained by various research groups, it has been proposed that one decapeptide regulates the release of both LH and FSH (Schally *et al.*, 1971; Yen *et al.*, 1972; Kastin *et al.*, 1972). Analogs of LRF with inhibitory action of greater potency than the natural decapeptide hormone have been synthesized (Monahan *et al.*, 1973).

Analogs of the LRF molecule, a decapeptide, were synthesized by Coy *et al.* (1973). They concluded that the N-terminal sequence pyro-Glu-His-Trp is probably the functional part of the LRF molecule and that the remaining amino acids are most likely involved only in binding to receptor sites and/or in transport in blood. Two zones containing LRF and FRF activity have been distinguished by Burgus *et al.* (1973). An LRF-A zone corresponded to the mobility of the LRF decapeptide, and LRF-B was eluted earlier and well separated from the LRF-A zone.

It is still questionable whether LRF is the sole hypothalamic hormone regulating the release of LH and FSH, whether the feedback action of gonadal steroids is expressed via a single pituitary cell type, and whether LH and FSH are synthesized by one or by two distinct cell types.

Recently, it was shown that estrogen significantly inhibits LH and FSH release throughout the infusion period (Orias and Libertun, 1973). LRF can induce the release of LH when injected into the third ventricle (Weiner *et al.*, 1972).

However, two pituitary cells were distinguished using different cytological methods: one produces FSH and the other LH (Nakayama and Nickerson, 1973a).

Weiner *et al.* (1972) showed that intraventricular administration of LRF is less potent in causing LH release than intravenous LRF, even when the latter was diluted by the systemic circulation. Intracerebral implantation of cortisol inhibited the reproductive system in immature rats (Smith *et al.*, 1971).

The catecholaminergic and cholinergic systems have a stimulatory effect, whereas the indolaminergic system has an inhibitory effect on the release of LRF, FRF, and PIF from the neurosecretory elements of the hypothalamus. Relatively small doses of dopamine and large doses of norepinephrine, epinephrine, or acetylcholine induced the release of LH and FSH and inhibited LTH. The release of these hormones is mediated through the hypothalamus, secondary to a discharge of LRF, FRF, and PIF (Kamberi, 1973).

It has been shown that both cAMP and cGMP may be involved in the hypothalamic control of LH and LTH secretion (Ojeda and McCann, 1973).

A role of prostaglandins in mediating the action of LRF on LH secretion was suggested (Dowd *et al.*, 1973). The influence of prostaglandins on pituitary secretion of LH and LTH is expressed at the hypothalamic level, thus altering the output of RF and/or IFs (Harms *et al.*, 1973).

Recently, it was observed that colchicine significantly enhances the action of LRF and FRF by increasing the rate of release of LH and FSH (Sundberg *et al.*, 1973).

3. *The Feedback Action of Gonadal Steroids*

The sites of the feedback action of gonadal steroids may be the cells of the pituitary, hypothalamic nuclei, and extrahypothalamic nervous centers.

The control system for gonadotropins in the male rat is under the influence of steroids at a very early age; in females ovarian steroids do not appear to modify the secretion of FSH until the animals are 20 days old (Ojeda and Ramirez, 1972; Pantić and Genbačev, 1972; Pantić and Škaro, 1974; Pantić and Kosanović, 1973).

Since LRF and FRF have been isolated from bovine and ovine hypothalamus by Schally *et al.* (1971) and Amoss *et al.* (1971), there is no doubt that these hormones are synthesized in the cells of hypothalamic nuclei. However, the question is which nuclei are the sites of LRF and FRF synthesis.

4. *Gonadal Steroid Receptors*

Higher uptake of steroids in hypothalamic nuclei than in cerebellar nuclei was observed by Chader and Villee (1970a). They suggested that a specific estradiol receptor in the cytoplasmic soluble fraction of rabbit hypothalamus appears to be somewhat different from those described in the uterus of the rat (Kato and Villee, 1967a,b; Chader and Villee, 1970b). By measuring the uptake and retention of estradiol-^{3}H, Kato and Villee (1967a) showed that the anterior hypothalamus and ME have a characteristic pattern similar to that shown by the uterus and vagina. They mentioned that specific estradiol-binding receptors are of limited capacity in the anterior hypothalamus and anterior hypophysis. Two different processes in transferring estrogen from the cytoplasmic receptors to the nuclear fractions have been proposed by Musliner *et al.* (1970).

It is now generally accepted that receptors are involved in the mechanism of gonadal steroid action. The hormone–receptor complex apparently binds to specific chromatin acceptor sites. The chromatin acceptor proteins for steroid hormone receptors have the capacity to recognize, and may be located at specific DNA sequences within the cell genome which are the sites for initiating the nuclear events requisite for steroid hormone action in target tissue (Steggles and King, 1972).

The first step in mediation of the estrogen effect on gonadotropic release is the formation of a high-affinity complex between estrogen and cytoplasmic proteins in the neurons and pituitary cells. Some recent data concerning the estrogen receptors are mentioned here.

Estrogen receptors exist in a selected area of the neonatal brain as early as day 2. Selective *in vivo* uptake of estrogen-^{3}H by the central nervous system of the neonatal female rat showed that radioactivity was concentrated in nuclei of the following neurons: n. tractus diagonalis; n. preopticus medialis; n. periventricularis; n. interstitialis striae terminalis; n. paraventricularis, pars magnocellularis; n. arcuatus hypothalami; n. parmammilaris; n. ventromedialis, pars ventromedialis; n. ventralis amygdale; and n. corticalis amygdale (Sheridan *et al.*, 1973).

The estradiol-concentrating mechanism in the rat pituitary is both qualitatively and quantitatively identical in male and female animals (Korach and Muldon, 1973).

Six hours following the injection of estrogen into 5-day-old rats, radioactivity was not detected in the pituitary, but significant radioactivity was detected in the hypothalamus and the central nervous system. However, the uptake of ^{14}C-labeled estrogen was 10-fold higher in gonads than in the central nervous system, and very high in the uterus and the adrenals.

Pituitary cells concentrate estradiol-^{3}H rapidly at 37°C, and peak uptake occurred by 20 minutes. Nuclear uptake and retention increased gradually and peaked at 30 minutes (Leavitt *et al.*, 1973).

The number of specific estrogen-binding receptor sites per pituitary cell is estimated to be 19,000 (Leavitt *et al.*, 1973) and 12,000 (Notides *et al.*, 1972). Progesterone binding increased quantitatively after pretreatment of castrate animals with estradiol-17β (Rao *et al.*, 1973).

The number of high-affinity estrogen-binding sites in the female rat hypothalamus and pituitary increases from day 19 to day 28, being maximal in mature animals. The number of "receptors" varied during sexual cycle, and the data showed that their number was higher during metestrus than proestrus.

5. *Hypothalamic Nuclei and Gonadotropic Activity*

Two separate hypothalamic mechanisms controlling FSH and LH secretion and two levels of hypothalamic control of LH secretion were suggested by Baraclough and Gorski (1961) and Flerkó (1966). Other articles deal with the role of the MBH and other hypothalamic centers in the control of gonadotropin release (Halász and Pupp, 1967; Graber *et al.,* 1967; Stetson, 1969; Peter, 1970; Stern, 1972).

Watanable and McCann (1968) showed that localization of FRF is restricted to the stalk-ME region. Two years later Crighton *et al.* (1970) demonstrated *in vitro* the localization of the LH-releasing factor in "the medial, basal tuberal region, an area which included the median eminence and arcuate nucleus, and more rostrally in an area which included the suprachiasmatic nucleus and all but the most rostral portions of the preoptic nucleus."

A close relationship exists between the neurosecretory cells of fish hypothalamus, especially between the NLT neurons and pituitary gonadotropic cells (Zambrano, 1971; Polenov, 1950; Polenov and Garlov, 1973; Pavlović and Pantić, 1974; Pantić and Lovren, unpublished data). It has been shown that the NLT neurosecretory system in fish is homologous to the hypothalamic arcuate (infundibular) ME system of tetrapods (Zambrano and De Robertis, 1968).

The anterior hypothalamic-preoptic area has been suggested as a positive feedback site for estrogen in the immature rat (Davidson, 1969). A possible connection between the retina and preoptic nucleus has been suggested (Vullings, 1973). In addition, the MBH appears to be involved in the negative feedback action of estrogen (Ramirez and McCann, 1964; Dukić, 1972).

The experiments of Ravona *et al.* (1973) with electrolytic lesions in birds showed that completely atrophied testes and combs and the absence of spermatogenesis and dehydrogenase activity in testicular interstitial tissues occurs in birds with lesions located in mammillary nuclei and in the posterior part of ventromedial nuclei. They postulated complete impairment of FRF and LRF to be the cause of these phenomena.

Regression of both tubular and interstitial tissue of the testes and regression of the cloacal gland were induced in birds when small pellets of testosterone propionate were implanted in a medial or paramedial position in the ventral hypothalamus or adenohypophysis. This was considered to be a result of interruption of the release of LH or LH and FSH (Graber *et al.,* 1967; Stetson, 1972).

The MBH has also been suggested as the effective site of steroid action (Flerkó, 1966; Davidson, 1969). Implantation of small pellets of testosterone propionate–charcoal (1:1) in a medial or paramedial position in the ventral hypothalamus or adenohypophysis results in regression of testes and the cloacal gland (Stetson, 1972). The effect of castration on the ultrastructure of the arcuate nucleus and the outer zone of the ME has been described (Zambrano and De Robertis, 1968).

The MBH in the guinea pig is refractory to electrochemical stimulation for the induction of ovulation (Quinn, 1973). Following the localization of estradiol-^{3}H and catecholamine in hypothalamic neurons, it was observed that a simultaneous nuclear accumulation of silver grains and cytoplasmic catecholamine fluorescence occurred in neurons of the periventricular and arcuate nuclei at the level of the NVM (Grant and Strumpf, 1973). Akmayev *et al.* (1973) investigated the role of α and β tanocytes in hypothalamohypophysial transmission, with particular attention to the role of the nerve cells of the arcuate nucleus and to the tanocytes.

C. TSH Cells

TSH cells are usually elongated or stellate cells with specific granules which are clearly PAS-positive and stained blue by aldehyde thionine.

In the pituitary of Pacific salmon these cells are located in the anteroventral zone, mainly in the rostral pars distalis (Olivereau, 1972). In the avian pituitary they are present mostly in the cephalic region.

TSH cells in *Poecilia reticulata* are active from birth (Sage and Bromage, 1970a,b). In rat pituitary these cells have a granule size between 50 and 100 nm and first occur on the seventeenth fetal day; the formation of these granules precedes the appearance of gonadotropins (Fink and Smith, 1971). These specific granules are very small, and their diameter is about 100 nm to a maximum of 200 nm.

Two types of TSH cells have been identified in the adenohypophysis of the fish *Cichlasoma biocellatum:* (1) PAS-positive and (2) alcian blue–negative basophils (Mattheij *et al.*, 1971). However, in mammals more types of TSH cells were described (Tixier-Vidal *et al.*, 1972; and others).

TSH cells possess specific granules smaller than those of the other specific cells, and they can be identified early by using other criteria as well, such as cell size, long cytoplasmic extensions, and vesiculation of ER (Fig. 9). As are other pituitary cells, they are controlled by the hormones produced in the hypothalamus or in the corresponding target organs and they react specifically.

TSH cells are very sensitive to thyroid deficiency, in which cells hypertrophy and are transformed into so-called thyroidectomy cells. These thyroidectomy cells arise from so-called delta transformed TSH cells (Tixier-Vidal *et al.*, 1972). The properties of thyroidectomy cells after thyroidectomy or administration of radioiodine have been examined (Jovanović *et al.*, 1966; Pantić and Stošić, 1966; Tixier-Vidal *et al.*, 1971; and many others). As a result of gradual vacuolation and fusion of these vacuoles in the cytoplasm, large vacuoles with a "signet ring" are formed

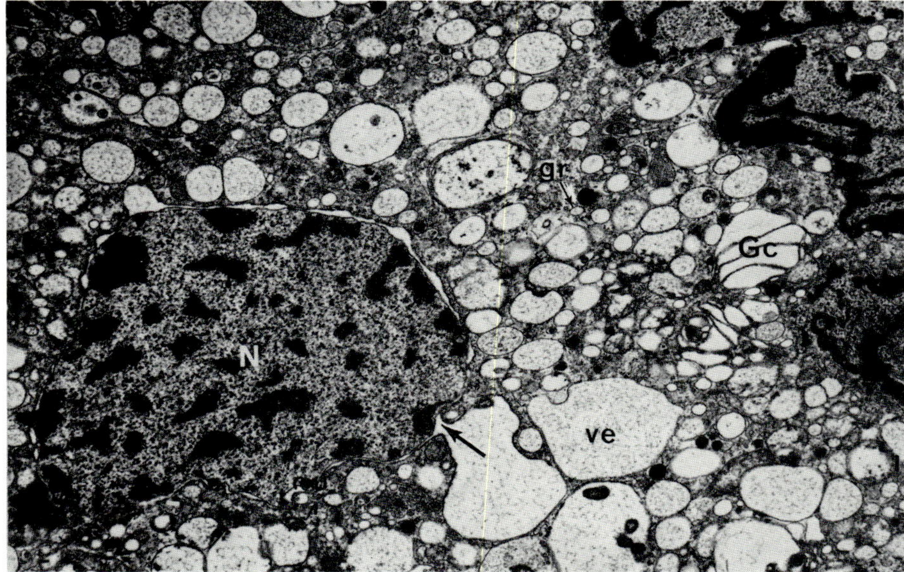

FIG. 9. TSH cell with nucleus (N). Note the continuity between the external nuclear membrane and the ER (arrow), granules (gr), Golgi complex (Gc), and large vesicles (ve) in the cytoplasm.

during hypertrophy. As hyperactive cells they lose the capacity to store TSH and may be devoid of specific granules. Dense bodies are located in the vesicles of ER (Fig. 10). In these hypertrophic cells colloid droplets, lipid granules, and lysosomal bodies are also present. As a result of prolonged stimulative activity, thyroidectomy cells undergo degeneration.

1. *TSH*

TSH is a glycoprotein consisting of two subunits (TSH-α and TSH-β). As mentioned earlier, the structures of bovine TSH-α and ovine ICSH-α are nearly identical (see Li, 1972). There are 67 homologous positions, or about 55–60% of either polypeptide chain of TSH-β and ICSH-β.

Radioimmunoassay and primary structure determination of TSH-β subunits show that human TSH-β is far more homologous with bovine TSH-β in sequence (12 differences) than is human ICSH-β with bovine ICSH-β (36 differences) (Shome and Parlow, 1973).

The molecular weight of TSH is 28,000, and its main role is to stimulate thyroid follicular cell activity. The role in the stimulation of thyroid cell activities is briefly reviewed by Pantić (1974).

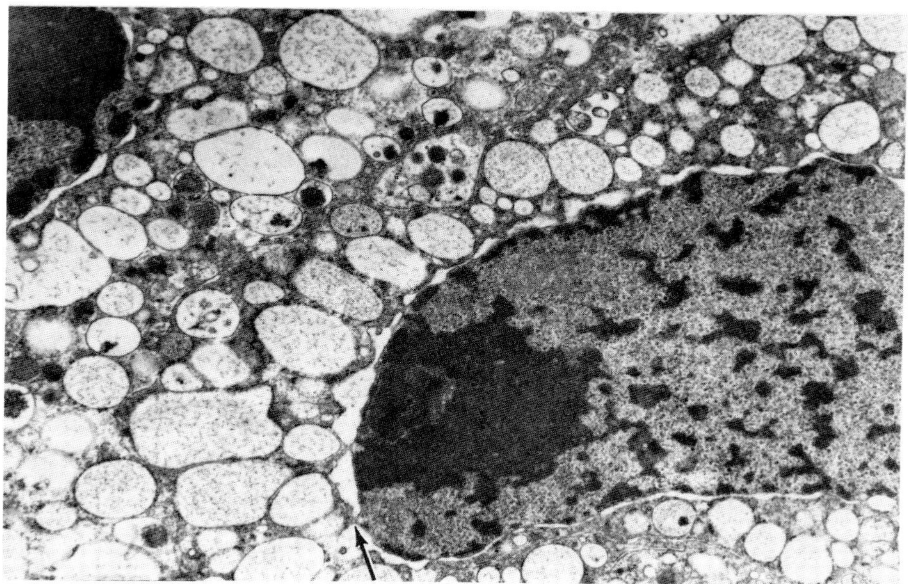

FIG. 10. Portion of hypertrophied TSH cells. Note large nucleolus and continuity between the karyoplasm and cytoplasm (arrow); large vesicles predominate in the cytoplasm, suggesting high activity.

2. *TRF*

The structure of TRF, a tripeptide obtained from ovine sources, has been established (Burgus *et al.,* 1969, 1970). It has been shown that the synthetic tripeptide has a stimulative effect on the release of TSH in all species of mammals examined so far (Blackwell and Guillemin, 1973). It was observed that TRF isolated from porcine hypothalamus has the same structure as ovine TRF: pGlu-His-Pro-NH_2 (Nair *et al.,* 1970). Synthetic TRF analog II, with specific activity 8 to 10 times the biological potency of native TRF, has been reported (Vale *et al.,* 1971).

The specificity, mechanism of action, and properties of TRF *in vivo* and *in vitro* have been reviewed (Blackwell and Guillemin, 1973). The first event in TRF action, analogous to that for larger protein hormones, is selective binding to receptors on the plasma membrane of TSH cells (Wilber and Seibel, 1973). TRF (10 ng/ml) *in vitro* causes prompt release of TSH. T_4 and T_3 added to the medium completely inhibited TRF mediator release (May and Donabedian, 1973).

Biological activity of TRF *in vitro* and *in vivo* was directly proportional to the affinity of membrane receptors for TRF (Vale *et al.,* 1973a,b).

Using ^{3}H-labeled TRF, Grant *et al.* (1971) described the specific binding of the labeled hormone to TSH tumor cell membranes, dispersed normal pituitary cells, or isolated membrane. This TRF binding was not affected by the presence of L-T_4 or L-T_3 (Labrie *et al.*, 1972).

TRF increases plasma prolactin and TSH in rams; the prolactin response was dose-related, while the TSH response was not. The magnitude of the TSH response decreased with successive TRF injections, while that of prolactin did not (Davis and Borger, 1973).

TRF in doses of 100, 500, or 1000 μg increased prolactin as much as 11-fold (Kelly *et al.*, 1973).

Ten nanograms or more caused larger increases in TSH (10-fold) and LTH (4-fold) and also stimulated growth hormone release (2-fold) in cells cultured for 4 days (Machlin and Jacobs, 1973).

Synthetic TRF *in vitro* significantly increased pituitary cAMP 25% in 20 minutes' incubation, with maximal effects (200%) at 90 minutes (Chayoth *et al.*, 1973).

The magnocellular part of the NPV in the hypothalamus of *Bufo vulgaris* has been suggested as a site of TRF synthesis and release (Milin, 1963). The NPV, NVM, and ME are involved and are most responsible for the regulation of thyroid follicular cell activity (Stošić *et al.*, 1965).

An assay for the binding of ^{3}H-labeled TRH has been described by Labrie *et al.* (1972). They observed that K^+ and Mg^{2+} increase formation of the receptor–TRF complex at optimal concentrations of 5–25 mM and 0.2–2.5 mM, respectively; inhibition occurred at higher concentrations tested.

Burgus *et al.* (1973) suggested the following mechanism for the action of TRF and its analogs on TSH cells: binding to the receptors of the TSH cell plasma membrane. This binding must be dependent on its three-dimensional conformation which determines the juxtaposition of important functional groups to the surface of the receptor site. They pointed out that "each residue in the TRF molecule plays an important role in determining the conformation of the molecule and that the conformation of the hystidyl residue is important for the binding of TRF to its receptor site."

It has been reported that synthetic TRF *in vitro* significantly increases cAMP with maximal effects (200%) at a 90-minute incubation (Chayoth *et al.*, 1973).

Pituitary responsiveness to TRF is not modulated by T_3 as the primary factor, and thyroid hormone economy may not be the sole regulatory determinant of TSH function (Clifton-Bligh *et al.*, 1973).

The stimulative effect of TRF on rat pituitary TSH cells leads to the extrusion of specific granules by exocytosis both *in vivo* and *in vitro*

(Pelletier *et al.,* 1971). It was observed that extrusion of granules by exocytosis occurs after appropriate stimulation in ACTH and STH cells as well, but this process was not seen in gonadotropic cells.

3. *The Feedback Action of Thyroid Hormones*

Various levels of the neuroendocrine system have been proposed as sites of thyroid hormone feedback actions. Some of them are noted.

The cells of the NPV, NVM, and ME are the most sensitive cells for thyroid feedback action (Knigge and Joseph, 1971; Knigge and Silverman, 1972; Stošić and Pantić, 1970, 1973a,b).

Thyroid hormones (T_3 and T_4) have negative feedback action on pituitary TSH cells.

Many articles deal with the examination of the pathways involved in the regulation of TSH synthesis and release. However, the mechanisms of thyroid hormone feedback action at the level of brain centers and pituitary cells are still far from being understood (see Pantić, 1974).

VI. Concluding Remarks

The isolation, purification, and determination of the primary structure of pituitary hormones, hypothalamic factors, and their analogs represent an exciting development in the field of neuroendocrinology.

Many cytological, cytochemical, immunohistochemical, and biochemical methods have been improved and used to increase our knowledge of the biology of pituitary cells, including the mechanisms regulating the synthesis of specific hormones and the ability to accumulate their products in specific granules and to release hormones at the appropriate time. It has become clear that the activity of pituitary cells is controlled by the corresponding hypothalamic factors and by the feedback action of target hormones.

The differentiation of the pituitary cells at the level of the genome is influenced by many processes such as inductive interaction between the pituitary cells and the cells originating from the diencephalon and the concentration of hormones in fetal circulation. However, many other internal as well as external factors can influence and even alter the development of the pituitary gland and of hypothalamus-pituitary target organs and their balanced interrelationship. Fetal and postnatal periods of development are the most sensitive, so that many factors and agents, especially hormones, can temporarily or permanently alter differentiation of the pituitary cells.

The main internal factors controlling pituitary cell activities are: hormones produced by the cells of hypothalamic nuclei, target and pituitary cells, bioamines, and other agents such as prostaglandins and cAMP.

In order to facilitate further successful investigation, more attention should be paid at the molecular level to the intracellular events that lead to increased cell complexity and the differentiation of specific cell types in the pituitary.

Many attempts have been made to establish the specific sites in the hypothalamic nuclei responsible for the synthesis and release of the corresponding hypothalamic hormones (RF and IFs), and to elucidate the mechanism by which target hormones regulate pituitary cells. More data are necessary to understand the role of extrahypothalamic nervous centers and of external factors in the control of the hypothalamopituitary system.

In considering the mechanisms regulating hormone release and synthesis, two main high-affinity binding sites, termed receptors, for hormones are proposed: (1) receptors for the protein hormones located at the surface of the plasma membrane of endocrine cells, and (2) receptors for steroids that are components of target cell cytoplasm and nuclei.

If it is assumed that the mode of hormone action is expressed via a receptor, the following questions arise: What is the nature of the receptors involved in the initial steps of a hormone effect? What is the difference in the number of receptors per cell during ontogenesis, various stages of cell activities, and tumorigenesis? And what are the pathways of the mechanisms regulating hormone biosynthesis, accumulation, and release?

As research data become available, we expect advances in the elucidation of the mechanisms of hormone action, which are of major interest for further progress in neuroendocrinology and cell biology.

Acknowledgments

This manuscript was prepared during my residence in the United States as a visiting professor under an arrangement between the Council of Academies of Sciences and Arts of Yugoslavia and the U. S. National Academy of Sciences. It was completed in the Department of Biochemistry, Harvard Medical School, Boston (1973).

The author thanks Dr. Olivera Pantić, Veterinary Faculty, Belgrade for his cooperation, and Dr. Jonathan Li, Department of Biochemistry, Harvard Medical School, Boston, for reading the manuscript. Thanks are also due to Miss Dusica Havelka for technical assistance in preparation of the electron micrographs.

References

Abraham, M. (1971). *Gen. Comp. Endocrinol.* **17**, 334.

Akmayev, I. G., Fidelina, O. V., Kabolova, Z. A., Popov, A. P., and Schitkova, T. A. (1973). *Z. Zellforsch. Mikrosk. Anat.* **137**, 493.

Aler, G. M. (1970). *Acta Zool. (Stockholm)* **51**, 149.

Aler, G. M. (1971). *Acta Zool. (Stockholm)* **52**, 275.

Alešin, B. V. (1971). "Medicina," 440 pp. Moskva.

Allanson, M., Foster, C. L., and Cameron, E. (1969). *J. Reprod. Fert.* **19**, 121.

Allen, J. P. (1973). *Endocrine Soc. Annu. Meet., 55th, Chicago*, A-80, Abstr. No. 64.

Amoss, M., Burgus, R., Blackwell, R., Vale, W., Fellows, R., and Guillemin, R. (1971). *Biochem. Biophys. Res. Commun.* **44**, 205.

Aubert, M. L., Grumbach, M. M., and Kaplan, S. L. (1973). *Endocrine Soc., Annu. Meet., 55th, Chicago*, A-49, Abstr. No. 1.

Baker, B. I. (1972). *Gen. Comp. Endocrinol.* **19**, 515.

Ball, J. N., and Baker, B. I. (1969). *In* "Fish Physiology" (W. S. Hoar and D. J. Randall, eds.), Vol. 2, pp. 1–110. Academic Press, New York.

Barraclough, C. A., and Gorski, R. A. (1961). *Endocrinology* **68**, 68.

Beamer, W. G., Murr, S. M., and Geschwind, I. I. (1972). *Endocrinology* **90**, 823.

Bern, H. A., and Nicoll, C. S. (1968). *Recent Progr. Horm. Res.* **24**, 681.

Birk, Y., and Li, C. H. (1964). *J. Biol. Chem.* **239**, 1048.

Blackwell, R. E., and Guillemin, R. (1973). *Annu. Rev. Physiol.* **35**, 357.

Bohus, B., and Strashimirov, D. (1970). *Neuroendocrinology* **6**, 197.

Bower, A., and Hadley, M. E. (1972). *Gen. Comp. Endocrinol.* **19**, 147.

Bowers, C. Y., Friesen, H. G., Hwang, P., Guyda, H. J., and Folkers, K. (1971). *Biochem. Biophys. Res. Commun.* **45**, 1033.

Brasch, M., and Betz, T. W. (1971). *Gen. Comp. Endocrinol.* **16**, 241.

Braun, T., and Sepsenwol, S. (1973). *Endocrine Soc., Annu. Meet., 55th, Chicago*, A-98, Abstr. No. 99.

Brazeau, P., Vale, W., Burgus, R., Ling, N., Butcher, M., Rivier, J., and Guillemin, R. (1973). *Science* **179**, 77.

Burgus, R., Dunn, T. F., Desiderio, D., Guillemin, R. (1969). *C. R. Acad. Sci., Ser. D* **269**, 1870.

Burgus, R., Dunn, T. F., Desiderio, D., Ward, D. V., Vale, W., and Guillemin, R. (1970). *Nature (London)* **226**, 321.

Burgus, R., Butcher, M., Ling, M. M. N., Monahan, M., Rivier, J., Fellows, R., Amoss, M., Blackwell, R., Vale, W., and Guillemin, R. (1971). *C. R. Acad. Sci., Ser. D* **273**, 1611.

Burgus, R., Butcher, M., Amoss, M., Ling, N., Monahan, M., Rivier, J., Fellows, R., Blackwell, R., Vale, W., and Guillemin, R. (1972). *Proc. Nat. Acad. Sci. U. S.* **69**, 278.

Burgus, R., Rivier, J., and Amoss, M. (1973). *Endocrine Soc., Annu. Meet., 55th, Chicago*, A-144, Abstr. No. 191.

Campantico, E., Giunta, C., Guardabassi, A., and Vietti, M. (1972). *Gen. Comp. Endocrinol.* **18**, 396.

Celis, M. E., Taleisnik, S., and Walter, R. (1971). *Proc. Nat. Acad. Sci. U. S.* **68**, 1428.

Chader, G. J., and Villee, C. A. (1970a). *Proc. Int. Soc. Psychoneuroendocrinol., Brooklyn, N. Y.* pp. 17–24.

Chader, G. J., and Villee, C. A. (1970b). *Biochem. J.* **118**, 93.

Chretien, M., and Li, C. H. (1967). *Can. J. Biochem.* **45**, 1163.

Chayoth, R., Zor, U., Kaneko, T., and Field, J. B. (1973). *Endocrine Soc., Annu. Meet., 55th, Chicago*, A-148, Abstr. No. 200.

Christiansen, R. O., and Desantel, M. (1973). *Endocrine Soc., Annu. Meet., 55th, Chicago*, A-100, Abstr. No. 104.

Clifton-Bligh, P., Silverstein, G. E., and Burke, G. (1973). *Endocrine Soc., Annu. Meet., 55th, Chicago,* A-61, Abstr. No. 25.

Cole, R. D., Li, C. H., Harris, J. I., and Pon, N. G. (1956). *J. Biol. Chem.* **219,** 903.

Coy, D. H., Coy, E. J., and Schally, A. V. (1973). *Endocrine Soc., Annu. Meet., 55th, Chicago,* A-145, Abstr. No. 193.

Crighton, D. B., Schneider, H. P. G., and McCann, S. M. (1970). *Endocrinology* **87,** 323.

Davidson, J. (1969). *In* "Frontiers in Neuroendocrinology" (W. F. Ganong and L. Martini, eds.), p. 343. Oxford Univ. Press, London and New York.

Davis, S., and Borger, M. L. (1973). *Endocrine Soc., Annu. Meet., 55th, Chicago,* A-210, Abstr. No. 324.

Dharmamba, M., and Nishioka, R. S. (1968). *Gen. Comp. Endocrinol.* **10,** 409.

Dharmamba, M., Handin, R. I., Nandi, J., and Bern, H. A. (1967). *Gen. Comp. Endocrinol.* **9,** 295.

Dingemans, K. P., and Feltkamp, C. A. (1972). *Z. Zellforsch. Mikrosk. Anat.* **124,** 387.

Dixon, J. S., and Li, C. H. (1960). *J. Amer. Chem. Soc.* **82,** 4568.

Doerr-Schott, J., and Follenius, E. (1970). *Z. Zellforsch. Mikrosk. Anat.* **111,** 427.

Dowd, A. J., Hoffman, D. C., and Speroff, L. (1973). *Endocrine Soc., Annu. Meet., 55th, Chicago,* A-135, Abstr. No. 173.

Dukić, S. (1972). Ph.D. Dissertation, p. 1. Belgrade Univ., Belgrade.

Elekes, K., and Peczely, D. (1972). *Gen. Comp. Endocrinol.* **18,** 589. (Abstr.)

Ellsworth, L. R., and Armstrong, D. T. (1973). *Endocrinology* **92,** 840.

Emmart, E. W. (1969). *Gen. Comp. Endocrinol.* **12,** 519.

Emmart, E. W., and Mossakowski, M. J. (1967). *Gen. Comp. Endocrinol.* **9,** 391.

Emmart, E. W., Spicer, S. S., and Bates, R. W. (1963). *J. Histochem. Cytochem.* **11,** 365.

Emmart, E. W., Bates, R. W., and Turner, W. A. (1965). *J. Histochem. Cytochem.* **13,** 182.

Emmart, E. W., Pickford, G. E., and Wilhelmi, A. E. (1966). *Gen. Comp. Endocrinol.* **7,** 571.

Enemar, A., Falck, B., and Iturriza, F. C. (1967). *Z. Zellforsch. Mikrosk. Anat.* **77,** 325.

Everett, J. W., and Sawyer, C. H. (1949). *Endocrinology* **45,** 581.

Farquhar, M. G. (1961). *Trans. N. Y. Acad. Sci.* **23,** 346.

Farquhar, M. G. (1969). *In* "Lysosomes in Biology and Pathology" (J. T. Dingle and H. B. Fell, eds.), Vol. 2, pp. 462–482. North-Holland Publ., Amsterdam.

Farquhar, M. G. (1971). *Mem. Soc. Endocrinol.* **19,** 79–124.

Farquhar, M. G., and Rinehart, J. F. (1954). *Endocrinology* **55,** 857.

Fawcett, D. W., Long, J. A., and Jones, A. L. (1969). *Recent Progr. Horm. Res.* **25,** 315.

Fernholm, B. (1972). *Z. Zellforsch. Mikrosk. Anat.* **132,** 451.

Fink, G., and Smith, G. C. (1971). *Z. Zellforsch. Mikrosk. Anat.* **119,** 208.

Flerkó, B. (1966). *In* "Neuroendocrinology" (L. Martini and W. F. Ganong, eds.), Vol. 1, p. 613. Academic Press, New York.

Fontaine, M., and Fontaine, Y. A. (1962). *Gen. Comp. Endocrinol., Suppl.* **1,** 63.

Fontaine, Y. A. (1969). *Acta Endocrinol.* (*Copenhagen*), *Suppl.* **136,** 1.

Forbes, M. S. (1972). *Gen. Comp. Endocrinol.* **18,** 146.

Foster, D. L., Roach, J. F., Karsch, F. J., Norton, H. W., Cook, B., and Nalbandov, A. V. (1972). *Endocrinology* **90,** 102.

Fuxe, K., and Hokfelt, T. (1969). *In* "Frontiers in Neuroendocrinology" (W. F. Ganong and L. Martini, eds.), p. 47. Oxford Univ. Press, London and New York.
Ganong, W. F., Scapagnini, U., Cuello, A. C., and Shoemaker, W. J. (1973). *Endocrine Soc., Annu. Meet. 55th, Chicago,* A-80, Abstr. No. 64.
Ganong, W. F., Biglieri, E. G., and Mulrow, P. J. (1966). *Recent Progr. Horm. Res.* **22,** 381.
Genbačev, O., and Pantić, V. (1972). *Yugoslav. Physiol. Pharmacol. Acta* **8**(3), 309.
Geschwind, I. I. (1969). *In* "Frontiers in Neuroendocrinology" (W. F. Ganong and L. Martini, eds.), p. 389. Oxford Univ. Press, London and New York.
Geschwind, I. I. (1972). *Bien. Symp. Anim. Reprod., 10th,* **34,** Suppl. 1, p. 19.
Geschwind, I. I., Li, C. H., and Barnafi, L. (1956). *J. Amer. Chem. Soc.* **78,** 4494.
Geschwind, I. I., Li, C. H., and Barnafi, L. (1957). *J. Amer. Chem. Soc.* **79,** 1003.
Gewirtz, G., and Yalow, R. S. (1973). *Endocrine Soc., Annu. Meet., 55th, Chicago,* A-53, Abstr. No. 9.
Ginsburg, J. (1971). *Annu. Rev. Pharmacol.* **11,** 387.
Goluboff, L. G., and Ezrin, C. (1969). *J. Clin. Endocrinol. Metab.* **29,** 1533.
Gosbee, J. L., Kraicer, J., Kastin, A., and Schally, A. (1970). *Endocrinology* **86,** 560.
Graber, J. W., Frankel, A. I., and Nalbandov, A. V. (1967). *Gen. Comp. Endocrinol.* **9,** 187.
Grant, G., Vale, W., and Guillemin, R. (1971). *Biochem. Biophys. Res. Commun.* **46,** 28.
Grant, L. D., and Stumpf, W. E. (1973). *J. Histochem. Cytochem.* **21,** 404.
Green, M. A., and Erwin, H. (1954). *J. Clin. Invest.* **33,** 938.
Guedent, S. L. (1972). *J. Microsc.* **14,** 152.
Guillemin, R. (1971). *Advan. Metab. Disord.* **5,** 1.
Halász, B., and Pupp, L. (1967). *Endocrinology* **80,** 608.
Halász, B., and Gorski, R. A. (1967). *Endocrinology* **80,** 608.
Hallahan, C., and Orsi, B. A. (1972). *Gen. Comp. Endocrinol.* **18,** 428.
Harms, P. G., Ojeda, S. R., and McCann, S. M. (1973). *Endocrine Soc., Annu. Meet., 55th, Chicago,* A-134, Abstr. No. 172.
Harris, J. I., and Lerner, A. B. (1957). *Nature (London)* **179,** 1346.
Hendrickson, A. E., Wagoner, N., and Cowan, W. M. (1972). *Z. Zellforsch. Mikrosk. Anat.* **135,** 1.
Herbert, D. C., and Hayashida, T. (1970). *Science* **169,** 378.
Herlant, M., Ectors, F., and Dessy, C. (1972). *C. R. Acad. Sci., Ser. D* **274,** 1183.
Hirono, M., Igarashi, M., and Matsumoto, S. (1970). *Neuroendocrinology* **6,** 274.
Hofmann, K., and Yajima, H. (1962). *Rec. Progr. Horm. Res.* **18,** 41.
Howell, S. L. (1972). *Excerpta Med. Found. Int. Congr. Ser.* No. 256, Abstr. 625.
Ishikawa, H., Ohtsuka, Y., Soyama, F., and Yoshimura, F. (1972). *Endocrinol. Jap.* **19,** 215.
Jacobs, L. S., Snyder, P. J., Wilber, J. F., Utiger, R. D., and Daughaday, W. H. (1971). *J. Clin. Endocrinol. Metab.* **33,** 996.
Jørgensen, C. B., and Vijayakumar, S. (1971). *Gen. Comp. Endocrinol.* **17,** 575.
Jovanović, M., Pantić, V., and Djurdjević, Dj. (1962). *Int. Congr. Mundial Vet., 16th, Madrid,* Abstr. No. 17, p. 45.
Jovanović, M., Pantić, V., Mihallović, M., Sinadinović, J., and Krainćanić, M. (1966). *Congr. Radiobiol., Cortina d'Ampeco,* Abstr. No. 467.
Kalra, P. S., Fawcett, C. P., Krulich, L., and McCann, S. M. (1973). *Endocrinology* **92,** 1256.
Kamberi, I. A. (1973). *Progr. Brain Res.* **39,** 276.

Kamberi, I. A., Schneider, H. P., and McCann, S. M. (1970). *Endocrinology* **86**, 278.
Kamberi, I. A., Mical, R. S., and Porter, J. C. (1971a). *Endocrinology* **88**, 1003.
Kamberi, I. A., Mical, R. S., and Porter, J. C. (1971b). *Endocrinology* **88**, 1294.
Kanematsu, S., and Mikami, S. (1970). *Gen. Comp. Endocrinol.* **14**, 25.
Karsch, F. J., Dierschke, D. J., Weick, R. F., Yamaji, T., Hotchkiss, J., and Knobil, E. (1973). *Endocrinology* **92**, 799.
Kastin, A. J., Viosca, S., and Schally, A. V. (1970). *In* "Hypophysiotropic Hormones of the Hypothalamus" (J. Meites, ed.), p. 171. Williams & Wilkins, Baltimore, Maryland.
Kastin, A. J., Arimura, A., and Schally, A. V. (1971). *Nature (London), New Biol.* **231**, 29.
Kastin, A. J., Schally, A. V., Gual, C., and Arimura, A. (1972). *J. Clin. Endocrinol. Metab.* **34**, 753.
Kato, J., and Villee, C. (1967a). *Endocrinology* **80**, 567.
Kato, J., and Villee, C. (1967b). *Endocrinology* **80**, 1133.
Kelly, P. A., Bedirian, K. N., Baker, R. D., and Friesen, H. G. (1973). *Endocrinology* **92**, 1289.
Knigge, K. M., and Joseph, S. A. (1971). *Neuroendocrinology* **8**, 273.
Knigge, K. M., and Silverman, A. J. (1972). *In* "Brain-Endocrine Interaction; Median Eminence: Structure and Function" (K. M. Knigge, D. E. Scott, and A. Weindl, eds.), p. 350. Karger, Basel.
Knowles, F., and Vollrath, L. (1966). *Z. Zellforsch. Mikrosk. Anat.* **75**, 317.
Korach, K. S., and Muldoon, T. G. (1973). *Endocrine Soc., Annu. Meet., 55th, Chicago,* A-237, Abstr. No. 377.
Kurosumi, K. (1968). *Arch. Histol. Jap.* **29**, 329.
Labrie, F., Barden, N., Poirier, G., and DeLean, A. (1972). *Proc. Nat. Acad. Sci. U. S.* **69**, 283.
Lagios, M. D. (1973). *Gen. Comp. Endocrinol.* **20**, 362.
Leavitt, W. W., Kimmel, G. L., and Friend, J. P. (1973). *Endocrinology* **92**, 94.
Levina, S. E. (1972). *Gen Comp. Endocrinol.* **19**, 242.
Li, C. H. (1962). *Recent Progr. Horm. Res.* **18**, 1.
Li, C. H. (1972). *Proc. Amer. Phil. Soc.* **116**(5), 365.
Li, C. H., and Dixon, J. S. (1956). *Science* **124**, 934.
Li, C. H., and Evans, H. M. (1944). *Science* **99**, 183.
Li, C. H., and Starman, B. (1964). *Nature (London)* **202**, 291.
Li, C. H., and Yamashiro, D. (1970). *J. Amer. Chem. Soc.* **92**, 7608.
Li, C. H., Simpson, M. E., and Evans, H. M. (1940). *Science* **92**, 355.
Li, C. H., Evans, H. M., and Simpson, M. E. (1942). *J. Biol. Chem.* **146**, 627.
Li, C. H., Geschwind, I. I., Levy, A. L., Harris, J. I., Dixon, J. S., Pon, N. G., and Porath, J. O. (1954). *Nature (London)* **173**, 251.
Li, C. H., Meienhofer, J., Schnabel, E., Chung, D., Lo, T., and Ramachandran, J. (1960). *J. Amer. Chem. Soc.* **82**, 5760.
Li, C. H., Meienhofer, J., Schnabel, E., Chung, D., Lo, T., and Ramachandran, J. (1961). *J. Amer. Chem. Soc.* **83**, 4449.
Li, C. H., Barnafi, L., Chretien, M., and Chung, D. (1965). *Nature (London)* **208**, 1093.
Li, C. H., Barnafi, L., Chretien, M., and Chung, D. (1966). *Proc. Pan-Amer. Congr. Endocrinol., 6th, Mexico City, 1965* p. 349.
Li, C. H., Dixon, J. S., Lo, T. B., Schmidt, K. D., and Pankov, Y. A. (1970). *Arch. Biochem. Biophys.* **141**, 705.

Licht, P. (1972). *Gen. Comp. Endocrinol.* **19,** 273, 282.
Licht, P., and Papkoff, H. (1972). *Gen. Comp. Endocrinol.* **19,** 102.
Machlin, L., and Jacobs, L. (1973). *Endocrine Soc., Annu. Meet., 55th, Chicago,* A-243, Abstr. No. 389.
Martinović, P., Pantić, V., Žgurić, M., Martinović, J., and Ivanisević, O. (1969). *Proc. Conf. Eur. Comp. Endocrinol., 5th, Utrecht* p. 108.
Matsuo, H., Baba, Y., Nair, R. M. G., Amimura, A., and Schally, A. V. (1971). *Biochem. Biophys. Res. Commun.* **43,** 1334.
Mattheij, J. A. M. (1970). *Z. Zellforsch. Mikrosk. Anat.* **104,** 337.
Mattheij, J. A. M., Kingma, F. J., and Stroband, H. W. J. (1971). *Z. Zellforsch. Mikrosk. Anat.* **121,** 82.
May, P. B., and Donabedian, R. K. (1973). *J. Clin. Endocrinol. Metab.* **36,** 605.
Mendoza, D., Arimura, A., and Schally, A. V. (1973). *Endocrinology* **92,** 1153.
Mikami, S., Hashikawa, T., and Farner, D. S. (1973). *Z. Zellforsch. Mikrosk. Anat.* **138,** 299.
Milin, R. (1963). *Ann. Endocrinol.* **24,** 255.
Mirecka, J., and Pearse, A. G. E. (1971). *Folia Histochem. Cytochem.* **9,** 365.
Mitnick, M. A., Valverde, R. C., and Reichlin, S. (1973). *Exp. Biol. Med.* **143,** 418.
Monahan, M., Rivier, J., Burgus, R., Amoss, M., Blackwell, R., Vale, W., and Guillemin, R. (1971). *C. R. Acad. Sci., Ser. D* **273,** 508.
Monahan, M., Vale, W., Rivier, C., Grant, G., and Guillemin, R. (1973). *Endocrine Soc., Annu. Meet., 55th, Chicago,* A-145, Abstr. No. 194.
Multon, H. S., and Erickson, J. E. (1970). *Gen. Comp. Endocrinol.* **15,** 484.
Musliner, T. A., Chader, G. J., and Villee, C. (1970). *Biochemistry* **9,** 4448.
Nadler, R. D. (1972a). *Neuroendocrinology* **9,** 349.
Nadler, R. D. (1972b). *Trans. N. Y. Acad. Sci.* **34,** 572.
Nagahama, Y., Nishioka, R. S., and Bern, H. A. (1973). *Z. Zellforsch. Mikrosk. Anat.* **136,** 153.
Nair, R. M. G., Barrett, J. F., Bowers, C. Y., and Schally, A. V. (1970). *Biochemistry* **9,** 1103.
Nair, R. M. G., Kastin, A. J., and Schally, A. V. (1971). *Biochem. Biophys. Res. Commun.* **43,** 1376.
Nakane, P. K. (1970). *J. Histochem. Cytochem.* **18,** 9.
Nakayama, I., and Nickerson, P. (1972). *Amer. J. Anat.* **135,** 93.
Nakayama, I., and Nickerson, P. (1973a). *Amer. J. Pathol.* **71,** 279.
Nakayama, I., and Nickerson, P. (1973b). *Endocrinology* **92,** 516.
Nakayama, I., Nickerson, P., and Skelton, F. R. (1969). *Lab. Invest.* **21,** 169.
Nakayama, I., Nickerson, P., and Skelton, F. R. (1970). *Amer. J. Pathol.* **58,** 377.
Newman-Taylor, A., Branch, B. J., Casady, R. L., and Turner, B. B. (1973). *Endocrine Soc., Annu. Meet., 55th, Chicago,* A-81, Abstr. No. 65.
Nicoll, C. S., and Bern, H. A. (1971). *Lactog Hormones, Ciba Found. Symp.* p. 299.
Nicoll, C. S., and Licht, P. (1971). *Gen. Comp. Endocrinol.* **17,** 490.
Nicoll, C. S., Mena, F., Sanguannoi, H., Tai, M., and Green, S. (1973). *Endocrine Soc., Annu. Meet., 55th, Chicago,* A-250, Abstr. No. 403.
Notides, A. C. (1970). *Endocrinology* **87,** 987.
Notides, A. C., Hamilton, D. E., and Rudolph, J. H. (1972). *Biochim. Biophys. Acta* **271,** 214.
Ojeda, S. R., and McCann, S. M. (1973). *Endocrine Soc., Annu. Meet., 55th, Chicago,* A-136, Abstr. No. 176.
Ojeda, S. R., and Ramirez, V. D. (1972). *Endocrinology* **90,** 466.

Olivereau, M. (1966). *Amer. J. Zool.* **6**, 598.
Olivereau, M. (1969). *Gen. Comp. Endocrinol., Suppl.* **2**, 32.
Olivereau, M. (1972). *Z. Zellforsch. Mikrosk. Anat.* **128**, 175.
Olivereau, M., and Dimovska, A. (1969). *Gen. Comp. Endocrinol.* **13**, 523.
Orias, R., and Libertun, C. (1973). *Endocrine Soc., Annu. Meet., 55th, Chicago,* A-147, Abstr. No. 197.
Pantić, V. (1965). *Proc. Congr. Biol. Gibier, Belgrade,* p. 87.
Pantić, V. (1968a). *Symp. Cell Biol., Belgrade* No. 49.
Pantić, V. (1968b). *Excerpta Med. Found. Int. Congr. Ser.* No. 166, p. 70. Abstr. No. 120.
Pantić, V. (1971). *Glas, Srp. Akad. Nauka Umet., Od. Med. Nauka,* CCLXXXI, **24**, 273.
Pantić, V. (1972). *Symp. Electron Microsc., Istanbul.* p. 133. Abstr. No. 95.
Pantić, V. (1974). *Int. Rev. Cytol.* **38**, 153.
Pantić, V., and Djuričić, V. (1966). *Arch. Biol. Sci.* **18**, 115.
Pantić, V., and Genbačev, O. (1969). *Z. Zellforsch. Mikrosk. Anat.* **95**, 280.
Pantić, V., and Genbačev, O. (1970). *Electron Microsc., Proc. Int. Congr., 7th, Grenoble* p. 569.
Pantić, V., and Genbačev, O. (1972). *Z. Zellforsch. Mikrosk. Anat.* **126**, 41.
Pantić, V., and Kosanović, M. (1973). *Gen. Comp. Endocrinol.* **21**, 108.
Pantić, V., and Škaro, A. (1974). *Cytobiologie* **91**, 72.
Pantić, V., and Stošić, N. (1966). *Citologia* **8**(1), 17.
Pantić, V., and Stošić, N. (1974). *Glas, Srp. Akad. Nauka Umet., Od. Med. Nauka* (in press).
Pantić, V., Stošić, N., Genbačev, O., and Žgurić, M. (1967). *Conf. Eur. Comp. Endocrinol., 4th, Carlsbad.*
Pantić, V., Ožegović, B., Genbačev, O., and Milković, S. (1971a). *J. Microsc. (Paris)* **12**, 225.
Pantić, V., Genbačev, O., Milković, S., and Ožegović, B. (1971b). *J. Microsc. (Paris)* **12**, 405.
Papkoff, H., Gospodarowicz, D., Candiotti, A., and Li, C. H. (1965). *Arch. Biochem. Biophys.* **111**, 431.
Papkoff, H., Gospodarowicz, D., and Li, C. H. (1967a). *Arch. Biochem. Biophys.* **120**, 434.
Papkoff, H., Mahlmann, L. J., and Li, C. H. (1967b). *Biochemistry* **6**, 3976.
Papkoff, H., Sairam, M. R., and Li, C. H. (1971a). *J. Amer. Chem. Soc.* **93**, 1531.
Papkoff, H., Solis-Wallckermann, J., Martin, M., and Li, C. H. (1971b). *Arch. Biochem. Biophys.* **143**, 226.
Pavlović, M., and Pantić, V. (1974). *Arch. Biol. Sci.* **25**(1–2).
Pelletier, G., Peillon, F., and Vila-Procine, E. (1971). *Z. Zellforsch. Mikrosk. Anat.* **115**, 501.
Peter, R. E. (1970). *Gen. Comp. Endocrinol.* **14**, 334.
Pieters, A., and Herlant, M. (1972). *C. R. Acad. Sci., Ser. D* **274**, 3002.
Polenov, A. L. (1950). *Dokl. Akad. Nauk SSSR* **73**(5), 1025.
Polenov, A. L., and Garlov, P. E. (1973). *Z. Zellforsch. Mikrosk. Anat.* **136**, 461.
Quinn, D. L. (1973). *Endocrine Soc., Annu. Meet., 55th, Chicago,* A-254, Abstr. No. 412.
Rajaniemi, H., and Vanha-Perttula, T. (1972). *Endocrinology* **90**, 1.
Ramirez, V. D., and McCann, S. M. (1964). *Endocrinology* **74**, 814.
Rao, B. R., Wiest, W. G., and Allen, W. M. (1973). *Endocrinology* **92**, 1229.

Ratner, A., and Peake, G. T. (1972). *Excerpta Med. Found. Int. Congr. Ser.* No. 256, Abstr. 627.

Ravona, H., Snapir, N., and Perek, M. (1973). *Gen. Comp. Endocrinol.* **20,** 112.

Romeis, B. (1940). *In* "Handbuch der mikroskopischen Anatomie des Menschen" (W. v. Mollendorff, ed.), Vol. 6, Part 3, p. 1. Springer-Verlag, Berlin and New York.

Roux, M. (1971). *Arch. Anat. Microsc. Morphol. Exp.* **60,** 107.

Sage, M., and Bern, H. A. (1971). *Int. Rev. Cytol.* **31,** 339.

Sage, M., and Bromage, N. R. (1970a). *Gen. Comp. Endocrinol.* **14,** 127.

Sage, M., and Bromage, N. R. (1970b). *Gen. Comp. Endocrinol.* **14,** 137.

Sage, M., and Bromage, N. R. (1971). *Gen. Comp. Endocrinol.* **14,** 137.

Saint Girons, H. (1961). *Arch. Biol.* **72,** 211.

Salem, M. H. M., Norton, H. W., and Nalbandov, A. V. (1970a). *Gen. Comp. Endocrinol.* **14,** 270.

Salem, M. H. M., Norton, H. W., and Nalbandov, A. V. (1970b). *Gen. Comp. Endocrinol.* **14,** 281.

Sandow, J., and Locke, W. (1973). *Endocrine Soc., Annu. Meet., 55th, Chicago,* A-119, Abstr. No. 141.

Sawyer, C. H. (1972). *In* "The Neurobiology of the Amygdala" (E. B. Eleftherion, ed.), p. 745. Plenum, New York.

Sawyer, C. H., Markee, J. E., and Hollinshead, W. H. (1947). *Endocrinology* **41,** 395.

Sawyer, C. H., Everett, J. W., and Markee, J. E. (1949). *Endocrinology* **44,** 218.

Schally, A. V., Arimura, A., Bowers, C. Y., Kastin, A. J., Sawano, S., and Redding, T. W. (1968). *Recent Progr. Horm. Res.* **24,** 497.

Schally, A. V., Redding, T. W., Bowers, C. Y., and Barrett, J. F. (1969). *J. Biol. Chem.* **244,** 4077.

Schally, A. V., Baba, Y., Nair, R. M. G., and Bennett, C. D. (1971). *J. Biol. Chem.* **246,** 6647.

Schally, A. V., Arimura, A., and Kastin, A. J. (1973). *Science* **179,** 341.

Scharrer, B. (1967). *Amer. Zool.* **7,** 161.

Schilling, R. M., and Blobel, R. (1973). *Acta Endocrinol.* (*Copenhagen*) **73,** 1.

Schneider, B., Gewirtz, G., Krieger, D., and Yalow, R. S. (1973). *Endocrine Soc., Annu. Meet., 55th, Chicago,* A-52, Abstr. No. 8.

Sheridan, P. J., Sar, M., and Stumpf, W. E. (1973). *Endocrine Soc., Annu. Meet., 55th, Chicago,* A-67, Abstr. No. 38.

Shome, B., and Parlow, A. F. (1973). *Endocrine Soc., Annu. Meet., 55th, Chicago,* A-60, Abstr. No. 23.

Siperstein, E. R. (1963). *J. Cell Biol.* **17,** 521.

Siperstein, E. R., and Miller, K. J. (1970). *Endocrinology* **86,** 451.

Smith, R. E., and Farquhar, M. G. (1966). *J. Cell Biol.* **31,** 319.

Smith, E. R., Johnson, J., Weick, R. F., Levine, S., and Davidson, J. M. (1971). *Neuroendocrinology* **8,** 94.

Spies, H. G., and Clegg, M. T. (1971). *Neuroendocrinology* **8,** 205.

Stachura, M. E., and Frohman, L. A. (1973). *Endocrine Soc., Annu. Meet., 55th, Chicago,* A-119, Abstr. No. 142.

Steggles, A. W., and King, R. J. B. (1972). *Eur. J. Cancer* **8,** 323.

Steinberg, A., and Chowdhury, M. (1971). *Endocrinology* **88,** Suppl., A74.

Stern, M. J. (1972). *Gen. Comp. Endocrinol.* **18,** 439.

Stetson, M. H. (1969). *Amer. Zool.* **9,** 1078.

Stetson, M. H. (1972). *Gen. Comp. Endocrinol.* **19,** 37.

Stetson, M. H., and Erickson, J. E. (1971). *Gen. Comp. Endocrinol.* **17,** 105.

Stoeckel, M. E., Dellmann, H. D., Porte, A., and Gertner, C. (1971). *Z. Zellforsch. Mikrosk. Anat.* **122,** 310.

Stoeckel, M. E., Dellmann, H. D., Porte, A., Klein, M. J., and Stutinsky, F. (1973). *Z. Zellforsch. Mikrosk. Anat.* **136,** 97.

Stošić, N., and Pantić, V. (1966). *Yugoslav. Physiol. Pharmacol. Acta* **2/3,** 231.

Stošić, N., and Pantić, V. (1969). *Annu. Meet. Eur. Soc. Radiat. Biol., 7th, Ulm* No. 139.

Stošić, N., and Pantić, V. (1970). *Arh. Biol. Nauka* **22**(1–4), 67.

Stošić, N., and Pantić, V. (1973a). *Yugoslav. Physiol. Pharmacol. Acta* **9,** 3, 335.

Stošić, N., and Pantić, V. (1973b). *Zb. Biotehn. Fak. Vet.* **10**(1), 45.

Stošić, N., Pantić, V., Pavlović-Hournac, M., and Radivojević, D. (1965). *Yugoslav. Physiol. Pharmacol. Acta* **1**(2/3), 194.

Stošić, N., Pantić, V., Pavlović-Hournac, M., and Radivojević, D. (1969). *Z. Zellforsch. Mikrosk. Anat.* **102,** 554.

Sundberg, D. K., Krulich, L., Fawcett, C. P., Illner, P., and McCann, S. M. (1973). *Proc. Soc. Exp. Biol. Med.* **142,** 1097.

Szentagothai, J., Flerkó, B., Mess, B., and Halász, B. (1968). "Hypothalamic Control of the Anterior Pituitary," 353 pp. Akadémiai Kiadó, Budapest.

Teleisnik, S., and Orias, R. (1965). *Amer. J. Physiol.* **208,** 293.

Tixier-Vidal, A., Kerdelhue, B., Berault, A., Picart, R., and Jutisz, M. (1971). *Gen. Comp. Endocrinol.* **17,** 33.

Tixier-Vidal, A., Chandola, A., and Franquelin, F. (1972). *Z. Zellforsch. Mikrosk. Anat.* **125,** 506.

Vale, W., Rivier, J., and Burgus, R. (1971). *Endocrinology* **89,** 1485.

Vale, W., Brazeau, P., Grant, G., Nussey, A., Burgus, R., Rivier, J., Ling, N., and Guillemin, R. (1972a). *C. R. Acad. Sci., Ser. D* **275,** 2913.

Vale, W., Grant, G., Amoss, M., Blackwell, R., and Guillemin, R. (1972b). *Endocrinology* **91,** 562.

Vale, W., Grant, G., and Guillemin, R. (1973a). *In* "Frontiers in Neuroendocrinology" (W. F. Ganong and L. Martini, eds.). Oxford Univ. Press, London and New York.

Vale, W., Brazeau, P., Rivier, C., Rivier, J., Grant, G., Burgus, R., and Guillemin, R. (1973b). *Endocrine Soc., Annu. Meet., 55th, Chicago,* A-118, Abstr. No. 139.

Van Oordt, P. G. W. J. (1965). *Gen. Comp. Endocrinol.* **5,** 131.

Van Oordt, P. G. W. J. (1968). *In* "Perspectives in Endocrinology" (E. J. W. Barrington and C. B. Jørgensen, eds.), p. 405. Academic Press, New York.

Vellano, C., Mazzi, V., and Lodi, G. (1973). *Gen. Comp. Endocrinol.* **20,** 177.

Vila-Porcile, E. (1972). *Z. Zellforsch. Mikrosk. Anat.* **129,** 328.

Villee, C. A. (1964). *Horm. Steroids; Biochem., Pharmacol., Ther., Proc. Int. Congr., 1st, Milan, 1962* **1,** 375–380.

Villee, C. A. (1968). *In* "Perspectives in Reproduction and Sexual Behavior" (M. Diamond, ed.), p. 237. Indiana Univ. Press, Bloomington, Indiana.

Vullings, H. G. B. (1973). *Z. Zellforsch. Mikrosk. Anat.* **136,** 355.

Watanabe, S., and McCann, S. M. (1968). *Endocrinology* **82,** 664.

Weiner, R., Gorski, R. A., and Sawyer, C. H. (1972). *Int. Symp., Munich, 1971* p. 236.

Weitzman, E., Fukushima, D., Nogeire, C., Roffwargtt, H., Gallagher, T. F., and Hellman, L. (1971). *J. Clin. Endocrinol. Metab.* 33, 14.

Wilber, J. F., and Seibel, M. J. (1973). *Endocrinology* **92,** 888.
Woods, J. E., deVries, G. W., and Thommes, R. C. (1971). *Gen. Comp. Endocrinol.* **17,** 407.
Wright, D. R., and Goodman, A. D. (1973). *Endocrine Soc., Annu. Meet., 55th, Chicago,* A-120, Abstr. No. 143.
Yang, W. H., Sairam, M. R., Papkoff, H., and Li, C. H. (1972). *Science* **175,** 637.
Yates, F. E., Russell, S. M., Dallman, M. F., Hedge, G. A., McCann, S. M., and Dhariwal, A. P. S. (1971). *Endocrinology* **88,** 3.
Yen, S. S. C., Rebar, R., Vandenberg, G., Naftolin, F., Ehara, Y., Engblom, S., Ryan, K. J., Benirschke, K., Rivier, J., Amoss, M., and Guillemin, R. (1972). *J. Clin. Endocrinol. Metab.* **34,** 1108.
Zambrano, D. (1970a). *Z. Zellforsch. Mikrosk. Anat.* **110,** 9.
Zambrano, D. (1970b). *Z. Zellforsch. Mikrosk. Anat.* **110,** 496.
Zambrano, D. (1971). *Gen. Comp. Endocrinol.* **17,** 164.
Zambrano, D., and De Robertis, E. (1968). *Z. Zellforsch. Mikrosk. Anat.* **87,** 409.
Zambrano, D., and Iturriza, F. C. (1973). *Gen. Comp. Endocrinol.* **20,** 256.
Zeleznik, A. J., and Midgley, A. R. (1973). *Endocrine Soc., Annu. Meet., 55th, Chicago,* A-69, Abstr. No. 41.
Žgurić, M., and Pantić, V. (1969). *Bull. Ass. Anatomistes, Congr. 54th, Sofia, Bulgaria.* No. 146, p. 679.
Žgurić, M., and Pantić, V. (1970a). *Yugoslav. Physiol. Pharmacol. Acta* **6**(3), 471.
Žgurić, M., and Pantić, V. (1970b). *Proc. Int. Congr. Radiat. Res., 4th, Evian,* p. 104. Abstr. No. 398.
Žgurić, M., and Pantić, V. (1970c). *Arh. Biol. Nauka* **22**(1–4), 87.
Žgurić, M., Pantić, V., and Bogović, B. (1968). *Arch. Biol. Sci.* **20,** 15.
Žgurić-Hristić, M., and Pantić, V. (1972). *Annu. Meet. Eur. Soc. Radiat. Biol., 9th,* p. 187. Rome.
Žgurić-Hristić, M., and Pantić, V. (1974). *Int. J. Rad. Biol.* **25** (in press).

Fine Structure of the Thyroid Gland

HISAO FUJITA

Hiroshima University, School of Medicine, Hiroshima, Japan

I. Introduction

The thyroid gland (*glandula thyreoidea*, Schilddrüse) was first briefly described by Galen (130–201 A.D.) in *De Voce* (quoted in Rolleston, 1936), and the name thyroid was proposed by Thomas Wharton (1614–1673), in *Adenographia* (quoted in Singer and Underwood, 1962) to describe the organ similar to a shield (*thyreos*, Greek) in shape. The function of this organ was a riddle for a long time, though exophthalmic goiter had been described in Assyrian bas reliefs and in medical scripts of ancient Egypt, China, India, and Rome (pre-Christian era), and a harmful quality of water was thought to be the cause of goiter in Rome (Plinus, 23–79 A.D., *Naturalis Historia*, quoted in Hertz, 1943; Vitruvius, 1st century A.D., On Architecture, quoted in Castiglioni, 1958). The thyroid has been thought to be an organ providing a fluid for the lubrication of the larynx (Galen, 130–201 A.D., quoted in Singer and Underwood, 1962), an organ for keeping the beautiful form of the neck (Leonardo da Vinci, 1452–1519, quoted in Bargmann, 1939), and a gland secreting certain materials into the circulation in an emergency (Ling, 1836). Henle (1841, quoted in Bargmann, 1939), Stannius (1846, quoted in Bargmann, 1939), and Gerlach (1850) thought this organ was a blood vessel gland similar to a lymph node, which was considered to be a lymph vessel gland.

In the nineteenth century the characteristic microscopic structure of this gland was first noted (King, 1836) but, until the middle of that century, the follicles filled with gummy fluid (which were called cells by King) were thought to communicate with one another. Bardleben (1841, quoted in Singer and Underwood, 1962) was the first investigator who stated that each follicle is an independent structure without any intercommunication.

Now it is well known that the thyroid gland secretes thyroid hormones containing iodide. The relationship between iodide and goiter had been considered by many investigators for a long time. Rogerius (1170, *Practica Chirvrgiae*, quoted in Castiglioni, 1958), Prosser (1769), Coindet (1820), Prout (1834), and von Basedow (1840) tried to use iodine preparations such as seaweed and burnt sponges for the treatment of goiter or thyroid disorders. Since Baumann (1895) isolated an iodine-binding amorphous compound from the thyroid gland and called it *Thyrojodin*, iodine has been regarded as an important component of the thyroid tissue. One of the hormones secreted by the thyroid gland was isolated as a crystal containing 65% iodine in December 1914, and called thyroxine by Kendall (1915, 1919). However, the chemical structure proposed ($C_{11}H_{10}O_3NI_3$, thyroxyindol) by him was wrong, and in 1926 Harrington clarified it to be a derivative of tyrosine ($C_{15}H_{11}O_4NI_4$). Thyroxine is also called tetraiodothyronine (T_4), owing to its chemical structure.

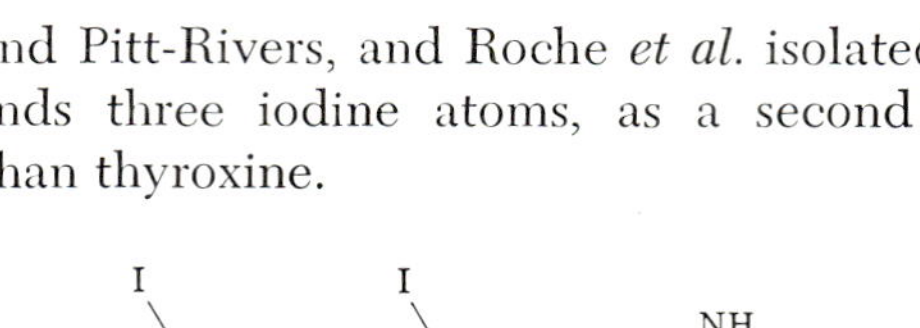

I I NH₂ HO O CH₂—CH—COOH I I

3, 5, 3′, 5′-Tetraiodothyronine (T_4)
= thyroxine

In 1952, Gross and Pitt-Rivers, and Roche *et al.* isolated triiodothyronine (T_3), which binds three iodine atoms, as a second hormone having stronger action than thyroxine.

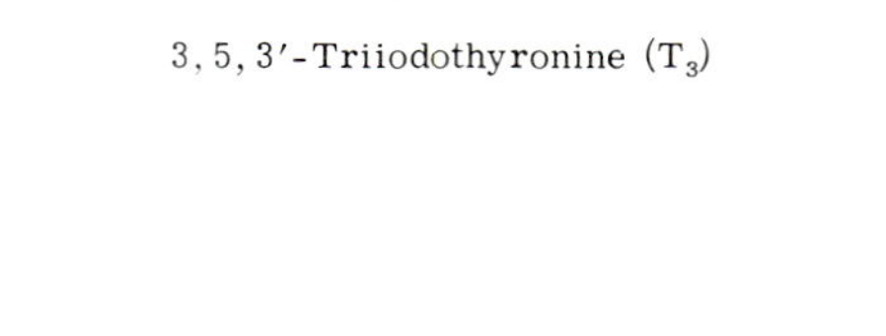

I I NH₂ HO O CH₂—CH—COOH I

3, 5, 3′-Triiodothyronine (T_3)

Although diiodothyronine (T_2) also has very weak hormonal action, little of this material is secreted, and now T_3 and T_4 are generally known to be thyroid hormones. In addition, another kind of thyroid hormone, thyrocalcitonin, has been discovered by Copp *et al.* (1962) and Hirsch *et al.* (1963). Now it is generally accepted that T_3 and T_4 are secreted by the follicular epithelial cell, and thyrocalcitonin by the parafollicular cells.

With the development of electron microscopy, knowledge of the functional morphology of the thyroid has been quickly advanced, and many difficult problems concerning the mechanism of synthesis and release of thyroid hormones have been solved. This article reviews the studies concerning the fine structure and its functional properties in the thyroid gland, except for the parafollicular cell.

II. Fine Structure and Its Functional Properties in the Thyroid of Mammals

A. General View

1. *Outline of the Thyroid Function*

The thyroid gland consists of numerous ball-like structures called follicles, and of interfollicular connective tissues with blood capillaries. Each follicle is composed of numerous follicular epithelial cells arranged as a simple cuboidal epithelium, a lumen surrounded by the epithelial cells, and a few parafollicular cells located singly or in groups in the basal part of the follicular epithelium. Each follicular lumen surrounded by the follicular epithelium is a completely enclosed area, storing colloid materials (thyroglobulin) secreted by the epithelial cells. The secretory process of T_3 and T_4 consists of several complicated steps: (1) synthesis of a large molecular glycoprotein called thyroglobulin in the follicular epithelial cells, (2) release of thyroglobulin into the follicular lumen, (3) iodination of the tyrosyl residues of thyroglobulin, (4) reabsorption of thyroglobulin from the follicular lumen into the follicular epithelial cells, (5) hydrolysis of thyroglobulin for the liberation of T_3 and T_4, and (6) release of T_3 and T_4 from the follicular epithelial cells into the connective tissue space.

Thyrocalcitonin, which is a calcium-lowering agent, is secreted by the parafollicular cells into the connective tissue space.

2. *Outline of the Fine Structure of the Follicular Epithelial Cell*

We present here an outline of the fine structure of the thyroid follicular epithelial cell of higher vertebrates (Figs. 1–3). Since the works of Braun-

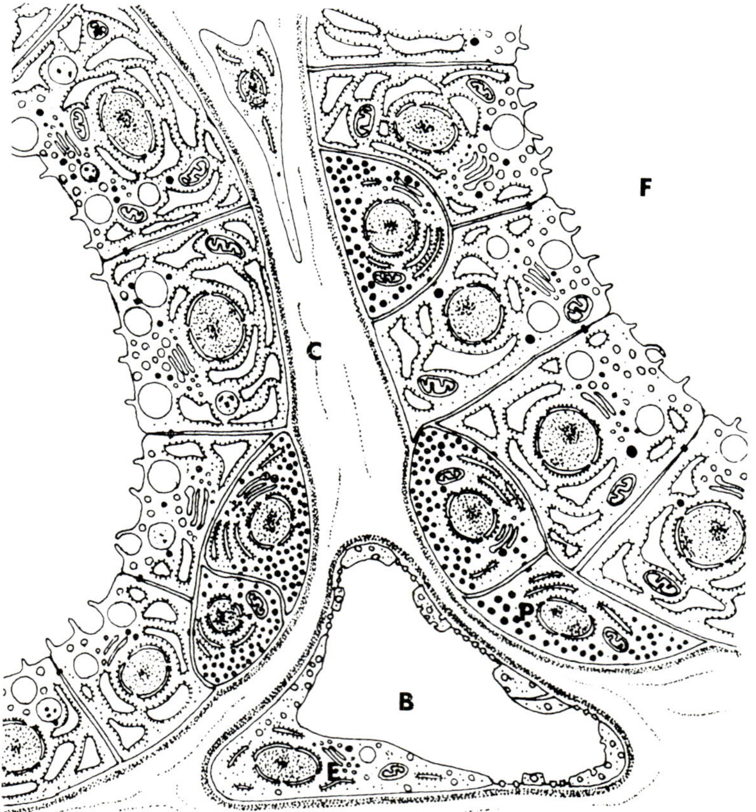

Fig. 1. A schematic representation of part of a mammalian thyroid. B, Blood capillary; C, connective tissue space; E, endothelial cell; F, follicular lumen; P, parafollicular cell.

steiner *et al.* (1953) and Monroe (1953), numerous articles have been published on the electron microscopy of the thyroid, and several excellent reviews deal with this problem (Ekholm, 1964; Wissig, 1964; Lupulescu and Petrovici, 1968; Fujita, 1968). We agree with almost all of these descriptions.

The occurrence of microvilli at the apical surface of the follicular epithelial cell was recognized in the earliest electron microscope studies of this gland made by Braunsteiner *et al.* (1953), Monroe (1953), and Dempsey and Peterson (1955). Recently, we obtained detailed data on the length, diameter, population, and surface area of the microvilli using freeze-etching electron microscopy, as shown in Table I (Fujita *et al.*, unpublished data).

TABLE I

MICROVILLI OF FOLLICULAR EPITHELIAL CELLS OF RATS AND RABBITS IN FREEZE-ETCH ELECTRON MICROGRAPHS[a]

	Number of microvilli per μm^2	Diameter of a microvillus (μm)	Length of a microvillus (μm)	Surface area of a microvillus (μm^2)	Ratio of surface area of plasma membrane including microvilli to that without microvilli
Rat	10.3 ± 1.25*	0.14 ± 0.057	0.44 ± 0.054	0.209	2.99
Rabbit	$9.2 + 1.12$	$0.13 + 0.048$	$0.72 + 0.095$	0.307	3.70

* Standard deviation.

[a] Fujita, Mishima, and Otsuka (unpublished data).

The cytoplasm is characterized by elements of well-developed rough endoplasmic reticulum with somewhat dilated cisternae, located mainly in the basal to lateral parts of the cytoplasm. Most of the basal and lateral cytoplasm is occupied by rough endoplasmic reticulum, and there is little cytoplasmic matrix in this region. Mitochondria, club-shaped or oval, with lamellar cristae, are distributed throughout the cytoplasm, and some of them are located among the elements of the rough endoplasmic reticulum. The Golgi apparatus, consisting of smooth-surfaced vacuoles, lamellae, and vesicles, is generally located in the supranuclear region of the cell. The membrane of the rough endoplasmic reticulum facing the Golgi apparatus is smooth-surfaced (Fig. 2). Small protrusions arising from it and small vesicles 50–100 nm in diameter, which may be derived from these protrusions, are sometimes observed. This phenomenon, which is now considered to be involved in the transportation of materials from the rough endoplasmic reticulum to the Golgi apparatus, has been termed blebbing or budding in many protein-secreting glands. In addition to these structures, cytoplasmic granules have been given attention since the work of Ekholm and Sjöstrand (1957). Near the Golgi apparatus and in the subapical region of the cytoplasm, there are at least three kinds of granules or vesicles: (1) small round, somewhat less dense vesicles 150–200 nm in diameter called subapical vesicles or subapical granules; (2) large colloid droplets 500–4000 nm in diameter; and (3) small highly electron-dense granules 100–300 nm in diameter. Some large colloid droplets are less dense, like the follicular colloid, and other droplets are

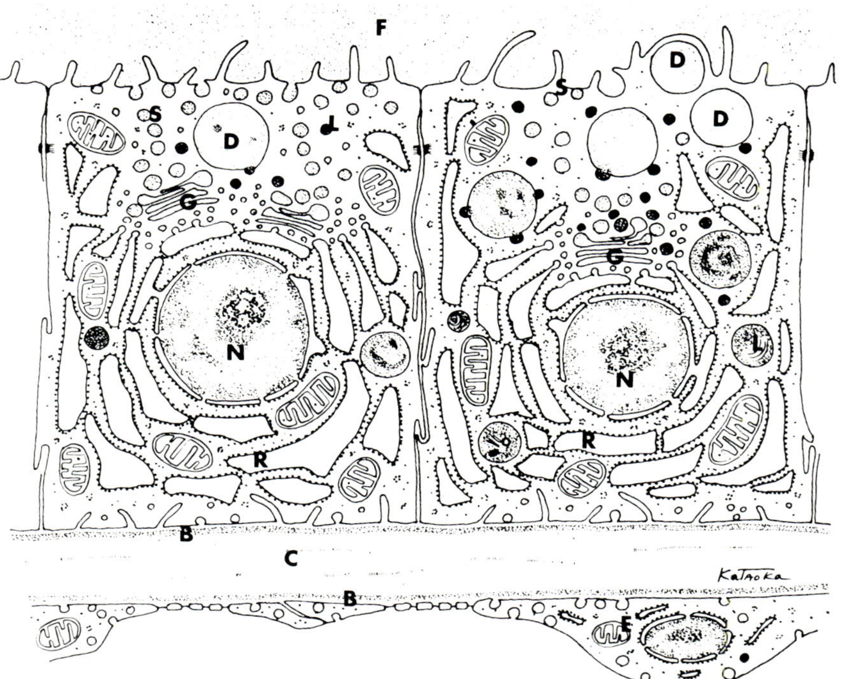

FIG. 2. A schematic representation of follicular epithelial cells of a mammalian thyroid. B, Basal lamina; C, connective tissue space; D, reabsorbed colloid; E, capillary endothelial cell; F, follicular lumen; G, Golgi apparatus; L, lysosome; N, nucleus; R, rough endoplasmic reticulum.

homogeneously or heterogeneously dense. These large colloid droplets are sometimes fused with the small dense granules which might be primary lysosomes. The nature and functional properties of these granules and droplets are described and discussed in the following sections.

Sometimes cytoplasmic processes and pseudopodia containing small and large colloid droplets, protruding from the apical part of the cell into the folliculur lumen, are seen, especially in experimental hyperfunctional conditions (Herman, 1960; Fujita, 1963; Wissig, 1963, 1964; Wetzel *et al.*, 1965; Seljelid, 1967a,b,c,d,e). These cytoplasmic processes and pseudopodia are known to be intimately related to the endocytosis of colloid from the folicular lumen into the cell. Microtubules and microfilaments sometimes found in the apical cytoplasm, especially near the Golgi apparatus, are thought to play a role in the reabsorption of colloid (Nève *et al.*, 1972).

Between the adjacent cells there are a typical tight junction, an inter-

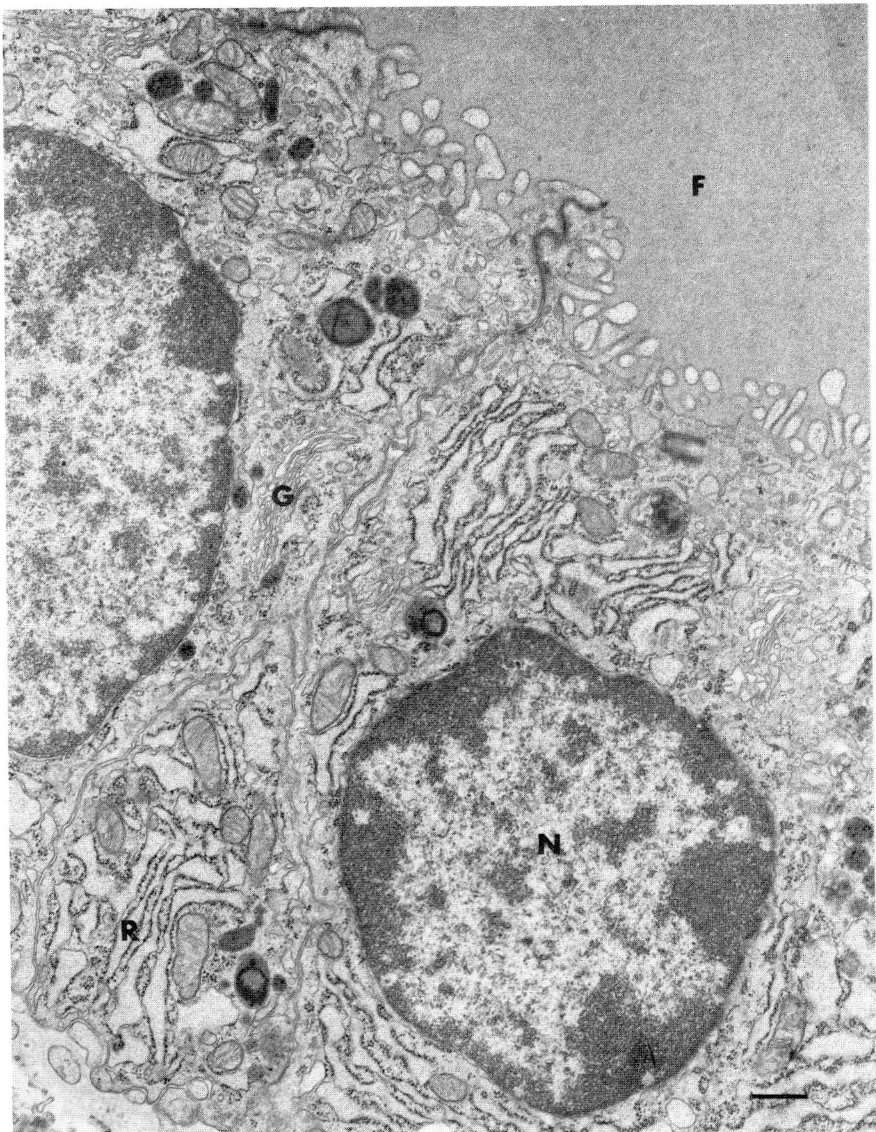

FIG. 3. Parts of follicular epithelial cells of a mouse thyroid. Note well-developed rough endoplasmic reticulum (R), small lysosomelike dense granules located near the Golgi apparatus (G), and large dense bodies. F, Follicular lumen; N, nucleus. ×7000.

mediate junction, and a few desmosomes. Interdigitations and basal infoldings are not well developed. In the following section, the functional properties of all these cytoplasmic structures are discussed in detail.

B. Cell Structure and Its Function

1. *Synthesis of Thyroglobulin (Figs. 4 and 5)*

T_3 and T_4 are not directly synthesized in the follicular epithelial cells. Complicated procedures are necessary to produce these thyroid hormones. First the cell has to synthesize a high-molecular-weight glycoprotein, called thyroglobulin by Oswald (1899).

There are several kinds of subunits of thyroglobulin: 3–6, 12, 19, 27, and 33S; and the 19S protein is thyroglobulin in a strict sense. The 19S thyroglobulin forms about 90% and the 27S about 7% of the synthesized protein. A molecular model for 19S thyroglobulin, the molecular weight of which is about 660,000, has been proposed by Edelhoch (1965). It consists of two equal subunits (12S) (Salvatore *et al.*, 1967), each made up of two 6S subunits of molecular weight 160,000, which are held together by one or a few disulfide bonds (S—S) (Edelhoch, 1965; De Crombrugghe *et al.*, 1966). Thus it is now believed that thyroglobulin has four peptide linkages. Thyroid hormones such as T_3 and T_4 are contained in thyroglobulin, and the hydrolysis of thyroglobulin is needed to liberate them.

As mentioned above, rough endoplasmic reticulum with dilated cisternae is well developed in the cytoplasm of all the follicular epithelial cells. Using numerous kinds of amino acids taken up into the follicular epithelial cells from the connective tissue space, the precursor protein of thyroglobulin is thought to be synthesized on the attached ribosomes of the rough endoplasmic reticulum and stored in its cisternae. Studies using electron microscope autoradiography and biochemistry have clarified that thyroglobulin is synthesized in the rough endoplasmic reticulum (Nadler *et al.*, 1964; Ekholm and Strandberg, 1966, 1967a,b, 1968; Olin *et al.*, 1970; Regard et Mauchamp, 1971; Fujita, 1970; Feeney and Wissig, 1972). Before these investigations this view had already been suggested by electron microscope studies of the thyroid (Dempsey and Peterson, 1955; Fujita *et al.*, 1958, 1963; Wissig, 1960, 1963). The role of mitochondria in the production of proteinaceous secretory substances is also a very interesting problem. From light microscopy some investigators had once thought that mitochondria might be transformed into secretory granules (Takagi, 1922; Kano, 1953). However, no one believes this now. Mitochondria in which ATP is produced are located among the elements of the rough endoplasmic reticulum, and it is clear that the mitochondria

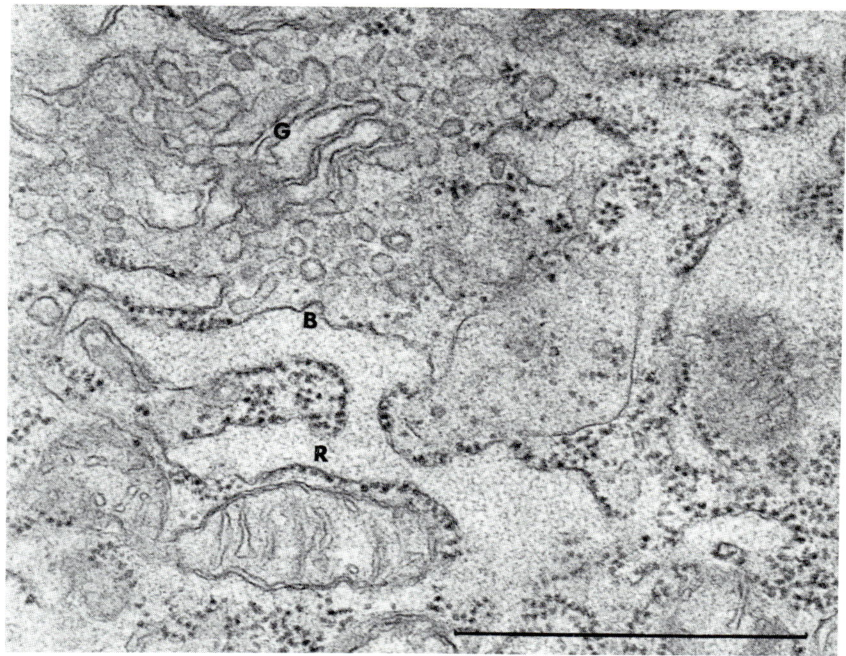

FIG. 4. Part of a follicular epithelial cell of a mouse thyroid. Note the relationship between well-developed rough endoplasmic reticulum (R) and dilated cisternae and Golgi apparatus (G). Note budding (B) from the rough endoplasmic reticulum to the Golgi apparatus. ×58,000.

are the important energy source for synthesizing the exportable protein in the rough endoplasmic reticulum.

The thyroglobulin precursor protein synthesized and stored in the cisternae of the rough endoplasmic reticulum is transferred to the Golgi apparatus. Before the relationship between the rough endoplasmic reticulum and Golgi apparatus was clear, there were many opinions as to the formation of secretory substances. Some investigators believed that the secretory granules arise directly from the rough endoplasmic reticulum by disappearance of attached ribosomes on the membrane of the endoplasmic reticulum (Dempsey and Peterson, 1955; Wang, 1958). Since Zeigel and Dalton (1962) and Palade *et al.* (1962) found blebs or buds arising from the smooth membrane of rough endoplasmic reticulum facing the Golgi apparatus in the exocrine pancreatic cell, the relationship between these organelles has suddenly become clear. The protein stored in the cisternae of the rough endoplasmic reticulum is known to be transported to the Golgi apparatus by the pinching off of small vesicles

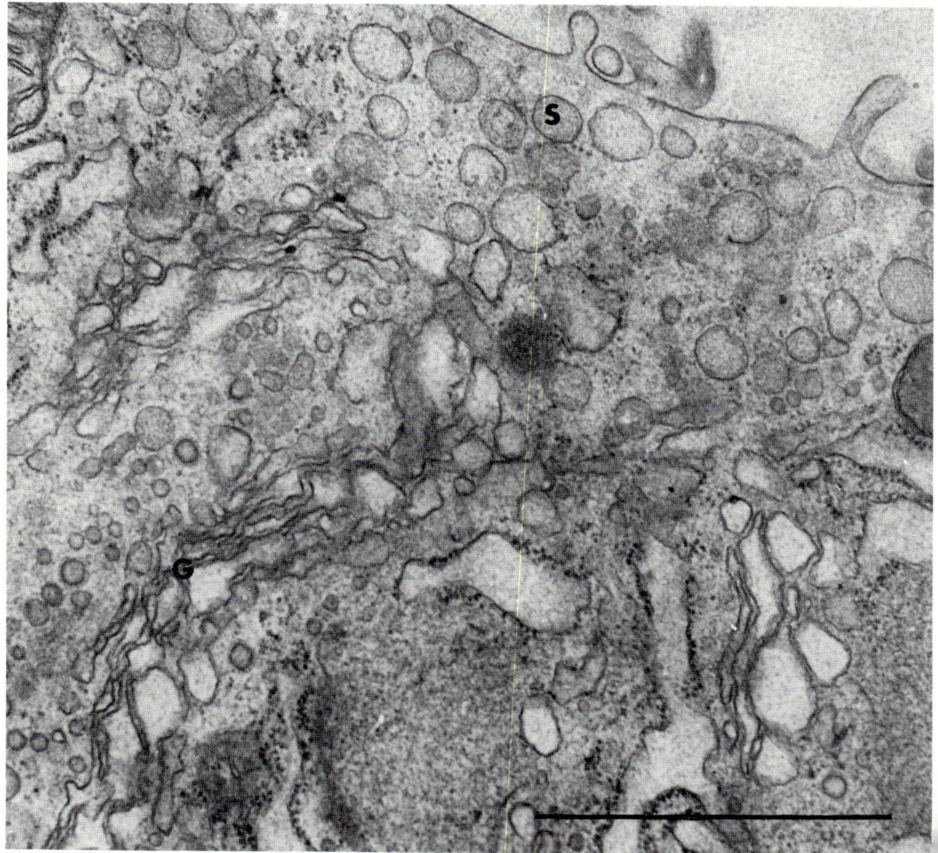

FIG. 5. Part of a follicular epithelial cell of a mouse thyroid. Secretory granules (S) derived from the Golgi apparatus (G) are seen. ×47,000.

from the budding element of rough endoplasmic reticulum facing the Golgi apparatus. The small vesicles are believed to move to the Golgi saccule for fusion with it. Haddad *et al.* (1971) called these small vesicles, 50–70 nm in diameter, the intermediate vesicles. Numerous data using the electron microscope autoradiography of amino-^{3}H acids have contributed to the solution of the problem concerning the synthesis and transportation of secretory substances in many kinds of proteinaceous glandular cells such as the exocrine pancreatic cell (Jamieson and Palade, 1967; van Heyningen, 1964), the parathyroid cell (Nakagami *et al.*, 1971), and the pancreatic B cell (Howell *et al.*, 1969). The protein transferred to the Golgi apparatus has been considered to mature to the secretory materials in the cisternae of the Golgi lamellae or vacuoles in most of the protein-secreting cells mentioned above (Palade *et al.*, 1961; Kurosumi, 1963). The intracellular route for synthesis and transportation of the exportable protein (thyroglobulin) in the thyroid follicular cell is the same as that of the other protein-secreting cells. Using electron microscope autoradiography, Nadler *et al.* (1964) reported that silver grains appear over the endoplasmic reticulum of the rat thyroid cell at 10 minutes after the injection of leucine-4,5-^{3}H, over the Golgi apparatus at 1 hour, and over the subapical vesicles at 3½ hours.

Table II lists data showing the speed of synthesis and transportation of protein in the mouse thyroid cell (Fujita, 1970). This table shows that amino acid has already begun to be incorporated into the synthesized protein in the rough endoplasmic reticulum within 15 minutes after the injection of leucine-^{3}H, and that the secretory granules (subapical vesicles) begin to be produced in the Golgi apparatus after 30–45 minutes and are released into the follicular lumen after 45 minutes.

The Golgi apparatus, which is composed of smooth vacuoles, lamellae, and vesicles, has at least two functions: condensation of proteinaceous substances transported, and binding of carbohydrate components to the protein. As Kurosumi (1963) reported for most protein-secreting cells, the small vesicles derived from the rough endoplasmic reticulum by budding are fused with the Golgi lamellae, and the substances transported by this mechanism are condensed and bound to some other materials as

TABLE II

PERCENT OF GRAIN COUNTS IN VARIOUS CELL ELEMENTS OF MOUSE THYROID EPITHELIAL CELLS AT VARIOUS TIMES AFTER INTRAPERITONEAL INJECTION OF 1 mCi LEUCINE-^{3}H[a]

	15 minutes	30 minutes	45 minutes	1 hour	2 hours	3 hours
Rough endoplasmic reticulum	67.2	55.7	46.5	44.2	34.2	32.3
Golgi apparatus	1.5	7.9	10.2	9.7	6.0	6.0
Subapical vesicles	0.3	6.4	14.3	14.5	14.2	9.7
Large droplets	0.0	0.0	0.1	0.0	0.0	0.4
Plasma membrane and cytoplasmic matrix	12.0	9.7	10.6	10.6	19.0	19.4
Nucleus	8.8	8.8	8.4	9.3	12.0	13.3
Mitochondria and lysosomes	10.2	11.5	12.0	11.7	14.1	18.9
Follicle lumen	−	±	+	+	++	++

KEY: −, negative; ±, trace; +, positive; ++, strongly positive.

[a] About 500 grains were counted, using about 100 electron micrographs of low magnification in each case. Fujita (1970).

a function of the Golgi membrane; then the secretory materials mature in the Golgi lamellae, to be pinched off from them and mature to the independent secretory granule.

It is well known that thyroglobulin is bound to carbohydrate elements to become a glycoprotein. In solving the problem concerning the site of incorporation of carbohydrates into thyroglobulin, the studies using electron microscope autoradiography made by Leblond and his coworkers, for example, Whur *et al.* (1969), Haddad *et al.* (1971), and Haddad (1972), have played an important role. Using rat thyroid lobes incubated in a medium with mannose-^{3}H and galactose-^{3}H, Whur *et al.* (1969) showed that mannose-^{3}H label localizes initially in the rough endoplasmic reticulum of the thyroid follicular cell, by 1–2 hours is transferred to the Golgi apparatus, and at 3 hours and subsequently is present over apical vesicles and follicular colloids; galactose-^{3}H label localizes initially in the Golgi apparatus, rapidly transfers to the apical vesicles, and then to the follicular colloid. In addition, Haddad *et al.* (1971) reported that fucose-^{3}H label appears chiefly over the Golgi saccules at 3–5 minutes, over the apical vesicles at 35 minutes to 1 hours, over the follicular colloid at 4 hours, and over the follicular colloid and reabsorbed colloid droplets at 30 hours after the injection of radioactive sugar into the rat. Obtaining similar results *in vitro*, these investigators concluded that mannose is incorporated into thyroglobulin in the rough endoplasmic reticulum, and galactose and fucose in the Golgi apparatus. The enzymes necessary for these reactions, mannosyl transferase and *N*-acetylglucosaminyl transferase, have already been demonstrated in rough microsomes of sheep thyroid (Bouchilloux *et al.*, 1970), and galactosyl transferase activity has been detected in the Golgi-rich fraction of thyroid cells (Bouchilloux *et al.*, 1969, 1970). The autoradiographic data of Leblond and his coworkers mentioned above agree well with these biochemical data.

2. *Secretory Granules* (*Fig. 6*)

The follicular epithelial cell secretes two different substances in different directions: thyroglobulin which is released into the follicular lumen, and the thyroid hormones such as T_3 and T_4 which are secreted into the connective tissue space and blood capillaries. This section deals with only the former. There are at least three kinds of granules in the follicular epithelial cell, as mentioned in Section II,A: the small less dense granules, the large colloid droplets, and the small dense granules. Before 1964, it was not known which of these was the secretory granule containing thyroglobulin. Wissig (1960, 1963) and Fujita (1963) first considered that the large less dense colloid droplets (500–4000 nm in diameter) might be secretory substances derived from the Golgi apparatus, while Nadler *et al.* (1962), Wollman *et al.* (1964), and Wetzel *et al.* (1965)

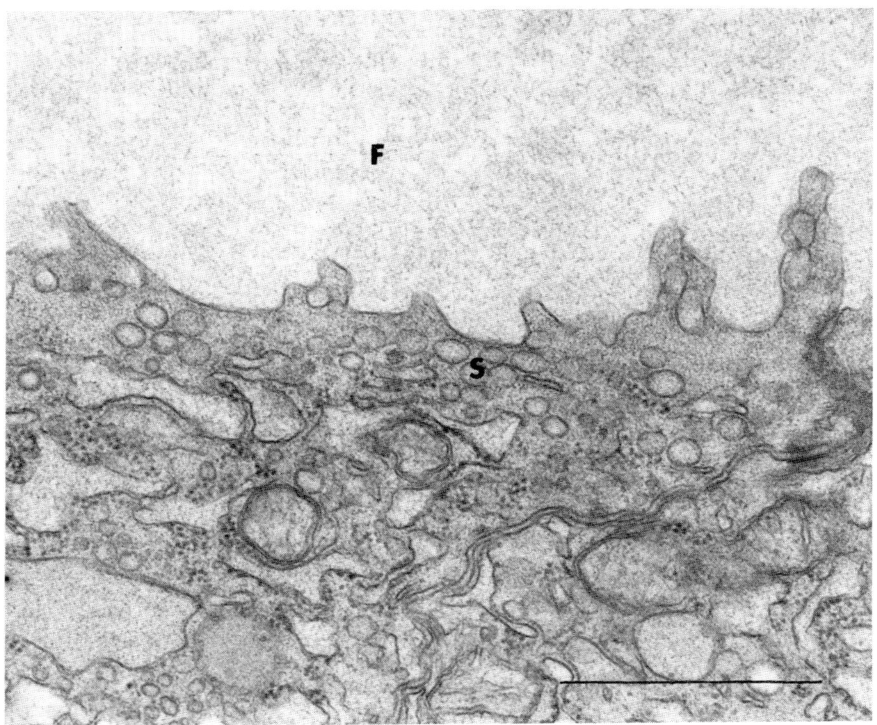

FIG. 6. An apical region of a follicular epithelial cell of a rat thyroid. Note many secretory granules (S) located in the subapical cytoplasm. F, Follicular lumen. ×35,000.

thought that the droplets were not secretory materials but reabsorbed ones. Studies using electron microscope autoradiography have played a great role in solving this problem. As mentioned above, Nadler *et al.* (1964) clearly showed, using the autoradiography of leucine-^{3}H, that the secretory substances consisting of thyroglobulin are present in the small vesicles (150–200 nm in diameter) located in the apical or subapical cytoplasm. In addition, studies using the electron microscope autoradiography of ^{125}I have made it clear that the large colloid droplets are not secretory substances but reabsorbed ones (Sheldon *et al.*, 1964; Stein and Gross, 1964; Bauer and Meyer, 1965; Ekholm and Smeds, 1966; Fujita, 1969). Details are described in Section II,B,4. Thus it is now generally agreed that the secretory substances are contained in the subapical vesicles.

Although it is a fact that most of the apical or subapical vesicles (150–200 nm in diameter) are secretory granules (Fig. 7) containing thyro-

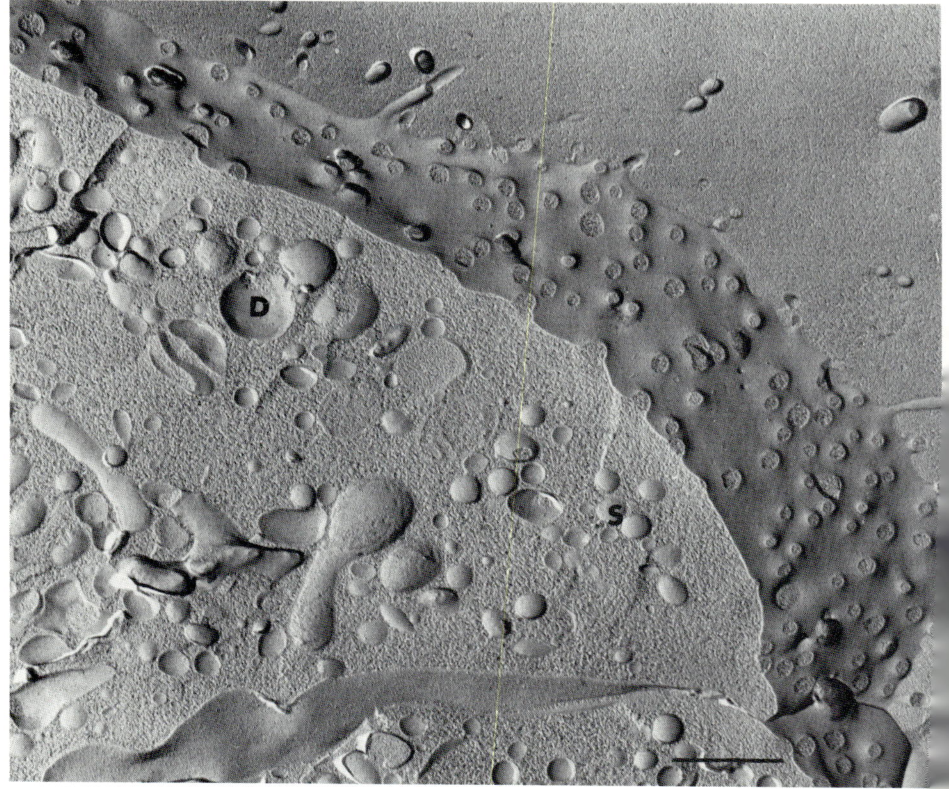

FIG. 7. A freeze-etch electron micrograph of part of a follicular epithelial cell of a rabbit thyroid. Note numerous microvilli at the apical surface and secretory granules (S) in the subapical cytoplasm. D, Colloid droplet. ×15,000.

globulin, it is difficult to determine whether all the vesicles contain secretory materials. The reason is mentioned in Section II,B,4. By freeze-etching electron microscopy also, many secretory granules 150–200 nm in diameter are easily recognized in the subapical and apical cytoplasm. They are sometimes in contact with the apical plasma membrane and are thought to be released into the follicular lumen by reverse pinocytosis, like those of other protein-secreting organs.

3. *Site of Iodination of Thyroglobulin*

The tyrosyl residue in thyroglobulin is iodinated in the thyroid gland. This reaction is termed iodination of thyroglobulin. The site of iodination has been one of the important problems in recent thyroid research, and we have published a detailed review on this subject (Fujita, 1972).

Before this problem is considered, uptake of inorganic iodide into the thyroid is discussed. It has been well known that inorganic iodide is quickly taken up into the thyroid gland from the blood capillaries. It is not clear why iodide ion in the blood capillaries is easily taken up into the follicular epithelial cell after passing through the capillary endothelium, pericapillary space, basal laminae, and plasma membrane. The uptake of iodide ion is called iodide trapping. Although the mechanism of iodide trapping is not clear, active transport (an ion pump) has been proposed as an explanation (Doniach and Logothetopoulos, 1955; Woodbury and Woodbury, 1963), and ouabain-sensitive ATPase seems to be implicated in the taking up of iodide ion into the cell (Wolff and Halmi, 1963). Fujita and co-workers (unpublished data) demonstrated the ATPase reaction in the basal as well as the apical plasma membrane in the follicular epithelial cell of guinea pigs (Fig. 8), using the medium of Farquhar and Palade (1966) and Laird and Yates (1973). The reaction for ATPase in the apical plasma membrane becomes markedly stronger after the injection of TSH. ATPase in the plasma membrane is thought to be necessary for active transport across the membrane, and it is easy to speculate that ATPase activity in the basal plasma membrane of the follicular epithelial cell has a role in taking up iodide ion into the cell. However, it is difficult to determine whether or not the enzymic activity in the apical plasma membrane is related to the transport of inorganic iodide into the follicular lumen. This enzyme might play a role in reabsorption of colloid from the follicular lumen into the cytoplasm, because the activity tends to increase after the injection of TSH. However, the ATPase reaction is specific not only to the plasma membrane of the thyroid cell, and many other kinds of cells show the positive reaction in their plasma membrane. The problem remains as to the mechanism by which iodide ion is accumulated selectively in the thyroid gland.

In studying the localization of inorganic and organic iodide taken up into the thyroid, dry-mounted autoradiography of radioactive iodine using freeze-dried, unfixed, and unembedded sections is a powerful tool. Pitt-Rivers and Trotter (1953) and Doniach and Logothetopoulos (1955) found silver grains for ^{131}I in the follicular cells, as well as in the follicular lumen, using freeze-dried sections, 15 or 30–50 minutes after the injection of ^{131}I into thiouracil-treated or propylthiouracil-treated animals, respectively. Andros and Wollman (1964) demonstrated that the accumulation of inorganic iodide was much higher in the cells of some follicles than in the follicular lumen 5 minutes after the injection of ^{131}I into propylthiouracil-treated mice, and they concluded that the basal plasma mem-

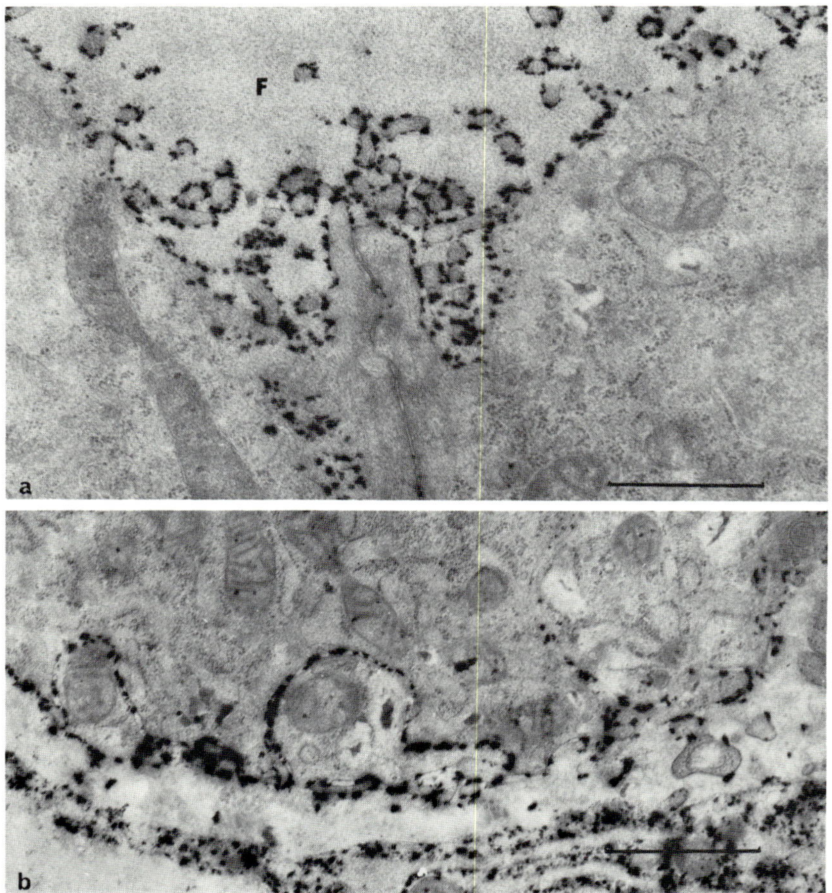

FIG. 8. ATPase reaction of the thyroid follicular epithelial cell of a guinea pig 1.5 hours after injection of TSH (4 IU). Reaction products are localized at the apical (a) and basal plasma membrane (b). F, Follicular lumen. (a) ×24,500; (b) ×15,000.

brane is responsible for iodine uptake and that the apical plasma membrane exerts some control over the passage of radioiodide from the cells into the follicular lumen. Our data, from light microscope autoradiography of freeze-dried sections, show that silver grains for iodide are accumulated chiefly over the follicular lumen 2 hours after the injection of ^{131}I into mercaptoimidazole-treated rats; these silver grains disappear in the fixed, dehydrated, and paraffin-embedded sections obtained from the same animal. This means that the silver grains seen in these freeze-dried autoradiographic sections reveal inorganic iodide. Inorganic iodide taken up into the follicular epithelial cells is transported to the follicular lumen gradually, and so silver grains for inorganic iodide are located over the follicular lumen as well as over the follicular epithelial cells a few minutes after the injection of ^{131}I, as shown by Andros and Wollman (1964); they are localized almost entirely over the follicular lumen a few hours after injection of radioactive iodine, as shown by our studies.

The site of the iodination of thyroglobulin is now discussed. The autoradiography of radioactive iodine, using fixed, dehydrated, and embedded materials, is one of the most effective and important methods of examining the morphological aspects of this problem, and the detection of iodoproteins in cellular fractions of follicular cells from normal or cultured thyroids has been applied for this purpose in biochemical studies. Light microscope autoradiography of radioactive iodine was first performed by Hamilton *et al.* (1940) and Gorbman and Evans (1941). They only recognized radioactive iodine accumulated in the thyroid colloid. Since the studies of Leblond and Gross (1948) and Doniach and Pelc (1949), which were planned to detect the site of iodination of thyroglobulin, numerous investigators have tried to use autoradiography at the light as well as electron microscope level to determine the site of the iodination of thyroglobulin. During the usual autoradiographic procedures, inorganic iodide is almost completely washed away by fixatives and alcohol, and only organic iodide bound to thyroglobulin is detected by this method. So this technique is suitable for localizing only organic iodide in the tissue. Although ^{131}I is the preferred isotope at the electron microscope level; ^{131}I emits high-energy particles, while the energy radiation of ^{125}I is low and more useful for determining fine-structural localization.

a. *Light Microscope Autoradiography.* Leblond and Gross (1948), who were the first to use light microscope autoradiography of ^{131}I to investigate the site of iodination of thyroglobulin, showed that the radioactivity is present mostly in the epithelium, especially in the apical portion of the cell, 1 hour after the injection of radioisotope into rats receiving nonradioactive iodide or into hypophysectomized rats, while radioactive iodine is demonstrated mostly in the follicular lumen as early as 2 minutes after injection into iodine-deficient animals. Although the resolving power of autoradiography was not good at that time, their data seem to suggest the possibility that the iodination of thyroglobulin takes place in the follicular epithelial cells as well as in the follicular lumen. However, Doniach and Pelc (1949), who found radioactive iodide only in the follicular lumen of normal rat thyroids within 10 minutes after injection, showed that iodination might occur in the follicular lumen. Later, Leblond and his co-workers made many excellent studies of this problem, and they believe that iodination occurs mostly in the follicular lumen. With improvements in the autoradiographic technique at the light microscope level, many investigators, such as Nadler and Leblond (1954), Nadler *et al.* (1954), Wollman and Wodinsky (1955), Levenson (1960), van Heyningen and Sandborn (1963), Pitt-Rivers *et al.* (1964), and Lowenstein and Wollman (1967a,b, 1970), detected the site of iodination of thyroglobulin. Most of their articles reported that the follicular lumen is the main site of iodination of thyroglobulin, although there were a few exceptions. The data obtained by Wollman and Wodinsky (1955) are very interesting. They found silver grains always localized in the colloid between 11 seconds and 1 hour after radioiodine injection, especially in the narrow ring at the edge of the luminal colloid at 2 minutes and earlier, and concluded that iodination takes place in the luminal, especially the peripheral colloid. An accumulation of silver grains over the peripheral region of the follicular lumen after the injection of radioiodine is called a ring reaction. Nadler and Leblond (1954) and Nadler *et al.* (1954) also noted the same pattern. Since then many workers have regarded the edge of the luminal colloid as an important area for iodination of thyroglobulin.

However, Levenson (1960), who removed the luminal colloid from normal rat thyroid labeled with ^{131}I *in vivo* for 1 hour, demonstrated that cellular organic iodide was present in the remaining epithelium and also found the ring reaction in the peripheral zone of the luminal colloid of the rat 30 minutes to 1 hour after the administration of ^{131}I. He concluded that the iodination of thyroglobulin begins in the epithelial cells and continues in the follicular lumen. Pitt-Rivers *et al.* (1964), who showed that silver grains are present over the follicular epithelial cells of normal rats 2 hours after the injection of ^{125}I, and of nonradioactive iodide-treated rats 10–15 minutes after the injection of ^{131}I, concluded that thyroglobulin is iodinated in the follicular epithelial cells.

b. *Electron Microscope Autoradiography* (*Fig.* 9). In determining the site of iodination of thyroglobulin at the fine-structural level, electron microscope autoradiography is needed. Since the work of Kayes *et al.* (1962), many investigators have used this method for solving this problem (Stein and Gross, 1963, 1964; Lupulescu *et al.*, 1964; Ibrahim and Budd, 1965; Lupulescu and Petrovici, 1965; Simon and Droz, 1965; Ekholm, 1966; Takano and Honjin, 1968; Fujita, 1969; Tixier-Vidal *et al.*, 1969; Nadler, 1971). Most of them have concluded that the main site of iodination of thyroglobulin is the follicular lumen. Some of their reports are discussed here.

Stein and Gross (1963, 1964), who demonstrated silver grains for organic iodide chiefly over the colloid cell boundary overlying the microvilli of the cells in T_4-treated *mice* 15 minutes after the injection of ^{125}I, and autoradiographic reactions evenly distributed over the follicular colloid in normal or TSH-stimulated mice, postulated that iodination of

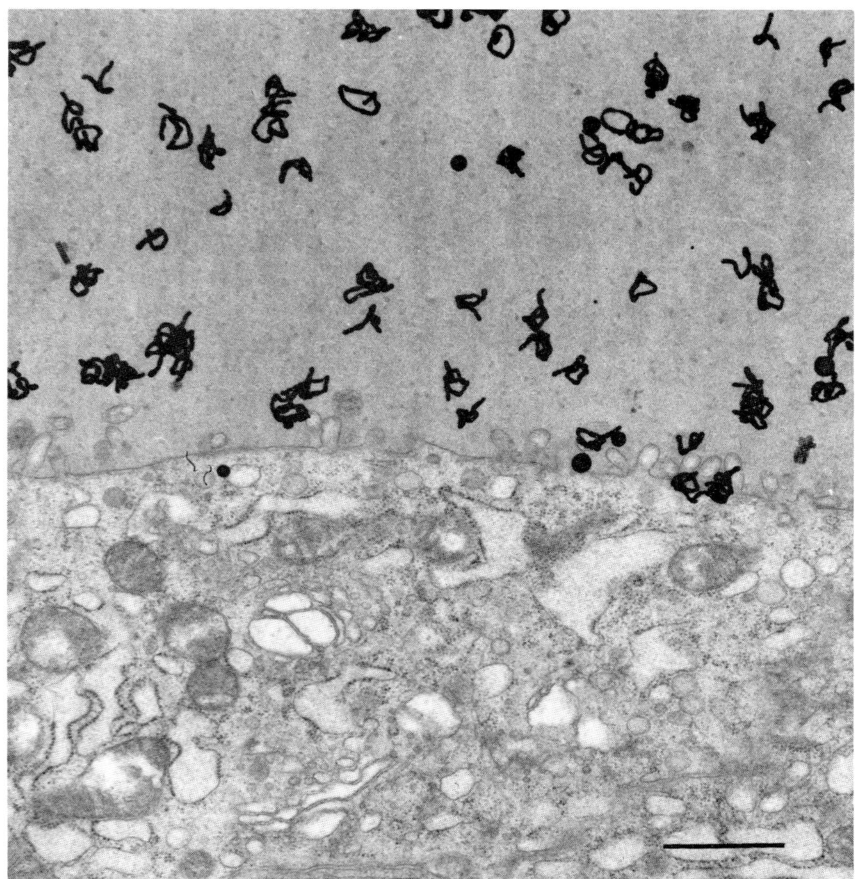

FIG. 9. An electron microscope autoradiograph of a mouse thyroid 1 hour after injection of 200 μCi of ^{125}I. Silver grains are localized over the follicular lumen and the apical plasma membrane region. ×16,000.

thyroglobulin occurs in the follicular colloid in association with the microvilli. Ibrahim and Budd (1965) found the grains randomly distributed over the follicular lumen and not over the cytoplasm in the rat thyroid 3 minutes to 4 hours after the injection of ^{125}I, and concluded that iodine binding occurs initially in the extracellular colloid pool. Their opinions coincide in regard to iodination of thyroglobulin taking place in the follicular lumen. Similar results and views have been reported by Lupulescu and Petrovici (1965), Simon and Droz (1965), Nadler, (1965, 1971), and Fujita (1969). We wish to introduce our data as follows. Silver grains appeared over the follicular lumen of the mouse thyroid

3 minutes after intraperitoneal injection of 200 μCi of ^{125}I and increased in number with time. No grains were detectable over most follicular cells of the mouse, although a few cells showed occasional grains over the cytoplasm, especially over the Golgi apparatus and apical cytoplasm. In some of the follicles, grains were markedly numerous at the apical plasma membrane or peripheral luminal region. From these observations it was generally concluded that the site of iodination of thyroglobulin is not in the cytoplasm but in the follicular lumen and apical plasma membrane region. Although he does not deny the possibility that iodination also takes place in the cytoplasm, Nadler (1965) concluded that the protein moiety of thyroglobulin is secreted by the follicular epithelial cells into the periphery of the follicular lumen, where it is iodinated. However, Ekholm (1966) and Takano and Honjin (1968) stated that iodination of thyroglobulin occurs within the cells, especially in the rough endoplasmic reticulum of guinea pigs and mice, respectively, although numerous grains were seen over the follicular lumen. In spite of the very few silver grains localized over the cytoplasm and the large number of them over the follicular lumen in the micrographs shown by Takano and Honjin (1968), they refused to consider the numerous grains over the lumen. Although Lupulescu and Petrovici (1965) also found quite a few silver grains over the rough endoplasmic reticulum, and Fujita (1969) over the Golgi apparatus, they attached importance to the numerous grains over the follicular lumen. Tixier-Vidal *et al.* (1969) also believe that iodination of thyroglobulin occurs in the elements of the rough endoplasmic reticulum on the basis of their data from sheep thyroid cells isolated by trypsinization and incubated in the presence of ^{125}I. They found a few silver grains autoradiographically over the elements of the rough endoplasmic reticulum in the incubated free thyroid cells without follicle structure, and labeled thyroglobulin by ultracentrifugation of these cells in a sucrose density gradient and by gel filtration on Sephadex G-200. Using 10-, 13-, and 18-day-old chick embryos, whose follicular lumens are not very large in volume, Fujita (1969) found several silver grains over the cytoplasm, especially over the Golgi apparatus, over the apical cytoplasm showing small vesicles, over the rough endoplasmic reticulum, and over the follicular lumen, and considered the possibility that some iodination occurs in the cytoplasm.

To conclude, we wish to emphasize the following. In adult vertebrates whose thyroid follicles are completed, the main site of iodination of thyroglobulin is generally the follicular luminal colloid, especially its peripheral region. However, iodination also takes place partly in the cytoplasm. Data from chick embryos, whose follicle lumens are not very large (Fujita, 1969), and from incubated and dissociated thyroid cells

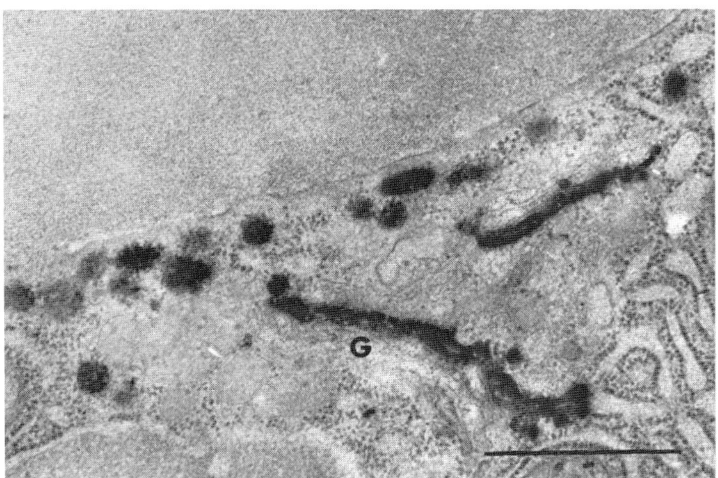

FIG. 10. Peroxidase reaction in the frog (*R. nigromaculata*) thyroid. Reaction products are localized in the Golgi lamellae (G) and subapical vesicles. ×26,000. (From Nanba, 1972b.)

without folliclular lumen (Tixier-Vidal *et al.* 1969), indicate the possibility that iodination also occurs in the cytoplasm, for example, in the rough endoplasmic reticulum, Golgi apparatus, and subapical vesicles. As to the reason why iodination takes place almost entirely in the follicular lumen, we are of the opinion that there is much more thyroglobulin in the follicular lumen than in the cell cytoplasm and that numerous molecules of thyroglobulin in the lumen have not yet been completely iodinated and therefore injected radioiodine combines with luminal colloid preferentially. The histochemical data on the peroxidase reaction discussed in the next section also agree with this view.

c. *Peroxidase Activity* (*Figs. 10 and 11*). It is now believed that the enzyme responsible for iodination of thyroglobulin is a peroxidase (Alexander, 1961; Suzuki *et al.*, 1961; Hosoya *et al.*, 1962; De Groot *et al.*, 1965; Taurog, 1970). Since the works of Serif and Kirkwood (1958) and Alexander (1959), the oxidation of trapped iodide ion occurs using hydrogen peroxide and the enzymic action of peroxidase. Oxidized iodide, whose exact ionic state is unknown, iodinates the tyrosyl residues in thyroglobulin, the reaction being catalyzed by peroxidase. Although another enzyme, tyrosine iodinase, was proposed for this reaction (Fawcett and Kirkwood, 1954; Serif and Kirkwood, 1958; Yip, 1964, 1965), this enzyme has never been purified, and iodination of thyroglobulin takes place through the enzymic action of crystalline chloroperoxidase (Taurog and Howell, 1966), of horseradish peroxidase (Nunez *et al.*, 1966a,b), and of purified peroxidase (Hosoya, 1968). Reactions are summarized as follows.

$$\text{NADH (or NADPH)} + H^+ + O_2 \xrightarrow{\text{peroxidase}} \text{NAD}^+ \text{ (or NADP}^+\text{)} + H_2O_2$$

$$H_2O_2 + I^- \xrightarrow{\text{peroxidase}} \text{Oxidized iodide} + H_2O_2$$

$$\text{Oxidized iodide} + \text{tyrosine} \xrightarrow{\text{peroxidase}} \text{Iodotyrosine}$$

From these observations, the fine-structural localization of peroxidase activity in the thyroid is an important subject in considering the site of iodination of thyroglobulin.

Endogenous peroxidase was first histochemically demonstrated in rat follicular epithelial cells by Dempsey (1944), and in rat follicular luminal colloid by De Robertis and Grasso (1946). The electron microscope histochemistry method for localizing peroxidase using 3,3′-diaminobenzidine tetrahydrochloride (DAB) and hydrogen peroxide, demonstrated by Graham and Karnovsky (1966) and Essner (1969), has been applied to thyroid research. The reaction products of peroxidase activity have been demonstrated in the perinuclear cisternae, cisternae of the rough endoplasmic reticulum, a few Golgi lamellae, and subapical vesicles in the thyroid follicular cell of the rat by Strum and Karnovsky (1970), Nakai and Fujita (1970), and Shin *et al.* (1970). In addition, Strum

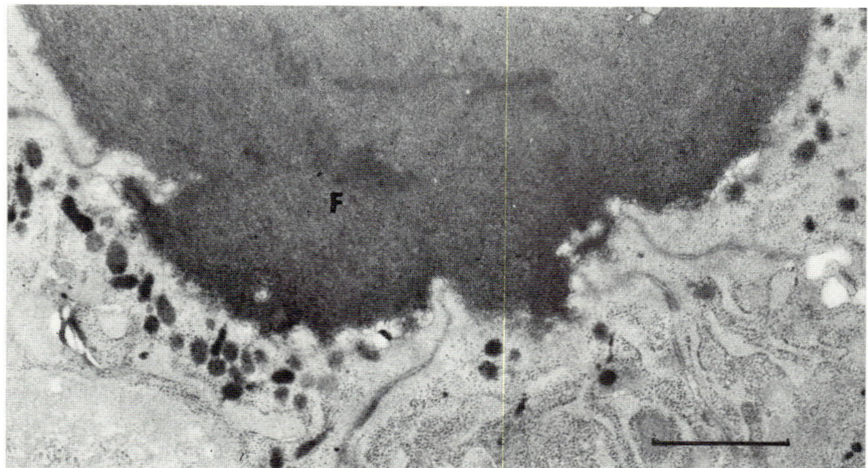

Fig. 11. Same as Fig. 10. Reaction products are localized in the follicular lumen (F) and subapical vesicles. ×18,000.

and Karnovsky (1970), Shin *et al.* (1970), and Fujita (1972) found the positive reaction around the external surface of the microvilli projecting from the apical cytoplasm or in the periluminal region. In the frog (*Rana nigromaculata*), the reaction is positive in the Golgi lamellae, in some apical vesicles, and in the peripheral colloid of the follicular lumen or the whole colloid of the small lumen (Fujita, 1972; Nanba, 1972b, 1973).

However, Hosoya *et al.* (1972) reported that the reaction product is observed only in the membrane of the rough endoplasmic reticulum in pig thyroid cells. Tice and Wollman (1972), who tried preoxidation of thyroid tissue using hydrogen peroxide, glucose-glucose oxidase, or potassium ferricyanide before immersion in DAB solution, showed that the reaction product does not occur in the cisternae of cytoorganelles but is associated with the membrane of cytoorganelles and the apical plasma membrane. These are problems to be solved in the future.

Strum and Karnovsky (1970) believe that the peroxidase protein is synthesized in the rough endoplasmic reticulum and transported to the Golgi apparatus to be condensed, and that the peroxidase activity exists chiefly in the apical vesicles and on the external surface of the microvilli. The peroxidase-positive granules are thought to be released into the follicular lumen by reverse pinocytosis, although it is not clear whether the peroxidase-positive granules are identical or not with the thyroglobulin-containing secretory granules. In rats with goiter experimentally induced by aminotriazole, the peroxidase activity decreases progressively, particularly within the most expanded cisternae of the endoplasmic reticulum, although a few follicles display the activity (Strum and Karnovsky, 1971). Using the thyroid from tadpoles kept in water containing ^{125}I for 2 or 50 hours, Nanba (1973) performed electron microscope autoradiography, after the cytochemical technique for peroxidase localization, for simultaneous demonstration of the peroxidase reaction and silver grains of ^{125}I. She observed that grains were mostly localized over the colloid lumen and a few were over the peroxidase-positive vesicles in the cytoplasm. This suggests that the apical plasma membrane region and the peripheral colloid lumen are the major sites of iodination of thyroglobulin.

It has been shown biochemically that the enzyme for iodine binding is present in the subcellular fractions (or microsome-mitochondrial fractions) (Kondo, 1961; Suzuki *et al.*, 1961; Alexander and Corcoran, 1962; De Groot and Davis, 1962; Hosoya *et al.*, 1962; Klebanoff *et al.*, 1962; Maloof and Soodak, 1964; Alexander, 1965). This microsomal fraction might correspond to the rough endoplasmic reticulum, Golgi vesicles, and subapical vesicles. Benabdeljlil *et al.* (1967), who succeeded in purifying apical particles (which may contain microvilli) by using sheep thyroid, demonstrated the existence of iodide–peroxidase activity in these particles.

Based on all these data from electron microscope cytochemistry and biochemistry, we believe that peroxidase is localized in the rough endoplasmic reticulum, Golgi lamellae, subapical vesicles, and peripheral follicular lumen, and that these might play a role in iodination of thyroglobulin. Nevertheless, autoradiographic studies have shown that iodination of thyroglobulin takes place almost entirely in the follicular lumen, especially in its peripheral region, as mentioned above. As an explanation for the inconsistency between the localization of peroxidase and the site of iodination of thyroglobulin, we emphasize the following. Thyroglobulin in the follicular lumen occurs in a far larger quantity than in the cell cytoplasm, and numerous molecules of thyroglobulin in the follicular lumen may not yet be completely iodinated; therefore injected iodine combines with luminal colloid preferentially. Tice and Wollman (1972) have considered three possibilities: (1) intracellular thyroglobulin is segregated from the enzyme; (2) no hydrogen peroxide or no iodide is available to the enzyme at intracellular sites; (3) intracellular inhibitors are present.

4. *Reabsorption of Colloid* (*Figs. 12–16*)

Thyroglobulin, stored in the follicular lumen and called follicular colloid, is reabsorbed into the follicular epithelial cells by the action of TSH secreted from the adenohypophysis. The reabsorption mechanism has been one of the most important problems in thyroid morphology. As mentioned in Section II,A, there are at least three kinds of granules in the apical cytoplasm of the follicular epithelial cell: small vesicles, large droplets, and small dense granules. It is well known that the large colloid droplets increase in number and accumulate in the apical region of the cell after the injection of TSH (Herman, 1960; Nadler *et al.*, 1962; Wissig, 1963; Fujita, 1963; Wollman *et al.*, 1964; Wetzel *et al.*, 1965; Seljelid, 1967a,b,c). The large droplets were first described by Biondi (1892) light microscopically as a secretion product of the follicular epithelial cells containing a material similar to luminal colloid. They are positive to the periodic acid–Schiff (PAS) reaction (McManus, 1946), low in electron density, and homogeneous like the colloid in the follicular lumen; it is clear that they contain colloid materials (Gersh and Caspersson, 1940). However, it was difficult to determine whether they are derived from the luminal colloid or are released into the follicular lumen. Wissig (1963) and Fujita (1963) considered them secreted materials, while Wollman and Spicer (1961), Nadler *et al.* (1962), and Wollman *et al.* (1964) considered them substances reabsorbed from the follicular lumen into the cytoplasm of the follicular epithelial cells.

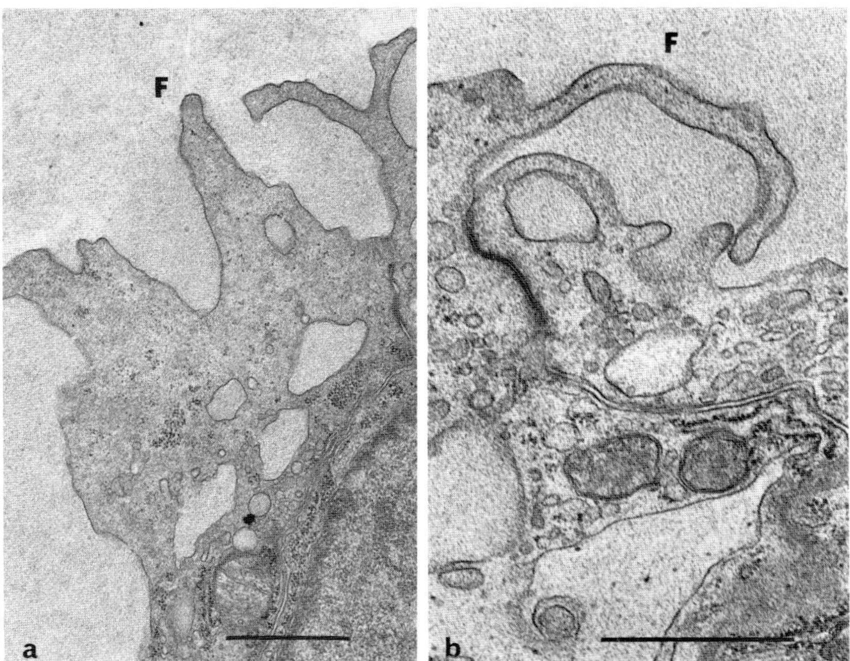

FIG. 12. Pseudopods in the apical region of the thyroid follicular epithelial cell of a bat awakened for 3 hours from long-term hibernation. (a) 3 hours after injection of TSH (1 IU). (b) 3 hours after injection of TSH (1 IU). F, Follicular lumen. (a) ×16,000; (b) ×26,000.

Electron microscope autoradiography has played a great role in solving this problem. The studies done by Sheldon *et al.* (1964), Stein and Gross (1964), Bauer and Meyer (1965), Ekholm and Smeds (1966), Fujita (1969), and Seljelid *et al.* (1970) have shown that the large droplets are not secreted material but reabsorbed material. Injected ^{125}I accumulates in the follicular lumen of the thyroid after several hours to several days. If T_4 is given to this animal after that, the follicle colloid stored is not used for liberation of the thyroid hormone, and radioactive iodine is completely confined to the follicular lumen. When TSH is injected, luminal colloid must be reabsorbed into the cell to be hydrolyzed for the liberation of T_3 and T_4. All the above investigators, who applied this characteristic to determine the origin of intracellular colloid droplets, found silver grains over the large colloid droplets as well as over the follicular lumen a few minutes to 1 hour after the injection of TSH.

A microinjection study made by Seljelid (1967b) has also contributed toward solving this problem. A solution of equine ferritin was instilled

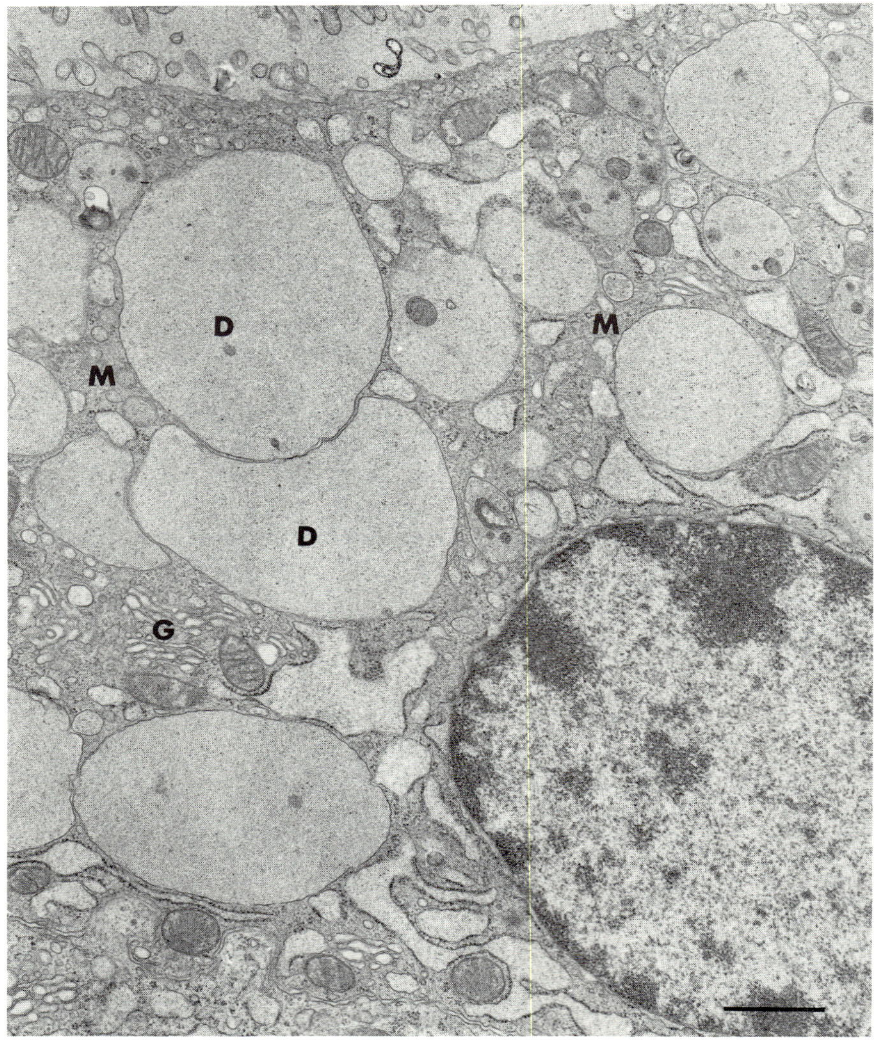

FIG. 13. An accumulation of many reabsorbed colloid droplets (D) in the follicular epithelial cell of a rat 1.5 hours after injection of TSH (1 IU). G, Golgi apparatus; M, microtubules. ×13,500.

into rat thyroid follicles using the technique of microinjection, and then TSH was administered. Ferritin particles were observed in the intracellular colloid droplets 10–20 minutes after the TSH injection. From the data of electron microscope studies using autoradiography as well as

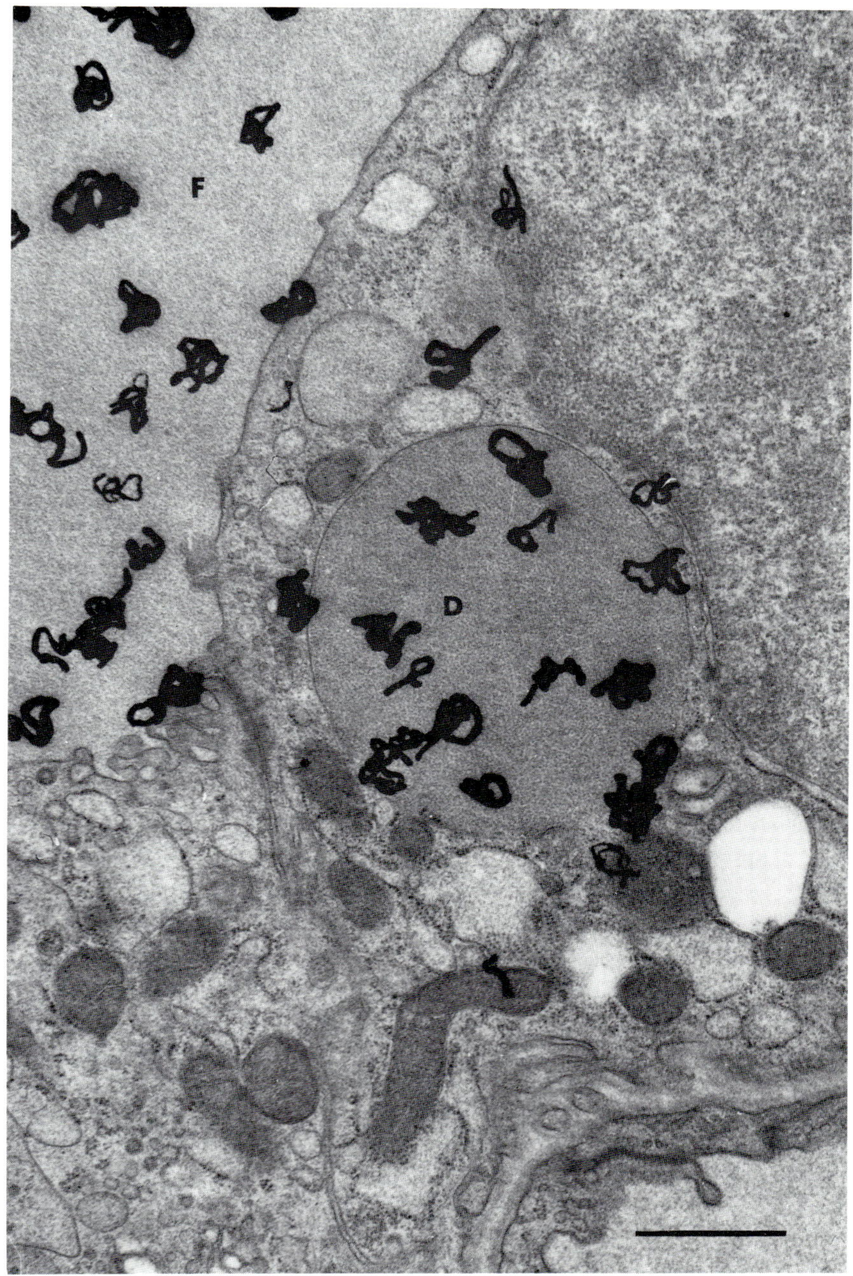

FIG. 14. An electron microscope autoradiograph of a part of the thyroid follicle from the thyroid of a mouse that received 200 μCi of ^{125}I 24 hours prior to fixation and 1 IU of TSH 1 hour before fixation. Silver grains are seen over the follicle colloid (F) and intracellular large droplet (D). ×20,000. (From Fujita, 1969.)

microinjection, it is concluded that intracellular colloid droplets contain material reabsorbed from the follicular lumen.

Problems have arisen concerning the reabsorption mechanism of the colloid. It is well-known that cytoplasmic protrusions like pseudopods appear in the apical cytoplasm of the follicular epithelial cells of the thyroid after the injection of TSH (Ekholm, 1960; Herman, 1960; Wissig, 1963; Fujita, 1963; Wetzel *et al.*, 1965; Lupulescu *et al.*, 1966; Ekholm and Smeds, 1966; Seljelid, 1965, 1966, 1967b,c,d,e). Pseudopods projecting into the follicular lumen are rarely observed in normal animals, often in animals receiving TSH 5 minutes to 2 hours before sacrifice, and in patients suffering from hyperthyroidism. This structure, quite irregular in outline and shape, 0.3–4.0 μm in width, and 0.5–10.0 μm in length, possesses large colloid droplets 0.5–5.0 μm in diameter, small vesicles 0.1–0.5 μm in diameter, and a cytoplasmic matrix containing free ribosomes. Sometimes long, narrow projections extend from the surface of the pseudopod and are though to function in engulfment of colloid. Detailed descriptions of the formation of pseudopods have been presented by Wetzel *et al.* (1965) and Seljelid (1967b,c). Although it was once thought that these might be involved in the release of colloid droplets into the follicular lumen (Wissig, 1963; Fujita, 1963), most investigators now believe them to be involved in engulfment of colloid. Silver grains are localized over the colloid droplets contained in the pseudopod.

However, there is a problem whether or not the colloid is usually reabsorbed only by a mechanism of engulfment, like phagocytosis. Seljelid (1967b) and Seljelid *et al.* (1970) showed, using the microinjection method and autoradiography at the electron microscope level, the possibility that luminal colloid is reabsorbed both by micropinocytosis and phagocytosis. It has been noted in Seljelid's article (1967b) that ferritin particles instilled into the follicular lumen appear, 10–20 minutes after TSH administration, in vacuoles located in pseudopods protruding into the follicle lumen, and in vacuoles, recognizable as colloid droplets, deeper in the cytoplasm; they also appear in small vesicles in the apical cytoplasm, and in bristle-coated pits in the plasma membrane, indicating a micropinocytotic process. In addition, based on the data from electron microscope autoradiography of serial sections and microinjection experiments introducing Thorotrast and colloidal gold into individual follicle lumina, Seljelid and his co-workers (1970) reported that luminal colloid is taken up in small endocytic vesicles which may fuse later to form larger colloid droplets. It was speculated by them that the mass engulfment of colloid by pseudopods cannot be excluded, but that this process plays no major role under physiological conditions. If so, it is not so easy to discriminate between the reabsorbed vesicles and the secretory

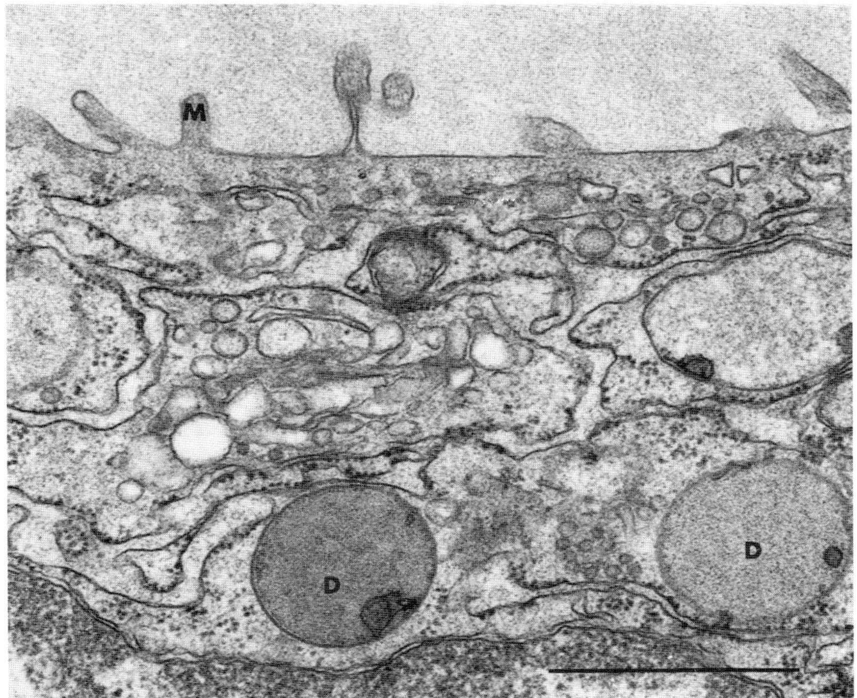

FIG. 15. Large, electron-dense droplets (D) containing heterogeneous substances, in the follicular epithelial cell of a rat thyroid 1 hour after injection of TSH (1 IU). These droplets are considered to be in the process of hydrolysis of the reabsorbed colloid. M, Microvilli. ×39,000.

granules only by their shape, although the large droplets are undoubtedly reabsorbed material. Their ideas are also clear from the data we obtained using electron microscope autoradiography. Silver grains are located over the large colloid droplets, as well as over some of the small vesicles after the injection of TSH, and the many features showing the fusion of small or large colloid droplets with one another are observed. After hypophysectomy, intracellular colloid droplets markedly decrease in number and at last disappear (Wetzel *et al.*, 1965; Seljelid, 1967a; Schwarz, 1967; Fujita and Suemasa, 1968). This means that TSH secreted by the adenohypophysis is necessary for reabsorption of colloid.

The mechanism of TSH action on the reabsorption of colloid is also an interesting problem. An excellent review has been published by Wollman (1969) on the biochemical activity of TSH in the induction of colloid droplets. The concentration of adenosine 3′,5′-monophosphate

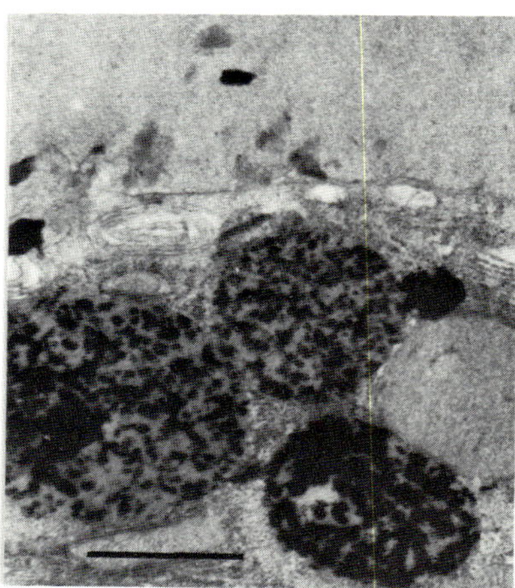

Fig. 16. Acid phosphatase reaction in the follicular epithelial cell of the frog (*R. nigromaculata*). Dense reaction deposits are seen on the large droplets. ×16,600.

(cAMP) in the thyroid gland is promptly elevated after TSH is administered (Klainer *et al.*, 1962; Gilman and Rall, 1966, 1968a,b). Sutherland *et al.* (1965) reported that cAMP is an intracellular mediator responsible for the activation of cells stimulated by hormones. N^6-2′-*O*-dibutyryl adenosine 3′,5′-monophosphate (DB-cAMP), having an action similar to that of cAMP and better penetration of the cell membrane, induces the formation of colloid droplets (Pastan and Wollman, 1967; Nève and Dumont, 1970a) and glucose oxidation (Pastan, 1966; Pastan and Macchia, 1967) in the cells of incubated dog thyroid slices. Based on these data, TSH is thought to stimulate the thyroid cell to increase adenyl cyclase activity and production of cAMP, which might have a function in the reabsorption of colloid. However, it is not clear whether the reabsorption of colloid is controlled by the cell nucleus or not. The data of Ekholm and Elmquist (1967) and Fujita and Suemasa (1968) have shown that actionomycin D may inhibit the endocytic activity of follicular cells induced by TSH injection. Rats treated with T_4 for 4 days for suppression of thyroid (Ekholm and Elmquist, 1967), and hypophysectomized rats (Fujita and Suemasa, 1968), were used to study the appearance of intracellular colloid droplets after the injection of TSH into animals pretreated with actinomycin D or not pretreated. It was shown by Fujita and Suemasa (1968) that cytoplasmic large droplets appeared in hypophysectomized rat thyroid cells within 30 minutes after intraperitoneal injection of TSH and increased in number with time, but these droplets were seldom seen in rats pretreated with actinomycin D. Quite similar data were published by Ekholm and Elmquist (1967). They considered two possibilities in the interpretation of this phenomenon: (1) TSH might act on the nuclear DNA, thereby inducing synthesis or activation of the enzymes responsible for endocytosis, (2) TSH exerts its effect directly on the apical plasma membrane of the follicular cells by activation of an enzyme system residing in the membrane. In addition to these, Fujita and Suemasa (1968) considered another possibility: mRNA is necessary for formation of the cytoplasmic structural protein of the follicular cell cytoplasm which was reduced in quantity by hypophysectomy, and actinomycin D, by blocking the synthesis of this mRNA, suppresses formation of the structural protein and vital activity of the thyroid follicular cells, and so the endocytotic activity induced by TSH treatment is also inhibited. If so, the inhibition of endocytosis is not a direct and specific reaction caused by actinomycin D. The effect of TSH on iodine release from the thyroid gland is also inhibited by pretreatment with actinomycin D 2 days before TSH injection (Halmi *et al.*, 1966; Kriss *et al.*, 1966) and not inhibited by pretreatment shortly before TSH injection (Halmi *et al.*, 1966). Based on these facts, Wollman (1969) concluded that mRNA synthesis is not needed for both droplet formation and iodine release, but that a factor required for the process has a short half-life and presumably the mRNA that codes for it also has a short half-life.

Recently, the function of microtubules and microfilaments has received attention in the transportation and release of the secretory granules in some protein-secreting cells. An article dealing with the relationship between microtubules and microfilaments and the reabsorption of colloid in the thyroid has been published (Nève *et al.*, 1972). In the follicular epithelial cells, the occurrence of microtubules (Seljelid, 1967a; Nève and Wollman, 1971) and microfilaments associated with one another is well known (Nève *et al.*, 1972). According to Nève *et al.* (1972), two kinds of microfilaments, 50 and 100 A wide, occur in the dog thyroid. The former are located especially in the microvillous and subcortical region and in the pseudopods after the injection of TSH, while microtubules are sometimes localized at the base of and inside the pseudopod. It was demonstrated by them (Nève *et al.*, 1972) that, in TSH- or DB-cAMP-stimulated slices, the release of butanol-extractable ^{131}I (iodine and thyroid hormones) and the phagocytosis of colloid are inhibited by cytochalasin B which is the inhibitor of the contractile function of microfilaments, and by colchicine, vinblastine, vincristine, ethanol, and deuterium oxide, which are microtubule inhibitors, and they speculated that microfilaments might be required for the new formation of pseudopods and that microtubules act as scaffolding in the cells and in the pseudopods.

5. *Liberation and Release of Hormones* (*Figs. 17 and 18*)

Reabsorbed colloid droplets must be hydrolyzed for the liberation of the thyroid hormones T_3 and T_4. Lysosomes containing enzymes necessary for hydrolysis play a major role in this reaction. Small dense granules 100–300 nm in diameter, located chiefly in the apical or subapical cytoplasm, are believed to be primary lysosomes derived from the Golgi apparatus. The acid phosphatase reaction often used for identification of lysosomes is positive in these small dense bodies, in heterogeneously or homogeneously dense droplets (Wollman *et al.*, 1964; Wetzel *et al.*, 1965; Seljelid, 1965, 1967a,c; Kosanovic *et al.*, 1968), and in inner Golgi lamellae (Seljelid, 1967a), while less dense droplets considered to be newly reabsorbed are negative to this reaction. In addition, esterase and aryl sulfatase activities have also been demonstrated in small and large dense granules and bodies by Wollman *et al.* (1964) and Seljelid and Helminen (1968), respectively. After the injection of TSH, the small dense granules (primary lysosomes) move to the apical part of the cell rapidly (within a few minutes) to fuse with the reabsorbed colloid droplets, and the lysosomal enzymes mix with the colloid materials (Wetzel *et al.*, 1965; Seljelid, 1967d; Wollman, 1969; Seljelid *et al.*, 1971). Although features showing the intimate contact between the limiting membranes of the colloid droplet and the dense granule are easily recognized, the fusion of both is seldom seen (Wetzel *et al.*, 1965; Ekholm and Smeds, 1966; Seljelid, 1967b), and it might occur very rapidly (Wollman, 1969). Then the colloid droplets change from low to high in electron density, and from negative to positive in acid phosphatase reaction (Wetzel *et al.*, 1965; Seljelid, 1967d; Wollman, 1969). These colloid droplets belong to secondary lysosomes: phagolysosomes.

Seljelid (1967e) reported that the movement of primary lysosomes to the apical cytoplasm is a direct effect of TSH, from the observation that lysosomal movement occurs even if the reabsorption of colloid droplets is inhibited by the intraluminal microinjection of cysteine. However, Kosanovic *et al.* (1968), who found that 1 hour of TSH action does not measurably change the enzyme activities using the cytochemical method for detecting thyroglobulin-hydrolyzing activities, speculated that the mechanism for the prompt increase in hormone release by TSH is not an accelerated biosynthesis of hydrolytic enzymes but an induction of endocytosis, although 24 hours of TSH stimulation resulted in an increase in enzyme activities. Lysosomal bodies are present even in animals

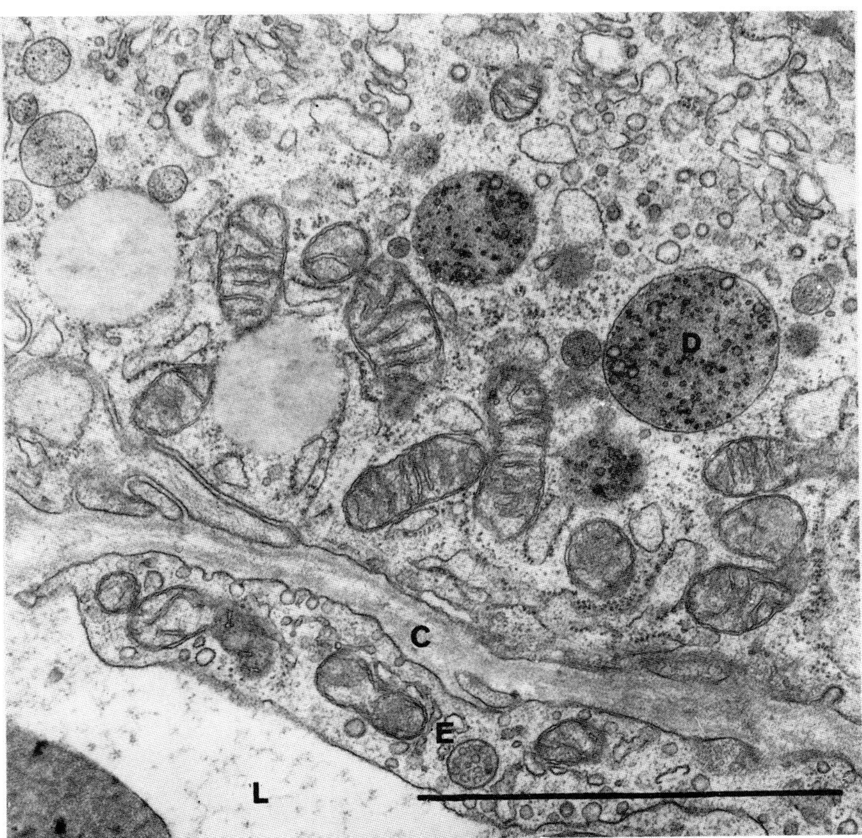

FIG. 17. A basal part of the follicular epithelial cell of the mouse thyroid 1 hour after injection of TSH. Note two heterogeneously dense bodies (D) which might be the reabsorbed colloids in the process of hydrolysis. Small dense granules suggesting primary lysosomes are scattered. C, Connective tissue space; E, capillary endothelial cell; L, capillary lumen. ×56,000.

under experimental hypofunctional conditions. In the suppressed thyroid cells of prolonged T_4-treated rats (Ekholm and Elmquist, 1967; Seljelid *et al.*, 1971) and of long-term hypophysectomized rats (Fujita and Suemasa, 1968), lysosomal dense bodies are seen.

Through hydrolysis of thyroglobulin taking place in the dense colloid droplets fused with lysosomes, T_3, T_4, and other amino acids are liberated. Dense membranous fragments, lamellar structures, and vacuoles are often seen in some dense colloid droplets and might be products of the hydrolyzing process.

Liberated hormones such as T_3 and T_4 are released into the connective tissue space from the basal part of the cell, and parts of monoiodotyrosine, diiodotyrosine, and other amino acids produced by hydrolysis of thyroglobulin are reutilized for new synthesis of thyroglobulin. T_3 and T_4, which are amino acid derivatives and 651 and 777 in molecular weight, respectively, do not seem to have any granular structures, because they are too small in molecular size as compared with other proteinaceous secretory materials having granules. So it is difficult to detect the pathway of the liberating hormones in the cytoplasm. They are speculated to be transported through the cyoplasmic matrix located in the well-developed rough endoplasmic reticulum. At 1 hour after the injection of TSH into a mouse preinjected with ^{125}I 24 hours before, silver grains are seen over the follicular lumen, over the reabsorbed colloid droplets, over the cytoplasmic matrix, over the basal plasma membrane, and over the pericapillary space (Fujita, 1969). Similar data were reported by Sheldon *et al.* (1964) and Stein and Gross (1964). This suggests the possibility that the silver grains found over the cytoplasmic matrix, over the basal plasma membrane, and over the pericapillary space mean the occurrence of T_3 and T_4, although they are amino acids which are easily washed away and moved by the procedures of fixation, dehydration, and embedding (Sheldon *et al.*, 1964; Stein and Gross, 1964; Fujita, 1969). If T_3 and T_4 are present in the cytoplasmic matrix without any granular structures and are released through the plasma membrane without making any visible structural change, the release mechanism of the thyroid hormones from the thyroid cell to the pericapillary space might correspond to a specific diacrine or transfusion type, namely, Kurosumi's fifth type (Kurosumi, 1961). The release mechanism of the thyroid hormone is a problem to be solved in the future.

The fate of the altered colloid droplet is also a problem for the future. The droplets become smaller in size and higher in electron density and migrate toward the basal cytoplasm, and the heterogeneous contents of many dense bodies may consist of the residuum from catabolic processes. Wetzel *et al.* (1965) speculated that the enzymes involved with these altered lysosomes, which may correspond to residual bodies, are reutilized by freshly incorporated colloid droplets.

We now describe the crystals appearing in the colloid droplets in some normal and experimental animals. By light microscopy needlelike cytoplasmic crystals were found in normal opossum thyroid by Bensley (1914) and Bargmann (1939). The same structures were seen electron microscopically in thyroidal colloid droplets of normal rats and chicks, of TSH-, vasopressin-, oxytocin-, and dicarbamate-treated rats (Yoshi-

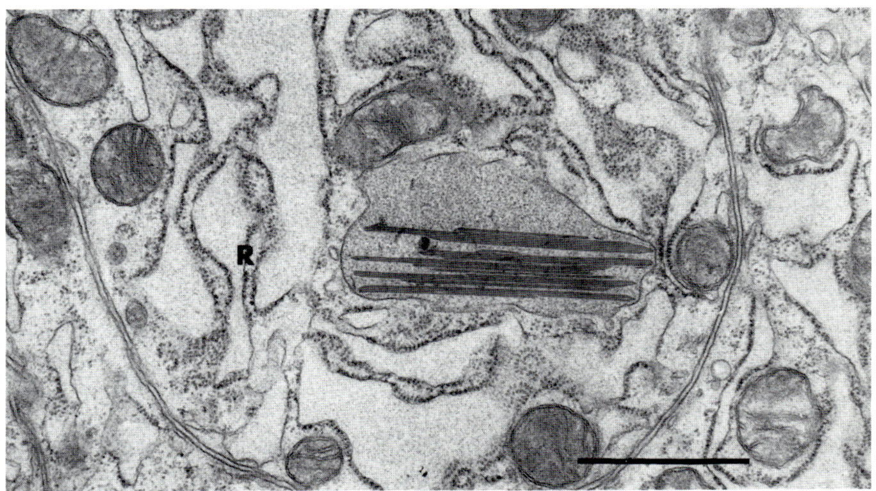

FIG. 18. Crystals in the colloid droplet of the follicular epithelial cell of a normal mouse thyroid. R, Rough endoplasmic reticulum. ×23,000.

mura and Irie, 1961), of TSH-treated hibernating bats (Fujita, 1971), of normal mice (Fujita, 1968; Elfvin, 1971), of old rats (Youson and van Heyningen, 1968), of normal teleosts (Fujita and Machino, 1965), and of TSH-treated chimaeroid fishes (Nakai and Gorbman, 1969).

The fine structures of these crystals are somewhat different from one another. Those in the teleost *Seriola quinqueradiata* consist of aggregates of regularly arranged needle-shaped filaments, and each filament is about 110–120 A in thickness and composed of three layers with two dense layers, each about 35–40 A in width and separated by a less dense layer of the same width (Fujita and Machino, 1965). The crystals in mice consist of a row of parallel dense and light lines of about the same thickness, and the periodicity of the line pattern is about 70 A (Elfvin, 1971). The hibernating bat thyroid a few hours after receiving TSH shows a similar structure (Fujita, 1971), and the chimaeroid fish *Hydrolagus colliei* has aggregates of irregularly arranged, curved fine filaments in the colloid droplets 1–3 days after injection of TSH.

The crystals appear mostly in the case of injection of a thyroid stimulating agent. Elfvin (1971) noted that the reaction product for acid phosphatase is present around but not directly over regions that show the crystalline structure. Similar results were obtained by Nakai and Gorbman (1969). These observations suggest that this structure might be a modified colloid droplet (Yoshimura and Irie, 1961) or altered thyroglobulin derived from a hydrolytic process (Nakai and Gorbman, 1969; Fujita, 1971), although the fine structural analysis offers no information about the origin of the crystalline material (Elfvin, 1971).

C. Blood Capillaries and Nerve Fibers

Among numerous follicles are interfollicular loose connective tissue elements containing blood vessels and nerve elements.

It is well known that endocrine glands are supplied with rich vascularization for taking up raw materials necessary for hormone synthesis and for transporting the released hormone. The outer surface of each thyroid follicle is encapsulated by a capillary network (Billroth, 1882). A scanning electron microscope study using corrosion casts prepared with methylmethacrylate (Murakami, 1971) reveals that the blood capillaries derived from the interlobular or interfollicular arteries make a complex network encapsulating each follicle like a basket (Fujita and Murakami, 1974). In the dog and rat the wall of each basket, consisting of the capillary network, is often common with that of the adjacent follicles, while in the monkey the capillary network of one basket is completely independent of that of the adjacent ones (Fujita and Murakami, 1974). (See Fig. 19.)

The fine structure of the thyroid blood capillary has been described by many investigators (Ekholm, 1957; Wissig, 1960; Fujita, 1963; and others). Between the basal plasma membrane of the follicular epithelial cell and that of the capillary endothelial cell, there are two layers of basal lamina (basement membrane) and a pericapillary space (connective tissue space) containing connective tissue cells, fibrils, and tissue fluid. The endothelial wall shows numerous fenestrations (pores) 400–500 A in diameter, which are closed by a thin diaphragm. The diaphragm is not trilaminar but monolayered, and there is a ringlike structure in its central area. The overall diameter of the ringlike structure is about 200 A, and the light center is about 70 A in diameter. This structure is common in the thyroid as well as in all other endocrine glands of higher vertebrates. Elfvin (1965) and Fujita and Kataoka (1969) described the fine structure of the diaphragm in the rat adrenal medulla and adenohypophysis, respectively. The tissue fluid in the connective tissue space is derived from the blood plasma, which might be transferred through the capillary endothelium, and the thyroid hormones released from the basal part of the follicular epithelial cell into the pericapillary space through the basal lamina must enter the capillary lumen through the basal lamina and endothelium. Hashimoto and Hama (1968) reported that exogenous peroxidase particles injected into the blood vessel pass through the central ringlike structure of the endothelial fenestrae and

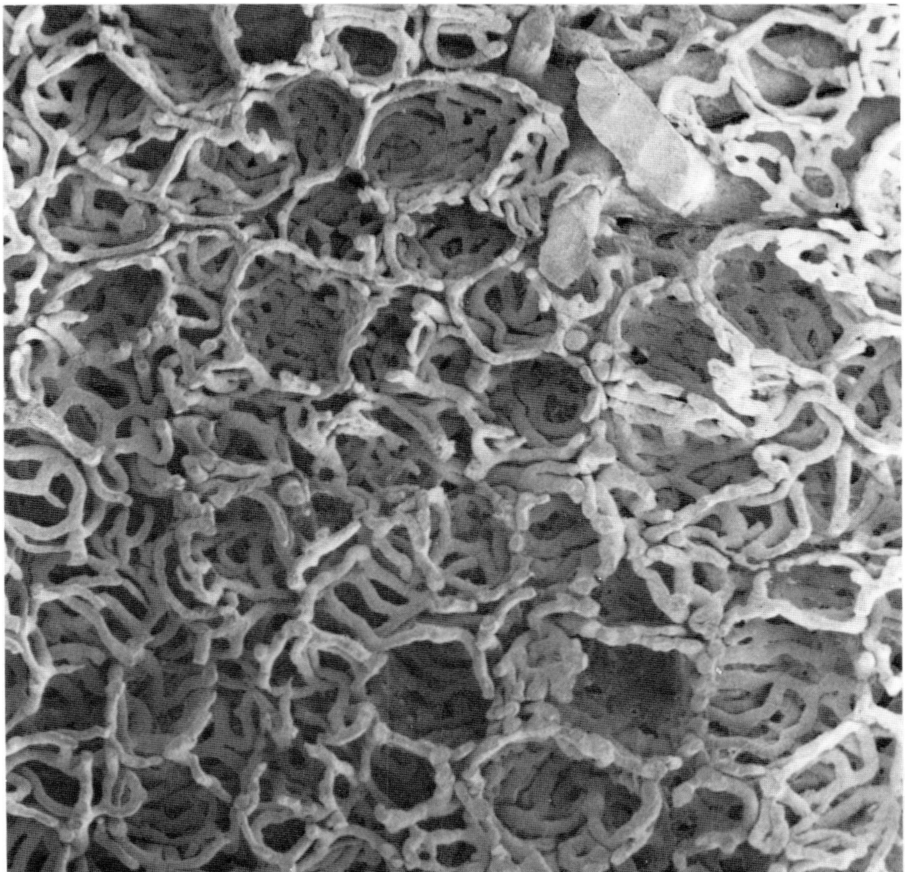

FIG. 19. Vascular cast of the dog thyroid seen under the scanning electron microscope. The interfollicular blood capillaries make a complex network encapsulating each follicle. ×180. (From Fujita and Murakami 1974.)

enter the pericapillary space across the basal lamina in the choroid plexus and area postrema of the cerebrum. Based on this study, it is possible that the endothelial fenestrae are one of the transendothelial pathways and that the central ringlike structure has a special function. Many pinocytotic vesicles found in the endothelial cell might also have a role in transportation of the hormone through the endothelium.

When acid dyes such as trypan blue or lithium carmine are injected into animals, phagocytes belonging to the reticuloendothelial system are stained blue or red, respectively. In the thyroid, some of the connective tissue cells (Fig. 20) located in the pericapillary space are stained well

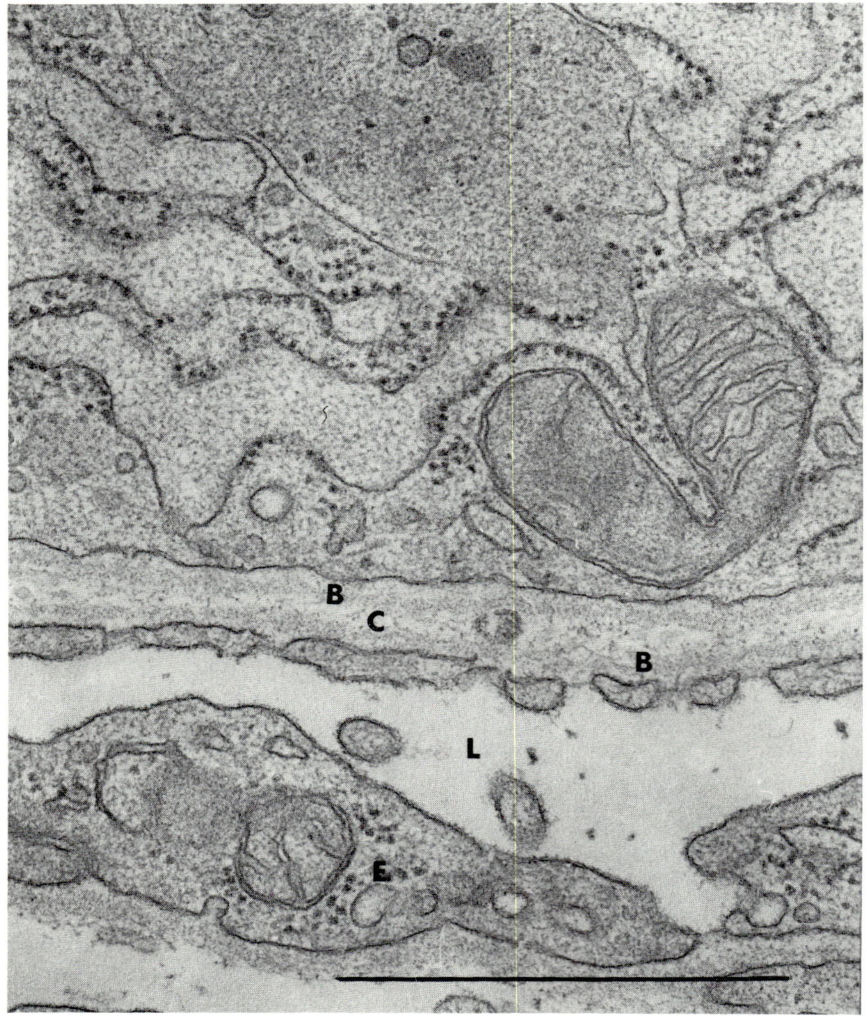

FIG. 20. Basal part of the follicular epithelial cell and a blood capillary of the mouse thyroid 1 hour after injection of TSH. B, Basement membrane; C, connective tissue space; E, endothelium; L, capillary lumen. ×60,000.

vitally and show many lysosomes in their cytoplasm, while the endothelial cells do not react so markedly. This phenomenon is observed in all other endocrine organs. These pericapillary connective tissue cells are histiocytes having phagocytic activity, and connective tissue fibrils such as reticular and collagen fibrils are thought to be produced by pericapillary fibrocytes. The capillaries are sometimes located in the basal part of the follicular endothelial cell, and are termed endocellular or endoepithelial capillaries in the mouse (Bucher, 1940; Matthaes, 1972) and in the rat (Bucher and Riedel, 1967). In this case there are a narrow pericapillary space and basal laminae between the plasma membranes of the follicular epithelial cell and of the capillary endothelium. In some parts of the thyroid the pericapillary space is discontinuous and the two basal laminae are fused. (See Fig. 21.)

The thyroid is also supplied with lymph vessels, whose origin is not clear. The tissue fluid in the pericapillary space might enter the lymph vessel after passing through its endothelium. There is a possibility that the thyroid hormones enter partly into the lymph vessel (Kotani *et al.*, 1968).

Nonmyelinated nerve fibers, each of which is enclosed by Schwann cell cytoplasm, are sometimes seen in the pericapillary space (Brettschneider, 1963). These are thought to be sympathetic fibers derived from the cervical nerve, or parasympathetic fibers from the vagus nerve. It is very difficult to find the nerve terminal that is in direct contact with

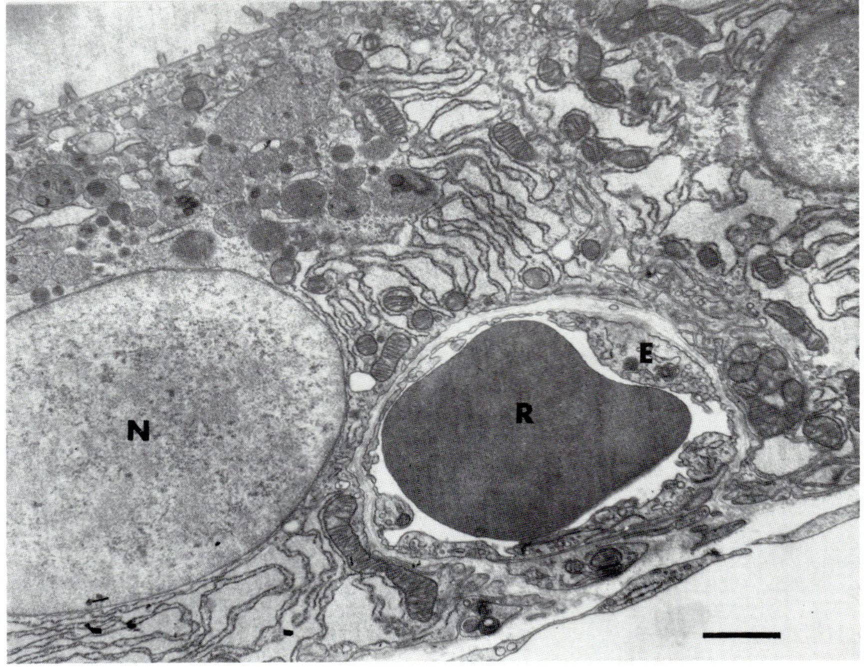

FIG. 21. Intraepithelial blood capillary in the mouse thyroid. E, Endothelium; N, nucleus of the follicular epithelial cell; R, erythrocyte. ×9800.

the follicular epithelial cell, like that in the adrenal cortex, while numerous synaptic contacts are seen in the adrenal medulla and pancreatic islet. The thyroid stimulated humorally by TSH and the adrenal cortex stimulated by ACTH do not seem to be directly controlled by autonomic nerve fibers. Nerve terminals containing noncored or cored synaptic vesicles are sometimes found in the pericapillary space. Although the smooth muscle of the interfollicular arteriole is in contact with the nerve terminal, the direct synaptic contact at the capillary endothelium is also very hard to locate. There are a narrow space and basal laminae between them, although Brettschneider (1963) regarded this relationship as the synaptic contact. The transmitter might be released into the pericapillary tissue fluid from the nerve terminal, and then indirectly stimulate the follicular epithelial cell and the capillary endothelium. (See Fig. 22.)

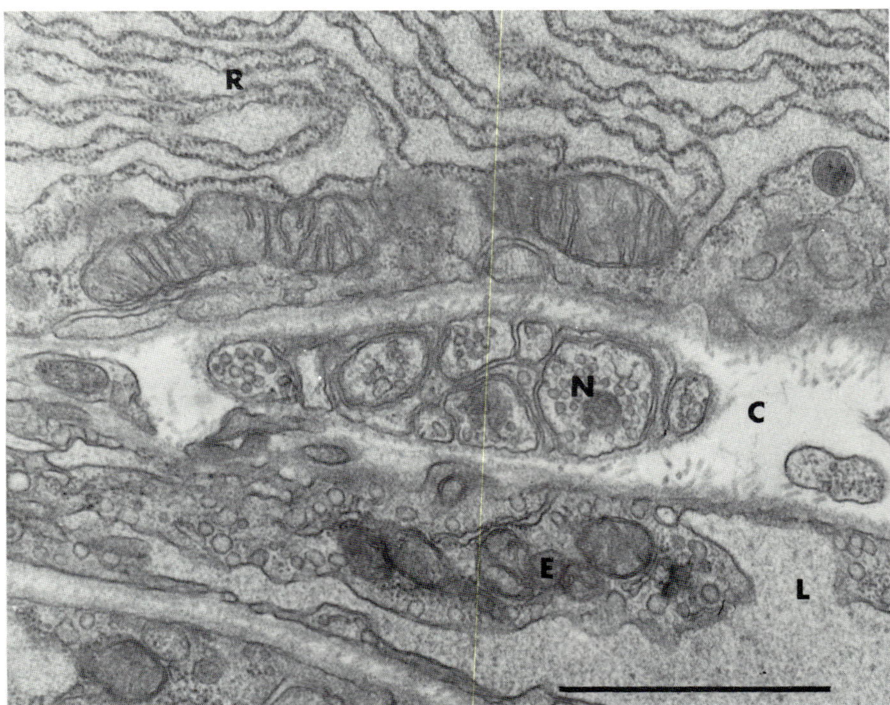

FIG. 22. Nerve terminals (N) containing noncored pinocytotic vesicles, in the pericapillary connective tissue space (C) of the mouse thyroid. R, Rough endoplasmic reticulum of the follicular epithelial cell; E, endothelial cell; L, capillary lumen. ×36,000.

III. Phylogenetic Aspects of the Fine Structure of the Thyroid Gland

Several excellent reviews have been published on the comparative aspects of the functional morphology of the thyroid (Bargmann, 1939; Gorbman, 1959; Barrington, 1959, 1964, 1968; Gorbman and Bern, 1962; Dodd and Matty, 1964; Fujita, 1971). However, none of them has dealt with the fine structure of this gland except for that of Fujita (1971). The typical thyroid, consisting of numerous follicle structures, is seen throughout the Pisces, Amphibia, Reptilia, Aves, and Mammalia, except for the larval lamprey, while the thyroid hormones (T_3 and T_4) are demonstrated in protochordates such as amphioxus and ascidians. The larval lamprey and protochordates have an endostyle which has been considered to be homologous to the thyroid of the vertebrate. Some of the endostylar cells of the larval lamprey become the follicular epithelial cells, forming follicle structures at metamorphosis.

A. Protochordates

An endostyle homologous to the thyroid has been found in ascidians and amphioxus, which are included in the phylum Chordata. Ascidians belong to the subphylum Tunicata, and amphioxus to the subphylum Cephalochordata. The endostyle of these animals (Fig. 23) is a groove

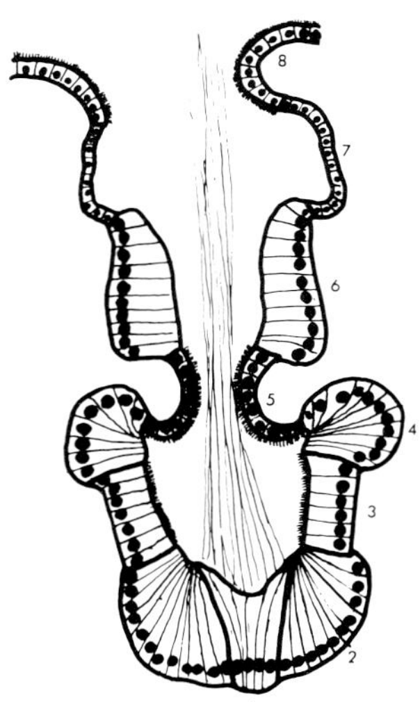

FIG. 23. A schematic representation of the transverse section of an endostyle of the ascidian *C. intestinalis* 1–8, zone-1 to -8 cells. (From Fujita 1972.)

in the floor of the gill chamber. The main function of the endostyle in these animals is thought to be gathering food, mixing it with mucoproteinaceous substances secreted by certain kinds of cells, and sending it into the digestive canal using numerous cilia.

The light microscope structure of the endostyle of the ascidian *Ciona intestinalis* has been described by Barrington (1957, 1959, 1964, 1965), Barrington and Barron (1960), Barrington and Thorpe (1965a,b), Olsson (1963), and others. The endostyle of *C. intestinalis* is a groove in the floor of the gill chamber. This organ is easily seen by the naked eye because of its white threadlike structure running longitudinally through the gill region. Eight zones are easily distinguished in the transverse section. Zone 1 is a median region forming the bottom of the hollow. Zones 2 through 8 consist of two symmetrical rows arranged in order, forming the lateral wall of the endostyle. Each zone is composed of simple cylindrical, cuboidal, or flattened epithelial cells. (See Fig. 24.)

The fine structure of the endostyle of the ascidian has been described by Aros and Viragh (1969), Fujita and Nanba (1971), and Thorpe *et al.*

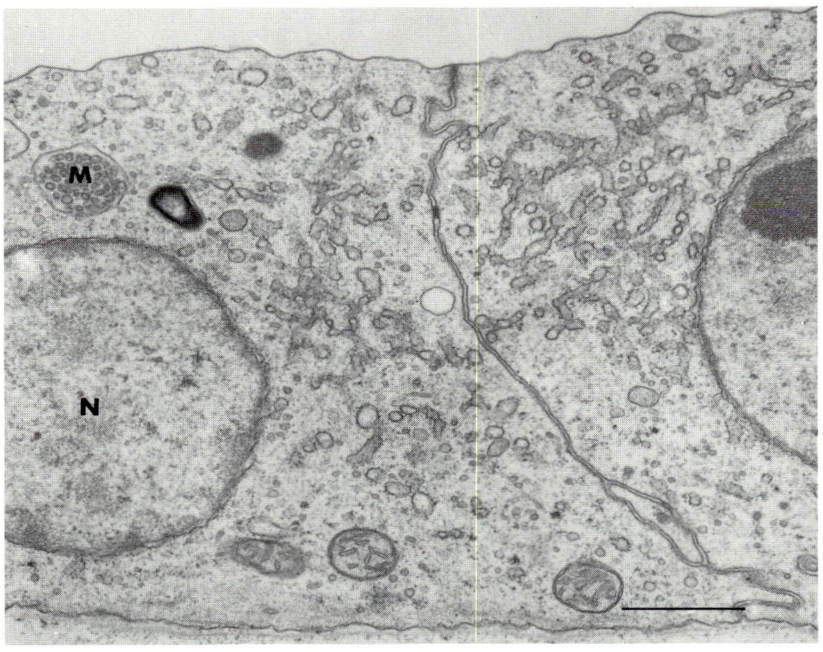

FIG. 24. Parts of zone-7 cells of an ascidian endostyle. M, Multivesicular body; N, nucleus. ×19,000.

(1972). Zone-1, -3, and -5 cells, which are characterized by numerous long cilia and show no indications of protein secretion but numerous small vesicles and cytoplasmic filaments, are thought to play a role in catching and transporting food, absorption of liquid, and supporting the endostylar construction (Fujita and Nanba, 1971). Zone-2, -4 and -6 cells, which are characterized by well-developed rough endoplasmic reticulum and numerous secretory granules derived from the Golgi apparatus, are believed to secrete proteins or mucoproteinaceous substances which might be necessary in the digestion of food (Fujita and Nanba, 1971). This observation coincides with the histochemical data of Olsson (1963) indicating that these cells are pyroninophile. However, the cells from zones 1 to 6 do not have any function for iodine binding. Zone-7 (Fig. 25) and -8 cells are considered homologous to the thyroid cell (Olsson, 1963; Barrington and Thorpe, 1965b; Fujita and Nanba, 1971; Thorpe *et al.*, 1972). Barrington and Thorpe (1965b) and Thorpe *et al.* (1972) considered zone-7 cells more important for the thyroidal function. Zone-7 and -8 cells (Fig. 26) contain poorly developed rough endoplasmic reticulum, a small Golgi apparatus, a few multivesicular bodies, a few lysosomes, and numerous small vesicles. In addition, zone-8 cells bear numerous cilia on their apical surface. As to the fine-structural findings of the ascidian endostyle, three articles published by Aros and Viragh (1969), Fujita and Nanba (1971), and Thorpe *et al.* (1972) are almost in agreement with one another.

In electron microscope autoradiography of ^{125}I, numerous silver grains are observed over the apical plasma membrane region of zone-7 and -8 cells, especially zone-8 cells, 1, 4, 6, 16, and 24 hours after immersion in seawater containing ^{125}I (Fujita and Nanba, 1971). Similar results have been published by Thorpe *et al.* (1972). Based on this observation, we believe that iodination of thyroglobulinlike protein occurs in the apical plasma membrane region of these cells. Since the endostylar lumen is open to the pharyngeal lumen, the materials in the endostylar lumen are washed away almost completely during the fixation, dehydration, and embedding of the tissue, and the possibility of the iodination of protein taking place within the endostylar lumen cannot be ruled out (Fujita and Nanba, 1971). If so, the main site of iodination of thyroglobulinlike protein is the apical plasma membrane region of zone-7 and -8 cells and the endostylar lumen. This is principally a pattern similar to that of the thyroid cell in the higher vertebrate. At 4, 6, 16, and 24 hours, especially 16 and 24 hours, after immersion in sea water containing ^{125}I, silver grains are localized over the multivesicular bodies and lysosomes in addition to the apical plasma membrane region of zone-7 and -8 cells (Fujita and Nanba, 1971). Thorpe *et al.* (1972) also reported similar data. The multi-

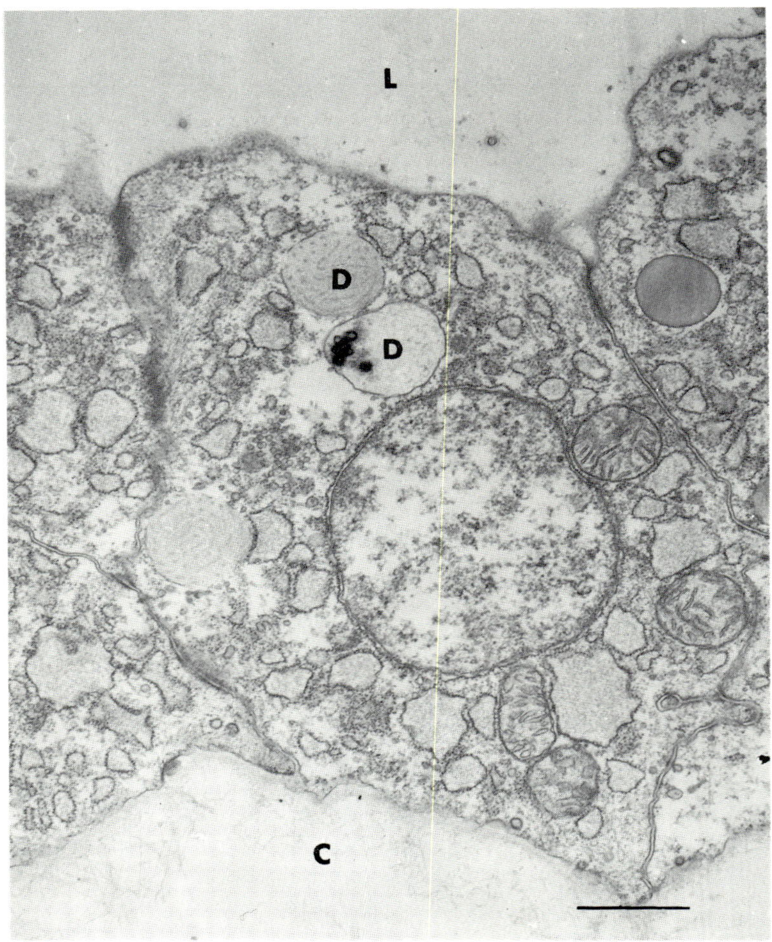

FIG. 25. Parts of zone-7 cells of the endostyle of the ascidian *C. intestinalis* immersed for 5.5 hours in seawater containing TSH (1 IU/liter). Large droplets (D) appear in the cytoplasm. L, Endostylar lumen; C, connective tissue. ×15,000.

vesicular bodies and lysosomes are thought to be involved with the iodinated thyroglobulinlike proteins, which might be reabsorbed into the cell. We wish to emphasize that organic iodide could be reabsorbed into the cytoplasm, to be contained in the multivesicular bodies and lysosomes. Then T_3 and T_4 liberated by hydrolysis of the protein might be released from the basal part of the cell into the connective tissue, although the possibility should not be denied that thyroglobulinlike protein

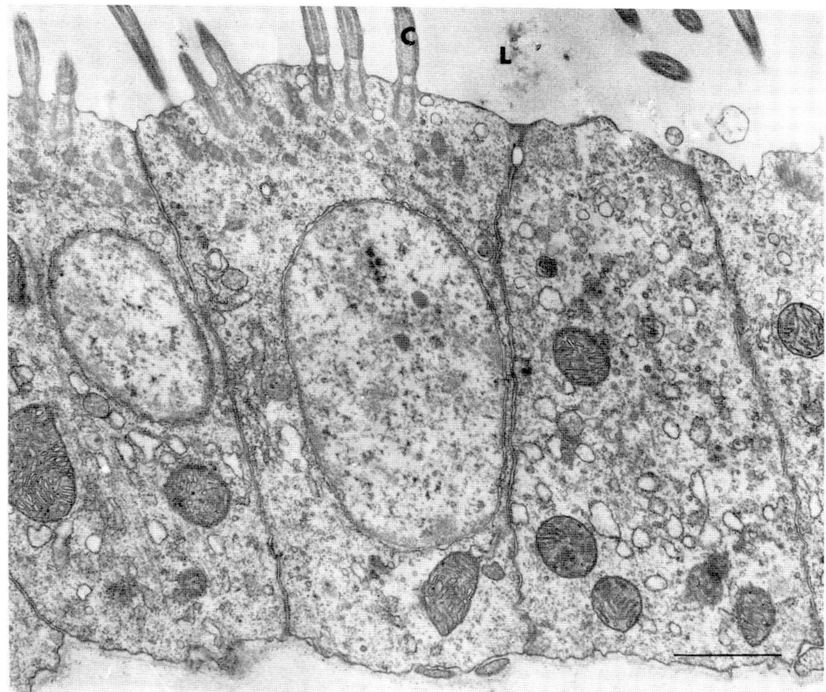

FIG. 26. Parts of zone-8 cells of an ascidian endostyle. C, Cilia; L, endostylar lumen. ×16,500.

is partly transported to the digestive canal with the food. T_4 has been demonstrated, using radiochromatography, in the ascidian endostyle by Barrington and Thorpe (1965a) and Barrington (1968). In addition, Kennedy (1966) reported the occurrence of T_3 and T_4 in ascidian blood.

The light microscope structure of the amphioxus endostyle has been described by Krause (1923), Franz (1927), Pietschmann (1929), Thomas (1956), Barrington (1958, 1959), Olsson (1963) and others. There are at least seven kinds of cells: zone-1, -2, -3, -4, -5, -6, and -7 cells, and zone-5 cells are known to have iodine-binding activity, according to Thomas (1956), Barrington (1958), and Olsson (1963). The following descriptions are based on observations we have made. Zone-1 cells are located in a median region forming the bottom of the floor of the endostyle, and zone-3 and -6 cells located on both lateral walls of this organ (about 1–3 μm in width and about 20–40 μm in height) have numerous long cilia on their apical surface. Rough endoplasmic reticulum and Golgi apparatus are not well developed and no secretory activities are observed, although lysosomelike dense bodies are scattered in the cytoplasm of these cells. Zone-2 and -4 cells, which are also narrow and tall, are characterized by the elements of well-developed rough endoplasmic reticulum and numerous electron-dense secretory granules 100–300 nm diameter, derived from the Golgi apparatus. Zone-5 cells, which have been regarded as the iodine-binding cells, also show well-developed rough endoplasmic reticulum and less dense secretory granules (200–500 nm) in diameter in their cytoplasm. These findings are almost in agreement with those published by Welsch and Storch (1969) for *Branchiostoma lanceolatum* (Pallas). Based on these data, it is clear that zone-2, -4, and -5 cells have secretory activity and that protein or mucoproteinaceous substances are released into the endostylar lumen. However, there are no similarities between the cytoplasmic structure of these cells and that of the endostylar cells of ascidians and larval lampreys and that of the thyroid cells of cyclostomes. Nothing is known about the iodine metabolism of the endostylar cells of amphioxus at the fine-structural level, although T_3 and T_4 have been demonstrated biochemically in the endostyle of an amphioxus by Covelli *et al.* (1960) and Tong *et al.* (1962). The functional morphology of the amphioxus endostyle is a subject to be studied in the future.

B. Cyclostomes

The cyclostome is the lowest and the most interesting vertebrate in regard to the evolutional meaning of the thyroid. One species of the cyclostome is the hagfish, and the other is the lamprey. The thyroids of both animals are located in the hypobranchial region and consist of widely scattered follicles with connective tissue elements. Larval lampreys, called ammocoetes, do not have a thyroid, but an organ termed an endostyle which is equivalent to the thyroid, and at metamorphosis some endostylar cells become follicular epithelial cells and form the follicle structure; in the hagfish a thyroid having many follicles is seen throughout its life. So morphologically the most primitive thyroid is that of the larval lamprey or of the hagfish, while functionally the most primitive organ homologous to the thyroid is that of protochordates such as ascidians and amphioxus.

The endostyle of the larval lamprey is an organ homologous to the thyroid (Müller, 1871; Dohrn, 1886; Kieckebusch, 1928; Bargmann, 1939). The endostyle of the larval lamprey (Fig. 27) consists of two hollow organs extending from the first to fifth gill arches and separated by a median septum. Each hollow communicates with the pharynx through a narrow duct, the ductus hypobranchialis, which is closed at metamorphosis. Epithelial cells of this organ are classified into five types; type-1

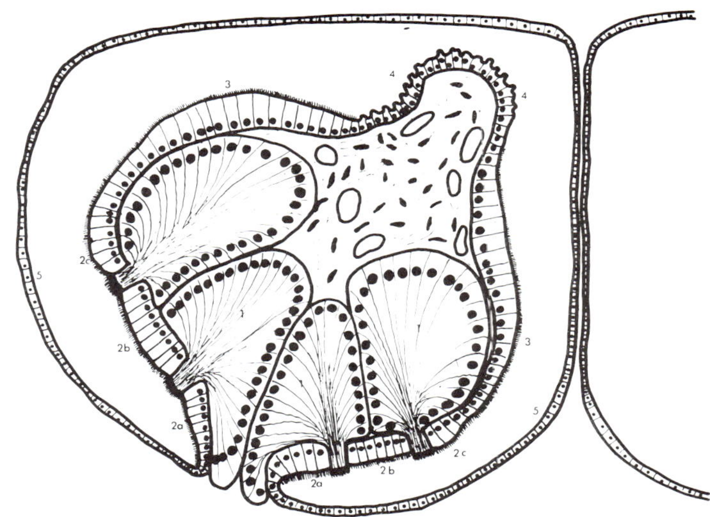

FIG. 27. Transverse section through one of the anterior chambers of the endostyle of a larval lamprey, ammocoete of *L. japonica*. 1–5, type-1 to -5 cells. (From Fujita, 1972.)

to -5 cells (Marine, 1913; Leach, 1939). Type-1 cells are subdivided into ventral type-1 and dorsal type-1 cells, and type-2 cells into type-2a, -2b, and -2c cells. Among them type-2c and -3 cells are considered homologous to the thyroid follicular epithelial cells, especially type-3 cells which are the main iodine-concentrating cells (Leloup, 1952, 1955; Sterba, 1953; Olivereau, 1955; Barrington and Franchi, 1956; Clements-Merlini, 1960a,b). At metamorphosis chiefly type-3 cells become follicular epithelial cells (Sterba, 1953; Clements-Merlini, 1960a,b; Honna, 1960).

The fine structure of the endostyle of the larval lamprey has been described by Egeberg (1965), Fujita and Honma (1968, 1969), and Hoheisel (1969, 1970) and reviewed by Barrington and Sage (1972). Their descriptions are principally almost coincident with one another. In the endostyle of larval lampreys, ammocoetes of *Lampetra japonica*, type-1 cells are characterized by elements of well-developed rough endoplasmic reticulum occupying most of the cytoplasm, numerous dark secretory granules 150–300 nm in diameter, and somewhat less dense ones 600–2000 nm in diameter derived from the Golgi apparatus. These seem to be proteinaceous secretory materials which might be not be involved with the thyroid hormone. Type-2a, and -2b cells, cylindrical in shape and with numerous microvilli and cilia, are poor in cytoorganelles and contain free ribosomes, glycogen particles, and a few lysomelike dense bodies. Type-2 and -3 cells, which have been considered homologous to thyroid cells, are characterized by a fairly well-developed rough endoplasmic reticulum. Although the elements of the endoplasmic reticulum do not show dilated cisternae, small stacks of flattened sacs, a few of which are continuous with the Golgi apparatus, are usually seen in the basal and lateral cytoplasm of these taller cells. In the infranuclear to the apical part of the cytoplasm are numerous small vesicles which might be derived from the Golgi apparatus. Some of them are thought to be secretory substances containing thyroglobulinlike protein. In addition, lysosomelike dense bodies, large or small, are scattered in the cytoplasm. At the apical surface of these cells, numerous cilia and microvilli projecting into the endostylar lumen are noted. In type-4 cells, which do not have any iodine-binding activity, rough endoplasmic reticulum is very poorly developed, and free glycogen particles and numerous vesicles are distributed throughout the cytoplasm. Large cytoplasmic processes projecting into the endostylar lumen are characteristic. Type-5 cells, forming the lining of the endostylar chamber are flattened and cuboidal in shape and contain small smooth vesicles, glycogen particles, and small and large lysosomelike dense bodies in their cytoplasm. Elements of rough endoplasmic reticulum, cilia, and microvilli are scarce in these cells.

By electron microscope autoradiography, it has been demonstrated that silver grains are localized chiefly over the apical plasma membrane region of type-2c and -3 cells 30 minutes and 1 and 2 hours after the intraperitoneal injection of sodium iodide-^{125}I (Fujita and Honma, 1969) (Fig. 28). As the endostylar lumen is open to the pharynx by a hypobranchial duct, materials in the lumen are easily washed away during the fixation, dehydration, and embedding of the tissue. So it is highly possible that the materials in the endostylar lumen are also labeled by radioactive iodine if they remain in the tissue. This suggests that the main site of iodination of thyroglobulinlike protein is the apical plasma membrane region and the endostylar lumen of these cells. Since a small number of dense granules in the cytoplasm of these cells is partly labeled 1–2 hours after the injection of radioiodine, we do not deny the possibility that iodination occurs in the cytoplasm slightly. At 6 and 24 hours (Figs. 29 and 30) after the injection of radioactive iodine, silver grains are localized over lysosomelike dense bodies and multivesicular bodies, in addition to the apical plasma membrane region. This means that the lysosomelike dense granules and bodies contain the iodinated thyroglobulinlike protein. Although it is difficult to determine whether these dense granules and bodies contain secreted material or reabsorbed material, the occurrence of reabsorption and hydrolysis of thyroglobulin in

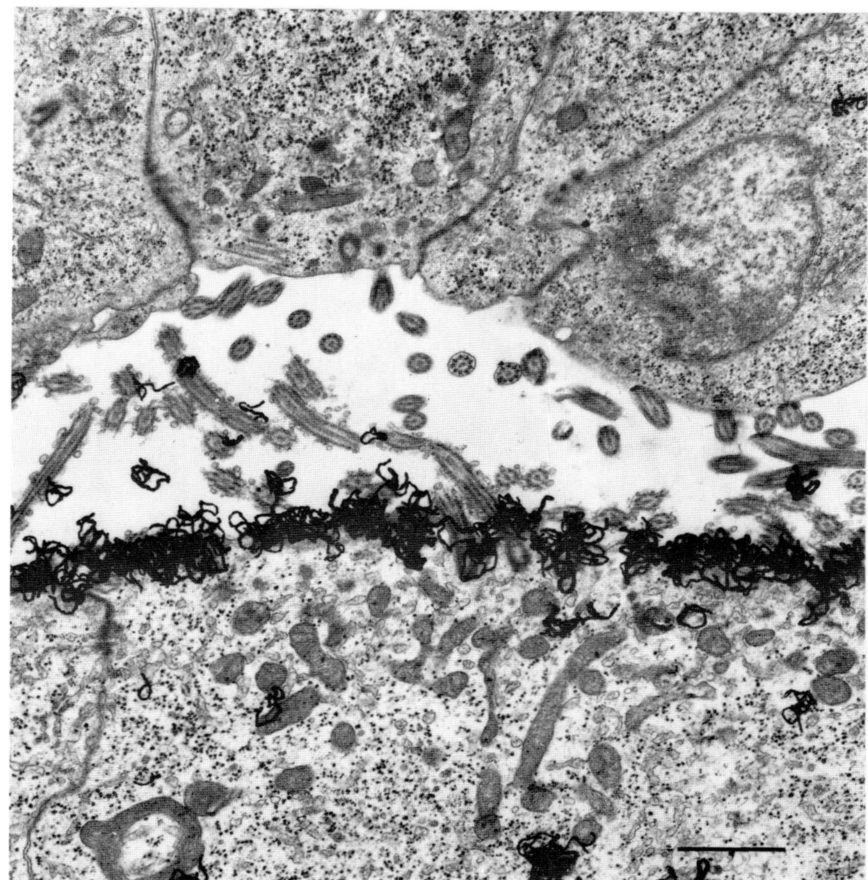

FIG. 28. Parts of type-4 (upper) and type-3 (lower) cells of a larval lamprey 2 hours after injection of ^{125}I (200 μCi). Silver grains are localized over the apical cell membrane region of a type-3 cell. ×14,500. (From Fujita and Honma, 1969.)

type-2c and -3 cells is considered highly possible. At 24 hours silver grains are often observed over the dense bodies localized in the subapical as well as basal cytoplasm, and over the dense connective tissue elements. It is probable that T_3 and T_4 might be liberated from the lysosomal dense bodies and released from the basal part of the cell into the connective tissue. It is interesting that some type-5 cells, whose rough endoplasmic reticulum is not so well developed, are also labeled. The type-5 cells, which lack activity in the synthesis of thyroglobulinlike protein, might be involved in the reabsorption of iodinated protein, as Egeberg (1965)

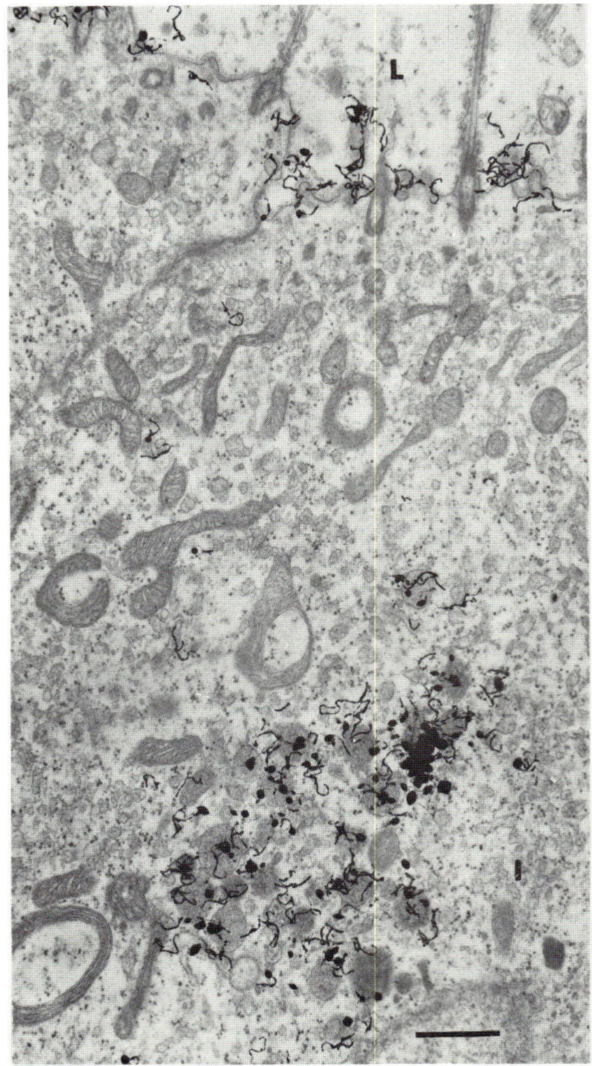

FIG. 29. Parts of type-3 cells of an endostyle of a larval lamprey 24 hours after injection of 200 μCi of ^{125}I. Silver grains are localized over the lysosomes, multivesicular bodies, and apical plasma membrane region. L, Follicular lumen. $\times 11,000$.

and Fujita and Honma (1969) suggested in summary, iodine metabolism in the endostyle of the larval lamprey shows a pattern almost similar to that of the higher vertebrate, and type-2c and -3 cells, especially type-3, may have a function similar to that of the thyroid cell. Although

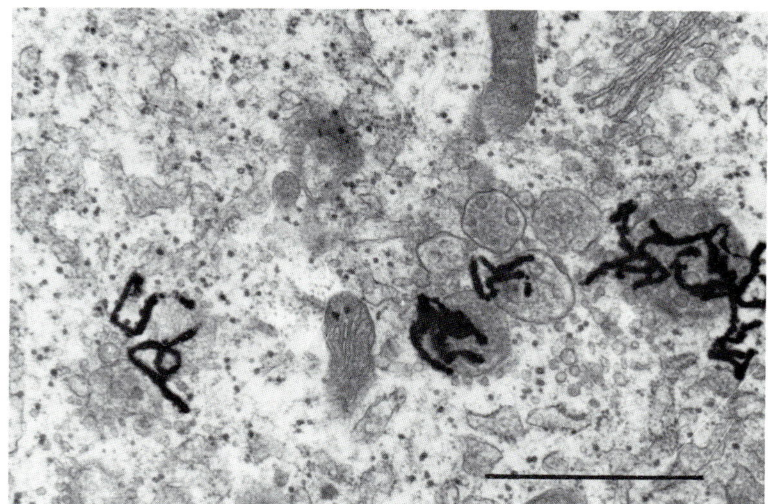

FIG. 30. Same as Fig. 30. Silver grains are localized over the lysosomes and multivesicular bodies. ×29,000.

Clements and Gorbman (1955), Gorbman (1959), and Thomas (1962) reported that thyroglobulinlike protein may be transported to and hydrolyzed in the alimentary canal and reabsorbed through the digestive epithelium, we wish to emphasize that the endostylar cells also reabsorb the iodinated thyroglobulinlike protein and hydrolyze it to liberate the thyroid hormone. T_3 and T_4 have been known to be present in extracts of the endostyle of a larval lamprey, ammocoetes of *Lampetra planeri* or of *Petromyzon marinus* (Leloup and Berg, 1954; Leloup, 1955; Salvatore *et al.*, 1959a,b; Roche *et al.*, 1961).

At metamorphosis of the larval lamprey, the hypobranchial duct is closed, many endostylar cells degenerate, and typical follicle structures are formed by some of the endostylar cells in the same region. The light microscope structure of the lamprey thyroid has been described by many investigators (Marine, 1913; Kraentzel, 1933; Eggert, 1938; Leach, 1939; Olivereau, 1952; Sterba, 1954; Lanzing, 1959; Honma, 1960).

The fine structure of the thyroid of the adult lamprey (Fig. 31) has been described by Fujita (1966), Fujita and Honma (1966), and Hoheisel (1969, 1970). The thyroid of the adult lamprey during the upstream migration period consists of many usual follicles containing the colloid in their lumen 20–100 × 700–400 μm in diameter, and an extremely large follicle without colloid called parafollicle. The thyroid function may be performed by common follicles containing the colloid. The follicle cells in the common follicular epithelium are classified into

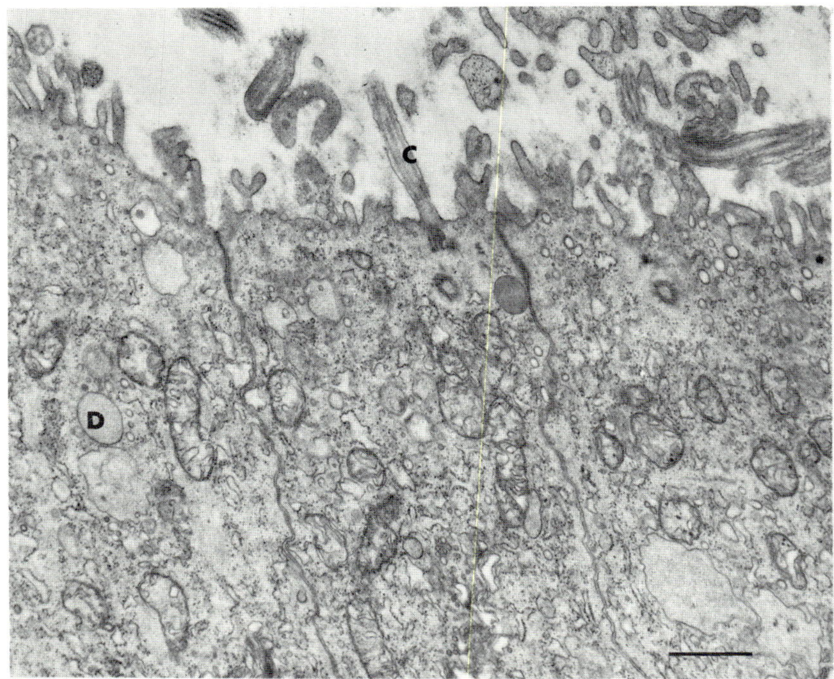

FIG. 31. Apical parts of the follicular epithelial cells of the adult lamprey thyroid. C, Cilia; D, colloid droplet. ×13,000.

three types: (1) nonciliated taller cells, (2) ciliated taller cells, and (3) nonciliated cuboidal cells (Fujita and Honma, 1966). It is thought that all these cells are principally identical, and differences among them are due to their functional state. All these cells, which are somewhat similar to type-2c and -3 cells in the fine structural appearance of the cytoplasm show fairly well-developed elements of rough endoplasmic reticulum, whose cisternae are not dilated but flattened (Fig. 32), being quite similar to those of the endostylar cells of the larval lamprey. The cytoplasm looks more compact as compared with that of the higher vertebrate thyroid. Small less dense granules (50–300 nm in diameter) and dense granules (200–600 nm in diameter), some of which might be derived from the Golgi apparatus, are seen in the subapical region of the taller cells, and a few large less dense droplets (500–2000 nm in diameter), which are similar to the reabsorbed colloid droplets of the higher vertebrate thyroid cell, are sometimes seen in the taller cells. In addition, lysosomelike, large dense bodies irregular in shape (1.0–7.0 μm in diam-

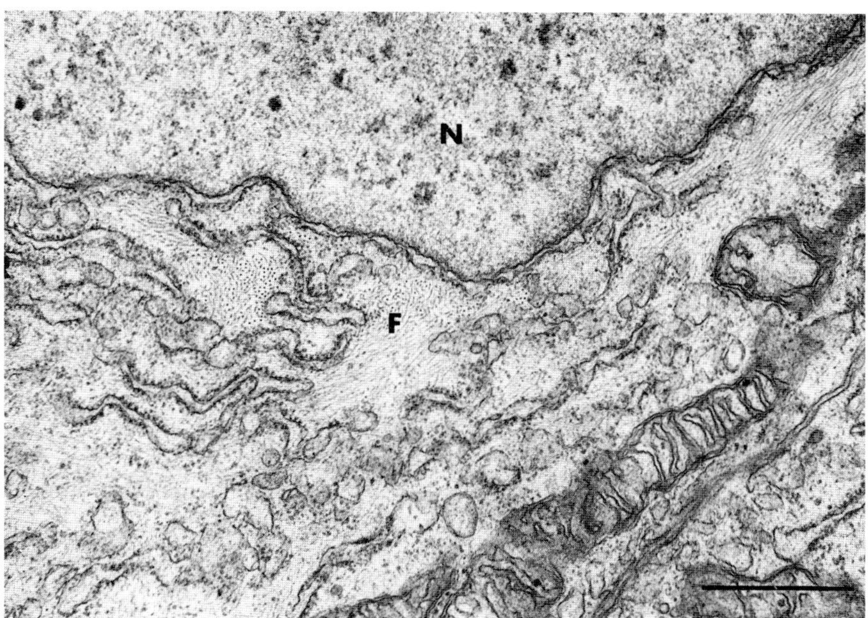

FIG. 32. Intranuclear region of the follicular epithelial cell of the adult lamprey thyroid. Note the cytoplasmic filaments (F) and rough endoplasmic reticulum (R) with flattened cisternae. N, Nucleus. ×20,000.

eter), heterogeneously electron-dense, and containing dense materials, lamellar structures, and less dense vacuoles are usually seen in these lamprey thyroid cells. These might correspond to the yellow pigments (Fig. 33) reported in light microscope studies (Lanzing, 1959; Honma, 1960). It is difficult to determine the function of the cytoplasmic inclusions, because no results have been published on the electron microscope autoradiography of radioactive iodine and radioactive amino acids. However, it is possible that the small less dense granule is a secretory substance containing thyroglobulinlike protein, the small dense granule is a primary lysosome, the large less dense droplet is a reabsorbed colloid, and a large heterogeneously dense body is a phagolysosome or autolysome; some of these may contain iodinated thyrogulobulin. However, the rough endoplasmic reticulum and the compact appearance of the cytoplasm of these cells tell us that the lamprey thyroid, as well as the ammocoete endostyle, is not as active as that of the higher vertebrate. Leloup and Fontaine (1960) stated, in a physiological study, that the uptake of ^{131}I by the thyroid gland and the rate of secretion of thyroid

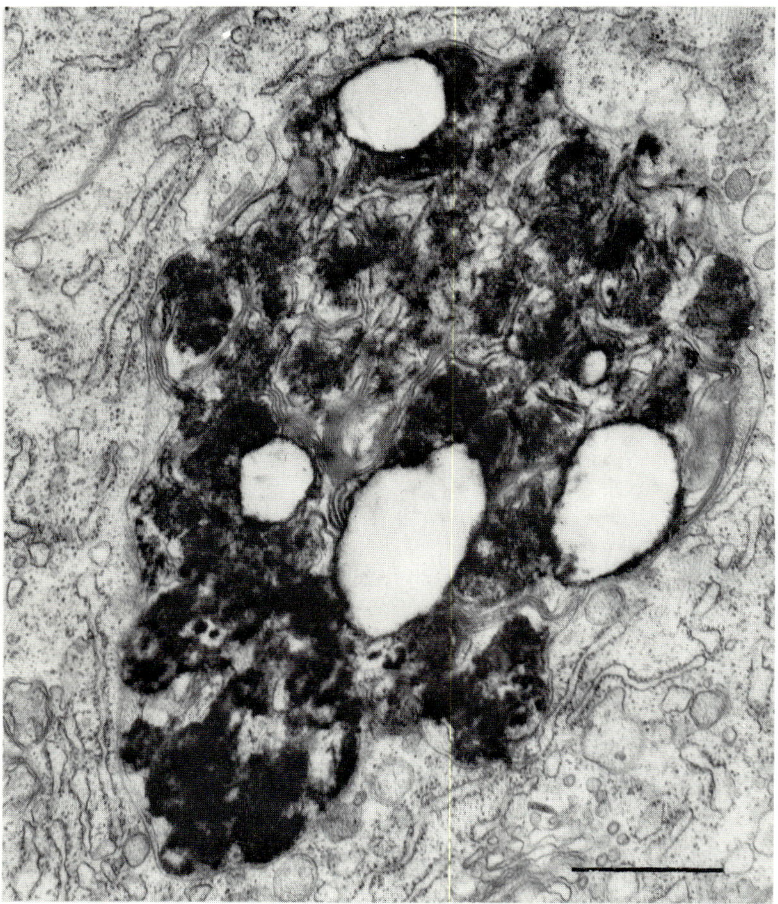

FIG. 33. A large dense body corresponding to the yellow pigment in the follicular epithelial cell of an adult lamprey thyroid. ×20,500. (From Fujita and Honma, 1966.)

hormone in a lamprey (*P. marinus*) are very low. The observation that the vascularization of the thyroid or its homologous organ in the adult as well as in the larval lamprey is not so rich and the interfollicular connective tissue elements are extremely dense in collagen fibrils, also suggests the hypofunction of the thyroid of these animals.

The thyroid gland of the hagfish (Fig. 34), embedded in the fatty tissue of the pharyngeal floor along the course of the ventral branchial aorta, also has an interesting structure. The light microscope structure

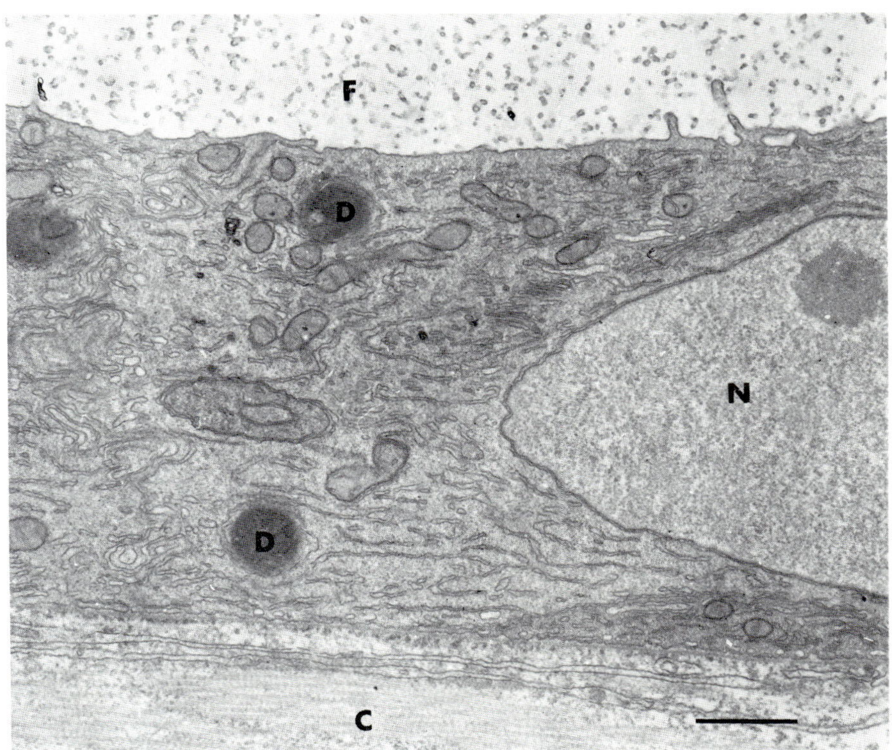

FIG. 34. Part of the thyroid follicular epithelial cell of a hagfish, *E. burgeri*. Elements of rough endoplasmic reticulum are flattened, and dense bodies (D) are seen. C, Connective tissue; F, follicular lumen; N, nucleus. ×13,500.

has been described by Cole (1905), Schaffer (1906), and Waterman and Gorbman (1963). The follicles are distributed more widely than those in the lamprey, and the interfollicular connective tissue consisting chiefly of fatty tissue is more abundant. The follicle lumen varies in size; it is sometimes extremely large and can be easily seen by the naked eye. We have also made electron microscope and electron microscope autoradiographic study of ^{125}I in *Eptatretus burgeri* (40–60 cm in length, obtained in July and August). Our data agree principally with those for *Eptatretus stouti* (37–41 cm in length, obtained in April) reported by Henderson and Gorbman (1971). The difference between the hagfish thyroid and the lamprey one is the occurrence of numerous cilia in the lamprey thyroid cell and their absence in the hagfish. Microvilli are also short in length and few in number, and the apical surface looks more flattened in the hagfish. The cytoplasm is generally compact, and the cytomembrane system is not so well developed as compared with that of the higher vertebrate thyroid cell; its structure is similar to that of the lamprey thyroid. Free ribosomes are scattered, and microfilaments are frequently observed in the cytoplasmic matrix. Rough endoplasmic reticulum consisting of tubular or lamellar elements is scattered in the cytoplasm. No dilated cisternae are seen. The Golgi apparatus composed of lamellae and vesicles is also small in size. These observations suggest that the hagfish thyroid cell is inactive, like that of the lamprey. Numerous small vesicles (30–80 nm in diameter), a few small dense granules (100–200 nm), less dense colloid droplets (500–1500 nm), and large homogeneously or heterogeneously dense bodies (500–3000 nm) are noted in the cytoplasm. The large dense bodies are sometimes encircled by a tubular element of smooth endoplasmic reticulum. Henderson and Gorbman (1971) speculated that the small vesicle is a secretory substance, and the dense granule is a lysosome which could fuse with reabsorbed colloid. Our electron microscope autoradiography shows a small number of silver grains appearing over the follicular lumen and the apical plasma membrane region 1 hour after the injection of 1 mCi of ^{125}I, and over a few small vacuoles and a few large dense bodies 6 hours after the injection of radioiodine and 1 hour after treatment with TSH. This suggested that the main site of thyroglobulin in the hagfish thyroid is the follicular lumen and the apical plasma membrane region, and that the small vacuoles and the large dense bodies contain reabsorbed materials. No phagocytotic features and pseudopods have been observed even in TSH-treated animals. This means that the luminal colloid is reabsorbed by micropinocytosis, as suggested by Henderson and Gorbman (1971). The large dense bodies are thought to be phagolysosomes containing iodinated colloid. All these fine-structural findings suggest that the hagfish thyroid is not as active, and that the iodine uptake activity is also not as strong as compared with that in the higher vertebrate. Waterman and Gorbman (1963) showed that the accumulation of iodine is very slow until an apparent maximum is reached at 144 hours in *Myxine glutinosa.* Among the follicles are abundant connective tissue containing poorly distributed blood capillaries. Although the endothelial cells are not fenestrated like those in the lamprey thyroid, numerous pinocytotic vesicles are usually observed in the endothelium.

C. Elasmobranchs to Mammals

The fine structure of the thyroid has been described by numerous investigators for various lower as well as higher vertebrates. All these animals show an almost similar pattern in the fine structure of the thyroid from elasmobranchs to mammals. A description of the fundamental pattern of the thyroid cell of these animals follows. The well-developed rough endoplasmic reticulum with dilated cisternae occupies a large part of the cytoplasm, and the Golgi apparatus is located in the supranuclear region. Small less dense or dense granules derived from the Golgi apparatus, large less dense droplets reabsorbed from the follicular lumen, and large dense colloid droplets which might be in the process of hydrolysis are seen in the supranuclear and the apical part of the cell. Mitochondria are distributed throughout the cytoplasm.

1. *Elasmobranchs*

The fine structure of the thyroid of elasmobranchs has been described by Nakai and Gorbman (1969) for the chimaeroid fish (ratfish) *Hydrolagus colliei.* The following discussion is based chiefly on their description. The ratfish thyroid gland is a solid encapsulated organ near the anterior extremity of the lower jaw, under the small tongue. The follicular epithelial cell is cylindrical and is characterized by well-developed rough endoplasmic reticulum with dilated cisternae, secretory granulelike small vesicles, large colloid droplets, and lysosomes of various sizes. The cell is stimulated by TSH treatment, as is that in the higher vertebrate. Iodine organification occurs in the follicular colloid, and the intracellular colloid droplets contain protein-bound iodine (Nakai and Gorbman, 1969). Large colloid droplets are often fused with small granules thought to be primary lysosomes. Lysosomes, heterogeneously or homogeneously dense, sometimes contain filamentous structures, crystalloids, and vesicular formations. Nakai and Gorbman (1969), showing the histochemical reaction deposits for acid phosphatase on the dense material but not on the filamentous structures, considered the filamentous structures and the crystalloids digestive or other reaction products resulting from interaction with lysosomes. It should be emphasized that there are fairly large differences between the chimaeroid fish and the cyclostome in the fine structure of the thyroid cell, especially in the elements of rough endoplasmic reticulum, and no striking differences between the chimaeroid fish and the higher vertebrate.

2. *Teleosts*

Descriptions of the fine structure of the teleost thyroid have been presented for *S. quinqueradiata* (Fujita and Machino, 1965), *Anguilla japonica* (Fujita *et al.*, 1966) (Fig. 35), *Semicossyphus reticulatus* (Suemasa *et al.*, 1968), and *Sebastiscus marmoratus* (Suemasa *et al.*, 1968). As Honma described light microscopically, the group of thyroid follicles lies on the dorsal surface or periphery of the ventral aorta, and

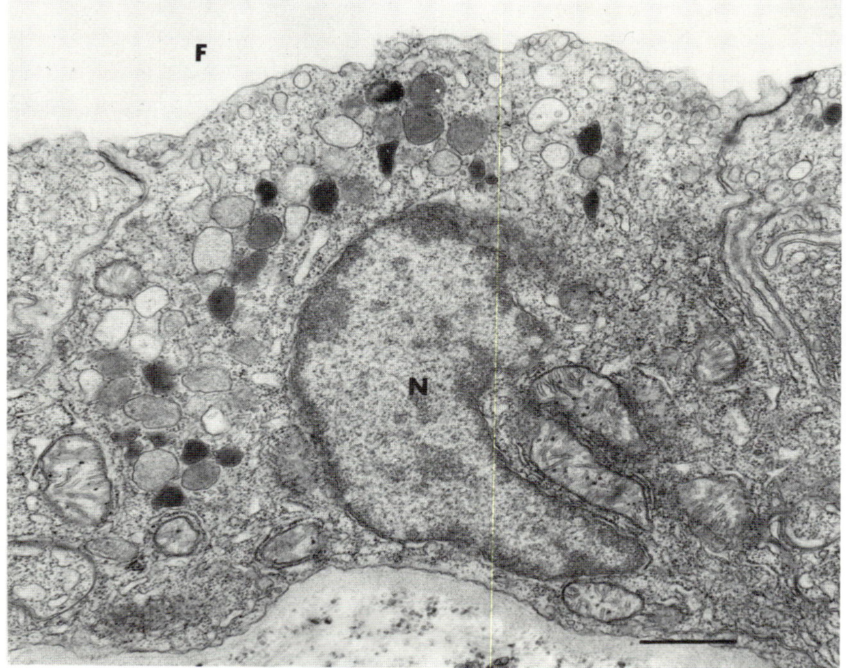

FIG. 35. Parts of thyroid follicular epithelial cells of a silver eel, *A. japonica.* Less dense and dense droplets, and small subapical vesicles, are seen in the cytoplasm. F, Follicular lumen; N, nucleus. ×15,000. (From Fujita *et al.*, 1966.)

the distribution of follicles ranges from the bifurcated regions of the first to second afferent branchial arteries, or to the third one. The thyroids of these fishes show a structure generally similar to those of higher vertebrates. In *Seriola, Semicossyphus,* and *Sebasticus* (Fig. 36), occurrences of crystals and of aggregates of fine filaments in certain large lysosomelike dense bodies are characteristic. Although no articles dealing with the electron microscope autoradiography of ^{125}I and amino-^{3}H acids have been publishd, the functional pattern of the teleost thyroid is probably similar to those of higher vertebrates and chimaeroid fish. The capillary endothelium of *S. marmoratus* (Fig. 37) shows relatively numerous fenestrations with a thin diaphragm, as do those of higher vertebrates.

3. *Amphibians*

Electron microscope studies of the amphibian thyroid have been made by several investigators. The thyroid follicular epithelial cells of the frog (Coleman *et al.*, 1968a,b,c; Nakai *et al.*, 1970; Regard and Mauchamp,

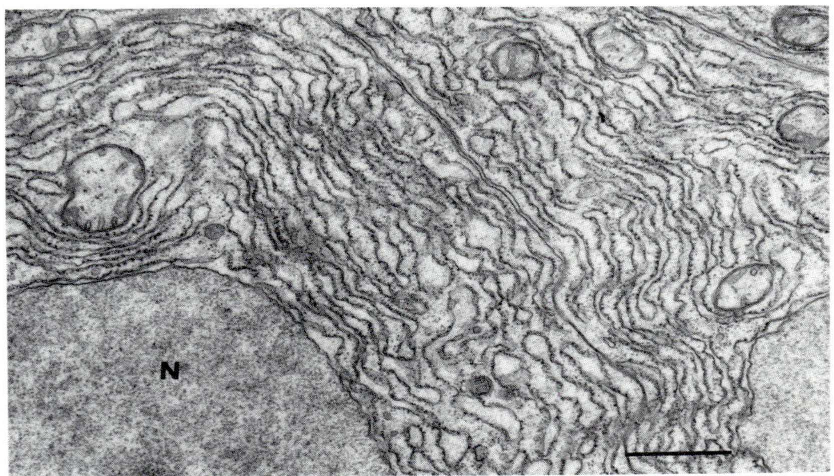

FIG. 36. Parts of the thyroid follicular epithelial cells of the teleost *S. marmoratus*. Note well-developed rough endoplasmic reticulum. N, Nucleus. ×17,000.

1971; Nanba, 1972a; Neuenschwander, 1972), of the salamander (Herman, 1960; Larsen, 1968; Setoguti, 1973a,b,c), and of the newt (Hearing and Eppig, 1969) are electron microscopically similar to those of the mammal and fish, although there are a few differences in the description of their details. Elements of the rough endoplasmic reticulum with dilated cisternae are well developed, especially in the basal part of the cytoplasm of all these animals. According to most of the above workers, small vesicles presumed to be derived from the Golgi apparatus and to contain secretory materials are seen in the apical cytoplasm, and large colloid droplets and dense granules are also located chiefly in the supranuclear cytoplasm. The small dense granules and large dense droplets are positive for the acid phosphatase reaction (Nakai *et al.*, 1970), and fusion of the large less dense droplets with the small dense granules, which might be primary lysosomes, sometimes occurs (Nakai *et al.*, 1970). Although it is somewhat difficult to distinguish clearly which of these granules is a secretory one, a reabsorbed one, or a lysosome, their functional properties are thought to be principally the same as those of the higher vertebrate. Lamellar structures, vacuoles, crystals, and filamentous structures are often seen in the large, heterogeneously dense bodies (Herman, 1960; Larsen, 1968; Nakai *et al.*, 1970; Setoguti, 1973a,b,c), and are similar to those seen in some teleosts and chimaeroid fish. The occurrence of a central cilium in the follicular epithelial cell has been reported for the frog (Coleman *et al.*, 1968a) and for the salamander (Larsen, 1968; Setoguti, 1973a), although its function is obscure.

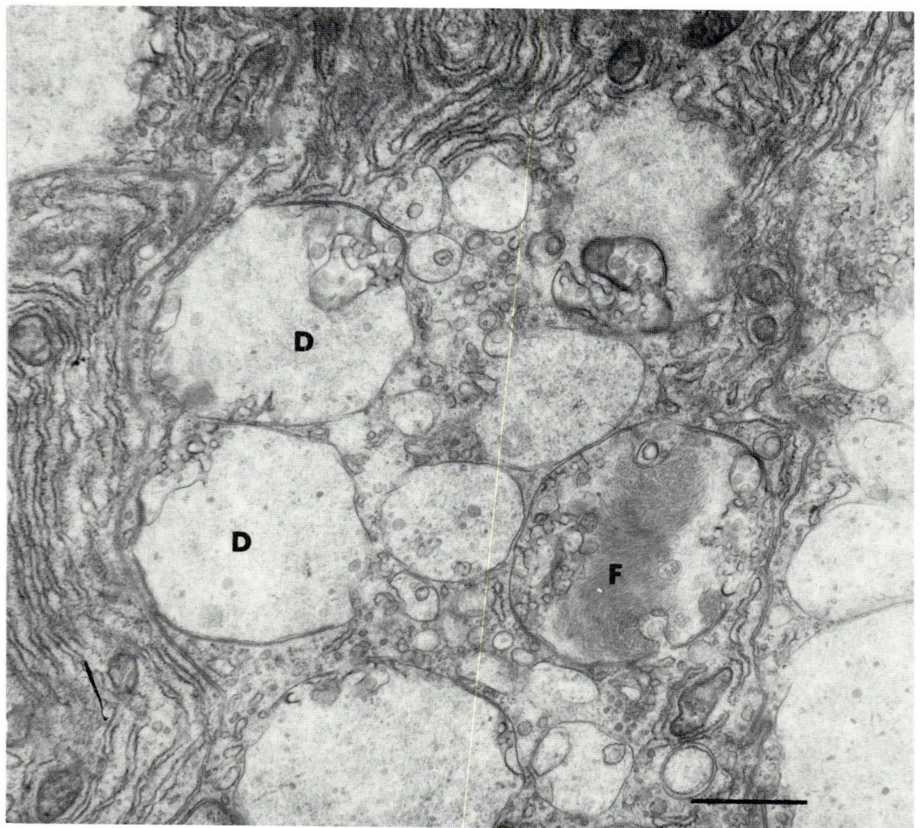

FIG. 37. An accumulation of colloid droplets (D) in the thyroid follicular epithelial cell of the teleost *S. marmoratus*. Note filamentous structures (F) in the droplet. ×15,000.

The iodination of thyroglobulin in the frog is considered to take place chiefly in the follicular lumen and in the apical plasma membrane region, as in other vertebrates (Nakai *et al.*, 1970). The amphibian thyroid reacts to TSH, and pseudopod-containing reabsorbed colloid droplets often appear after the injection (Herman, 1960; Setoguti, 1973b).

The capillary endothelium shows fenestrated structures similar to those in some teleosts and higher vertebrates.

4. *Reptiles, Birds, and Mammals*

Studies on the ultrastructure of the reptilian thyroid have been described for the snake (Watari, 1962) and the tortoise (Sakai, 1960; Muramoto,

1964; Lupulescu and Petrovici, 1968); those of the bird thyroid for the domestic fowl (Fujita *et al.*, 1958; Fujita, 1963; Hilfer, 1964; Fujita and Tanizawa, 1966), duck (Sakai, 1960), and pigeon (Sakai, 1960; Muramoto, 1964); and those of the mammalian thyroid for the rat (Braunsteiner *et al.*, 1953; Monroe, 1953; Weber *et al.*, 1954; Dempsey and Peterson, 1955; Walthard, 1955; Walthard and Roos, 1959; Irie, 1960; Lever, 1960; Roos, 1960; Waller, 1961; Wissig, 1960, 1963, 1964; Stoll *et al.*, 1961; Brettschneider, 1963; Fujita *et al.*, 1963; Tashiro and Sugiyama, 1964; Bauer and Meyer, 1965; Ibrahim and Budd, 1965; Wetzel *et al.*, 1965; Ekholm and Smeds, 1966; Seljelid, 1967a,b,c,d,e; Fujita and Suemasa, 1968; Lietz, 1968; Lupulescu and Petrovici, 1968; Youson and van Heiningen, 1968; Seljelid *et al.*, 1970, 1971; Feeney and Wissig, 1971), mouse (Ekholm and Sjöstrand, 1957; Ekholm, 1957, 1960, 1964; Irie, 1960; Ekholm *et al.*, 1963; Sheldon *et al.*, 1964; Stein and Gross, 1964; Lupulescu and Petrovici, 1968; Fujita, 1969, 1970), hamster (Irie, 1960; Nève and Wollman, 1971; Lietz, 1973), guinea pig (Braunsteiner *et al.*, 1953; Sakai, 1960; Ekholm, 1964; Fabre and Marescaux, 1968; Kosanović *et al.*, 1968; Lupulescu and Petrovici, 1968), rabbit (Sakai, 1960), beaver (Trolldenier, 1965), bat (Nunez and Becker, 1970; Nunez, 1971; Fujita, 1972), common seal (Harrison *et al.*, 1962), common dolphin (Harrison and Young, 1970), cat (Irie, 1960), dog (Irie, 1960; Sakai, 1960; Tashiro and Sugiyama, 1964; Nève and Dumont, 1970a,b; Nunez *et al.*, 1972; Nève *et al.*, 1972), sheep (Trolldenier, 1967; Nève *et al.*, 1968; Tixier-Vidal *et al.*, 1969), deer (Pantic, 1967), roebuck (Pantic, 1967), cow (Irie, 1960; Trolldenier, 1967), pig (Trolldenier, 1967), horse (Irie, 1960), and human (Noseda, 1954; Garnier, 1956; Kanaya, 1960; Sakai, 1960; Irvine and Muir, 1963; Lupulescu, 1965; Nève, 1965; Heimann, 1966; Greene *et al.*, 1966; Lupulescu and Petrovici, 1968; Toujas and Guelfi, 1969; Klinck *et al.*, 1970; Matthaes, 1972). Although there are a few differences in detail, the descriptions of the fine structure of the thyroid gland of these animals in these articles are essentially similar to one another. No characteristic differences in fine structure of the thyroid cell are found among reptiles, birds, and mammals. General descriptions of the fine structure of the thyroid and its functional properties in these animals are as mentioned in Section II,A.

The occurrence of a central cilium (or a central flagellum) should be noted in the thyroid cell. This element has been reported in the follicular epithelial cell of many kinds of animals. Microtubules of the 9 + 2 form are arranged in the central cilium of the follicular epithelial cell as in other organs. Although the function of this structure in the thyroid is not clear, we believe that all or most of the thyroid follicular epithelial cells have a central cilium on their apical surface. As a central cilium is usually seen not only in the thyroid follicular epithelial cell but also in other epithelial cells or glandular cells, this structure is not specific and not characteristic of the thyroid. As mentioned above, in the adult and larval lamprey, thyroid cells and their homologous cells bear numerous cilia on their apical surface. It is not known whether an evolution relationship exists between the numerous cilia in the lamprey thyroid cells and the central cilium in the thyroid cells of other vertebrates.

The occurrence of characteristic follicular epithelial cells called colloid cells is now discussed. Colloid cells, which are described by Langendorff (1889) in the dog and calf thyroid as cells having colloidlike clear or hyaline cytoplasm, are sometimes called Langendorff's cells. By electron microscopy the colloid cell is occasionally found in various species of normal and pathological animals, in metamorphosing salamanders (Setoguti, 1973a), normal mice (Nève *et al.*, 1970), in monkeys with experimental thyroiditis (Themann *et al.*, 1968), and in humans with autoimmune thyroiditis (Nève, 1966, 1969). Colloid cells are classified into two kinds: cells having very dilated cisternae of the rough endoplasmic reticulum, and cells showing disruption of the plasma membrane and extensive replacement of the cytoplasm by material resembling colloid (Nève *et al.*, 1970). Setoguti (1973a) reported a similar view classifying colloid cells.

IV. Ontogenetic Aspects of the Fine Structure of the Thyroid Gland

Since the first description of His (1885), it has been well known that the thyroid gland is chiefly derived from the endodermal epithelium of the ventral wall of the pharynx at the first branchial pouch level. The cells of this area begin to proliferate in the 2- to 3-mm human embryo. This is called a median thyroid anlage, which is later bifurcated to make up the left and right lobes and the isthmus. In addition, as described by His (1885), the lateral thyroid anlages (ultimobranchial bodies) located in the caudal part of the fourth branchial pouches may join the thyroid formation. The lateral thyroid anlages have been thought to fuse with the median thyroid to form parts of the left and right lobes. However, the role of the lateral anlages in forming the thyroid gland has not yet been completely defined. This article deals only with the development of the follicular cells and the formation of the follicle structure.

Fine-structural studies on the ontogenetic development of the thyroid gland have been made in the frog (Coleman *et al.*, 1968a; Nanba, 1972a; Neuenschwander, 1972), chick (Fig. 38) (Stoll *et al.*, 1957; Fujita and Machino, 1961; Hilfer, 1964; Fujita and Honma, 1965; Fugita and Tanizawa, 1966), rat (Feldman *et al.*, 1961; Ishikawa, 1965), and human (Shepard, 1966, 1967, 1968; Lietz *et al.*, 1971; Garcia-Bunuel *et al.*, 1972).

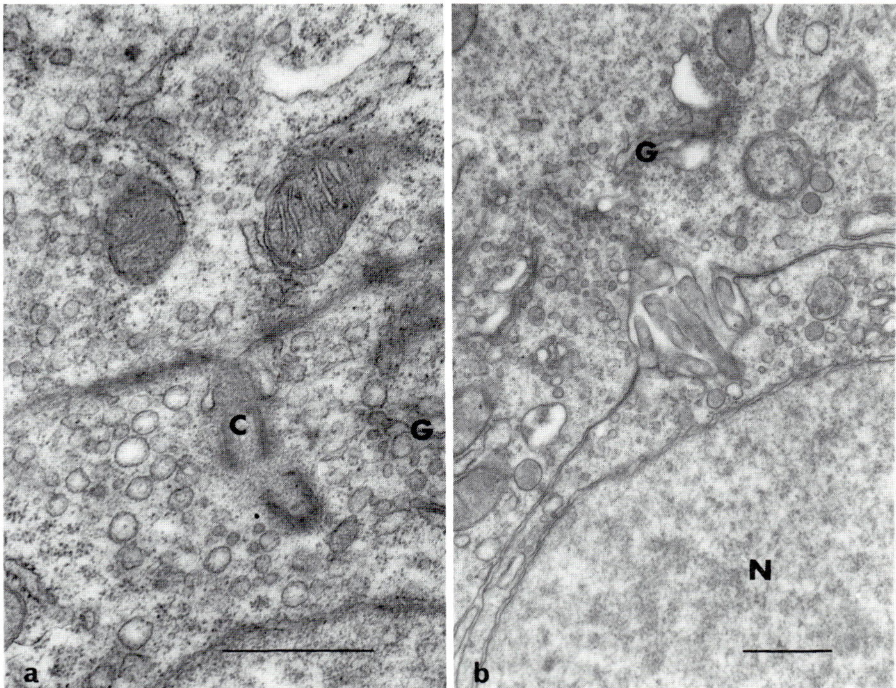

FIG. 38. First appearance of the primitive follicular lumen in an 8-day-old chick embryo. (a) A lumen begins to be made between two epithelial cells. C, Development of central cilium; G, Golgi apparatus. ×20,000. (b) A primitive follicle lumen appearing between two epithelial cells. G, Golgi apparatus; N, nucleus. ×12,000.

Ontogenetic and phylogenetic observations of the thyroid at the fine-structural level are very important and useful in understanding the functional morphology of the gland. Mechanisms of the formation of the secretory granule, of the follicular lumen (Fig. 39), and of the follicle structure have been clarified by these studies.

The fine structure of the thyroid cell without any secreting activity is, as in all the above animals, characteristic of poorly developed cytomembranes and abundant free ribosomes distributed throughout the cytoplasm. Elements of the rough endoplasmic reticulum are lacking, and the Golgi apparatus is very small in size; mitochondria having usual cristae are distributed throughout the cytoplasm. Neither secretory granules nor colloid droplets are observed and the epithelial cells aggregate to make up a cell cord without a follicular lumen. In addition, many lipid droplets are seen in functionally differentiated thyroid cells of the tadpole.

It has been reported that the follicular lumen appears electron micro-

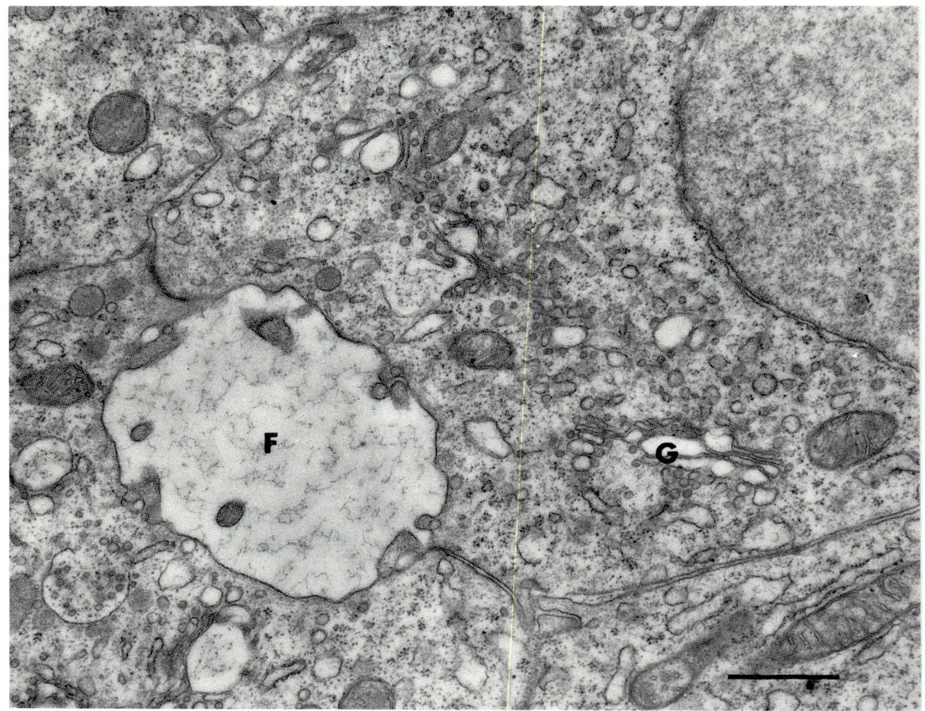

FIG. 39. A primitive follicle lumen (F), which is somewhat developed as compared with Fig. 38, in an 8-day-old chick embryo thyroid. Small vesicles around the lumen, which might be derived from the Golgi apparatus (G), are considered to be secretory granules containing thyroglobulin. ×14,500.

scopically in the 25-stage tadpole of *R. nigromaculata* (Nanba, 1972a), 8-day-old embryo of the chick (Fig. 40) (Hilfer, 1964; Fujita and Tanizawa, 1966), 17- to 18-day-old embryo of the rat (Feldman *et al.* 1961), and 60-mm embryo of the human (Olin *et al.*, 1970; Lietz *et al.*, 1971). Stoll *et al.* (1957, 1958) suggested a major role for the cytoplasmic vacuole (α cytomembrane) corresponding to that of the rough endoplasmic reticulum in the differentiation of thyroid cells. Detailed descriptions have been made by Hilfer (1964) and Fujita and Tanizawa (1966) of the functional differentiation of the thyroid cell and the mechanism of follicle formation in the chick embryo. Their data are summarized as follows (Fujita and Tanizawa, 1966). In the 8-day-old (rarely 7-day-old) chick embryo, the primitive follicular lumen about 1 μm in diameter appears between two epithelial cells. Fairly long microvilli projecting

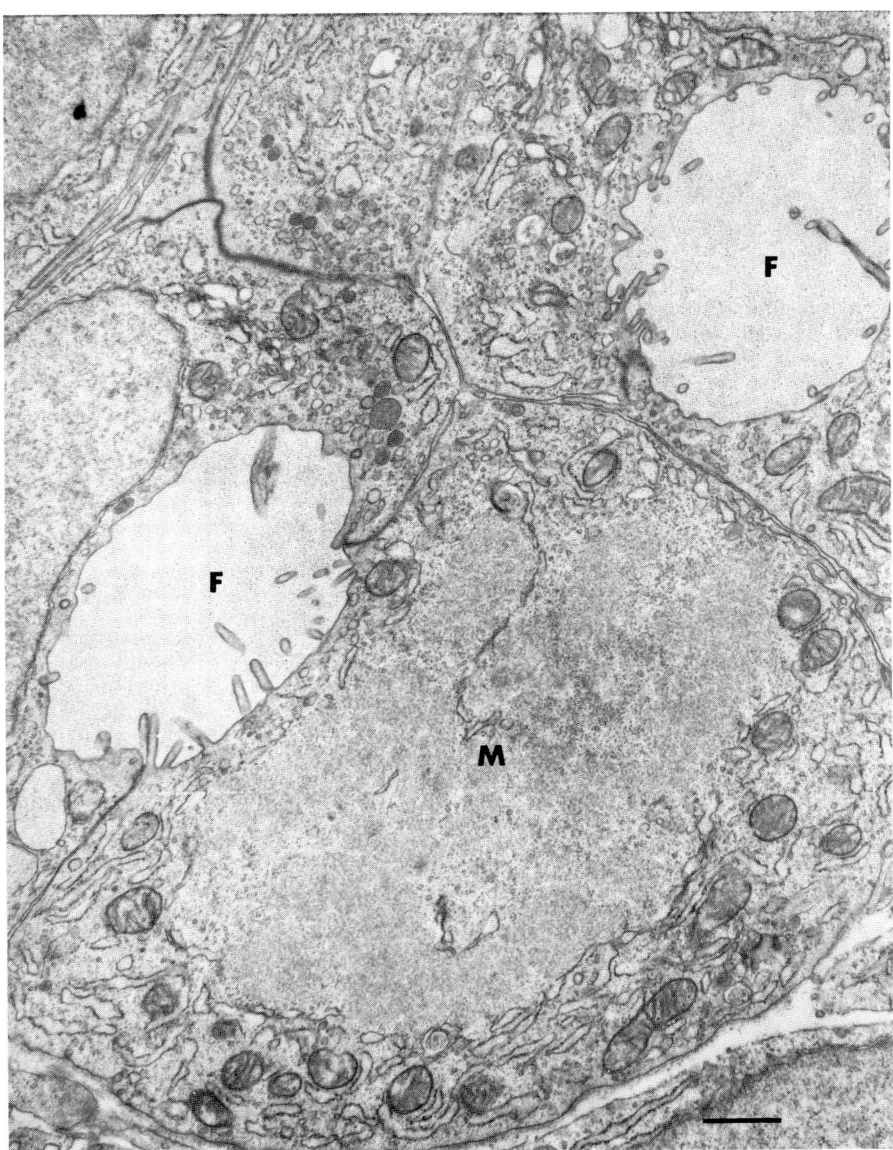

Fig. 40. Part of a thyroid of a 10-day-old chick embryo. Two follicle lumens (F) and a mitotic epithelial cell (M) are present. Rough endoplasmic reticulum is well developed as compared with that of an 8-day-old embryo. ×10,000. (From Fujita and Tanizawa, 1966.)

into the small lumen and a terminal bar forming a ring at the edge of the lumen are seen. Around the lumen are numerous small vesicles 50–150 nm in diameter, which might be secretory granules derived from the Golgi apparatus. The development of rough endoplasmic reticulum, the Golgi apparatus lying near the lumen, small vesicles derived from the Golgi apparatus, a central cilium projecting from the epithelial cell, and the appearance of the terminal bar are considered important factors in the formation of the primitive follicular lumen. A few small dense granules suggesting primary lysosomes also appear near the Golgi apparatus at 8 days of incubation. By mitosis of the epithelial cells and the increase in quantity of the colloid secreted into the lumen, the primitive follicular lumen enlarges with time. The daily development of the rough endoplasmic reticulum is especially notable, and at 16–17 days the follicular cell is almost as completely matured in its cytoplasmic fine structure as that of the adult animal. The loose mesenchyme begins to penetrate into the epithelial cell cord at 10–11 days to form each follicle unit, and the typical independent follicle unit is completed at 14 days of incubation.

The follicle formation of the frog embryo (tadpole) thyroid is almost similar to that of the chick embryo (Nanba, 1972a).

Hilfer and Stern (1971), using cultured thyroid cells from an 8-day-old chick embryo with or without spreading of mesenchyme elements, reported the necessity of mesenchymal tissue for cytodifferentiation, especially in the development of rough endoplasmic reticulum of the epithelial cell. In this experimental study, it was shown that the capsular mesenchymal tissues from an 8-day-old embryo are not effective in arrangement of the epithelial cells and in follicle formation, but those from a 16-day-old embryo are effective (Hilfer and Stern, 1971). The activity of lysosome has been reported to be important for rearrangement and reformation of the follicle structure in dissociated thyroid cells from a 16-day-old chick embryo (Hilfer *et al.*, 1968). Lysosomes are also necessary for hydrolysis of the reabsorbed colloid, as mentioned in Section B,5. We also found lysosomes in the thyroid cell of the 8-day-old chick embryo when its functional differentiation starts.

In the frog, the lysosomes appearing in the thyroid cell at stage IV (stage 25) are also useful in the digestion of numerous lipid droplets which disappear at functional differentiation (Nanba, 1972a). The first appearance of peroxidase activity is also an interesting aspect of tadpole embryology. The enzyme is first recognized histochemically in the vesicles near the Golgi complex or in the apical cytoplasm, and sometimes in the cisternae of the inner lamellae of the Golgi apparatus in early premetamorphosis (stage 25). This period coincides with that of the first concentration of ^{131}I uptake into the thyroid as shown by Kaye (1961). Peroxidase activity in the thyroid seems to parallel ^{131}I uptake into the thyroid throughout metamorphosis and reaches its maximum during climax metamorphosis (Nanba, 1973).

The development of the rat thyroid has been described by Feldman *et al.* (1961), who demonstrated very little thyroglobulin or immunologically similar protein in the thyroid of the 17-day-old rat embryo and slight ^{131}I accumulation in the 17- to 18-day-old rat. They did not find any direct correlation between the cytoplasmic structure and the presence of thyroglobulin, although the rough endoplasmic reticulum was noted to be responsible for the synthesis of thyroglobulin. In the 17-day-old embryo, they observed a very small follicular lumen between the two epithelial cells, and a few elements of rough endoplasmic reticulum and a small Golgi apparatus in the cytoplasm.

In regard to the ontogeny of the human thyroid, fine-structural studies of functional morphology have been made by several investigators (Shepard *et al.*, 1964; Shepard, 1966, 1967, 1968; Olin *et al.*, 1970; Lietz *et al.*, 1971; Garcia-Bunuel *et al.*, 1972). The mechanism of follicle formation is somewhat different from that in the chick embryo and in the tadpole. The intracellular lumina, which may communicate with the extracellular space, first appear in the 11-week or 60-mm human embryo, and are fused with one another to become typical follicle lumina (Shepard, 1966, 1968; Olin *et al.*, 1970). The relationship between activity for thyroglobulin synthesis and the appearance of rough endoplasmic reticulum and of cytoplasmic vesicles has been considered by Shepard (1966, 1968), and interesting biochemical and electron microscope studies have been performed in solving the details of this problem by Olin *et al.* (1970). They stated that in human fetuses of CR 47–54 mm the thyroid produces only noniodinated protein of low molecular weight (structural protein) and lacks rough endoplasmic reticulum but is rich in free ribosomes, and that in fetuses of CR 60 mm the thyroid synthesizes noniodinated or iodinated 17–19S thyroglobulin and shows a fairly well-developed rough endoplasmic reticulum. Also Lietz *et al.* (1971) did not find colloid in embryos less than 60 mm long. The appearances of rough endoplasmic reticulum, the Golgi apparatus, subapical vesicles, and intracellular lumina filled with dense materials are regarded as important signs of the functional differentiation of the thyroid cell.

The important role of the development of lysosomes in the onset of functional maturation of the human fetal thyroid has been reported (Garcia-Bunuel *et al.*, 1972). A rather abrupt increase in the number of lysosomes occurring at 10–12 weeks of gestation coincided chronologically with the appearance of follicular lumen and with the appearance of the other signs of functional differentiation mentioned above.

Characteristic thyroid cells which might originate from the ultimobranchial organ are now described. In addition to the usual follicle, Dunn,

(1944), Gorbman (1947a,b), and Wetzel and Wollman (1969) found a second kind of follicle in the normal mouse thyroid. It consisted of foamy colloid-containing cell debris and several kinds of unusual epithelial cells. According to Wetzel and Wollman (1969), the epithelial cells are classified into the following types: ciliated epithelial cells, agranular reticulum cells (named by Wetzel and Wollman, 1969), usual follicular epithelial cells, and parafollicular cells. Recently a fifth type of cell was added by Nève and Wollman (1972).

The ciliated cell bears numerous cilia on its apical surface, and its cytoplasm contains numerous mitochondria and poorly developed rough endoplasmic reticulum with nondilated cisternae (Wetzel and Wollman, 1969). The agranular reticulum cell is characterized by well-developed smooth endoplasmic reticulum, numerous large mitochondria, glycogen particles, and secretory granules (Wetzel and Wollman, 1969). Although this article does not deal with the parafollicular cell, we mention that this type cell has numerous secretory granules and is known to secrete thyrocalcitonin. The fifth type of cell, which is characterized by a relatively little cytoplasm containing clusters of fiber, vesicles near the basal plasma membrane, and half-desmosomes in the basal plasma membrane, resembles the ultimobranchial cell (Nève and Wollman, 1972). Recently, Calvert (1972) described three kinds of follicles in 2- to 4-week-old rats: usual follicles, ultimobranchial follicles, and mixed follicles. The ultimobranchial follicles possess a wall made of stratified squamous epithelium, and mixed follicles are lined by usual follicular epithelial cells and lumen-bordering cells, which might be forms transitional between ultimobranchial and parafollicular cells. Although numerous articles have been published on the fate of the ultimobranchial body, many questions remain to be answered in the future.

V. Fine Structure of the Thyroid Gland under Experimental Conditions

Several articles have been published on the histological and functional reactions of the thyroid under various experimental conditions or under certain pathological conditions. Among these only studies dealing with cytological alterations of the gland induced by some experimental condition are reviewed in this article.

A. Hypophysectomy and T_4 Treatment

Nièpce (1851) was probably the first investigator who noticed the functional relationship between the pituitary and thyroid gland. He found hypertrophy of the hypophysis in cretinous patients, and Smith (1921) recognized atrophy of the thyroid after hypophysectomy in the tadpole.

Since then numerous studies have been carried out on effects of hypophysectomy on various endocrine organs. Fine-structural changes in the rat thyroid after hypophysectomy (Fig. 41) have been observed by Dempsey and Peterson (1955), Wetzel *et al.* (1965), Seljelid (1967a), Schwarz (1967), Lupulescu and Petrovici (1968), and Fujita and Suemasa (1968). It is well known that TSH secreted by the beta cell of the adenohypophysis is necessary for performance of the thyroid function. TSH secretion is almost stopped, and the thyroid becomes free of this hormonal control after hypophysectomy. The descriptions presented by all these investigators generally agree with one another.

After hypophysectomy, thyroid follicular epithelial cells are attenuated; follicular lumens become larger in size, intracellular reabsorbed colloid droplets decrease in number, and elements of rough endoplasmic reticulum and Golgi apparatus are reduced in size. These reactions become stronger during the days after the operation. In 20- to 30-day hypophysectomized rats, no intracellular colloid droplets and no secretory granules are observed, elements of rough endoplasmic reticulum are extremely sparse, the Golgi apparatus located at the supranuclear region is very small, the cytoplasmic matrix is electron-dense and has free ribosomes, and a few small mitochondria and a few lysosomes are scattered through-

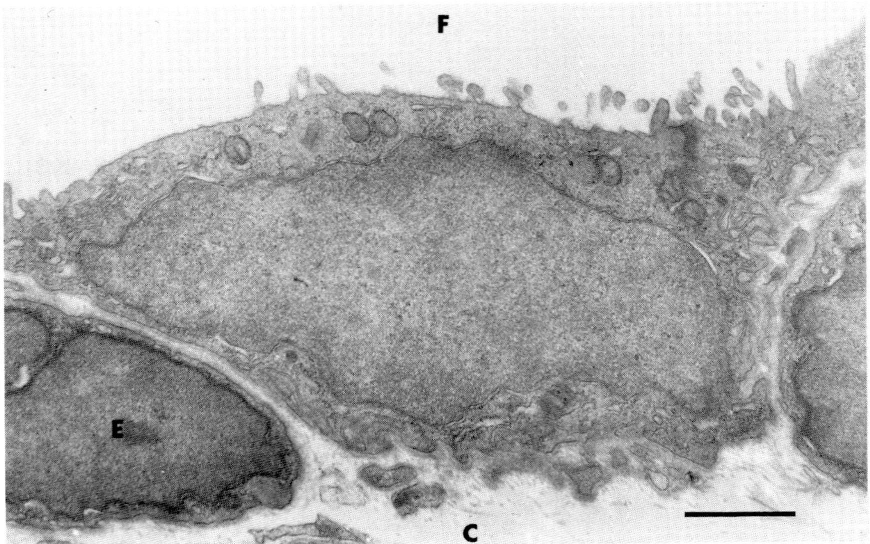

FIG. 41. An attenuated follicular epithelial cell of a 20-day-hypophysectomized rat. Cytomembranes are very poorly developed, and the cytoplasmic matrix looks compact. C, Connective tissue space; E, endothelial cell; F, follicular lumen. ×17,000.

out the cytoplasm (Fujita and Suemasa, 1968). Accumulation of radioiodine is markedly reduced in the hypophysectomized animal. These observations suggest that the reabsorption and hydrolysis of colloid and protein synthesis are extremely inhibited by the inhibition of TSH secretion. Parafollicular cells, which have been known to secrete thyrocalcitonin, are not affected by hypophysectomy (Fujita and Suemasa, 1968).

Similar changes are induced in the rat by prolonged treatment with T_4 for 5–15 days (Seljelid *et al.*, 1971). The follicular epithelial cells and their nuclei become flattened, and rough endoplasmic reticulum, Golgi apparatus, mitochondria, and polysomes are reduced in number and extent. Intracellular colloid droplets disappear, autophagic vacuoles are easily seen, and the acid phosphatase activity in the thyroid rapidly decreases. An animal in which the thyroid gland is suppressed by treatment with T_4 is sometimes called a T_4-blocked animal. Hypophysectomized or T_4-blocked animals were often used for experimental studies detecting the action of various chemical agents on the thyroid gland.

B. Hibernating Animals

Concerning the fine-structural changes in the thyroid during hibernation, reports have been published on the Chiroptera (Azzali, 1967), and on the bat (Fig. 42) (Nunez and Becker, 1970; Nunez, 1971; Fujita, 1971).

During prehibernation (late autumn) of the bat (Nunez and Becker, 1970), the thyroid follicular epithelial cells show proliferation of rough endoplasmic reticulum, development in size of the Golgi apparatus, and an increase in the number of subapical vesicles, of multivesicular bodies, and of dense granules as compared with those of active summer animals. These structures show a hyperactivity of the cell in synthesizing thyroglobulin.

Cytoorganelles such as the rough endoplasmic reticulum and the Golgi apparatus are reduced in amount and size with advancing days of hibernation. In midhibernation (late December to early January) and late hibernation (late January to late March) (Nunez and Becker, 1970; Nunez, 1971; Fujita, 1971), the follicular epithelial cells of the bat become markedly attenuated, rough endoplasmic reticulum and Golgi apparatus are very poorly developed, and mitochondria, being somewhat larger in size but rather fewer in number, are distributed in the cytoplasm. Intracellular colloid droplets are difficult to find, while many dense granules and dense bodies are observed throughout the cytoplasm. These changes, being fairly similar to those occurring after hypophysectomy, are understood to represent a hypofunctional state of the cell. The appearance of typical junctional complexes in the late-hiberating bat has

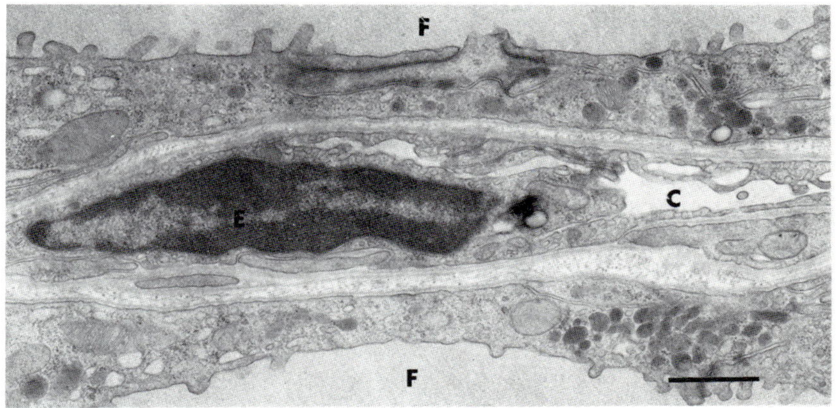

FIG. 42. Attenuated follicular epithelial cells of the thyroid of a hibernating bat. C, Blood capillary; F, follicular lumen; E, endothelial cell. ×14,000. (From Fujita, 1971.)

been reported by Nunez (1971). He observed that these complexes are characterized by a localized dense thickening of the apposed lateral plasma membranes, separated by an intercellular space 75 A wide, and that there is a close apposition of rough endoplasmic reticulum and mitochondria. The functional significance of this structure is not clear. The accumulation of radioiodine in the thyroid gland is also markedly reduced (Fujita, 1971). The follicular cells of these hibernating bats react to injected TSH (Fujita, 1971).

In 15-day aroused bats (24°C) in midwinter, although most cells are flattened and the elements of rough endoplasmic reticulum sparse, a few cells show fairly well-developed rough endoplasmic reticulum with somewhat dilated cisternae. Sometimes large colloid droplets occur in a few cells, and lysosomal granules are reduced in number.

In 60-day aroused bats, all the follicular epithelial cells show the same features as the active thyroid cells of the nonhibernating bat.

C. TSH Treatment

TSH secreted by beta cells of the anterior pituitary is the main factor controlling thyroid function. It is well known that this agent stimulates the gland to induce a hyperfunctional state and an elevation of T_4 release into the blood vessels. The hormone-releasing action of TSH is rapid, within 10–30 minutes (Campbell *et al.*, 1960; Taurog *et al.*, 1964; Rosenberg *et al.*, 1965).

Cytological alterations of the thyroid gland after the injection of TSH have been observed in various animals: in the salamander (Herman, 1960), tadpole (Neuenschwander, 1972), chick (Fujita, 1963), mouse (Ekholm, 1960), rat (Roos, 1960; Wissig, 1963; Wetzel *et al.*, 1965; Seljelid, 1965, 1967a,b,c,d,e; Lupulescu and Petrovici, 1968; Fujita and Suemasa, 1968), guinea pig (Sobel and Geller, 1965; Ekholm and Smeds, 1966; Kosanović *et al.*, 1968), and dog (Tashiro and Sugiyama, 1964; Nève and Dumont, 1970). The fine-structural changes of the follicular epithelial cell induced by this hormone are generally similar in all these animals, although there are some species differences in the speed of reaction. As mentioned in Section B,4 the first cytological reactions occurring after TSH injection are: pseudopod formation, appearance of intracellular colloid droplets, movement of lysosomes to the apical cytoplasm, and fusion of colloid droplets with the lysosomes.

Five minutes after the intravenous injection of TSH, pseudopods and reabsorbed colloid droplets appear in the apical part of the follicular epithelial cell of the rat (Wetzel *et al.*, 1965; Wollman, 1965; Seljelid, 1967b,c), while these reactions are observed in the dog 15 minutes after intravenous injection (Nève and Dumont, 1970b). The number of intracellular colloid droplets reaches a maximum 2 hours after TSH injection in the rat, and after 3–4 hours in the dog (Nève and Dumont, 1970b). The reabsorbed colloid is hydrolyzed by the lysosome to liberate T_3 and T_4, and the thyroid hormone level in the blood vessels is simultaneously elevated. The activity of the hydrolytic enzymes is markedly increased by repeated TSH treatment (Kosanović *et al.*, 1968; Seljelid *et al.*, 1971).

Other changes induced by the injection of TSH are marked dilation of the cisternae of rough endoplasmic reticulum and development of the Golgi apparatus (Herman, 1960; Wissig, 1963; Fujita, 1963). These reactions are observed somewhat later than the appearance of pseudopods and of intracellular colloid droplets (Nève and Dumont, 1970a,b). Biochemical data have shown the increase in RNA content in the thyroid after the injection of TSH (Fiala *et al.*, 1957; Creek, 1965; Kerkof and Tata, 1967). This observation means that an elevation of protein synthesis occurs, although it is not clear whether this is a direct or indirect action of TSH. Subapical vesicles (secretory granules) derived from the Golgi apparatus increase in number in the apical cytoplasm of rat follicular cells 1–3 hours after the injection of TSH. Some cells show extremely dilated cisternae of rough endoplasmic reticulum, which occupy most of the basal cytoplasm (Fujita, 1963; Tashiro and Sugiyama, 1964). Based on these observations, formation of the secretory granules containing thyroglobulin is thought to be stimulated by injected TSH. The relationship between lysosome formation and the increase in RNA content should also be considered, because large amounts of lysosomal enzymes are necessary for hydrolyzing the reabsorbed colloid. Even in elasmobranchs, the thyroid follicular epithelial cells react in a manner similar to those in the higher vertebrate (Nakai and Gorbman, 1969). The reaction to injected TSH in the hagfish thyroid is neither prompt nor clear. Zone-7 and -8 cells in the ascidian endostyle show large intracellular droplets a few hours after immersion in seawater containing TSH, according to our observations.

TSH-sensitive ATPase of the thyroid tissue has been demonstrated biochemically (Turkington, 1962). Recently, we found that the ATPase reaction becomes strongly positive in the apical plasma membrane of guinea pig follicular cells after the injection of TSH. Although its functional properties are obscure, there is a possibility that ATPase appearing in the apical plasma membrane after injection of TSH might be related to the endocytosis of colloid droplets.

The acute reaction of the thyroid induced to a single injection of TSH returns almost to normal after 24 hours in the rat (Lupulescu and Petrovici, 1968), although the dilation of cisternae of rough endoplasmic reticulum remains in some cells in the dog (Nève and Dumont, 1970b).

After prolonged treatment with TSH, a marked decrease in the volume of the follicular lumen, a remarkable increase in follicular cell height, proliferation of rough endoplasmic reticulum with dilated cisternae, hypertrophy of the Golgi apparatus, an increase in free ribosomes in number, an increase in mitochondria in number and size, an increase in intracellular colloid droplets in number and size, elongation of microvilli, and hyperemia of the sinusoid capillary are observed (Lupulescu and Petrovici, 1968). Similar results were obtained by Nève and Dumont (1970a,b) in the dog and, in addition to these, an increase in hydrolytic activity as a result of treatment with TSH in the rat thyroid was shown biochemically by Seljelid *et al.* (1971). These changes indicate elevation of release of thyroid hormone and of synthesis of thyroglobulin in the thyroid gland.

D. Treatment with Other Thyroid Stimulants

TSH is now believed to stimulate adenyl cyclase in the plasma membrane of the follicular epithelial cell to produce cAMP, which is the mediator of hormonal activity (Gilman and Rall, 1966). TSH elevates the thyroidal concentration of cAMP *in vivo* and *in vitro* (Gilman and Rall, 1966, 1968a,b). Since exogenous cAMP passes through the plasma membrane only with difficulty, DB-cAMP, which can penetrate the plasma membrane, is used in experimental studies instead of cAMP. Pastan and Wollman (1967) and Nève and Dumont (1970a) reported the formation of intracellular colloid droplets in slices of dog thyroid cultured in a medium containing DB-cAMP. A similar action of prostaglandin E_1 was also shown by Nève and Dumont (1970a,b).

The effect of catecholamines on the thyroid function is very complicated. It has been reported by Ericson *et al.* (1970) that *l*-epinephrine, *l*-isoproterenol, 5-hydroxytryptamine, and other aromatic monoamines induce endocytosis of thyroglobulin and release of the thyroid hormone. Using hypophysectomized or T_4-blocked mice, they noted the appearance of intracellular colloid droplets within 10 minutes (reaching a maximum in number at 30 minutes), their disappearance at 60 minutes, and an increase in blood ^{131}I level which reached a peak value at 2 hours after a single injection of these agents. However, Pastan and Wollman (1967) and Onaya and Solomon (1969) did not find any morphological changes in dog thyroid tissue cultured in a medium containing epinephrine, isoproterenol, or 5-hydroxytryptamine. On the contrary, inhibiting effects of these agents on the thyroid function have been reported (Szánto *et al.*, 1966; Cavazzuti and Chigi, 1969).

Melander and Sundler (1972) reported the following physiological data. The stimulatory effects of amines on the thyroid are induced only in hypophysectomized or T_4-blocked mice, and the amines and TSH produce additive responses when administered simultaneously. They are mutually antagonistic when administered at an interval of 2 hours. With an interval of 5 minutes between the administration of the amines and TSH, epinephrine and norepinephrine cause a diminished response to TSH, while 5-hydroxytryptamine and isoproterenol are additive with TSH (Melander and Sundler, 1972). It was speculated that all these amines probably activate the thyroid cell, but that epinephrine and norepinephrine inhibit the TSH-stimulated secretion of thyroid hormone by causing vasoconstriction in the thyroid. Parafollicular cells and mast cells which release 5-hydroxytryptamine, and adrenergic nerve terminals which release catecholamine, may have a role in controlling the function of thyroid follicular cells (Ericson *et al.*, 1970, 1972; Melander and Sundler, 1972).

E. Thiouracil Treatment

Thiouracil and its derivatives, such as methylthiouracil and propylthiouracil, used for the treatment of hyperthyroidism since Astwood (1944), are known to be blocking agents in the iodination of thyroglobulin. Noniodinated thyroglobulin has been obtained in thiouracil-treated hogs (Tarutani and Ui, 1968, 1969). After treatment with these chemicals, release of T_3 and T_4 into the blood vessels is markedly inhibited.

By a negative feedback mechanism, the release of thyrotropin-releasing factor (TRF) is strongly stimulated, and consequently TSH secretion by the beta cells of the adenohypophysis becomes high. Then the thyroid gland reacts to highly concentrated TSH released by the anterior pituitary. However, iodinated thyronine is not secreted by the thyroid, as iodination of thyroglobulin is inhibited by thiouracil; consequently, this reaction circle is strongly stimulated. Since long-term treatment with thiouracil results in experimental goiter, this agent is called a goitrogen. Numerous articles have been published on the cytological changes in the thyroid after prolonged or single injection of thiouracil or its derivatives (Dempsey and Peterson, 1955; Irie, 1960; Fujita *et al.*, 1963; Wissig, 1964; Langer *et al.*, 1967; Lupulescu and Petrovici, 1968; Coleman *et al.*, 1968b,c). Fine-structural alterations of the follicular epithelial cell are fairly similar to those occurring after TSH injection. The decrease in size of the follicular lumen, increase in cell height, dilation of cisternae of rough endoplasmic reticulum, hypertrophy of the Golgi apparatus, and elongation and irregularization of microvilli in length and shape are remarkable. In addition, characteristic lysosomal inclusion bodies increase in number in the cytoplasm of 6-month-treated rats (Fujita *et al.*, 1963).

F. Miscellaneous

Numerous reports have been published on the pathological alterations of the thyroid gland induced by various other experimental conditions, such as the administration of iodine (Waller, 1961; Lupulescu and Petrovici, 1968), fluoride (Waller, 1961), potassium thiocyanate (Lupulescu and Petrovici, 1968) or lithium (Heltne and Ollerich, 1972), deficiencies of iodine (Feldman, 1961; Lupulescu and Petrovici, 1963, 1964) and of tocopherol and ubiquinone (Blähser and Schnorr, 1971), cold stress (Dempsey and Peterson, 1955; Ehrenbrand, 1966; Krstić and Bucher, 1971), irradiation with radio-iodine (Wang, 1960; Sobel, 1964), and autoimmune injury (Sobel and Geller, 1965; Flax and Billote, 1965; Themann *et al.*, 1968; Karsensen, 1970; Nève, 1971). This article does not deal with the details of pathological changes in the thyroid.

References

Alexander, N. M. (1959). *J. Biol. Chem.* **234**, 1530.

Alexander, N. M. (1961). *Endocrinology* **68**, 671.

Alexander, N. M. (1965). *In* "Current Topics in Thyroid Research" (C. Cassano and M. Andreoli, eds.), pp. 43–54. Academic Press, New York.

Alexander, N. M., and Corcoran, B. J. (1962). *J. Biol. Chem.* **237**, 243.

Andros, G., and Wollman, H. (1964). *Proc. Soc. Exp. Biol. Med.* **115**, 775.

Aros, B., and Viragh, S. (1969). *Acta Biol. (Budapest)* **20**, 281.

Astwood, E. B. (1944). *J. Clin. Endocrinol. Metab.* **4**, 229.

Azzali, G. (1967). *Acta Bio-Med.* **38**, 5.
Bargmann, W. (1939). *In* "Möllendorffs Handbuch der mikroskopischen Anatomie des Menschen" (W. von Möllendorff, ed.), Vol. VI, Part 2. Springer-Verlag, Berlin.
Barrington, E. J. W. (1957). *J. Mar. Biol. Ass. U.K.* **36**, 1.
Barrington, E. J. W. (1958). *J. Mar. Biol. Ass. U.K.* **37**, 117.
Barrington, E. J. W. (1959). *In* "Comparative Endocrinology" (A. Gorbman, ed.), pp. 250–285. Wiley, New York.
Barrington, E. J. W. (1964). "Hormones and Evolution." English Univ. Press, London.
Barrington, E. J. W. (1965). "The Biology of Hemichordata and Protochordata." Oliver & Boyd, Edinburgh.
Barrington, E. J. W. (1968). *In* "Perspective in Endocrinology" (E. J. W. Barrington and C. B. Jorgensen, eds.), pp. 1–46. Academic Press, New York.
Barrington, E. J. W., and Barron, N. (1960). *J. Mar. Biol. Ass. U.K.* **39**, 513.
Barrington, E. J. W., and Franchi, L L. (1956). *Quart. J. Microsc. Sci.* **97**, 393.
Barrington, E. J. W., and Sage, M. (1972). *In* "The Biology of Lampreys" (M. W. Hardisty and I. C. Potter, eds.), pp. 105–134. Academic Press, New York.
Barrington, E. J. W., and Thorpe, A. (1965a). *Proc. Roy. Soc., Ser. B* **163**, 136.
Barrington, E. J. W., and Thorpe, A. (1965b). *Gen. Comp. Endocrinol.* **5**, 373.
Bauer, W., and Meyer, J. (1965). *Lab. Invest.* **14**, 1795.
Baumann, E. (1895). *Z. Physiol. Chem.* **21**, 319.
Benabdeljlil, C., Michel-Béchet, M., and Lissitzky, S. (1967). *Biochem. Biophys. Res. Commun.* **27**, 74.
Bensley, R. R. (1914). *Anat. Rec.* **8**, 431.
Billroth, T. (1882). "Die allgemeine chirurgische Pathologie und Therapie," 10th Ed. Reimer, Berlin.
Biondi, D. (1892). *Arch. Ital. Biol.* **17**, 475.
Blähser, S., and Schnorr, B. (1971). *Z. Zellforsch. Mikrosk Anat.* **114**, 493.
Bouchilloux, S., Ferrand, J., Gregoire, J., and Chabaud, O. (1969). *Biochem. Biophys. Res. Commun.* **37**, 538.
Bouchilloux, S., Chabaud, O., Michel-Béchet, M., Ferrand, M., and Athouel-Haon, A. M. (1970). *Biochem. Biophys. Res. Commun.* **40**, 314.
Braunsteiner, H., Fellinger, K., and Pakesch, F. (1953). *Endocrinology* **53**, 123.
Brettschneider, H. (1963). *Z. Mikrosk.-Anat. Forsch.* **69**, 630.
Bucher, O. (1940). *Z. Zellforsch. Mikrosk. Anat.* **30**, 432.
Bucher, O., and Riedel, B. (1967). *Z. Zellforsch. Mikrosk. Anat.* **79**, 364.
Calvert, R. (1972). *Anat. Rec.* **174**, 341.
Campbell, H. J., George, R., and Harris, G. W. (1960). *J. Physiol. (London)* **152**, 527.
Castiglioni, A. (1958). "A History of Medicine" (E. B. Krumbhaar, transl.) Knopf, New York.
Cavazzuti, G. R. R., and Chigi, C. (1969). *Acta Anat., Suppl.* **56**, 107.
Clements, M., and Gorbman, A. (1955). *Biol. Bull.* **108**, 258.
Clements-Merlini, M. (1960a). *J. Morphol.* **106**, 337.
Clements-Merlini, M. (1960b). *J. Morphol.* **106**, 357.
Coindet, J. F. (1820). *Ann. Chim. Phys.* **15**, 49.
Cole, F. J. (1905). *Anat. Anz.* **27**, 323.
Coleman, R., Evennett, R. J., and Dodd, J. M. (1968a). *Gen. Comp. Endocrinol.* **10**, 34.
Coleman, R., Evennett, R. J., and Dodd, J. M. (1968b). *Z. Zellforsch. Mikrosk. Anat.* **84**, 490.

Coleman, R., Evennett, R. J., and Dodd, J. M. (1968c). *Z. Zellforsch. Mikrosk. Anat.* **84**, 497.
Copp, D. H., Cameron, E. C., Cheney, B. A., Davidson, A. G. F., and Henze, K. G. (1962). *Endocrinology* **70**, 638.
Covelli, I., Salvatore, G., Sena, L., and Roche, J. (1960). *C. R. Soc. Biol.* **154**, 1165.
Creek, R. D. (1965). *Endocrinology* **76**, 1124.
De Crombrugghe, B., Pitt-Rivers, R., and Edelhoch, H. (1966). *J. Biol. Chem.* **241**, 2766.
De Groot, L. J., and Davis, A. M. (1962). *Endocrinology* **70**, 492.
De Groot, L. J., Thompson, J. E., and Dunn, A. D. (1965). *Endocrinology* **76**, 632.
Dempsey, E. W. (1944). *Endocrinology* **34**, 27.
Dempsey, E. W., and Peterson, R. R. (1955). *Endocrinology* **56**, 46.
De Robertis, E., and Grasso, K. (1946). *Endocrinology* **38**, 137.
Dodd, J. M., and Matty, A. J. (1964). *In* "Thyroid Gland" (R. Pitt-Rivers and W. R. Trotter, eds.), Vol. 1, pp. 303–356. Butterworth, London.
Dohrn, A. (1886). *Mitt. Zool. Stat. Neapel* **6**, 49.
Doniach, I., and Logothetopoulos, J. H. (1955). *J. Endocrinol.* **13**, 65.
Doniach, I., and Pelc, S. R. (1949). *Proc. Roy. Soc. Med.* **42**, 957.
Dunn, T. B. (1944). *J. Nat. Cancer Inst.* **4**, 555.
Edelhoch, H (1965). *Recent Progr. Hormone Res.* **21**, 1.
Egeberg, J. (1965). *Z. Zellforsch. Mikrosk. Anat.* **68**, 102.
Eggert, B. (1938). "Berblingers Zwanglose Abhandlungen aus dem Gebiet der inneren Sekretion," Vol. 3, Johann Ambrosius Barth, Leipzig.
Ehrenbrand, F. (1966). *Z. Anat. Entwicklungsgesch.* **125**, 211.
Ekholm, R. (1957). *Z. Zellforsch. Mikrosk. Anat.* **46**, 139.
Ekholm, R. (1960). *Int. Conf. Electron, Microsc., Proc., 4th, Berlin, 1958* **2**, 378–381, Springer, Berlin.
Ekholm, R. (1964). *In* "Electron Microscopic Anatomy" (S. Kurtz, ed.), pp. 221–237. Academic Press. New York.
Ekholm, R. (1966). *J. Ultrastruct. Res.* **14**, 419.
Ekholm, R., and Elmquist, L.-G. (1967). *Exp. Cell Res.* **48**, 640.
Ekholm, R., and Sjöstrand, F. (1957). *J. Ultrastruct. Res.* **1**, 178.
Ekholm, R., and Smeds, S. (1966). *J. Ultrastruct. Res.* **16**, 71.
Ekholm, R., and Strandberg, U. (1966). *J. Ulstrastruct. Res.* **16**, 71.
Ekholm, R., and Strandberg, U. (1967a). *J. Ultrastruct. Res.* **17**, 184.
Ekholm, R., and Strandberg, U. (1967b). *J. Ultrastruct. Res.* **20**, 103.
Ekholm, R., and Strandberg, U. (1968). *J. Ultrastruct. Res.* **22**, 252.
Ekholm, R., Helander, H., and Zelander, T. (1963). *Z. Zellforsch. Mikrosk. Anat.* **59**, 467.
Elfvin, L.-G. (1965). *J. Ultrastruct. Res.* **12**, 687.
Elfvin, L.-G. (1971). *J. Ultrastruct. Res.* **34**, 345.
Ericson, L. E., Melander, A., Owman, C., and Sundler, F. (1970). *Endocrinology* **87**, 915.
Ericson, L. G., Hakanson, R., Melander, A., Owman, C., and Sundler, F. (1972). *Endocrinology* **90**, 795.
Essner, E. (1969). *J. Histochem. Cytochem.* **17**, 454.
Fabre, M., and Marescaux, J. (1968). *Z. Zellforsch. Mikrosk. Anat.* **84**, 328.
Farquhar, M. G., and Palade, G. E. (1966). *J. Cell Biol.* **30**, 359.
Fawcett, D. M., and Kirkwood, S. (1954). *J. Biol. Chem.* **209**, 249.
Feeney, L., and Wissig, S. L. (1971). *Tissue Cell* **3**, 9.

Feeney, L., and Wissig, S. L. (1972). *J. Cell Biol.* **53**, 510.
Feldman, J. D. (1961). *In* "Advances in Thyroid Research" (R. Pitt-Rivers, ed.), pp. 318–323. Pergamon, Oxford.
Feldman, J. D., Vazquez, J. J., and Kurtz, S. M. (1961). *J. Biophys. Biochem. Cytol.* **11**, 365.
Fiala, S., Sproul, E. E., and Fiala, A. E. (1957). *Proc. Soc. Exp. Biol.* **94**, 517.
Flax, M. H., and Billote, J. B. (1965). *Ann. N. Y. Acad. Sci.* **124**, 234.
Franz, V. (1927). *Ergeb. Anat. Entwicklungsgesch.* **27**, 464.
Fujita, H. (1963). *Z. Zellforsch. Mikrosk. Anat.* **60**, 615.
Fujita, H. (1966). *Int. Conf. Electron Microsc., Proc., 6th, Kyoto, 1966,* **2**, 529–530, Maruzen, Tokyo.
Fujita, H. (1968). *In* "Fine Structure of Cells and Tissues. Electron Microscopic Atlas" (E. Yamada, K. Uchizono, and Y. Watanabe, eds.), Vol. 5, pp. 86–113. Igaku Shoin, Tokyo.
Fujita, H. (1969). *Virchows Arch. B Cell Pathol.* **2**, 265.
Fujita, H. (1970). *Gunma Symp. Endocrinol.* **7**, 49.
Fujita, H. (1971). *Z. Zellforsch. Mikrosk. Anat.* **121**, 301.
Fujita, H. (1972). *Arch. Histol. Jap.* **34**, 109.
Fujita, H., and Honma, Y. (1965). *Gunma Symp. Endocrinol.* [*Proc.*] **3**, 121.
Fujita, H., and Honma, Y. (1966). *Z. Zellforsch. Mikrosk. Anat.* **73**, 559.
Fujita, H., and Honma, Y. (1968). *Gen. Comp. Endocrinol.* **11**, 111.
Fujita, H., and Honma, Y. (1969). *Z. Zellforsch. Mikrosk. Anat.* **98**, 525.
Fujita, H., and Kataoka, K. (1969). *Z. Anat. Entwicklungsgesch.* **128**, 318.
Fujita, H., and Machino, M. (1961). *Exp. Cell Res.* **25**, 204.
Fujita, H., and Machino, M. (1965). *Anat. Rec.* **152**, 81.
Fujita, H., and Murakami, T. (1974). *Arch. Histol. Jap.* **36**, 180.
Fujita, H., and Nanba, H. (1971). *Z. Zellforsch. Mikrosk. Anat.* **121**, 455.
Fujita, H., and Suemasa, H. (1968). *Arch. Histol. Jap.* **30**, 45.
Fujita, H., and Tanizawa, Y. (1966). *Z. Anat. Entwicklungsgesch.* **125**, 132.
Fujita, H., Kano, M., and Okamoto, S. (1958). *Arch. Histol. Jap.* **14**, 61.
Fujita, H., Machino, M., and Nakagami, K. (1963). *Okajimas Folia Anat. Jap.* **39**, 157.
Fujita, H., Suemasa, H., and Honma, Y. (1966). *Arch. Histol. Jap.* **27**, 153.
Garcia-Bunuel, R., Anton, E., and Brandes, D. (1972). *Endocrinology* **91**, 438.
Garnier, B. (1956). *Schweiz. Z. Allg. Pathol. Bakteriol.* **19**, 129.
Gerlach, J. (1850). "Handbuch der allgemeinen und speziellen Gewebelehre des meschlichen Körpers." Braumüller, Mainz.
Gersh, I., and Caspersson, T. (1940). *Anat. Rec.* **78**, 303.
Gilman, A. G., and Rall, T. W. (1966). *Fed. Proc., Fed. Amer. Soc. Exp. Biol.* **25**, 617.
Gilman, A. G., and Rall, T. W. (1968a). *J. Biol. Chem.* **243**, 5867.
Gilman, A. G., and Rall, T. W. (1968b). *J. Biol. Chem.* **243**, 5872.
Gorbman, A. (1947a). *Cancer Res.* **7**, 746.
Gorbman, A. (1947b). *Anat. Rec.* **98**, 93.
Gorbman, A. (1959). *In* "Comparative Endocrinology" (A. Gorbman, ed.), pp. 266–282. Wiley, New York.
Gorbman, A., and Bern, H. A. (1962). "Textbook of Comparative Endocrinology." Wiley, New York.
Gorbman, A., and Evans, H. M. (1941). *Proc. Soc. Exp. Biol. Med.* **47**, 103.
Graham, R. C., and Karnovsky, M. J. (1966). *J. Histochem. Cytochem.* **14**, 291.

Greene, R., Inman, D., and Newman, R. (1966). *In* "Tumors of the Thyroid Gland" (A. Appaix, ed.), pp. 110–116. Karger, Basel.
Gross, J., and Pitt-Rivers, R. (1952). *Lancet* **i**, 439.
Haddad, A. (1972). *J. Histochem. Cytochem.* **20**, 220.
Haddad, A., Smith, M. D., Herscovics, A., Hadler, N. J., and Leblond, C. P. (1971). *J. Cell Biol.* **49**, 856.
Halmi, N. S., Westra, J. P., and Polly, R. E. (1966). *Endocrinology* **79**, 424.
Hamilton, J. G., Soley, M. H., and Eichhorn, K. B. (1940). *Univ. Calif., Berkeley, Publ. Pharmacol.* **1**, 339.
Harrington, C. R. (1926). *Biochem. J.* **20**, 293.
Harrison, R. J., and Young, B. A. (1970). *J. Anat.* **106**, 243.
Harrison, R. J., Rowlands, I. W., Whitting, H. W., and Young, B. A. (1962). *J. Anat.* **96**, 3.
Hashimoto, P. H., and Hama, K. (1968). *Med. J. Oscka Univ.* **18**, 331.
Hearing, V. J., and Eppig, J. T., Jr. (1969). *J. Morphol.* **128**, 369.
Heimann, P. (1966). *Acta Endocrinol. (Copenhagen), Suppl.* **110**, 1.
Heltne, C. E., and Ollerich, D. A. (1972). *Amer. J. Anat.* **136**, 297.
Henderson, N. E., and Gorbman, A. (1971). *Gen. Comp. Endocrinol.* **16**, 409.
Herman, L. (1960). *J. Biophys. Biochem. Cytol.* **7**, 143.
Hertz, S. (1943). "On Goiter and Allied Diseases." Munksgaard, Copenhagen.
Hilfer, S. R. (1964). *J. Morphol.* **115**, 135.
Hilfer, S. R., and Stern, M. (1971). *J. Exp. Zool.* **178**, 293.
Hilfer, S. R., Iszard, L. B., and Hilfer, E. K. (1968). *Z. Zellforsch. Mikrosk. Anat.* **92**, 256.
Hirsch, P. F., Gauthier, G. F., and Munson, P. L. (1963). *Endocrinology* **73**, 244.
His, W. (1885). "Anatomie Menschlicher Embryonen." Vogel, Leipzig.
Hoheisel, G. (1969). *Morphol. Jahrb.* **114**, 204.
Hoheisel, G. (1970). *Morphol. Jahrb.* **114**, 337.
Honma, Y. (1960). "Studies on the Morphology and Role of the Important Endocrine Glands in some Japanese Cyclostomes and Fishes." Niigata Daigaku Seibutsugaku Kyoshitsu, Niigata.
Hosoya, T. (1968). *Gunma Symp. Endocrinol. [Proc.]* **5**, 219.
Hosoya, T., Kondo, Y., and Ui, N. (1962). *J. Biochem. (Tokyo)* **52**, 180.
Hosoya, T., Matsukawa, S., and Kurata, Y. (1972). *Endocrinol. Jap.* **19**, 359.
Howell, S. L., Kostianovsky, M., and Lacy, P. E. (1969). *J. Cell Biol.* **42**, 695.
Ibrahim, M. S., and Budd, G. C. (1965). *Exp. Cell Res.* **38**, 50.
Irie, M. (1960). *Arch. Histol. Jap.* **19**, 39.
Irvine, W., and Muir, A. (1963). *Exp. Cell Res.* **29**, 73.
Ishikawa, K. (1965). *Okajimas Folia Anat. Jap.* **41**, 295.
Jamieson, J. D., and Palade, G. E. (1967). *J. Cell Biol.* **34**, 597.
Kanaya, T. (1960). *Hiroshima Daigaku Igakubu Kaibogaku Kyoshitsu Gyosekishu* **7**, 1.
Kano, K. (1953). *Arch. Histol. Jap.* **4**, 245.
Karsensen, R. (1970). *Immunology* **19**, 301.
Kaye, N. Y. (1961). *Gen. Comp. Endocrinol.* **1**, 1.
Kayes, J., Maunsbach, A. B., and Ullberg, S. (1962). *J. Ultrastruct. Res.* **7**, 339.
Kendall, E. C. (1915). *J. Amer. Med. Ass.* **64**, 2042.
Kendall, E. C. (1919). *J. Biol. Chem.* **39**, 125.
Kennedy, G. R. (1966). *Gen. Comp. Endocrinol.* **7**, 500.
Kerkof, P. R., and Tata, J. R. (1967). *Biochem. Biophys. Res. Commun.* **28**, 111.

Kieckenbusch, H. H. (1928). *Z. Morph. Ökol. Tiere* **11**, 247.

King, T. W. (1836). *Guy's Hosp. Rep.* **1**, 429.

Klainer, L. M., Chi, Y. M., Friedberg, S., Rall, T., and Sutherland, E. (1962). *J. Biol. Chem.* **237**, 1239.

Klebanoff, S. J., Yip, C., and Kessler, D. (1962). *Biochim. Biophys. Acta* **58**, 563.

Klinck, G. H., Oertel, J. E., and Winship, T. (1970). *Lab. Invest.* **22**, 2.

Kondo, Y. (1961). *J. Biochem. (Tokyo)* **50**, 210.

Kosanović, M., Ekholm, R., Strandberg, U., and Smeds, S. (1968). *Exp. Cell Res.* **52**, 147.

Kotani, M., Seiki, K., Higashida, M., Imanishi, Y., Yamashita, A., Miyamoto, M., and Horii, I. (1968). *Endocrinology* **82**, 1047.

Kraentzel, F. (1933). *Arch. Biol.* **44**, 469.

Krause, R. (1923). "Mikroskopische Anatomie der Wirbeltiere in Einzeldarstellungen," Vol. 4. W. de Gruyter, Berlin u. Leipzig.

Kriss, J. P., Pleshakov, V., and Chien, J. R. (1966). *J. Clin. Endocrinol. Metab.* **24**, 1005.

Krstic̀, R., and Bucher, O. (1971). *Anat. Anz., Erg* **65**, 555.

Kurosumi, K. (1961). *Int. Rev. Cytol.* **11**, 1.

Kurosumi, K. (1963). *Saibo Kagaku Shimpojumu,* **13**, 309.

Laird, R., and Yates, R. (1973). *J. Ultrastruct. Res.* **44**, 339.

Langendorff, O. (1889). *Arch. Anat. Physiol., Physiol. Abt., Suppl.* **219.**

Langer, P., Lupulescu, A., and Petrovici, A. (1967). *Endocrinol. Exp.* **1**, 121.

Lanzing, W. J. R. (1959). "Studies on the River Lamprey, Lampetra fluviatilis, during its Anadromous Migration." Uitgeversmmaatschappi Neerlandia, Utrecht.

Larsen, J. H., Jr. (1968). *J. Ultrastruct. Res.* **24**, 190.

Leach, W. J. (1939). *J. Morphol.* **65**, 549.

Leblond, C. P., and Gross, J. (1948). *Endocrinology* **43**, 306.

Leloup, J. (1952). *J. Physiol. (Paris)* **44**, 284.

Leloup, J. (1955). *J. Physiol. (Paris)* **47**, 671.

Leloup, J., and Berg, O. (1954). *C. R. Acad. Sci.* **238**, 1069.

Leloup, J., and Fontaine, M. (1960). *Ann. N. Y. Acad. Sci.* **86**, 316.

Levenson, V. I. (1960). *Bull Exp. Biol. Med. (USSR)* **49**, 51.

Lever, J. (1960). *Int. Conf. Electron Microsc., Proc., 4th, Berlin, 1958,* 2, 229–232. Springer, Berlin.

Lietz, H. (1968). *Beitr. Pathol. Anat. Allg. Pathol.* **138**, 342.

Lietz, H. (1973). *Z. Zellforsch. Mikrosk. Anat.* **138**, 569.

Lietz, H., Wöhler, J., and Pomp, H. (1971). *Z. Zellforsch. Mikrosk. Anat.* **113**, 94.

Loewenstein, J. E., and Wollman, H. (1967a). *Endocrinology* **81**, 1063.

Loewenstein, J. E., and Wollman, H. (1967b). *Endocrinology* **81**, 1074.

Loewenstein, J. E., and Wollman, H. (1970). *Endocrinology* **87**, 143.

Lupulescu, A. (1965). *Experientia* **21**, 8.

Lupulescu, A., and Petrovici, A. (1963). *Folia Endocrinol.* **16**, 105.

Lupulescu, A., and Petrovici, A. (1964). *Acta Anat.* **57**, 294.

Lupulescu, A., and Petrovici, A. (1965). *In* "Current Topics in Thyroid Research" (C. Cassano and M. Andreoli, eds.), pp. 85–94. Academic Press, New York.

Lupulescu, A., and Petrovici, A. (1968). "Ultrastructure of the Thyroid Gland." Krager, Basel.

Lupulescu, A., Langer, P., and Petrovici, A. (1966). *Rev. Roum. Endocrinol.* **3**, 197.

Lupulescu, A., Merculiev, E., Nicolae, M. (1964). *Acta Anat.* **57**, 37.

Lupulescu, A., Nogoescu, I., Petrovici, A., Nicolae, M., Stoian, M., Balan, M., and Stancu, H. (1967). *Acta Anat.* **66**, 321.

McManus, J. F. A. (1946). *Nature (London)* **158**, 202.
Maloof, F., and Soodak, M. (1964). *J. Biol. Chem.* **239**, 1995.
Marine, D. (1913). *J. Exp. Med.* **17**, 379.
Matthaes, P. (1972). *Acta Endocrinol. (Copenhagen)* **69**, 459.
Melander, A., and Sundler, F. (1972). *Endocrinology* **90**, 188.
Monroe, B. (1953). *Anat. Rec.* **116**, 345.
Müller, W. (1871). *Jena. Z. Med. Naturwiss.* **6**, 428.
Murakami, T. (1971). *Arch. Histol. Jap.* **32**, 445.
Muramoto, K. (1964). *Okajimas Folia Anat. Jap.* **39**, 321.
Nadler, N. J. (1965). *In* "Current Topics in Thyroid Research" (C. Cassano and M. Andreoli, eds.), pp. 73–76. Academic Press, New York.
Nadler, N. J. (1971). *Anat. Rec.* **169**, 384.
Nadler, N. J., and Leblond, C. P. (1954). *Brookhaven Symp. Biol.* **7**, 40.
Nadler, N. J., Leblond, C. P., and Bogoroch, R. (1954). *Endocrinology* **54**, 154.
Nadler, N. J., Sarkar, S. K., and Leblond, C. P. (1962). *Endocrinology* **71**, 120.
Nadler, N. J., Young, B. A., Leblond, C. P., and Mitmaker, B. (1964). *Endocrinology* **74**, 333.
Nakagami, K., Warshawsky, H., and Leblond, C. P. (1971). *J. Cell Biol.* **51**, 596.
Nakai, Y., and Fujita, H. (1970). *Z. Zellforsch. Mikrosk. Anat.* **107**, 104.
Nakai, Y., and Gorbman, A. (1969). *Gen. Comp. Endocrinol.* **13**, 285.
Nakai, Y., Nanba, H., and Fujita, H. (1970). *Arch. Histol. Jap.* **31**, 421.
Nanba, H. (1972a). *Arch. Histol. Jap.* **34**, 277.
Nanba, H. (1972b). *Histochemie* **32**, 99.
Nanba, H. (1973). *Arch. Histol. Jap.* **35**, 313.
Neuenschwander, P. (1972). *Z. Zellforsch. Mikrosk. Anat.* **130**, 553.
Nève, P. (1965). *J. Microsc. (Paris)* **4**, 811.
Nève, P. (1966). *Pathol. Eur.* **1**, 234.
Nève, P. (1969). *Virchows Arch., A* **346**, 302.
Nève, P. (1971). *Horm. Metab. Res.* **3**, 426.
Nève, P., and Dumont, J. E. (1970a). *Exp. Cell Res.* **63**, 285.
Nève, P., and Dumont, J. E. (1970b). *Z. Zellforsch. Mikrosk. Anat.* **103**, 61.
Nève, P., and Wollman, S. H. (1971). *Anat. Rec.* **171**, 81.
Nève, P., and Wollman, S. H. (1972). *Anat. Rec.* **172**, 37.
Nève, P., Rodesch, F. R., Dumont, J. E. (1968). *Exp. Cell Res.* **51**, 68.
Nève, P., Eagleton, G. B., and Wollman, S. H. (1970). *J. Clin. Endocrinol. Metab.* **31**, 38.
Nève, P., Ketelbant-Balasse, P., Willems, C., and Dumont, J. E. (1972). *Exp. Cell Res.* **74**, 227.
Nièpce, B. (1851). "Traité du Goitre et du Cretinisme." Balliére et Fils, Paris.
Noseda, I. (1954). *Z. Mikrosk.-Anat. Forsch.* **60**, 192.
Nunez, E. A. (1971). *Amer. J. Anat.* **131**, 227.
Nunez, E. A., and Becker, D. V. (1970). *Amer. J. Anat.* **129**, 369.
Nunez, E. A., Belshaw, B. B., and Gershon, M. D. (1972). *Am. J. Anat.* **133**, 463.
Nunez, J., Mauchamp, J., Pommier, J., Cirkovic, T., and Roche, J. (1966a). *Biochem. Biophys. Res. Commun.* **23**, 761.
Nunez, J., Mauchamp, J., Pommier, J., Cirkovic, T., and Roche, J. (1966b). *Biochim. Biophys. Acta* **127**, 112.
Olin, P., Ekholm, R., and Almqvist, S. (1970). *Endocrinology* **87**, 1000.
Olivereau, M. (1952). *Arch. Anat. Microsc. Morphol. Exp.* **41**, 1.
Olivereau, M. (1955). *C. R. Ass. Anat.* 42, 1113.
Olsson, R. (1963). *Acta Zool. (Stockholm)* **44**, 299.

Onaya, T., and Solomon, D. H. (1969). *Endocrinology* **85**, 150.

Oswald, A. (1899). *Hoppe-Seyler's Z. Physiol. Chem.* **27**, 14.

Palade, G. E., Siekevitz, P., and Caro, L. G. (1962). *Exocrine Pancreas; Norm. Abnorm. Funct. Ciba Found. Symp.* pp. 23–55.

Pantić, V. (1967). *Z. Zellforsch. Mikrosk. Anat.* **81**, 487.

Pastan, I. (1966). *Biochem. Biophys. Res. Commun.* **25**, 14.

Pastan, I., and Macchia, V. (1967). *J. Biol. Chem.* **242**, 5757.

Pastan, I., and Wollman, S. H. (1967). *J. Cell Biol.* **35**, 262.

Pietschmann, V. (1929). *In* "Handbuch der Zoologie" (W. Kükenthal, ed.), Vol. **6**, pp. 1–123. Walter de Gruyter, Berlin u. Leipzig.

Pitt-Rivers, R., and Trotter, W. R. (1953). *Lancet* **ii**, 918.

Pitt-Rivers, R., Niven, J. S. F., and Young, M. R. (1964). *Biochem. J.* **90**, 205.

Prosser, T. (1769). "An Account and Method of Cure of Bronchocele." Derbyneck, London.

Prout, W. (1834). "Chemistry, Meterology and the Function of Digestion Considered with Reference to Natural Theology." Pickering, London.

Regard, E., and Mauchamp, J. (1971). *J. Ultrastruct. Res.* **37**, 664.

Roche, J., Lissitzky, S., and Michel, R. (1952). *C. R. Acad. Sci., Paris* **234**, 1228.

Roche, J., Salvatore, G., and Covelli, I. (1961). *Comp. Biochem. Physiol.* **2**, 90.

Rolleston, H. D. (1936). "The Endocrine Organs in Health and Disease." Oxford Univ. Press, London and New York.

Roos, B. (1960). *Pathol. Microbiol.* **23**, 129.

Rosenberg, I. N., Athans, J. C., and Isaacs, G. H. (1965). *Recent Progr. Horm. Res.* **21**, 33.

Sakai, Y. (1960). *Kagoshima Daigaku Igaku Zasshi* **11**, 297.

Salvatore, G., Covelli, I., Sena, L., and Roche, J. (1959a). *C. R. Soc. Biol.* **153**, 1686.

Salvatore, G., Macchia, V., Vecchio, G., and Roche, J. (1959b). *C. R. Soc. Biol.* **153**, 1693.

Salvatore, G., Aloj, S., Salvatore, M., and Edelhoch, H. (1967). *J. Biol. Chem.* **242**, 5002.

Schaffer, J. (1906). *Anat. Anz.* **28**, 65.

Schwarz, W. (1967). *Anat. Anz., Erg.* **120**, 127.

Seljelid, R. (1965). *J. Histochem. Cytochem.* **13**, 687.

Seljelid, R. (1966). *Exp. Cell Res.* **41**, 688.

Seljelid, R. (1967a). *J. Ultrastruct. Res.* **17**, 195.

Seljelid, R. (1967b). *J. Ultrastruct. Res.* **17**, 401.

Seljelid, R. (1967c). *J. Ultrastruct. Res.* **18**, 1.

Seljelid, R. (1967d). *J. Ultrastruct. Res.* **18**, 237.

Seljelid, R. (1967e). *J. Ultrastruct. Res.* **18**, 479.

Seljelid, R., and Helminen, H. J. (1968). *J. Histochem. Cytochem.* **16**, 467.

Seljelid, R., Reith, A., and Nakken, K. F. (1970). *Lab. Invest.* **23**, 595.

Seljelid, R., Helminen, H. J., and Thies, G. (1971). *Exp. Cell Res.* **69**, 249.

Serif, G. S., and Kirkwood, S. (1958). *J. Biol. Chem.* **233**, 109.

Setoguti, T. (1973a). *Z. Zellforsch. Mikrosk. Anat.* **137**, 177.

Setoguti, T. (1973b). *Z. Zellforsch. Mikrosk. Anat.* **137**, 195.

Setoguti, T. (1973c). *Arch. Histol. Jap.* **35**, 377.

Sheldon, H., McKenzie, J. M., Van Nimwegan, D. (1964). *J. Cell Biol.* **23**, 200.

Shepard, T. H. (1966). *In* "Organogenesis" (R. L. DeHaan and H. Ursprung, eds.), pp. 493–512. Holt, New York.

Shepard, T. H. (1967). *J. Clin. Endocrinol. Metab.* **27**, 945.

Shepard, T. H. (1968). *Gen. Comp. Endocrinol.* **10**, 174.
Shepard, T. H., Andersen, H., Andersen, H. J. (1964). *Anat. Rec.* **149**, 363.
Shin, W. X., Ma, M., Quintana, N., and Novikoff, A. B. (1970). *Congr. Int. Microsc. Electron. Grenoble* **3**, 79–80.
Simon, C., and Droz, B. (1965). *In* "Current Topics in Thyroid Research" (C. Cassano and M. Andreoli, eds.), pp. 77–84. Academic Press, New York.
Singer, C., and Underwood, E. A. (1962). "A Short History of Medicine." Oxford Univ. Press (Clarendon), London and New York.
Smith, P. E. (1921). *Anat. Rec.* **11**, 57.
Sobel, H. J. (1964). *Arch. Pathol.* **78**, 53.
Sobel, H. J., and Geller, J. (1965). *Amer. J. Pathol.* **46**, 149.
Stein, O., and Gross, J. (1963). *Exp. Cell Res.* **31**, 208.
Stein, O., and Gross, J. (1964). *Endocrinology* **75**, 787.
Sterba, G. (1953). *Wiss. Z. Friedrich-Schiller-Univ., Jena, Math.-Naturwiss. Reihe* **2**, 239.
Sterba, G. (1954). *Wiss. Z. Friedrich-Schiller-Univ., Jena, Math.-Naturwiss. Reihe* **3**, 239.
Stoll, R., Blanquet, P., Lachapèle, A. P., Maraud, R., and Magimel, R. (1957). *Electron Microsc. Proc. Stockholm Conf. 1956*, pp. 174–176.
Stoll, R., Maraud, R., Blanquet, P., Mounier, J., and Lachapèle, A. (1958). *Ann. Endocrinol.* **19**, 183.
Stoll, R., Maraud, R., and Lacourt, P. (1961). *Ann. Endocrinol.* **22**, 281.
Strum, J. M., and Karnovsky, M. J. (1970). *J. Cell Biol.* **44**, 655.
Strum, J. M., and Karnovsky, M. J. (1971). *Lab. Invest.* **24**, 1.
Suemasa, H., Honma, Y., and Fujita, H. (1968). *Arch. Histol. Jap.* **29**, 363.
Sutherland, E. W., Oye, I., and Butcher, R. W. (1965). *Recent Progr. Horm. Res.* **21**, 623.
Suzuki, M., Nagashima, M., and Yamamoto, K. (1961). *Gen. Comp. Endocrinol.* **1**, 103.
Szánto, L., Reviczky, A. L., and Grynaeus, T. (1966). *Acta Physiol.* **29**, 183.
Takagi, K. (1922). *Okajimas Folia Anat. Jap.* **1**, 69.
Takano, I., and Honjin, R. (1968). *Okajimas Folia Anat. Jap.* **44**, 173.
Tarutani, O., and Ui, N. (1968). *Biochem. Biophys. Res. Commun.* **33**, 733.
Tarutani, O., and Ui, N. (1969). *Biochim. Biophys. Acta* **181**, 136.
Tashiro, M., and Sugiyama, S. (1964). *Okajimas Folia Anat. Jap.* **40**, 131.
Taurog, A. (1970). *Recent Progr. Horm. Res.* **26**, 189.
Taurog, A., and Howell, E. M. (1966). *J. Biol. Chem.* **244**, 1329.
Taurog, A., Porter, J. C., and Thio, D. T. (1964). *Endocrinology* **74**, 902.
Themann, H., Andrara, J. A., Rose, N. R., Andrara, E. D., and Witebsky, E. (1968). *Clin. Exp. Immunol.* 3, 491.
Thomas, I. M. (1956). *J. Mar. Biol. Ass. U. K.* **35**, 203.
Thomas, I. M. (1962). *In* "The Evolution of Living Organisms" (G. W. Leeper, ed.), pp. 166–172. Melbourne Univ. Press, Melbourne.
Thorpe, A., Thorndyke, M. C., and Barrington, E. J. W. (1972). *Gen. Comp. Endocrinol.* **19**, 559.
Tice, L. W., and Wollman, S. H. (1972). *Lab. Invest.* **26**, 63.
Tixier-Vidal, A., Picart, R., Rappaport, L., and Nunez, J. (1969). *J. Ultrastruct. Res.* **28**, 78.
Tong, W., Kerkof, P., and Chaikoff, I. L. (1962). *Biochim. Biophys. Acta* **56**, 326.
Toujas, L., and Guelfi, J. (1969). *Z. Zellforsch. Mikrosk. Anat.* **94**, 118.

Trolldenier, H. (1965). *Z. Mikrosk.-Anat. Forsch.* **73**, 438.
Trolldenier, H. (1967). *Z. Mikrosk.-Anat. Forsch.* **77**, 29.
Turkington, R. W. (1962). *Biochim. Biophys. Acta* **65**, 386.
van Heyningen, H. E. (1964). *Anat. Rec.* **148**, 485.
van Heyningen, H. E., and Sandborn, E. (1963). *J. Appl. Phys.* **34**, 2525.
von Basedow, C. A. (1840). *Wochenschr. Heilk.* **6**, 197.
Waller, U. (1961). *Schweiz. Menatsschr. Zahnheilk.* **71**, 561.
Walthard, B. (1955). *Bull. Schweiz. Akad. Med. Wiss.* **11**, 346.
Walthard, B., and Roos, B. (1959). *Verh. Deut. Ges. Pathol.* **42**, 335.
Wang, R. (1958). *Denshi Kembikyo* **6**, 184.
Wang, R. (1960). *Nippon Naibumpi Gakkai Zasshi* **35**, 1309.
Watari, N. (1962). *Arch. Histol. Jap.* **23**, 21.
Waterman, A. J., and Gorbman, A. (1963). *Gen. Comp. Endocrinol.* **3**, 58.
Weber, G., Zampi, G., and Ignesti, C. (1954). *Arch. "De Vecchio" Anat. Pathol. Med. Clin.* **22**, 297.
Welsch, U., and Storch, V. (1969). *Z. Zellforsch. Mikrosk. Anat.* **102**, 432.
Wetzel, B. K., and Wollman, S. H. (1969). *Endocrinology* **84**, 563.
Wetzel, B. K., Spicer, S. S., and Wollman, S. H. (1965). *J. Cell Biol.* **25**, 593.
Whur, P., Herscovics, A., and Leblond, C. P. (1969). *J. Cell Biol.* **43**, 289.
Wissig, S. L. (1960). *J. Biophys. Biochem. Cytol.* **7**, 419.
Wissig, S. L. (1963). *J. Cell Biol.* **16**, 93.
Wissig, S. L. (1964). *In* "The Thyroid Gland" (R. Pitt-Rivers and W. Trotter, eds.), Vol. 1, pp. 32–70. Butterworth, London.
Wolff, J., and Halmi, H. S. (1963). *J. Biol. Chem.* **238**, 847.
Wollman, S. H. (1965). *In* "Current Topics in Thyroid Research" (C. C. Cassano and M. Andreoli, eds.), pp. 1–18. Academic Press, New York.
Wollman, S. H. (1969). *In* "Lysosomes" (J. T. Dingle and H. B. Fell, eds.), Vol. 2, pp. 483–512. North-Holland Publ., Amsterdam.
Wollman, S. H., and Spicer, S. S. (1961). *Fed. Proc., Fed. Amer. Soc. Exp. Biol.* **20**, 201.
Wollman, S. H., and Wodinsky, I. (1955). *Endocrinology* **56**, 9.
Wollman, S. H., Spicer, S. S., and Burstone, M. S. (1964). *J. Cell Biol.* **21**, 191.
Woodbury, D. M., and Woodbury, J. W. (1963). *J. Physiol (London)* **169**, 553.
Yip, C. C. (1964). *Biochim. Biophys. Acta* **90**, 216.
Yip, C. C. (1965). *Biochim. Biophys. Acta* **96**, 75.
Yoshimura, F., and Irie, M. (1961). *Z. Zellforsch. Mikroshk. Anat.* **55**, 204.
Youson, J., and van Heyningen, H. E. (1968). *Amer. J. Anat.* **122**, 377.
Zeigel, R. F., and Dalton, A. J. (1962). *J. Cell Biol.* **15**, 45.

Postnatal Gliogenesis in the Mammalian Brain

A. PRIVAT

Laboratoire de cultures de tissu nerveux, I.N.S.E.R.M. U-106, Paris, France

I. Introduction

The first account on neuroglia was probably that by Dutrochet (1824), who described small globules (*corpuscules globuleux*) he observed in great quantities on large globules (*cellules globuleuses*) in the central nervous system of two molluscs. Later, Virchow (1854) described neuroglia as a tissue per se, and Deiters recognized neuroglial cells as being different from neurons. His (1889) postulated that neurons and neuroglial cells arise from two different stem cells, whereas Schaper (1897) estimated that these two cells are indeed two forms of the same cell, both giving rise to undifferentiated apolar cells. In 1909, Ramón y Cajal, using silver impregnations, observed two cell types in the spinal cord of the chick embryo, the apolar neuroblast, and the neuroglial cell or astrocyte; the latter arises through the transformation of neuroepithelial cells, and intermediate stages were found. In 1913, Ramón y Cajal described a third cell, the so-called third element. With the silver carbonate technique, del Rio-Hortega (1919) showed that this third element was in fact made of two ramified cells, oligodendrocytes and microglial cells. He suggested that the oligodendrocyte is, as is the astrocyte, a neuroectodermal cell, whereas microglia is a mesodermal element. Some controversy remained for years on the origin of glial cells, as it was very difficult to obtain satisfactory metallic impregnations of very immature cells, although Penfield (1932) attempted to give a satisfactory identification on paraffin sections routinely stained. Kershman (1938) shed some new light on this problem with the description of the subependymal plate, a dense layer of transitional cells located between the matrix layer and the mantel layer. This layer contained, in addition to apolar neuroblasts, the precursors of neuroglial cells; he proposed (Kershman, 1939) that microglia consists of mesenchymal cells originating from meninges and blood vessels.

A new period in our knowledge of neuroglial cells began 15 years ago with the use of two new techniques. The electron microscope has provided us with precise criteria for the identification of glial cells, and has given several indications of their possible function. Radioautography has permitted investigators to label immature cells through incorporation of labeled nucleotides, and to follow their ultimate fate.

The purpose of this article is to summarize recent advances on postnatal gliogenesis obtained by these two techniques.

II. The Matrix Cell

A. The Subependymal Layer of the Lateral Ventricle

The protracted presence of mitotic figures around the lateral ventricles of the rat brain (Allen, 1912; Opalski, 1934; Kershman, 1938; Globus and Kuhlenbeck, 1944; cf. Fig. 1) made logical the investigation of the cell type involved in such a proliferative process. A light microscope study by Smart (1961) in the mouse conducted with the use of thymidine-^{3}H radioautography lead to the following results: Two cell types were distinguished, as seen with hematoxylin–eosin stain, on paraffin sections; the first type stained lightly and had an irregular nucleus with frequent indentations and chromatin granules clumped against the nuclear membrane. Cytoplasm was rare, and stained poorly. The second type had a smaller, darker nucleus which was very irregular in shape; the chromatin granules were coarse and distributed throughout the nucleus. The cytoplasm appeared as one or two tails attached to the nucleus. These two cell types had a tendency to occur in small groups of similar type. Intermediate cells were also found. These two cell types were found to

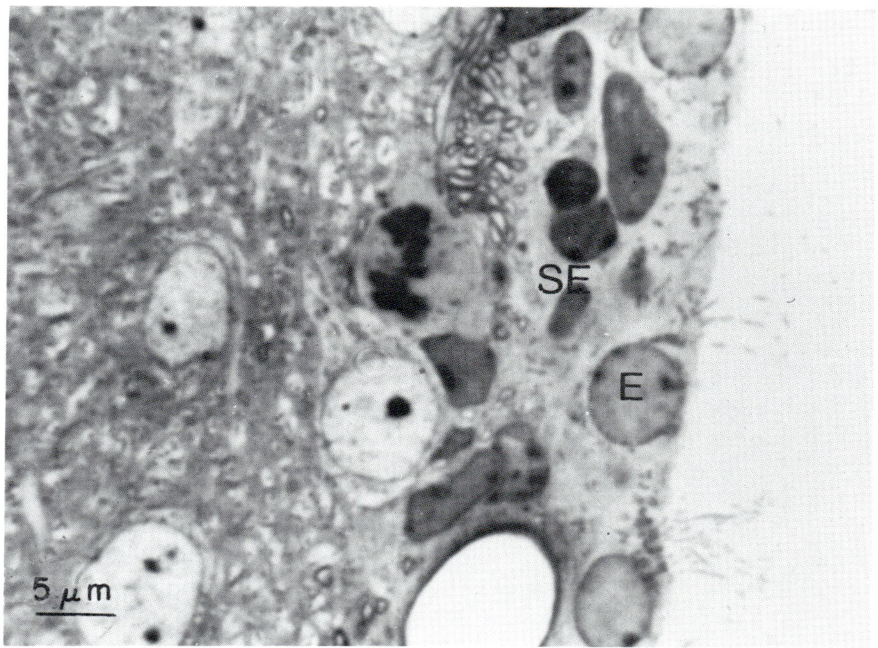

FIG. 1. External wall of the lateral ventricle of an adult rat. 1-Micrometer Epon section. The ependyma (E) is underlined by the subependymal layer (SE), where a few mitoses still occur. The caudate nucleus is at left. ×2000.

occur in variable proportion with age; 3 days after birth 90% were of the dark type, whereas the proportion was nearly 50:50 in the adult. After thymidine-^{3}H injection, dark-nucleated cells were found to be preferentially labeled; it was hypothesized, on morphological grounds, that they transformed into spongioblasts, precursors of glial cells, whereas the light-nucleated cells gave rise to neuroblasts (Angevine *et al.*, 1970).

However, these were only tentative conclusions, as it was difficult, at that time, and with those techniques, to give a fully reliable identification of neuronal and glial cell types. For the latter it would have been necessary to use metallic impregnation, which cannot be combined with radioautography. In a subsequent study, Altman (1966) labeled through intraperitoneal injection of thymidine-^{3}H the cells of the subependymal layer in a 13-day-old rat, and followed the fate of the labeled cells for two subsequent months. He found that substantial migration occurred through the corpus callosum and the fimbria, and that some cells reached the cortex. He concluded that these were essentially glial cells, and that this migration accounted for the increase in the glial cell/neuron ratio in the brain of the young rat. Almost a decade later most of these barriers were overcome, and it was possible to resume this study with the help of the electron microscope (Privat, 1970a). Indeed, the discrepancies in the first electron microscope description of mature glial types (Luse, 1956; Farquhar and Hartmann, 1957) were resolved with the works of Herndon (1964), Kruger and Maxwell (1966), Maxwell and Kruger (1965), and Caley and Maxwell (1968).

In the young rat (Privat, 1970a; Privat and Leblond, 1972), the proliferative subependymal layer is concentrated under the external wall of the lateral ventricle, with an extension between the corpus callosum and the head of the caudate nucleus. It appears, in the electron microscope (Fig. 2), as a collection of tightly packed cells underlying the ependyma, without any intervening basement membrane. The nuclei of subependymal cells appear darker and smaller than those of ependymal cells. They are ovoid or spindle-shaped, and usually run parallel to the ventricle surface. Their nuclear envelope is frequently indented, and the nucleoplasm exhibits a more-or-less patchy appearance due to many ill-defined clumps of chromatin. The cytoplasm, which is restricted to a rim around the nucleus and a somewhat larger accumulation at both poles, contains many free ribosomes, mostly arranged in small groups, or polysomes; cisternae of rough endoplasmic reticulum (ER) are very rare, and the few mitochondria present contain small dense granules; a few isolated microtubules but no filaments may be seen. These cells often have processes, running parallel to their main axis, which frequently arise from the two poles. These processes exhibit a content similar to that of the perikaryon. The only other cell type found in the subependymal layer is the mature microglial cell, whose features are identical to those described in the corpus callosum by Mori and Leblond (1969a). Mitoses were frequent among subependymal cells, especially in younger animals (20 days after birth). In the latter the subependymal layer appears as a compact, three- to five-cell-thick plate, separated from the underlying white or gray matter by the so-called border area (Fig. 3) which, in older animals, extends up to the ependyma, producing a fragmentation of the layer into several cell clumps. The border area is composed of scattered cells surrounded by a light-staining neuropil. The cells scattered within the border area are of many types. Besides an occasional neuroblast (Caley and Maxwell, 1968), found essentially in 20 DPN (days postnatal) rats, mature neurons are present, but the predominant cell type has an irregular nucleus with some clumps of chromatin, and a cytoplasm rich in ribosomes (Fig. 3). It is in many respects similar to the cells of the subependymal layer with, however, some differences;

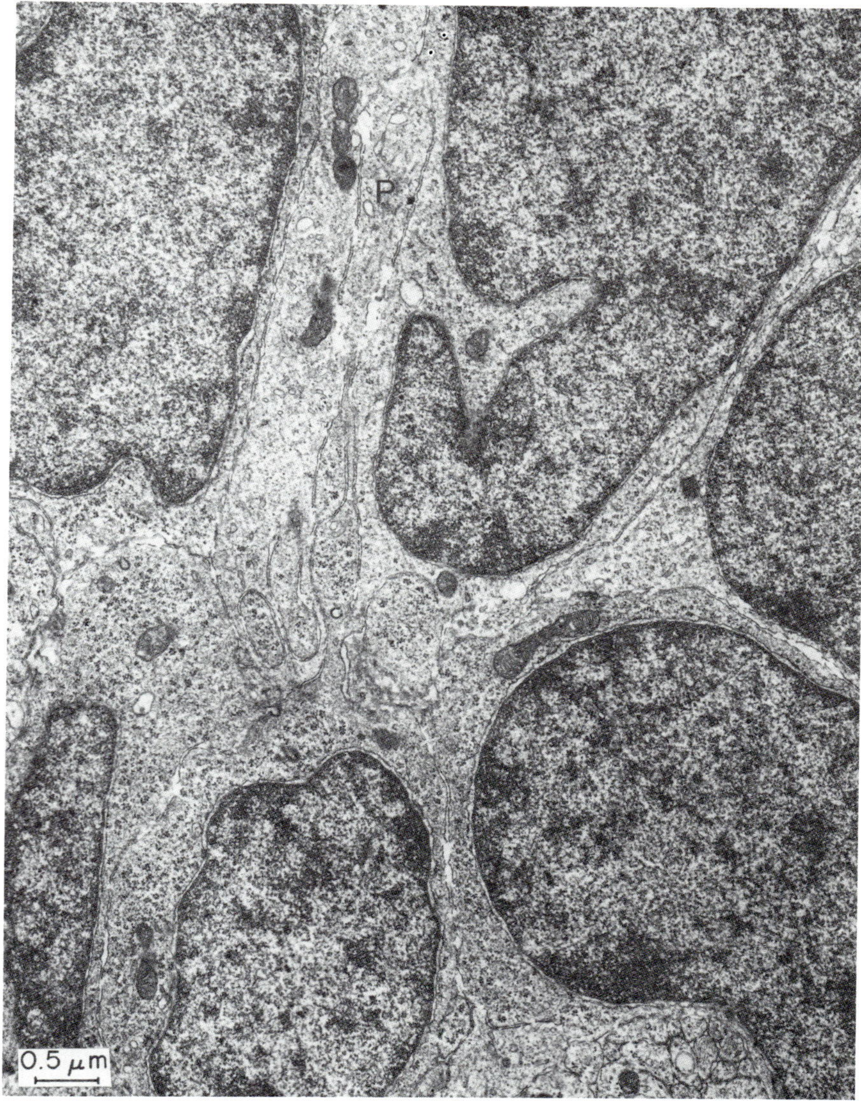

FIG. 2. Subependymal layer in a 40-gm rat. The nucleus of subependymal cells shows frequent indentation and contains irregular clumps of chromatin. The cytoplasm is scanty and contains numerous free ribosomes. A process (P) contains free ribosomes and a few microtubules. ×18,000. (From Privat and Leblond, 1972.)

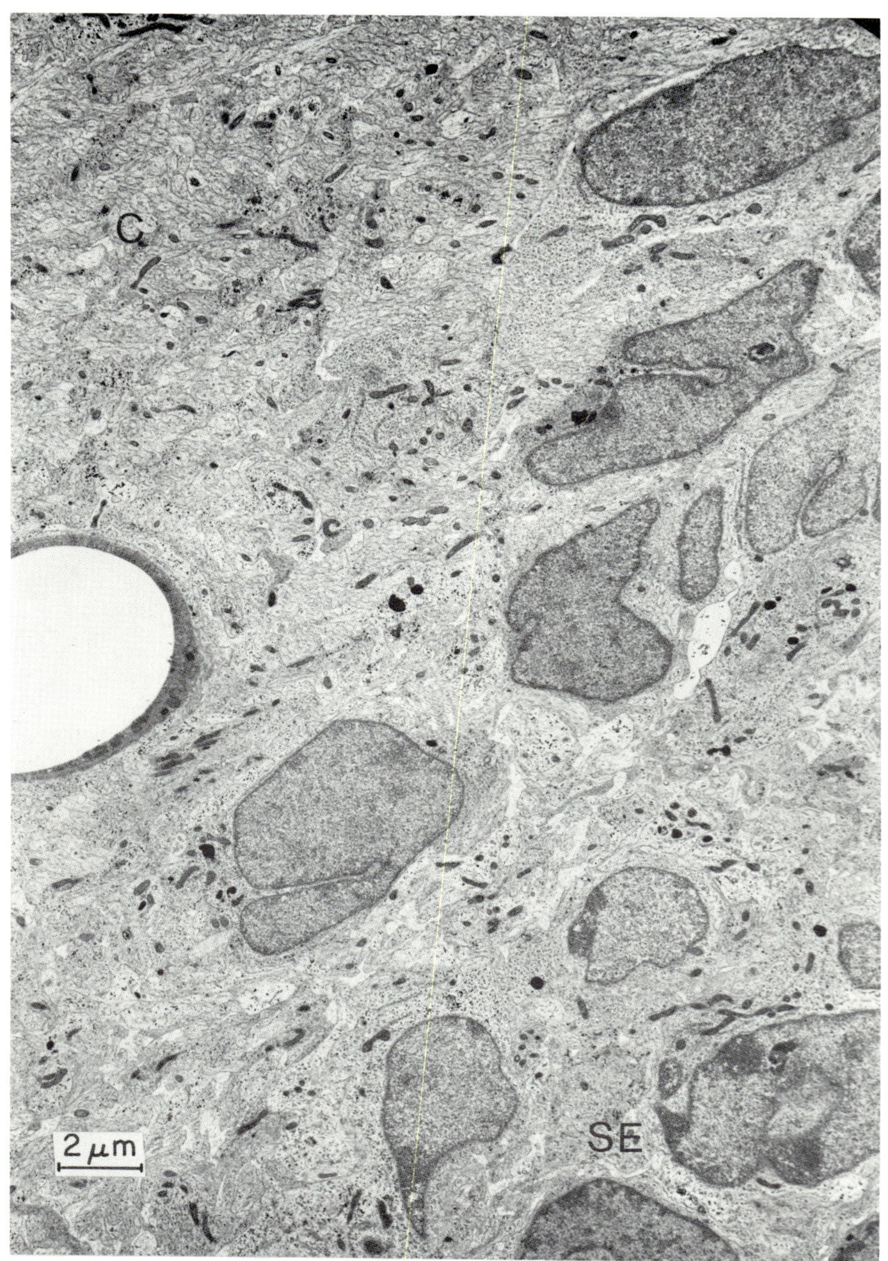

FIG. 3. Border area between subependymal layer (SE) and caudate nucleus (C) in a 80-gm rat. Cells are scattered on a light background of glial and neuronal processes. They have an irregular nucleus but appear lighter than those of the subependymal layer. ×5500. (From Privat and Leblond, 1972.)

the chromatin clumps are less numerous, thus giving the nucleus a lighter appearance, and the cytoplasm contains more organelles and ER cisternae. Other cells already have some characteristics of mature glial cells, and they are considered in the sections relevant to these cell types. Fully mature glial cells are rare, except for occasional microglial cells. Thus the subependymal layer appears to consist of a unique cell type which shows the characteristic features of immaturity, namely, an irregular nucleus and a high cytoplasmic content of ribosomes (Howatson and Ham, 1955; Munger, 1958).

Fujita (1963) came to the same conclusion, with continuous thymidine-^{3}H labeling experiments, for the matrix layer of the chick embryo.

The border area, containing light-nucleated cells, which tends to invade the subependymal layer, may account for the early distinction made by Smart (1961) between light- and dark-nucleated cells. The latter, present in large numbers in younger animals, correspond to true subependymal cells.

In adult rats (Blakemore, 1969), the situation appears only slightly different. The subependymal layer proper is reduced to a few clumps of immature cells which are separated by a neuropil made almost exclusively of astrocytic processes. Microglial cells and astrocytes appear far more numerous in this region than elsewhere in the brain.

In the newborn rabbit, Stensaas and Gilson (1972) found, with the help of serial sectioning, that subependymal cells had a large diversity of shape and size, ranging from small, simple cells to large, complex forms with multiple short processes. Nevertheless, transitional types were found which suggest that all these forms, together with mitotic elements, belong to the same population of cells. These investigators report that complex forms are much like the light-nucleated cells described by Smart (1961), whereas simple ones are similar to the dark-nucleated ones. In the adult cat (Demêmes and Marty, 1971), the subependymal layer appears substantially similar to that of the rodent; mitoses are still present in a population of immature-looking cells.

B. The Optic Nerve

The optic nerve was chosen for studies of gliogenesis since it represents a model without any neuronal production and, although it is a nerve, is truly part of the central nervous system; these studies were mainly performed on the rat (Peters, 1964; Peters and Vaughn, 1967; Vaughn and Peters, 1968; Vaughn, 1969), but also on the cat (Blunt *et al.*, 1972).

In the rat optic nerve, a true matrix layer is no longer present after birth; nevertheless proliferative activity is still present since, according to Vaughn (1969), mitotic cells make up 3–6% of the neuroglial population in the optic nerve during the first 2 weeks of postnatal growth. These mitotic cells are commonly found in the vicinity either of blood vessels, or of the glia limitans, at the periphery of the nerve. These most often are devoid of cytoplasmic processes. The main characteristic feature of these cells is the large amount of free ribosomes, with very rare cisternae of rough or smooth ER. Such cytoplasmic features are also characteristic of another population of cells, the so-called daughter cells (Vaughn, 1969), which appear smaller and more elongated than mitotic cells. These two cell types appear roughly similar to those of the subependymal layer, and they evidently belong to the same population of stem cells observed at different stages of their cycle (Fujita, 1963).

In the kitten optic nerve, 1–7 days after birth, a few cells were found (Blunt *et al.*, 1972) that mimic very closely the daughter cells described by Vaughn (1969). These, together with other cells apparently much more mature, were called glioblasts.

Besides mitotic and daughter cells, Vaughn (1969) has elaborated a complex series of intermediate forms (undifferentiated cells) leading to the precursors of mature neuroglial cells. All these are considered glioblasts. The first element of this series, the so-called small glioblast differs from the daughter cell mainly by the presence of cisternae of ER. They appear as long, thin, tortuous profiles winding irregularly through the cytoplasm, which shows larger protrusions and processes than those of daughter cells. In addition, the overall density of the cytoplasm is lighter, mainly because of a smaller content of ribosomes. Some of these cells show, in addition, a variable number of fat droplets and granular bodies. The interpretation of these inclusions is discussed in detail in Section V. Small glioblasts are said to transform into larger cells, large glioblasts which are characterized by a larger nuclear and cytoplasmic volume, and a pattern of chromatin suggesting that they are preprophase cells. The last of this series, the large glial precursor is distinguished by the presence of a nucleolus and an abundance of cytoplasmic organelles.

C. Cerebral Cortex

The superficial cerebral cortex of the newborn rat is packed with vertically arranged columns of undifferentiated cells; after the neuroblasts differentiate, a population of immature cells remains (Fig. 4) which later differentiate into neuroglial cells (Caley and Maxwell, 1968). The undifferentiated cell is characterized by a thin rim of cytoplasm containing scattered ribosomes, and few if any organelles. It quite often appears in a "semi-lunar" fashion on the convexity of a neuroblast (Fig. 4); the nuclear chromatin is clumped in small aggregates, often along the

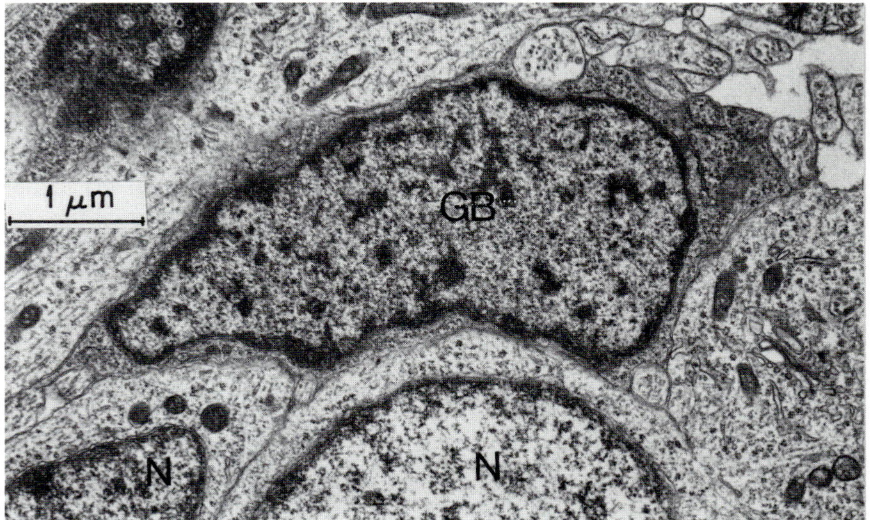

FIG. 4. Newborn rat cerebral cortex. The upper part of the cortical plate contains rows of maturing neurons (N) and a few immature-looking cells (GB) which are believed to transform into mature satellite glial cells. ×17,500.

nuclear membrane. These immature cells were no longer found in the cortex of ¾-month-old rats by Ling and Leblond (1973), in agreement with the previous study of Altman (1966) who stated that local multiplication and emigration of immature cells ceased almost completely about the twenty-fifth day after birth.

D. Cerebellar Cortex

The morphogenesis of the cerebellar cortex is characterized by a unique event in the central nervous system. An external matrix layer gives rise to granular and stellate neurons; the former migrate across the molecular layer and reach the internal granular layer, where they acquire mature neuronal characteristics; it was claimed, among others, by Fujita (1967; Fujita *et al.*, 1966) that the external matrix layer gives rise to a few glial cells after the production of neurons has ceased, between postnatal days 15 and 20 in the mouse.

In 16- to 20-day-old rats the external matrix layer appears as a collection of cell clusters separated by areas of neuropil, and principally by the large subpial end feet of Bergman glial cells. These are interposed between matrix cells and the basement lamina. Matrix cells appear most often as spindle-shaped, with the main axis parallel to the pial surface.

Their nuclear and cytoplasmic characteristics are similar to those of subependymal cells (Privat, 1972).

E. Persistence of Immature Cells in Adult Animals

The finding by Privat and Leblond (1972) of free subependymal cells in the corpus callosum of 1-month-old rats lead to quantitative estimation of the occurrence of these cells in the corpus callosum of adult rats (Ling and Leblond, 1973). Their number was found to decrease regularly from 1 to 5 months, but even at this age a few were still present. Similarly, Privat (1972) found in the molecular layer of the adult rat cerebellum a few cells that could not be classified as mature neuronal or glial cells. These were considered quiescent immature cells.

III. The Oligodendrocyte Line

The existence of several types of oligodendrocytes was recognized long ago by light microscopy of metallic impregnation (del Rio-Hortega, 1928; Cammermeyer, 1960) or in Nissl pictures (Kryspin-Exner, 1943). With the introduction of electron microscopy, a wide spectrum of descriptions were reported, from large pale cells (Luse, 1956; Dempsey and Luse, 1958) to a small dense cell with a round nucleus (Farquhar and Hartmann, 1957; Palay, 1958). The latter opinion was more or less shared by subsequent workers (De Robertis and Gershenfeld, 1961; Palay *et al.*, 1962; Malmfors, 1963; Mugnaini and Walberg, 1964; Schultz, 1964; Bodian, 1964; Wendell-Smith *et al.*, 1966; Stensaas and Stensaas, 1968; King, 1968), although Kruger and Maxwell (1966), and Caley and Maxwell (1968) admitted that these cells could display a wide range of nuclear and cytoplasmic densities, which tend to parallel each other.

Recent studies of the maturation of oligodendrocytes in various regions of the brain have cleared up some of these discrepancies.

A. Ultrastructural Evidence of Differentiation

1. *Subependymal Layer and Corpus Callosum*

Evidence of differentiation toward the mature oligodendrocyte was found in a few cells of the border area, in the vicinity of the subependymal layer (Privat, 1970a; Privat and Leblond, 1972). A few cells contained more ER cisternae than those of the subependymal layer, and these had a tendency to be grouped in stacks, as is the case in oligodendrocytes. Others had small, light processes emanating from the cytoplasm in a bushy fashion. Their nuclei appeared regularly shaped, with finely dispersed chromatin. Most of these cells were isolated in the border area,

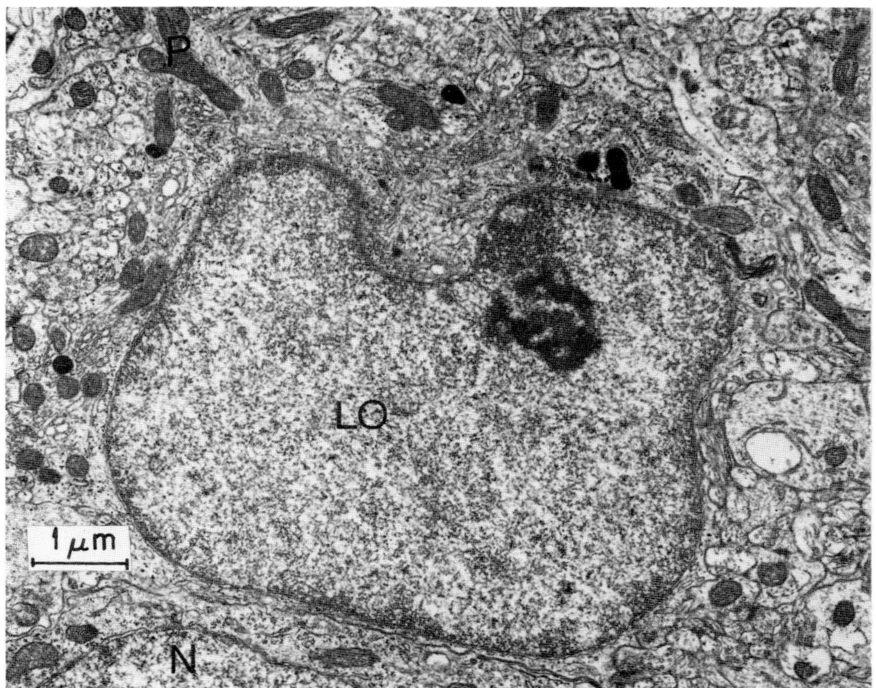

FIG. 5. Border area in a 80-gm rat showing a light oligodendrocyte (LO) in contiguity with a neuron (N). The cell has the typical large pale nucleus and prominent nucleolus. The cytoplasm contains numerous organelles and emits a process (P). ×12,500. (From Privat and Leblond, 1972.)

but others were found as satellites to a neuron, a common feature of the oligodendrocyte of the caudate nucleus (Figs. 5 and 6).

In the corpus callosum of 1-month-old rats, a complete spectrum of the oligodendrocyte cell line may be found (Hillebrand, 1966); in addition to the above-cited poorly differentiated cells (Privat and Leblond, 1972), three types of oligodendrocytes were found with the electron microscope (Mori and Leblond, 1970).

The *light oligodendrocyte* (Fig. 7) is characterized by a large, pale, nucleus and cytoplasm. The nucleus is as light as the axoplasm of neighboring myelinated axons and has a main diameter of 7 μm. The nuclear envelope is smooth, with inner and outer membranes running close to each other. There are no clear-cut chromatin masses, and the large nucleolus, usually located centrally, exhibits fibrillar and granular regions. The abundant cytoplasm is not as light as the nucleus; it contains

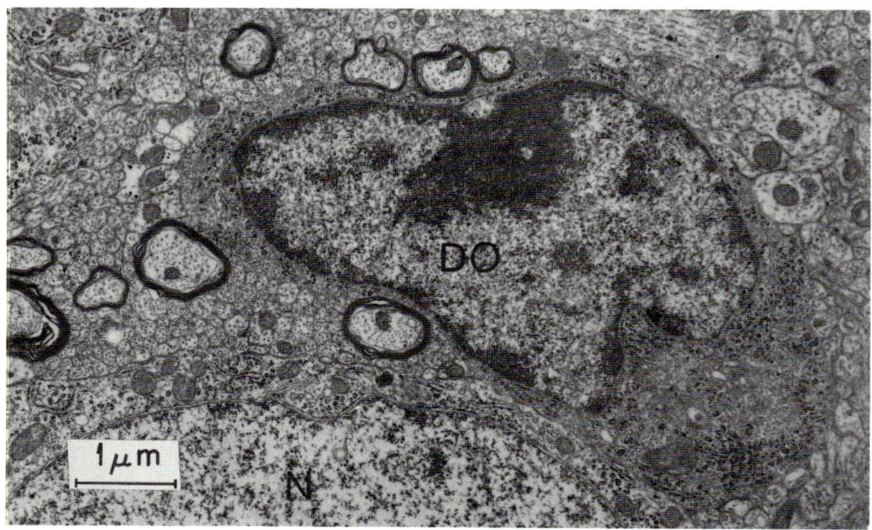

FIG. 6. Border area in a 80-mg rat showing a dark oligodendrocyte (DO) satellite of a neuron (N). Its dense cytoplasm contains numerous cisternae of rough ER, and a well-developed Golgi zone. The nuclear chromatin has a more patchy appearance when compared to the dark oligodendrocyte in Fig. 9. ×12,500.

numerous organelles: free ribosomes arranged in rosettes, a rather scanty rough ER, and numerous microtubules. The Golgi apparatus is made up of a few stacks and small saccules. Fine processes, found in large numbers in poorly differentiated cells, are prominent in light oligodendrocytes. They run fairly straight, with an uniform diameter, and contain numerous microtubules.

The line of demarcation between this cell and the second type, the *medium oligodendrocyte* (Fig. 8), is not a clear-cut one. They constitute a continuous spectrum of sizes and shades. A typical medium oligodendrocyte is denser and smaller than a light one. The oval, or more often spherical, nucleus has a diameter varying from 4 to 7 μm, and is denser than the axoplasm of neighboring axons. A few chromatin masses are stacked along the nuclear membrane: the nucleolus is less distinct than that of light cells. The cytoplasm is denser and less abundant when compared to that of the light cells. Rough ER is more abundant, and has larger, more regularly stacked cisternae. The Golgi zone is more prominent, with distended saccules. Microtubules are present. The ground cytoplasm appear denser than in light cells. Fine processes are still present, but few; they contain microtubules.

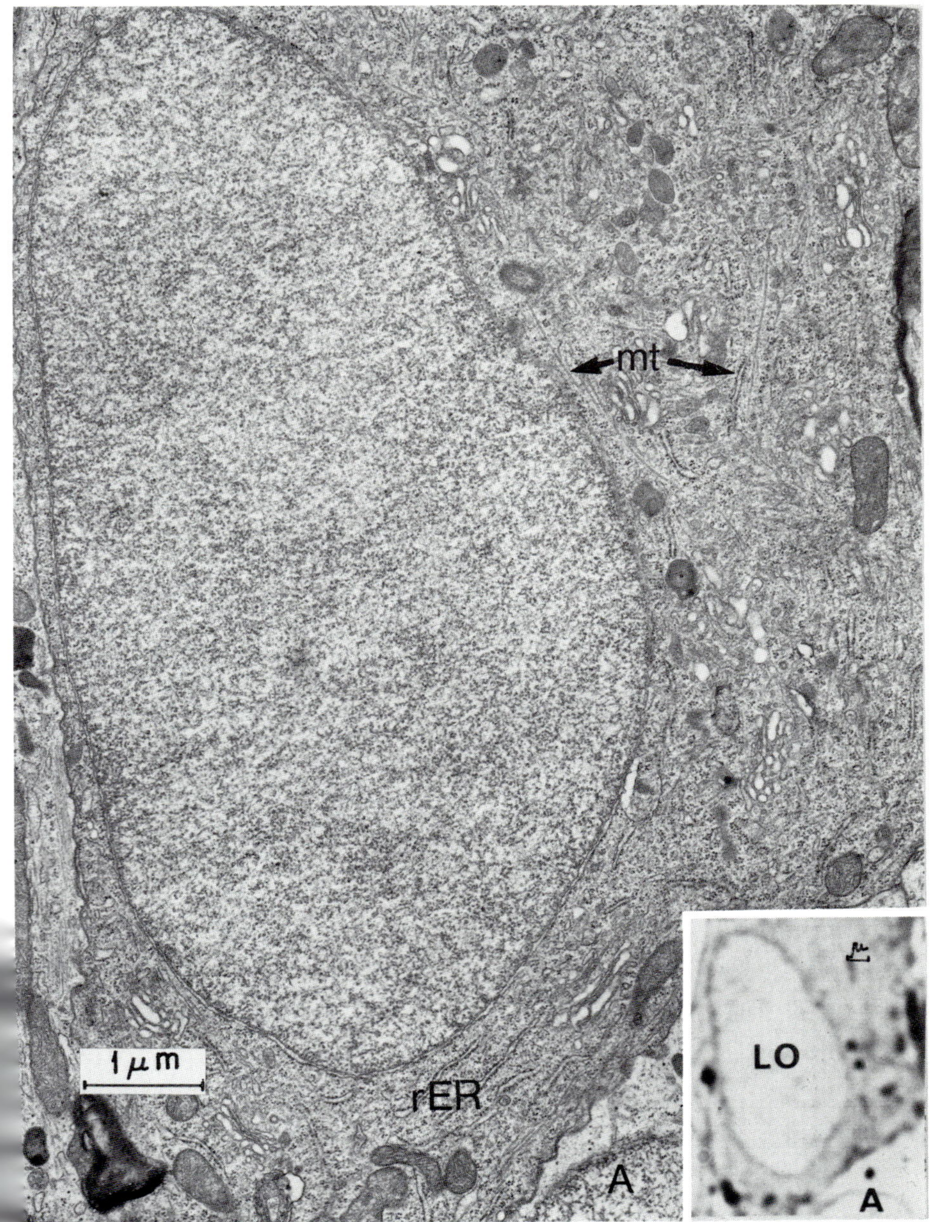

FIG. 7. Light oligodendrocyte (LO) in the corpus callosum of a 80-gm rat in an electron micrograph and in a serial light micrograph (inset). In both the nucleus is pale and has evenly dispersed chromatin; the cytoplasm is copious and contains numerous organelles, among which are microtubules (mt) and rough ER (rER). In the lower right corner is an astrocyte (A). ×15,000. Inset, ×3000. (From Ling *et al.*, 1973.)

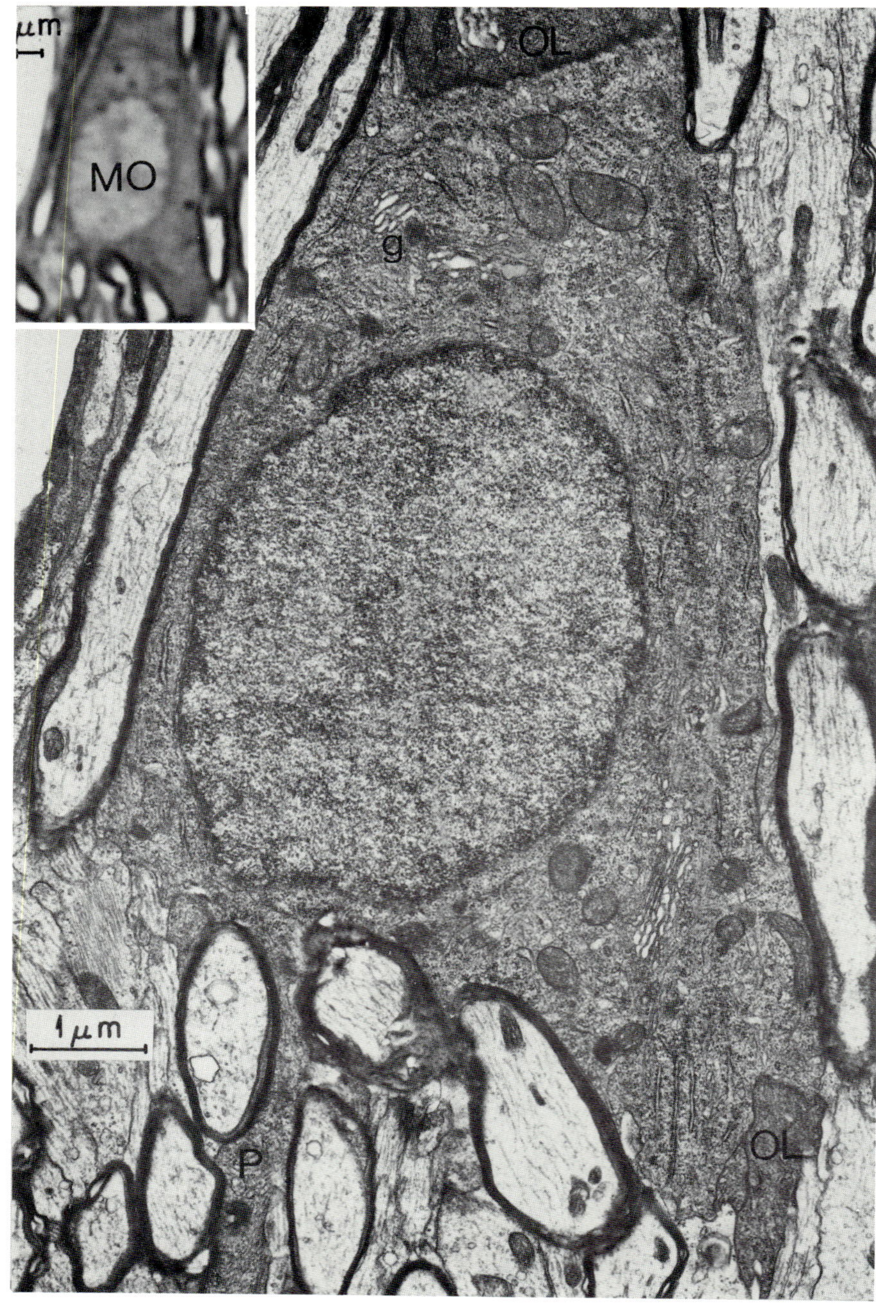
µm
MO
OL
g
1 µm
P
OL

Typical *dark oligodendrocytes* (Fig. 9) are cells whose cytoplasm and nucleus are much darker than the neighboring axons. Again, a large spectrum of shades and sizes exists, with no clear-cut demarcation from the medium oligodendrocyte. In typical cells the dense and relatively small nucleus is frequently eccentrically located. It is more-or-less spherical and has a diameter of 3.5–5 μm. Large chromatin masses are attached along the nuclear membrane. The eccentrically located nucleolus is sometimes poorly distinguished. In the rather scanty dark cytoplasm, the distended saccules of the Golgi zone appear light. The rough ER appears as stacks of rectilinear cisternae. The cytoplasm also contains *lamellar bodies* which are frequently associated with membranous structures present in the cell: smooth ER mitochondria, and so on. Mori and Leblond (1970) suspected that these three cell types represent a maturation series, and they indeed found that light and medium oligodendrocytes incorporated tritiated thymidine and underwent mitosis, and that dark ones did not, as if the latter were the true mature form of oligodendrocytes.

2. *The Optic Nerve*

Transitional stages between undifferentiated cells and mature oligodendrocytes were readily identified in the rat optic nerve (Vaughn, 1969). *Young oligodendrocytes* present soon after birth are characterized by a more abundant cytoplasm containing a well-developed Golgi apparatus and cisternae of rough ER. The nuclei frequently show a well-developed nucleolus, and the chromatin is still aggregated along the nuclear envelope.

Active oligodendrocytes (Fig. 10) appear about the fifth postnatal day, and become common 7–9 days after birth, corresponding to the onset of myelination in the optic nerve. Active oligodendrocytes are large cells with prominent cytoplasmic processes, a well-developed Golgi system, and rough ER. The nucleus shows a prominent nucleolus and evenly dispersed chromatin, a common feature of light oligodendrocytes found in the corpus callosum. But the distinctive feature is the presence of processes which in several instances have been seen closely associated with the outer loop of myelinated axons. This description is very similar

---

FIG. 8. Medium oligodendrocyte (MO). Same material and technique as Fig. 7. The nucleus is spherical and has some dense chromatin along the nuclear envelope. The cytoplasm is dense and contains outstanding Golgi zones (g) and several stacks of rough ER Microtubules are numerous. At lower left, a process (P) emanates from a cell. This cell is in contact with the cytoplasm of two other cells, probably medium or dark oligodendrocytes (OL). ×15,000. Inset, ×3000. (Courtesy of C. P. Leblond and J. Paterson.)

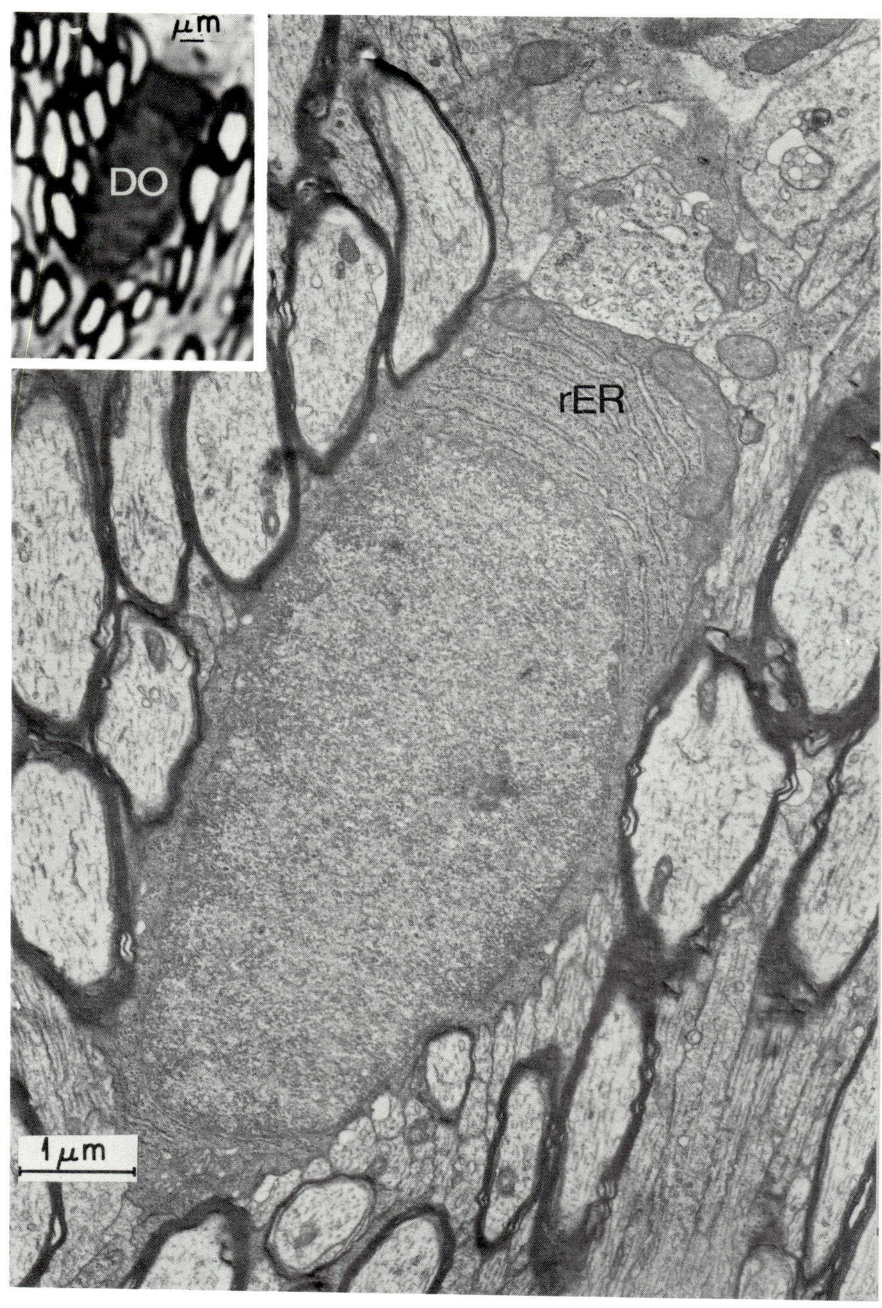
μm
DO
rER
1 μm

to that of light and medium oligodendrocytes in the corpus callosum (Mori and Leblond, 1970).

Mature oligodendrocytes (Fig. 11) are ovoid cells with few processes, exhibiting considerably more cytoplasmic and nuclear density than any other glial cell. Their nucleus is eccentrically located, and they exhibit a distinct clumping of chromatin along the nuclear envelope. The cytoplasm contains a well-developed Golgi apparatus and numerous short profiles of rough ER. They are in fact very similar to the medium and dark oligodendrocytes described by Mori and Leblond (1970) and the mature oligodendrocytes observed by several investigators (Mugnaini and Walberg, 1964; Kruger and Maxwell, 1966). Wendell-Smith *et al.* (1966), and Blunt *et al.* (1972) suggested that the oligodendrocyte is not the cell responsible for myelination, and they founded their assumption on the fact that myelination started in their material about 2–3 days after birth, whereas mature oligodendrocytes were found only much later. Therefore it is worth noting that cells described by these investigators as glioblasts and early oligodendrocytoblasts may correspond to young oligodendrocytes or immature oligodendroblasts of other workers who consider them myelin-forming cells. At variance, recent work by Hirose and Bass (1973) evidenced a close correlation between oligodendrocyte proliferation and deposition of cholesterol and cerebrosides in the optic nerve of the young rat.

3. *Cerebellar Cortex*

Ultrastructural study of the molecular layer of the cerebellar cortex of 16 to 25-day-old rats (Privat, 1972, 1973) revealed that, although most of the immature cells had the typical appearance of migrating granule neurons (Mugnaini and Forstronen, 1967; Rakic, 1971; Altman, 1972), some had a more spherical shape and a more abundant, less dense cytoplasm; a few were found as satellites to stellate neurons, as are some mature oligodendrocytes in the molecular layer of the cerebellar cortex in the adult rat (Fig. 12).

4. *Selective Impairment of Oligodendrocyte Maturation in the Jimpy Mouse Brain*

This neurological mutant, discovered by Phillips, was studied by Sidman and Hayes (1965) and Hirano *et al.* (1969), who found that this

---

Fig. 9. Dark oligodendrocyte (DO). Same material and technique as Fig. 7. Both nucleus and cytoplasm are denser than those of any glial cell. The cytoplasm is less abundant than in light and medium oligodendrocytes. Stacks of rough ER (rER) are prominent. ×15,000. Inset, ×3000. (Courtesy of C. P. Leblond and J. Paterson.)

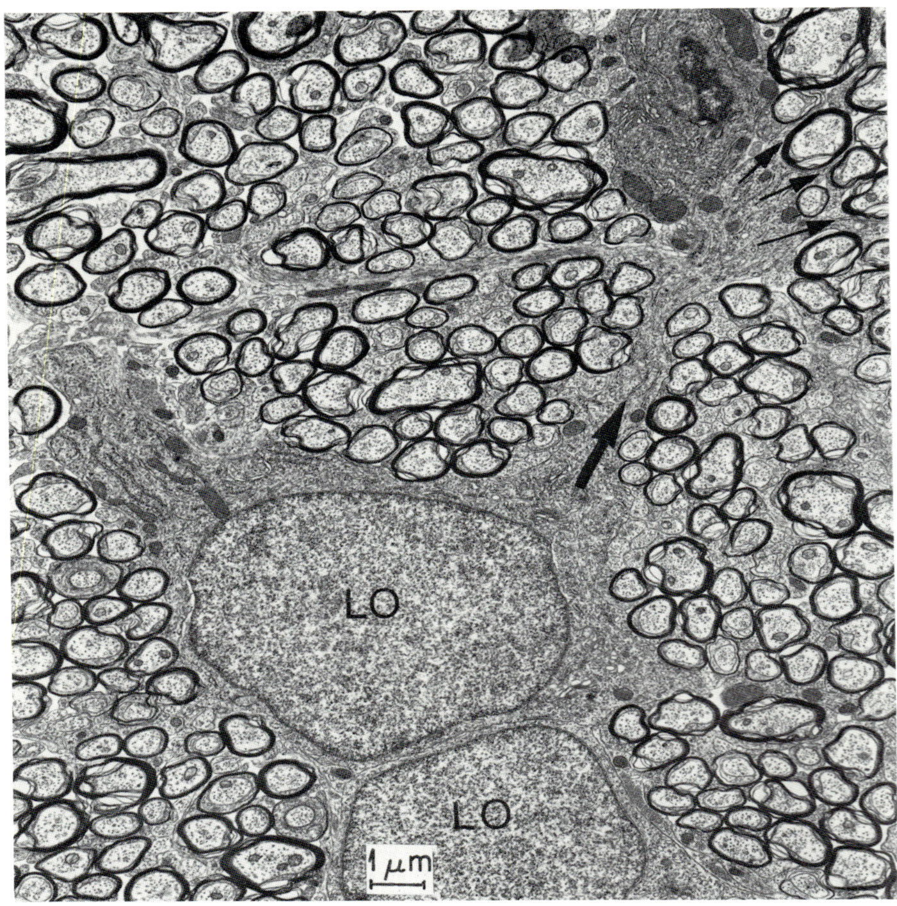

FIG. 10. Optic nerve of 20-day-old rat. Two light oligodendrocytes (LO) are presumed to be daughter cells of the same precursor. The upper one emits several processes, one of which (large arrow) is further subdivided into thin expansions (small arrows) which appear to be in continuity with the outer loop of myelinated axons. ×6500.

mutation was characterized by an almost total absence of myelin. As it is now well established that the oligodendrocyte is responsible for myelin deposition in the central nervous system (Peters, 1964; Bunge, 1968), it was decided to investigate the maturation of oligodendrocytes in the corpus callosum of Jimpy mice (Farkas-Bargeton *et al.*, 1972; Privat *et al.*, 1972).

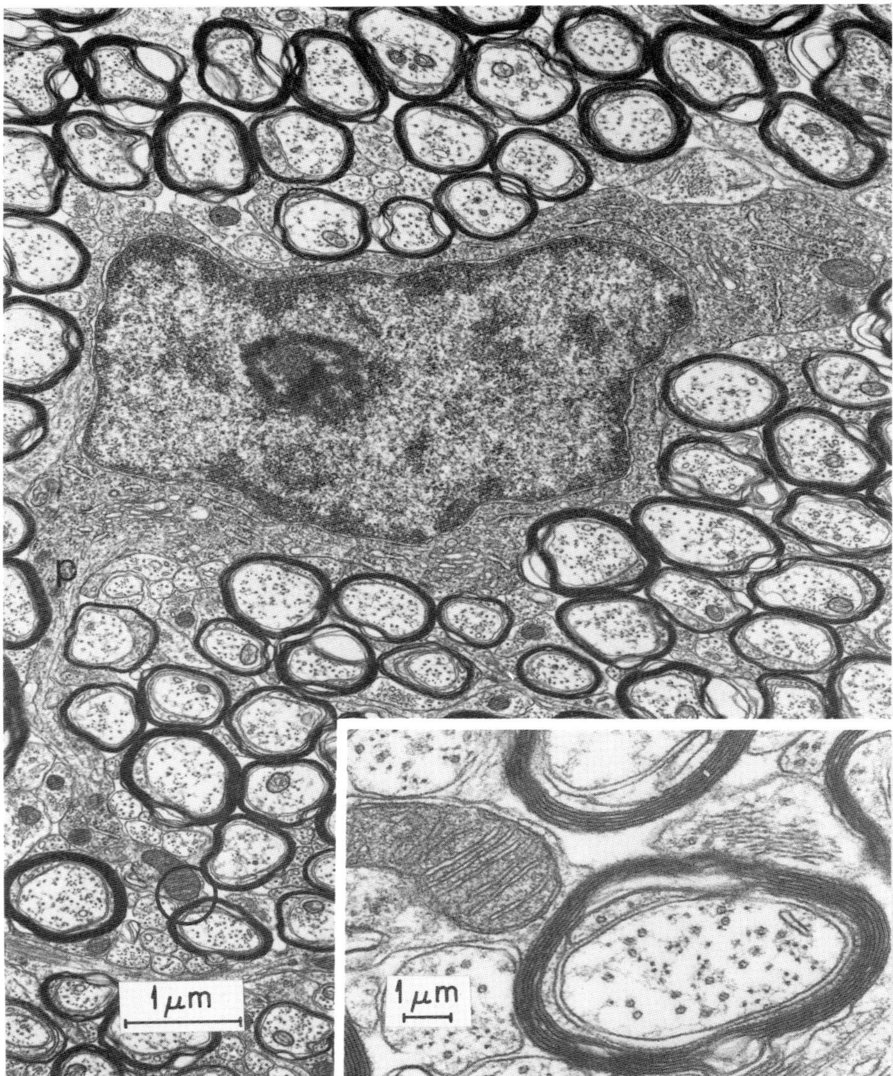

FIG. 11. Optic nerve of 20-day-old rat. A medium oligodendrocyte emits a process (P) containing numerous microtubules. This process abuts (inset) on a myelinated axon, where it appears to join the outer lamella of the sheath. Note that in both Figs. 10 and 11 the preterminal portion of the oligodendrocyte process contains rough ER and mitochondria. ×15,000. Inset, ×53,000.

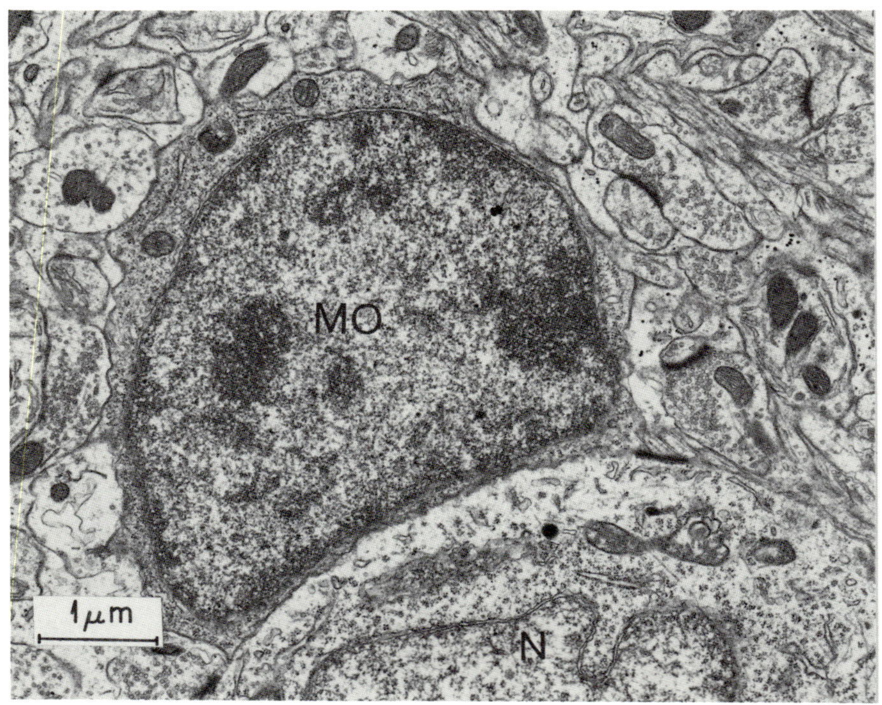

FIG. 12. Molecular layer of the cerebellum of a 1-month-old rat showing a medium oligodendrocyte (MO) satellite of a stellate neuron (N). See Fig. 8 for comparison. ×15,000.

When 14-day-old Jimpy mice were compared to controls, two differences were noted. First, the number of oligodendrocytes was drastically reduced; there were only light cells, compared to light and medium cells in the controls. The number of free subependymal cells, or glioblasts, was at least equal or increased compared to the controls. Second, the number of myelinated axons already appeared dramatically reduced, although myelination was far from being completed in the controls.

In 20-day-old and especially in 29-day-old controls, mature types of oligodendrocytes tended to predominate, whereas most axons became myelinated. Glioblasts were now present in small numbers. In Jimpy mice, at the same age, there was an almost total absence of myelin. A few myelinated fibers were found in the close vicinity of light oligodendrocytes, and these were few and usually isolated; the "domain" of myelination appeared in several instances well limited to a few fibers, and for each of them limited to an internode (Privat, unpublished). Medium and dark oligodendrocytes were absent. Glioblasts were present in large numbers throughout the corpus callosum. Pycnotic cells were found consistently more numerous in Jimpy mice than in the controls. The nature of these cells was, for most of them, difficult to determine, but in a few cases oligodendrocytes were positively identified. At all ages considered, the other glial cells—astrocytes and microglia cells—appeared unaffected by the mutation, except for some indices of reactivity, most likely due to the above-mentioned cell death; astrocytes showed more filaments and glycogen granules than usual, and microglial cells were packed with phagocyted products. In addition, the subependymal layer looked qualitatively and quantitatively normal. Thus it appeared with this study that the maturation of oligodendrocytes occurred through an independent line which could be affected specifically by a genetic disorder. Moreover, myelination was found to be dependent on the cells referred to as light oligodendrocytes, the two subsequent stages of their maturation, namely, the medium and dark cells presumed to be active in the maintenance of myelin, being absent.

B. Dynamic Evidence

The above results gave strong evidence in support of the differentiation of oligodendrocytes from immature neuroectodermal cells, but the building of a sequence from static images is largely dependent on the subjectivity of the observer. The ultimate proof could only arise from experiments in which labeling of immature cells was performed, and their fate followed at subsequent intervals.

1. *Subependymal Region*

This study (Privat, 1970a,b; Paterson *et al.*, 1972, 1973) required two conditions: first, that the immature cells of the subependymal layer be labeled specifically, and no other cell in the region. Second, to study a cell population large enough to make a quantification possible. The first condition was realized by introducing the label (thymidine-^{3}H) in the lateral ventricle of 1-month-old rats. The diffusion of the tracer was restricted to a limited area around the ventricular system, so that an almost perfect specificity was attained. The quantity injected (5 or 10 μCi in 50 μl) and the slow rate of injection made unlikely an important leakage into the brain parenchyme and the vascular system.

The second condition was realized by the use, for radioautography, of 1-μ plastic sections stained with toluidine blue, which allowed accurate cell identification and significative cell counts. It was then necessary to transpose the criteria of identification given by electron microscopy

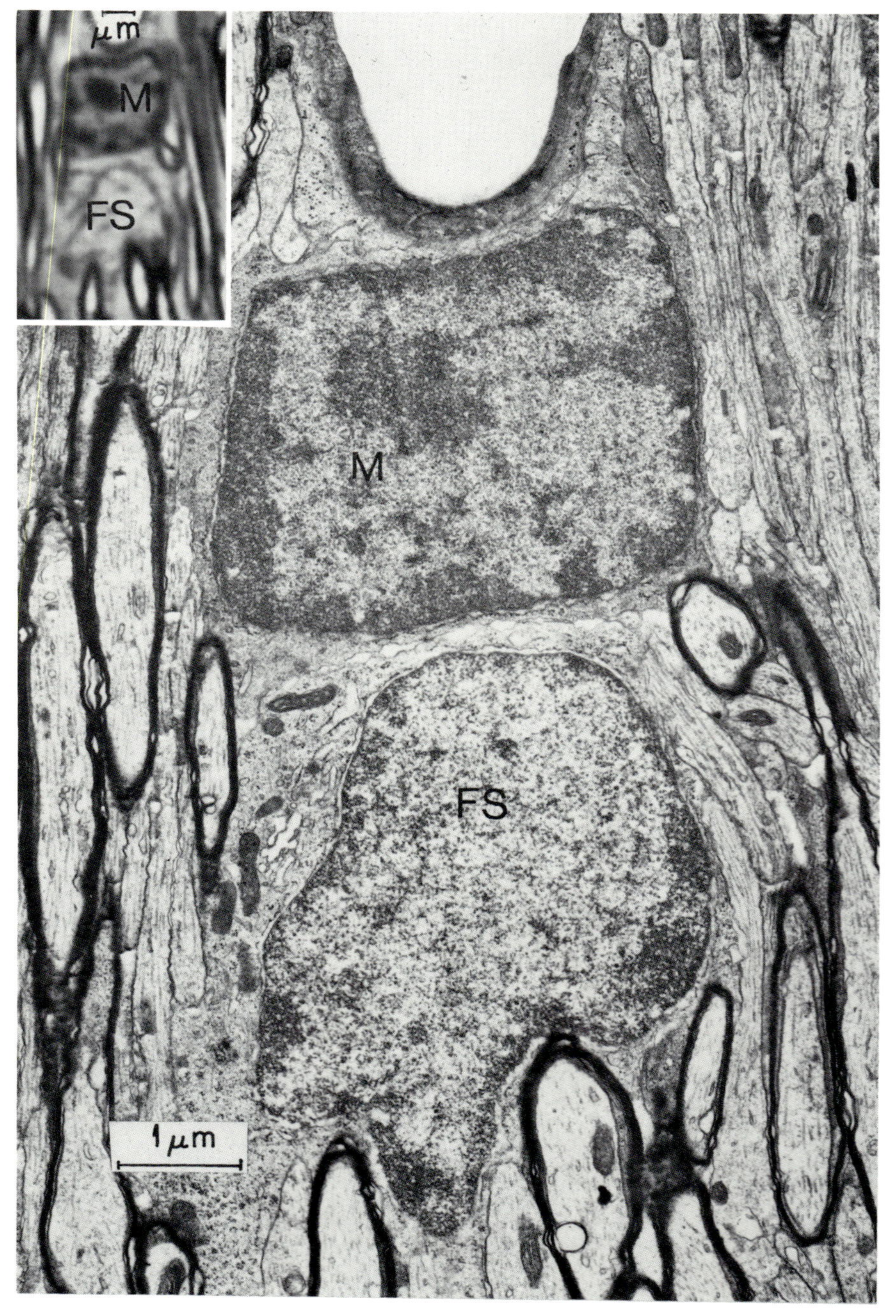
μm
M
FS
M
FS
1 μm

to light microscopy of 1-μ sections. For this purpose serial sectioning including thin and thick sections was performed, and the cells identified with the EM were looked for, with the aid of a few landmarks, on photomontages prepared from 1-μ sections (Ling *et al.*, 1973):

Subependymal cells (Fig. 1) have a small, irregularly shaped nucleus which appears angular and ovoid and shows indentations. The nucleus has a well-defined nuclear envelope and contains scattered ill-defined chromatin masses, with a few large masses. A patchwork appearance is fairly common, corresponding to prophase or telophase cells. The cytoplasmic boundaries are indistinct, except in poorly fixed material, where extracellular space is enlarged.

Free subependymal cells (Fig. 13) found in the corpus callosum usually have an ovoid or spindle-shaped nucleus, but less irregular shapes are not rare, especially near the border area. A distinct chromatin rim outlines the nucleus, but is more irregular than in astrocytes (see Section IV). A nucleolus is sometimes visible. The cytoplasm is always scanty and appears at one or both extremities of the nucleus as a narrow peak.

Light oligodendrocytes (Fig. 7), the largest of glial cells, display a large ovoid nucleus in abundant cytoplasm. The nucleolus is usually centrally located, but may be tangential to the nuclear envelope; when glutaraldehyde alone is used, instead of a mixture of aldehydes, the nucleoplasm displays some graininess.

In *medium oligodendrocytes* (Fig. 8), the nucleus and cytoplasm appear smaller and darker than in light oligodendrocytes. Nuclear chromatin is arranged in small clumps, sometimes along the nuclear envelope. The cytoplasm is well delineated from neighboring tissue.

Dark oligodendrocytes (Fig. 9) are the smaller and darker cells. The nucleoplasm has an overall density which parallels that of the cytoplasm. The nucleus is often eccentrically located in the cytoplasm. Its shape varies from oval to angular or crescent shape.

When thymidine-^{3}H was injected into the lateral ventricle (Privat, 1970a,b; Paterson *et al.*, 1972, 1973), a high percentage of the cells of the subependymal layer were found labeled 2 hours after the injection, as well as cells of the border area. By 2 days many labeled cells appeared in the corpus callosum, and a maximum was found at the same time in the subependymal layer. When the percentage of labeling was plotted

---

FIG. 13. Microglial cell (M) and free subependymal cell (FS) in the corpus callosum of a 80-mg rat. Technique as in Fig. 7. The microglial cell has an angular nucleus, with a sharp contrast between the large masses of dark chromatin and the light nucleoplasm, and a scanty cytoplasm with few organelles. The free subependymal cell has an irregularly shaped nucleus and cytoplasm rich in free ribosomes. ×15,000. Inset, ×3000. (Courtesy of C. P. Leblond and J. Paterson.)

against time for each cell type in the corpus callosum, it was found that the labeling of free subependymal cells reached a peak 2 days after the injection, light oligodendrocytes 4 days after the injection, medium oligodendrocytes between 4 and 18 days after the injection, and dark ones about 21 days. It was then inferred (Paterson *et al.*, 1973) that light oligodendrocytes occurred through transformation of free subependymal cells, and that they transformed into medium cells, which in turn gave rise to dark cells, the end product of the series. A statistical study of the relative number of glial cell types in the corpus callosum and neighboring cortex of the rat, from ¾ month to 6 months, confirmed these results (Ling *et al.*, 1973). It was found that in the corpus callosum the number of free subependymal cells decreased regularly from ¾ month onward, to become negligible at 5 months. This was also true of light and medium oligodendrocytes, whereas the number of dark ones increased regularly to reach at 5 months a sevenfold increase. In the neighboring cortex the events followed a parallel course, except that free subependymal cells were absent at all ages studied, and that the transformation into the mature cell, the dark oligodendrocyte, seemed to occur much more slowly.

2. *Cerebellar Cortex*

A similar method was used by Privat (1972, 1973) to study the fate of external granule cells in 16-day-old rats. Animals were sacrificed 2 hours to 25 days after intraperitoneal injection of thymidine, and radioautographs were prepared from 1-μ sections stained with toluidine blue.

Two hours after the injection, nearly 90% of the labeled cells were concentrated in the external granular layer and the upper portion of the molecular layer. They were identified as immature cells. Only 2% were located in the Purkinje and internal granular layers.

Four days after the injection, most of labeled cells were dispersed in the molecular and internal granular layers, and by 25 days 10% of the labeled cells could be positively identified as medium and dark oligodendrocytes distributed in the lower third of the molecular layer and the upper part of the internal granular layer.

It appeared then that the external granular layer was able to give rise, 16 days after birth, to a small although significant number of oligodendrocytes. Another finding was the persistence among labeled cells of a substantial percentage of immature-looking cells in adult rats.

C. Concluding Remarks

Studies made on the origin and maturation of oligodendrocytes in various animals and various structures are in general agreement on several points. These cells are generated mainly in the postnatal period, from undifferentiated cells of neuroectodermal origin. Their maturation parallels roughly the deposition of myelin in the various regions of the brain [in the spinal cord, where myelination is much more precocious, mature oligodendrocytes are present at birth (Phillips, 1973), and the sequence of maturation is very similar].

When oligodendrocyte maturation is disturbed, as in the Jimpy mouse, myelination does not take place, although other glial types are not affected.

The sequence of maturation is a protracted one, as most of the investigators consider three to five successive stages. Roughly three periods may be distinguished: first, a period of maturation *proprio sensu*, with cells of varying cytoplasm and nuclear shapes and sizes, characterized by a high content of free ribosomes; second, a period of active myelination, with large cells displaying a round or oval nucleus, a prominent nucleolus, and numerous large processes; third, a period of myelin support, with smaller, darker cells with few, if any, identifiable processes, a well-developed Golgi region, and rough ER. The last-mentioned cells are still actively metabolic, since it has been demonstrated that a turnover of lipoproteins occurs in the myelin sheath and, consequently, that the process of myelination continues throughout life (Banik *et al.*, 1968). The question of their duty is raised by the oligodendrocytes of the gray matter which, in regions such as the cerebral cortex, the caudate nucleus, and the molecular layer of the cerebellar cortex, do not seem to be associated with myelinated fibers; they often appear as neuronal satellites. Their maturation seems to follow the same steps, although perhaps more slowly (Ling *et al.*, 1973), and the mature cell has the same appearance.

IV. The Astrocyte Line

A. Morphological Evidence of Maturation

The identification of astrocytes was based for decades on a metallic impregnation technique devised by Ramón y Cajal (1913), namely, the gold sublimate method. This lead to the classic distinction between protoplasmic astrocytes, found in the gray matter, and fibrous astrocytes, present in the white matter. With ordinary stains, such as hematoxylin–eosin applied on paraffin sections, Penfield (1932) tentatively identified astrocytes as cells with a light, irregularly oval nucleus, larger than those of oligodendrocytes and microglial cells. These nuclear criteria were used by earlier electron microscopists (Luse, 1956; Farquhar and Hartmann, 1957) to identify astrocytes, and to describe their ultrastructural charac-

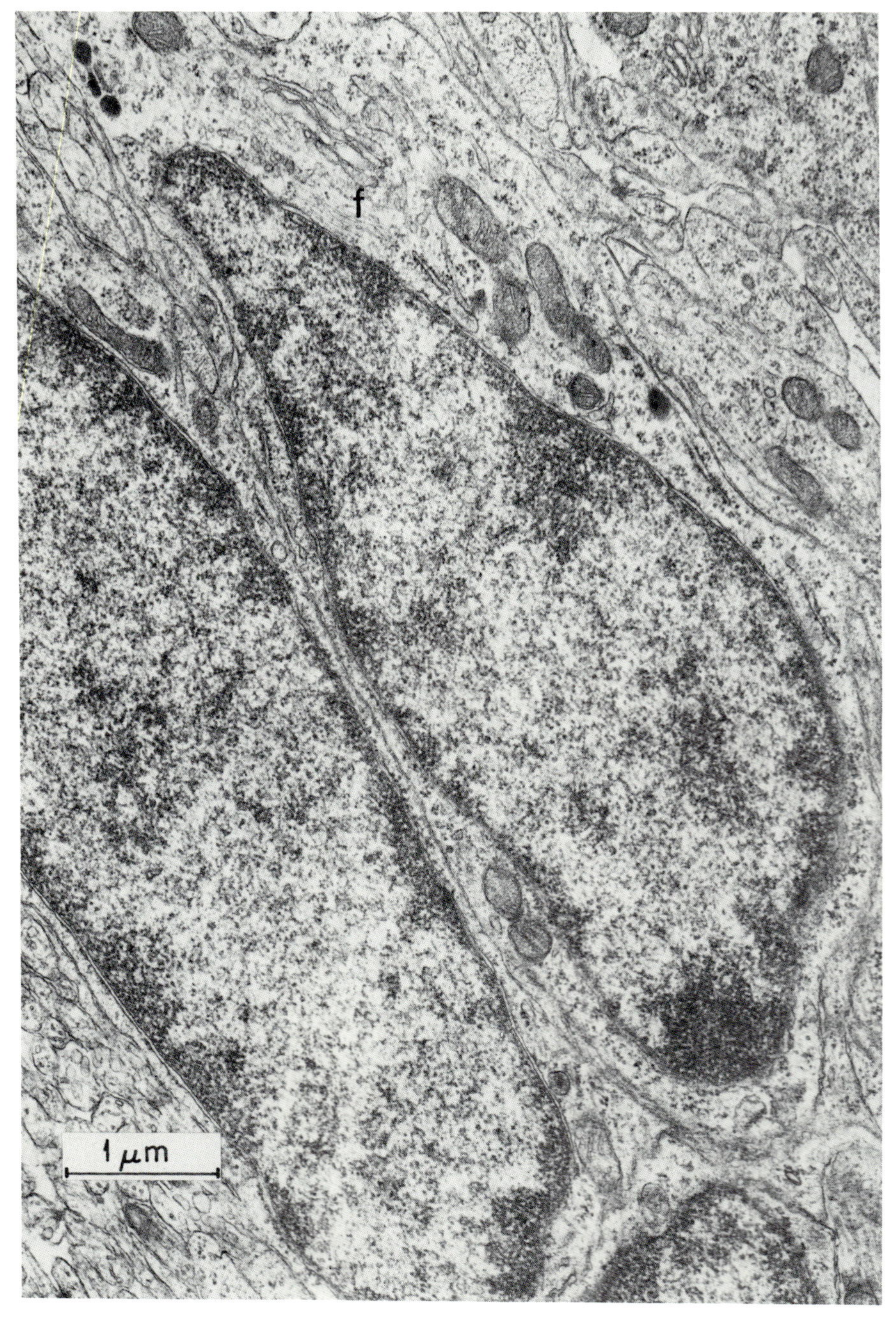
f
1 μm

teristics. The main criterion rapidly became the presence in the cytoplasm of bundles of filaments both in fibrous (Palay, 1958; De Robertis and Gershenfeld, 1961; Palay *et al.*, 1962; Malmfors, 1963; Schultz, 1964; Bodian, 1964) and in protoplasmic astrocytes (Herndon, 1964; Wolf, 1965; Kruger and Maxwell, 1966). Glycogen was also found as a regular feature of these cells (Kruger and Maxwell, 1966; Wendell-Smith *et al.*, 1966; Vaughn and Peters, 1968; Stensaas and Stensaas, 1968; King, 1968). Caley and Maxwell (1968) summarized the general consensus in describing astrocytes as the largest and lightest of the glial cells, with a few organelles in a light cytoplasm and filaments and glycogen in the pericaryon and processes. Finally, Vaughn and Pease (1967) and Mori and Leblond (1969b) filled the gap between light and electron microscopy in adapting the gold sublimate method of Ramón y Cajal (1913) for the electron microscope. In spite of the poor quality of the preparations, some cells were found to be overlaid by small dark metallic particles. These were essentially over bundles of filaments and spherical dense bodies, the so-called gliosomes described by Cajal. It appeared that these filaments and the glycogen particles were the two main ultrastructural criteria, with the nuclear morphology, for the identification of the first stages of astrocyte morphogenesis.

In the *subependymal region* of 1-month-old rats the first evidence of differentiation toward the mature astrocyte was found in a few cells of the border area (Fig. 14) (Privat, 1970a; Privat and Leblond, 1972). When nuclear criteria were considered, some free subependymal cells were found to differ from the cells of the subependymal layers only in the oval shape of the nucleus and a distinct rim of chromatin along the nuclear envelope. In others the cytoplasm appeared lighter, with a few short cisternae of rough ER and fewer free ribosomes. In others a few glycogen granules were found in the cytoplasm; finally, discrete bundles of filaments were found in some cells, with often isolated microtubules. Typical astrocytes, as described by Mori and Leblond (1969b) in the corpus callosum (Fig. 15), were also found in the border area. In older rats (Blakemore, 1969), astrocytes were found in the subependymal region in larger numbers than elsewhere beneath the ventricle, and these exhibited a high content of glycogen and gliofilaments.

In the *cerebral cortex* of young rats (Caley and Maxwell, 1968), astro-

---

FIG. 14. Border area of the subependymal layer in a 80-mg rat. The cell in the center has an elongated nucleus with patchy chromatin, a thin perinuclear rim of cytoplasm containing free ribosomes, a few cisternae of rough ER, some microtubules, and a few gliofilaments (f). This cell is believed to be a young astrocyte. At left and below are other cells with characteristics of immaturity. ×20,000.

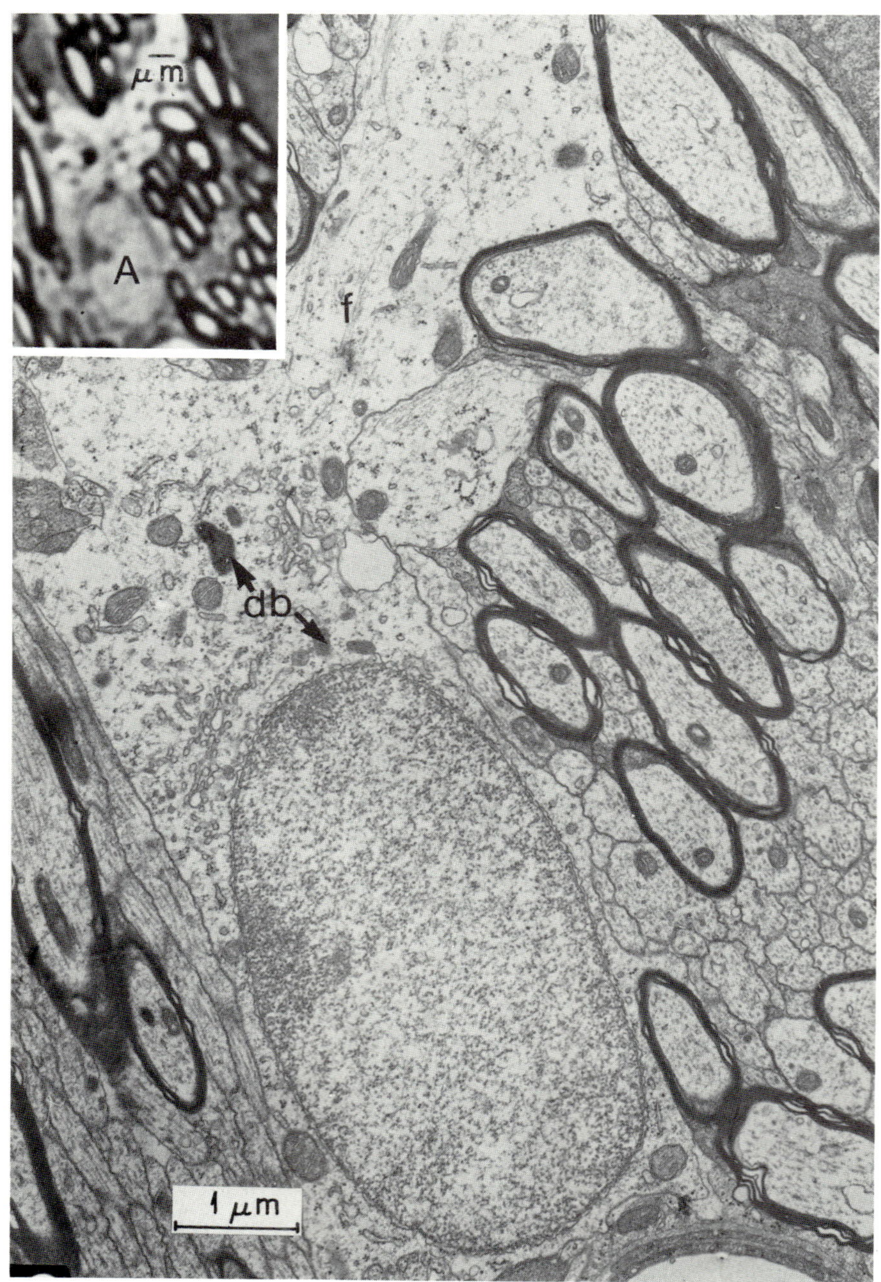
µm
A
f
db
1 µm

blasts were described as cells lighter than oligodendroblasts and oligodendrocytes, containing scattered ER, mitochondria, Golgi complexes, and vesicles. Their processes were broad, and frequently abutted onto the basal lamina of blood vessels. Glycogen granules were an inconstant feature, but cytoplasmic gliofilaments were considered a good criterion for the identification of astroblasts.

In the *optic nerve* of the young rat (Peters and Vaughn, 1967; Vaughn and Peters, 1968; Vaughn, 1969), two maturational stages were characterized. The young astrocyte differs from the undifferentiated precursor in the presence of radiating processes, often reaching the glia limitans, and the presence of short cisternae of rough ER. Immature astrocyes are characterized by the presence of a few filaments, together with microtubules. During maturation the number of filaments increases, while that of microtubules decreases (Peters and Vaughn, 1967). They are stellate cells with radiating processes, which are bifurcated to isolate bundles of axons and bear small, fingerlike projections which have some tendency to enclose unmyelinated axons.

The nuclei appear homogeneous, and have a thin rim of condensed chromatin beneath the nuclear envelope. The cisternae of rough ER are wide, regularly arranged, and filled with a flocculent, dense material. Some cells have the lucent cytoplasm characteristic of mature cells, but others are as dense as undifferentiated cells or young oligodendrocytes. In the kitten optic nerve (Wendéll-Smith *et al.*, 1966; Blunt *et al.*, 1972), early and late astrocytoblasts were found to be similar to the young and immature astrocytes described by Vaughn (1969) in the rat, except for greater nuclear and cytoplasmic density which may be accounted for by differences in tissue preparation.

In the *rat cerebellum*, as in other mammals, a special type of astrocyte is present, the so-called Bergman glia, or cell of Fananas, or epithelial cell of Golgi (Ramón y Cajal, 1909; Herndon, 1964; Petersen, 1969). Its perikaryon is adjacent to that of the Purkinje cells, and from it long, slender processes ascend to the surface of the folia where cytoplasmic end feet build up the glia limitans. Their nuclei are usually round or slightly oval, with a very discrete accumulation of chromatin along the nuclear membrane; their perikaryal cytoplasm is denser than is usual

---

FIG. 15. Astrocyte in the corpus callosum of a 80-mg rat. Technique as in Fig. 7. The light, oval nucleus has a thin condensation of chromatin along the nuclear membrane, which appears as a distinct outline in the semithin section. The light cytoplasm contains bundles of filaments (f) and a few dark bodies (db) which appear as dark spots on the toluidine blue-stained section. ×15,000. Inset, ×3000. (Courtesy of C. P. Leblond and J. Paterson.)

in astrocytes, and contains a larger amount of free ribosomes; tubules are sometimes present, but filaments are very rare in normal animals. Glycogen granules are an inconstant feature. The processes that ascend through the molecular layer contain bundles of gliofilaments.

These cells arise during fetal life from the matrix layer of the fourth ventricle; soon after birth they migrate toward the surface of the folium and grow up apical processes which join to form the glia limitans (Rakic and Sidman, 1973). Their postnatal maturation has been recently reinvestigated using the fluorescent antibody technique (Bignami *et al.*, 1972; Bignami and Dahl, 1973).

B. Dynamic Studies

Indications of transformation of undifferentiated neuroectodermal cells into astrocytes, obtained from static morphological data, needed to be confirmed by dynamic evidence from labeling experiments. It was found convenient (Privat, 1970b; Paterson *et al.*, 1973), as for oligodendrocytes (see Section III), to use radioautographic techniques on plastic sections stained with toluidine blue. Indeed, the thymidine-labeling studies of Altman (1966) and Lewis (1968), carried out on ordinary stained paraffin sections, although they detected migration from the subependymal region, were unable to determine with certainty which cell types were labeled.

Astrocytes were identified on ½-μm plastic sections stained with toluidine blue, according to the following criteria (Fig. 15). Both nucleus and cytoplasm appeared very light, with a sharp delineation of the nucleus by a thin condensation of chromatin. The nucleolus is sometimes visible as a small, dense spheroid. The cytoplasm is very pale, and may display large processes easily visible between myelinated axons and frequently approaching capillaries. When the initial fixation is carried out with glutaraldehyde alone, instead of the formaldehyde–glutaraldehyde mixture modified from the technique of Reese and Karnowsky (1967), the nucleoplasm has a ground-glass appearance (Ling *et al.*, 1973). When the cells of the subependymal layer were labeled with tritiated thymidine in 1-month-old rats, labeled astrocytes were found at subsequent time intervals in the corpus callosum (Privat, 1970a; Paterson *et al.*, 1973). Maximum labeling, 4 days after the injection, was always less than 1% of the cells of the corpus callosum, suggesting that the bulk of the astrocytes mature before the end of the first month after birth in the rat corpus callosum (Paterson *et al.*, 1973).

C. Proliferative Reactions

Several workers demonstrated the reactivity of astrocytes to a needle wound (del Rio-Hortega and Penfield, 1927; Sjöstrand, 1965; Adrian

1968; Cavanagh, 1970), or in the case of Wallerian degeneration (Nissl, 1892, 1894; de Vries, 1910; Brodal, 1939; Cammermeyer, 1955; Bignami and Ralston, 1969; Vaughn and Pease, 1970; Skoff and Vaughn, 1971; Cramer and Alpers, 1932). Most of them concluded that astrocytes are unable to undergo division in reaction to these various elements, but Cammermeyer (1955), Sjöstrand (1965), and Cavanagh (1970) were of the opposite opinion. Sjöstrand (1965) found in the hypoglossal nucleus, following a nerve crush, definite cell proliferation evidenced by thymidine-^{3}H labeling. Most of these cells were regarded as macrophages (see Section V), and a few were considered astrocytes. In Cavanagh's (1970) study, cell identification is at least dubious, as this investigator used paraffin sections stained with hematoxylin–eosin, and periodic acid–Schiff (PAS). On the one hand, it is well known that macrophages are able to take up PAS-positive debris and, on the other hand, Mori and Leblond (1969b) showed how glial cell identification in paraffin sections with ordinary stains was a source of error.

The opinion of Skoff and Vaughn (1971) is somewhat different; these investigators found that in young adult rats, following enucleation, astrocytes in the optic nerve take up tritiated thymidine. Furthermore, they demonstrated with electron microscopy mitoses in cells containing gliofilaments and microtubules. But conversely, they did not find any increase in the number of astrocytes of the optic nerve, although the number of multipotential glial cells increased markedly. They suggested that astrocytes, or astrocytelike cells, can transform into multipotential glial cells.

D. Concluding Remarks

As were oligodendrocytes, astrocytes were found according to both static and dynamic data to arise from neuroectodermal precursors. But, unlike oligodendrocytes, their maturation occurs mainly in the perinatal period, and few if any cells of this type appear to mature after 1 month in the species studied so far.

Astrocytes form a fairly stable population, and their reaction to various stimuli appears more to be a hypertrophy than a true proliferation.

V. Microglia

A. Introduction

Microglia was identified as a distinct element of the nervous tissue by del Rio-Hortega (1919), using the silver carbonate technic. Microglial cells were distinguished by their profusely branched processes, and possessed migratory as well as phagocytic activity. Del Rio-Hortega (1930,

1932) found that they appeared during ontogenesis under the pia and in the vicinity of the ventricles, where they constituted the so-called microglial fountains, and he suggested that they were of mesodermal origin. This opinion was shared among others, by Kershman (1939), who found that these mesenchymal cells migrated as round ameboid elements from the meninges and blood vessels and later developed into more complex branched forms.

However, several investigators believed that microglial cells, as other glial cell types, had a neuroectodermal origin. Pruijs (1927) found, in the rabbit, cells with characteristics intermediate between those of oligodendrocytes and microglial cells; Rydberg (1932) found, in the human fetus, cells intermediate between spongioblasts and microglial cells; Horstadius (1950), while studying neural crest derivatives, found that glial cells were of neuroectodermal origin. A common origin for all glial types was also postulated by Roussy and L'Hermitte (1930).

More recently, Cammermeyer (1970a,b) using the silver carbonate technique, expressed the view that microglial cells are extravascular cells possibly of hematogeneous origin, and that they are able to undergo mitosis.

During the last decade a considerable amount of information about microglia has been obtained with the electron microscope and with labeling experiments, and also with a combination of these two techniques. Most of this work was intended to shed some light on the origin of brain macrophages.

B. Ultrastructural Data

Earlier ultrastructural descriptions of microglia depicted it as consisting of small cells with dense nuclei (Luse, 1956; Farquhar and Hartmann, 1957; De Robertis and Gershenfeld, 1961; Blinzinger and Hager, 1962; Bodian, 1964; Herndon, 1964; Yasuzumi *et al.*, 1964), although major discrepancies existed between these descriptions; several investigators denied the existence of microglial cells in normal tissue (Malmfors, 1963; Kruger and Maxwell, 1966; Eager and Eager, 1966; King, 1968; King and Schwyn, 1970), which led to much controversy in the literature. Mori and Leblond (1969a) shed some light on this complicated issue; they adapted for electron microscopy the old technique of del Rio-Hortega (1919) in order to determine whether any identified cell type was selectively impregnated by silver carbonate, and to describe its characteristics. In spite of the poor preservation of the tissue after prolonged immersion in the formalin–ammonium bromide solution, these investigators were able to describe the impregnated cells as elongated, with a small nucleus bearing large patches of dark chromatin, scanty cytoplasm containing a few elongated tortuous cisternae of rough ER, a well-developed Golgi zone, and numerous dense bodies (Fig. 13).

They found this type of cell in two locations, pericytal and interstitial. The former are enclosed within an expansion of the basement lamina of a capillary, while the latter are scattered in the nervous parenchyme. Additional features of these cells, as found in routine electron microscope preparations, were the presence of irregular processes of various calibers whose density is greater than those of other glial cells, the absence of gliofilaments, and the rareness of microtubules (Fig. 16). These cells made up about 5% of the cell population of the corpus callosum.

This description was very similar to that of Blinzinger and Hager (1962) and Stensaas and Stensaas (1968). In addition, similar characters were described by Vaughn and Peters (1968) and Peters *et al.* (1970) for a cell called the third neuroglial element which was found in the optic nerve of fetal and postnatal rats, as well as in fully adult animals. These cells made up about 4–5% of the total number of cells in the optic nerve of adult rats but, according to Vaughn (1969), their number is much greater in the optic nerve of the fetal and early postnatal rat. Thus Vaughn (1969), Peters *et al.* (1970), and Vaughn *et al.* (1970) consider this cell an immature multipotential element, a precursor of all glial cell types in the optic nerve of the rat, which was called the small glial precursor (Vaughn, 1969); such third glial elements were also found in the red nucleus and substantia nigra of normal adult primates by King and Schwyn (1970), and in the hypoglossal nucleus of the rat after peripheral nerve crush by Fernando (1971). These discrepancies go far beyond a simple problem of terminology, since they imply a different ontogenic hypothesis. Mori and Leblond (1969a) found some evidence that pericytes could transform into interstitial microglial cells, as they demonstrated some of these cells crossing the basement lamina, thus favoring the opinion of Cammermeyer on extravascular cells of hematogeneous origin. Likewise, Privat and Leblond (1972), although they found with electron microscopy, as Cammermeyer (1965) did with metallic impregnation, numerous microglial cells in the vicinity of the subependymal layer or within it, were unable to find any morphological evidence of a transition between subependymal cells and microglial cells. The presence of numerous microglial cells in this region was also noted by Blakemore (1969), who discussed their possible origin *in situ*. In the same region, Stensaas and Reichert (1971) and Stensaas and Gilson (1972) found two types of microglial cells in the newborn rabbit, a round and ameboid form, and a ramified one, and they suggested that both were mature cell types. Moreover, they did not find, either with metallic impregnation, or with serial reconstruction through electron micrographs, any evidence of

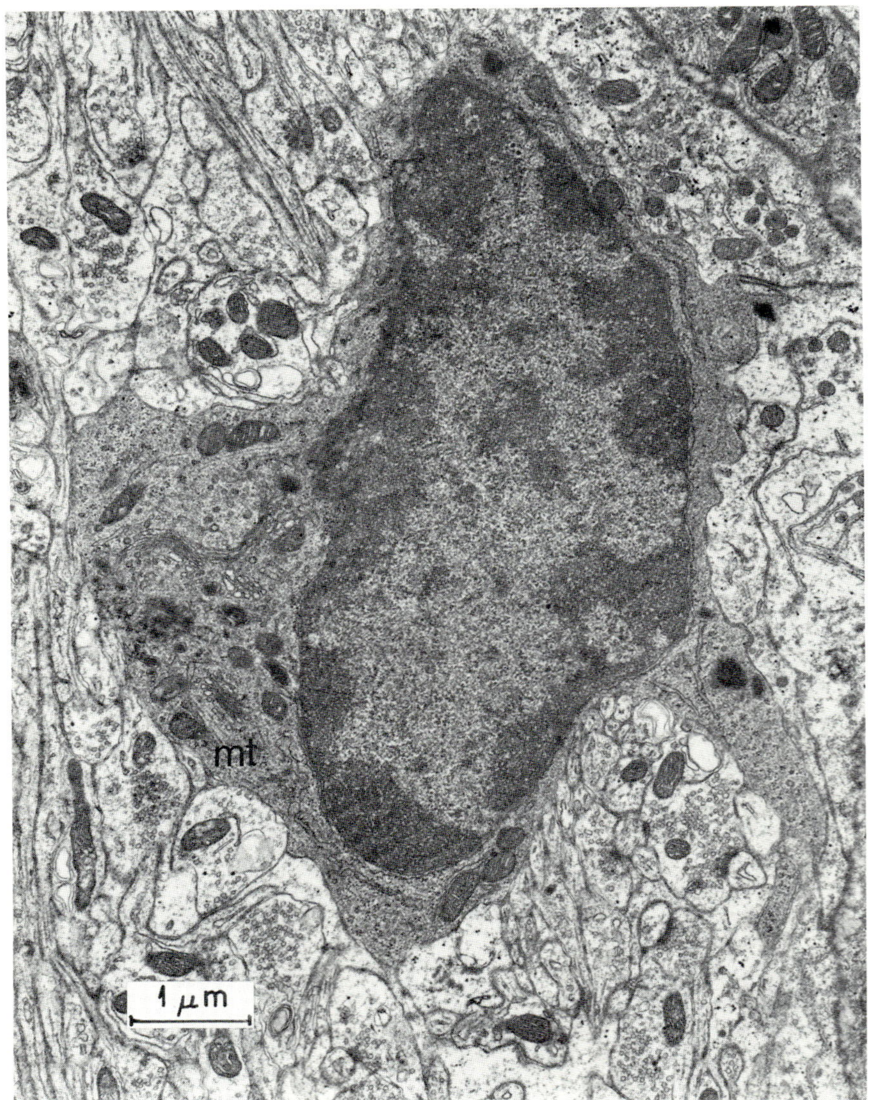

FIG. 16. Cerebellar cortex of 1-month-old rat. The typical microglial cell of the molecular layer exhibits an irregularly elongated nucleus, with large masses of dark chromatin, a relatively dense cytoplasm containing long, tortuous cisternae of rough ER, and a well-developed Golgi region. The few microtubules (mt) are a more unusual feature of this cell. ×15,000.

transitional forms between the cells of the matrix layer and these two types of microglial cells.

C. Experimental Data

1. *Dynamic Study*

In the course of their general evaluation of cell production of the subependymal layer of the rat brain, Privat (1970b), Ling *et al.* (1973), and Paterson *et al.* (1973) investigated the possibility that microglial cells could be generated by the subependymal cell, although Privat and Leblond (1972) did not find any ultrastructural evidence of such.

On the 1-μm Epon sections stained with toluidine blue used for this study (Ling *et al.*, 1973), microglial cells were easily distinguished from other cell types based on the following characters (Fig. 13). Their nucleus is smaller than that of any mature glial type and is characterized by a sharp contrast between dark clumps of chromatin and a slightly basophilic nucleoplasm. It is usually elongated, but may be globular or markedly irregular. The cytoplasm is not abundant, and appears grayish, with frequent dense bodies. Pericytes were found to display the same nuclear and cytoplasmic characters.

When tritiated thymidine was stereotaxically injected in the lateral ventricle (Privat, 1970b; Paterson *et al.*, 1973), no microglial cells were found labeled 2–4 hours after the injection, although numerous cells were labeled in the close vicinity of some of them, in the subependymal layer. Two days later, labeled microglial cells were found in the vicinity of the subependymal layer, in the corpus callosum, on the side of thymidine injection, and also on the opposite side, although in smaller numbers. Lewis (1968) had already reported such labeling, but the identification of microglial cells in routine paraffin sections made it doubtful. The number of labeled microglial cells in the corpus callosum reached a peak around the seventh day after injection, and decreased slowly thereafter. It was noted that, at every time interval after injection, the number of labeled microglial cells on the injected side was fourfold that on the contralateral side, and this was interpreted as a reaction to the trauma of injection. This was confirmed by the fact that, after intraperitoneal injection of thymidine, very few microglial cells were labeled, although the labeling pattern for other glial cells was similar to that obtained after intraventricular injection. These results were found to be consistent with a possible origin from subependymal cells, since these had enough label to admit a precursor–product relationship. However, one other origin, either from cerebrospinal fluid cells or, eventually, from labeled blood cells, could not be ruled out.

2. *Reactive Microglial Cells*

One of the main characteristics assigned to microglial cells by del Rio-Hortega (1919) was phagocytotic activity, and on that basis they were considered the source of brain macrophages. This led to much controversy (reviewed by Russel, 1962; Konigsmark and Sidman, 1963); briefly, workers using del Rio-Hortega's techniques suggested that oligodendrocytes are the only source of macrophages (Ferraro and Davidoff, 1928) or that they arise from both microglial and oligodendroglial cells (Field, 1957) or from endogenous mesenchymal elements (reviewed by Feigin, 1969) or, finally, from exogenous mesenchymal elements (Boggenstoss *et al.*, 1943).

A new insight into this problem was given by Adrian and Walker (1962) and by Konigsmark and Sidman (1963); both of these groups labeled blood leukocytes with tritiated thymidine before and after a wound located, respectively, in the spinal cord and the brain. Both groups concluded that the population of labeled cells in a wound was a complex one which included mononuclear leukocytes as well as indigenous elements. These indigenous elements were most likely microglial cells. Similar results were obtained by Huntington and Terry (1966), who nevertheless were unable to determine the exact nature of the endogenous element. More recently, Kitamura *et al.* (1972) found with electron microscope radioautography that, after a brain wound, macrophages originated exclusively from blood mononuclear leukocytes.

Another model is represented by glial reactions to retrograde degeneration. These have been studied by, among others, Cammermeyer (1955, 1965a,b), Rapos and Bakos (1959), Takano (1964), Watson (1965), Sjöstrand (1965, 1966, 1971), Kreutzberg (1966), Blinzinger and Kreutzberg (1968), Kirkpatrick (1968), Humbertson *et al.* (1969), Dixon (1969), Torvik and Skjorten (1971), Torvik (1972), Adrian and Smothermon (1970), Fernando (1971), Davidoff (1973), Wong-Riley (1972), Stenwig (1972), Matthews and Kruger (1973), Chow-Dewson (1966), and Eager and Eager (1966).

Sjöstrand (1965) demonstrated with thymidine-^{3}H radioautography a 10-fold increase in the labeling index of glial cells in the hypoglossal nucleus after nerve crush, compared to the contralateral nucleus. The majority of labeled cells were identified as microglial cells but, because of the use of paraffin section and routine stain, accurate identification could not be achieved, and this investigator could not rule out the possibility of leukocyte infiltration. This possibility was demonstrated by Adrian and Smothermon (1970), who labeled circulating leukocytes prior to hypoglossal injury. They found numerous labeled cells in the hypoglossal nucleus from 4 to 35 days after injury.

Torvik and Skjorten (1971) found, in facial nerve nuclei, different reactions according to the nature of the injury applied to the nerve: After crush lesions microglial cells were found, with the electron microscope, to proliferate and to cover restricted portions of the chromatolytic neuron. After complete nerve sections, microglial cells covered the whole surface of the cell, removed synaptic boutons, a feature already noted by Blinzinger and Kreutzberg (1968), and finally phagocytosed the whole dead cell. In a further experiment, Torvik (1972) positively identified microglial cells as the phagocytic elements, in a combined study with metallic impregnation and electron microscopy. With a different technique, Stenwig (1972) found no evidence of leukocyte infiltration after evulsion of the facial nerve and the same experimental procedure shown by Torvik (1972) to be followed by intensive proliferation of microglial cells.

After facial nerve crush, Sjöstrand (1971) found no evidence of invasion of labeled leukocytes. More recently, Davidoff (1973) observed that, after sectioning of the hypoglossal nerve, reactive microglial cells originated mostly from activated and proliferated oligodendroglial cells normally present in the nucleus.

However, Matthews and Kruger (1973), in a careful ultrastructural study of cellular reaction after neural degeneration in thalamic nuclei of the rabbit, found diapedesis of hematogeneous elements, and activation of endogenous pericytes. In an extensive discussion of the concept of microglia, these investigators concluded that the infiltration of hematogeneous elements is proportional to the extent of neuronal loss, as already pointed out by Torvik and Skjorten (1971) and Sjöstrand (1971).

Finally, Wallerian degeneration was also used as a model for the study of glial reactions by Bignami and Ralston (1969), van Crevel and Verhaart (1963), Vaughn *et al.* (1970), Vaughn and Pease (1970), Skoff and Vaughn (1971), Wong-Riley (1972), and Cook and Wisniewski (1973), among others.

Bignami and Ralston (1969) observed with the electron microscope massive phagocytic activity of cells that they classified as histiocytes, some of which had morphology similar to that of the classic microglial cells of light microscopy; the latter type was found only in late stages of Wallerian degeneration.

In their electron microscope study, Vaughn *et al.* (1970) and Vaughn and Pease (1970) noted a massive proliferation of multipotential glial cells or small glioblasts, which are morphologically similar to the microglial cells described by other investigators (Mori and Leblond, 1969a; Privat and Leblond, 1972; Wong-Riley, 1972). In a subsequent study, Skoff and Vaughn (1971) labeled circulating leukocytes with tritiated thymidine, both before and after eye enucleation, in order to detect leukocyte infiltration. This possibility was excluded, as well as the invasion of nervous tissue by proliferating pericytes. The proliferating cells, labeled with thymidine-^{3}H, appeared as undifferentiated cells and additional cells with ultrastructural features of microglial cells. In addition, these investigators concluded that neuroglial proliferation is shifted by Wallerian degeneration from the normal production of astrocytes to the formation of additional cells whose phagocytic function is likely. Moreover, the number of cells produced largely outnumbered the normal rate of proliferation in the optic nerve. In another region of the brain of another animal species, Wong-Riley (1972) found in the course of Wallerian degeneration the intervention of a phagocytic glial cell, similar in appearance to the microglial cell—or third glial cell—mentioned by other workers. Some of these cells were found enclosed in basal lamina of capillaries, as pericytes are.

At variance, Cook and Wisniewski (1973) estimated that oligodendroglial cells are mostly responsible for phagocytosis in Wallerian degeneration in cat and monkey optic nerve. This opinion was founded on the persistence, during degeneration, of the outer loop of oligodendrocytic cytoplasm. When degeneration is completed, these cells, which have a foamy cytoplasm, aggregate around blood vessels; no hematogeneous cell was found at any time. Likewise, Stenwig (1972) did not find any evidence of leukocytic invasion in the course of Wallerian degeneration in the dorsal column of the spinal cord in young rabbits.

D. Concluding Remarks

For many years after del Rio-Hortega (1919), microglial cells were generally looked upon as mesenchymal elements of the nervous tissue performing the task of phagocytosis. With the advent of electron microscopy, their existence in normal nervous tissue was questioned (Malmfors, 1963; Kruger and Maxwell, 1966), and the correspondence of any cell type identified, by electron microscopy, with the classic microglial cells of del Rio-Hortega is still an open question for some workers (Peters *et al.*, 1970), although Mori and Leblond (1969a) unequivocally demonstrated with electron microscopy a distinct cell type corresponding to the classic microglial cells of del Rio-Hortega. Another question closely related to the preceding was also reopened with the advent of labeling techniques combined with radioautography; the origin of brain phagocytes was extensively worked out in the last decade, and it was concluded that in some pathological processes (such as a brain wound) inducing extensive necrosis and vascular lesions, hematogeneous cells could account for a large proportion of phagocytes. In other experimental models in which extensive necrosis does not occur, indigenous elements are mostly involved. These appear morphologically similar to microglia as visualized by Mori and Leblond (1969a) in the electron microscopy of nervous tissue impregnated by the technique of del Rio-Hortega for microglial cells (1919). It remained to be determined if these elements originated from mesenchymal cells of the nervous system, that is, meningeal cells, pericytes, and so on, or from neuroectodermal cells. On the one hand, Vaughn (1969; Vaughn *et al.*, 1970) concluded the latter, on the basis of their extensive morphological analysis of the optic nerve in young rats. On the other hand, the group of McGill (Privat, 1970a,b; Privat and Leblond, 1972; Paterson *et al.*, 1973), working out the fate of subependymal cells with thymidine-^{3}H labeling, obtained results consistent with both hypotheses, although neuroectodermal origin appeared more likely. Finally, the possibility that mature macroglial cells could transform into microglial cells was advocated, among others, by Davidoff (1973) and is regarded as conceivable in many cases of human pathology (Foncin, personal communication).

VI. General Conclusions and Summary

Recent techniques developed in the two last decades, mainly electron microscopy and radioautography, have allowed some old opinions about gliogenesis to be discarded and some new facts established. Evidence of protracted gliogenesis throughout life, in several mammalian species, has been firmly established, and a common origin for astrocytes and oligodendrocytes from a neuroectodermal precursor has been demonstrated. It has also been shown that the production of either type from the matrix layer is not random, and that astrocytes are generated earlier than oligodendrocytes, although both are still produced, at different rates, in adult rats.

The origin of microglial cells is not so evident; nevertheless, several points have been clarified. Brain macrophages may originate from several sources and, when extensive necrosis occurs, blood leukocytes are always involved. In other cases indigenous elements are responsible for phagocytosis, and they most likely represent microglial cells. These might in turn originate from mesenchymal elements of the brain, or from neuroectodermal cells. This is still an open question, although several reasons argue for the latter.

A general issue that arose from all these studies, either experimental or purely descriptive, is the apparent plasticity of postnatal gliogenesis and, generally speaking, of glial cells throughout life. The persistence of immature cells in fully adult animals and the possibility that glial cell production in early adulthood is shifted from normal in reaction to various stimuli are factors of plasticity. Several questions remain to be solved in this area. The neuroectodermal origin of microglial cells needs to be unequivocally demonstrated. The possibility that apparently mature macroglial cell types transform into each other or into microglial cells needs to be explored.

Finally, almost nothing is known of the mechanisms that operate in the differentiation of the neuroectodermal precursor. Does the same cell give rise to neurons and neuroglial cells? Is one cell able to give rise to an astrocyte and an oligodendrocyte? The problems now shift from cytology to cell biochemistry and cytogenetics.

References

Adrian, E. K. (1968). *Amer. J. Anat.* **123**, 501.

Adrian, E. K., and Smothermon, R. D. (1970). *Anat. Rec.* **166**, 99.

Adrian, E. K., and Walker, B. E. (1962). *J. Neuropathol. Exp. Neurol.* **21**, 597.

Allen, E. (1912). *J. Comp. Neurol.* **22**, 547.

Altman, J. (1966). *Exp. Neurol.* **16**, 263.

Altman, J. (1972). *J. Comp. Neurol.* **145**, 353.

Angevine, J. B., Bodian, D., Coulombre, A. J., Edds, M. V., Hamburger, V., Jacobson, M., Lyser, K. M., Prestige, M. C., Sidman, R. L., Varon, S., and Weiss, P. A. (1970). *Anat. Rec.* **166**, 257.

Banik, N. L., Blunt, M. J., and Davison, A. N. (1968). *J. Neurochem.* **15**, 471.

Bignami, A., and Dahl, D. (1973). *Brain Res.* **49**, 393.

Bignami, A., and Ralston, H. (1969). *Brain Res.* **13**, 444.

Bignami, A., Eng, L. F., Dahl, D., and Uyeda, T. (1972). *Brain Res.* **43**, 429.

Blakemore, W. F. (1969). *J. Anat.* **104**, 423.

Blinzinger, K., and Hager, H. (1962). *Beitr. Pathol. Anat. Allg. Pathol.* **127**, 173.

Blinzinger, K., and Kreutzberg, G. W. (1968). *Z. Zellforsch. Mikrosk. Anat.* **85**, 145.

Blunt, M. J., Baldwin, F., and Wendell-Smith, C. P. (1972). *Z. Zellforsch. Mikrosk. Anat.* **124**, 293.

Bodian, D. (1964). *Bull Johns Hopkins Hosp.* **114**, 13.

Boggenstoss, A. H., Kernohan, J. W., and Drapiewski, J. F. (1943). *Amer. J. Clin. Pathol.* **13**, 333.

Brodal, A. (1939). *Z. Ges. Neurol. Psychiat.* **166**, 646.

Bunge, R. (1968). *Physiol. Rev.* **48**, 197.

Caley, D. W., and Maxwell, D. S. (1968). *J. Comp. Neurol.* **133**, 45.

Cammermeyer, J. (1955). *J. Comp. Neurol.* **102**, 133.

Cammermeyer, J. (1960). *Amer. J. Anat.* **106**, 197.

Cammermeyer, J. (1965). *Z. Anat. Entwicklungsgesch.* **124**, 543.

Cammermeyer, J. (1970a). *In* "Neurosciences Research" (S. Ehrenpreis and O. C. Solnitzky, eds.), Vol. 3, p. 43. Academic Press, New York.

Cammermeyer, J. (1970b). *Int. Congr. Neuropathol., 6th, Paris* p. 424.

Cavanagh, J. B. (1970). *J. Anat.* **106**, 471.
Chow, K. L., and Dewson, J. H. (1966). *J. Comp. Neurol.* **128**, 63.
Cook, R. D., and Wisniewski, H. M. (1973). *Brain Res.* **61**, 191.
Cramer, F., and Alpers, B. (1932). *Arch. Pathol.* **13**, 23.
Davidoff, M. (1973). *Z. Zellforsch. Mikrosk. Anat.* **141**, 427.
del Rio-Hortega, P. (1919). *Bol. Soc. Exp. Biol.* **9**, 69.
del Rio-Hortega, P. (1928). *Mem. Real Soc. Espan. Hist. Natur.* **14**, 5.
del Rio-Hortega, P. (1930). *Rev. Neurol.* **37**, 956.
del Rio-Hortega, P. (1932). *In* "Cytology and Cellular Pathology of the Nervous System" (W. Penfield, ed.), Sect. X, Vol. 2, pp. 483–534. Harper (Hoeber), New York.
del Rio-Hortega, P., and Penfield, W. (1927). *Bull. Johns Hopkins Hosp.* **41**, 278.
Demêmes, D., and Marty, R. (1971). *C. R. Acad. Sci., Ser. D* **272**, 2201.
Dempsey, A. W., and Luse, S. A. (1958). *In* "Biology of Neuroglia" (W. F. Windle, ed.), pp. 99–108. Thomas, Springfield, Illinois.
De Robertis, E., and Gerschenfeld, H. (1961). *Int. Rev. Neurobiol.* 3, 1.
de Vries, E. (1910). *Arb. Hirnanat.* **4**, 1.
Dixon, J. S. (1969). *Anat. Rec.* **163**, 101.
Dutrochet, H. (1824). "Recherches anatomiques et physiologiques sur la structure intime des animaux et des végétaux et sur leur motilité." Baillière, Paris.
Eager, R. P., and Eager, P. R. (1966). *Science* **153**, 553.
Farkas-Bargeton, E., Robain, O., and Mandel, P. (1972). *Acta Neuropathol.* **21**, 272.
Farquhar, M. G., and Hartmann, J. F. (1957). *J. Neuropathol. Exp. Neurol.* **16**, 18.
Feigin, I. (1969). *J. Neuropathol. Exp. Neurol.* **28**, 6.
Fernando, D. A. (1971). *Brain Res.* **27**, 365.
Ferraro, A., and Davidoff, L. M. (1928). *Arch. Pathol.* **6**, 1030.
Field, E. J. (1957). *J. Neuropathol. Exp. Neurol.* **16**, 48.
Fujita, S. (1963). *J. Comp. Neurol.* **120**, 37.
Fujita, S. (1967). *J. Cell Biol.* **32**, 277.
Fujita, S., Shimada, M., and Nakamura, T. (1966). *J. Comp. Neurol.* **128**, 191.
Globus, J. H., and Kuhlenbeck, H. (1944). *J. Neuropathol. Exp. Neurol.* **3**, 1.
Herndon, R. M. (1964). *J. Cell Biol.* **23**, 277.
Hillebrand, H. (1966). *Z. Zellforsch. Mikrosk. Anat.* **73**, 302.
Hirano, A., Sax, D. S., and Zimmermann, H. M. (1969). *J. Neuropathol. Exp. Neurol.* **28**, 388.
Hirose, G., and Bass, N. H. (1973). *J. Comp. Neurol.* **152**, 201.
His, W. (1889). *Arch. Anat. Entwicklungsgesch.*, p. 249.
Horstadius (1950). "The Neural Crest. Its Properties and Derivatives in the Light of Experimental Research." Oxford Univ. Press, London and New York.
Howatson, A. F., and Ham, A. W. (1955). *Cancer Res.* **15**, 62.
Humbertson, A., Zimmerman, E., and Leedy, M. (1969). *Z. Zellforsch. Mikrosk. Anat.* **100**, 507.
Huntington, H. W., and Terry, R. D. (1966). *J. Neuropathol. Exp. Neurol.* **25**, 646.
Kershman, J. (1938). *Arch. Neurol. Psychiat.* **40**, 937.
Kershman, J. (1939). *Arch. Neurol. Psychiat.* **41**, 24.
King, J. S. (1968). *Anat. Rec.* **161**, 111.
King, J. S., and Schwyn, M. (1970). *Z. Zellforsch. Mikrosk. Anat.* **106**, 309.
Kirkpatrick, J. B. (1968). *J. Comp. Neurol.* **132**, 189.
Kitamura, T., Hattori, H., and Fujita, S. (1972). *J. Neuropathol. Exp. Neurol.* **31**, 502.
Konigsmark, B. W., and Sidman, R. L. (1963). *J. Neuropathol. Exp. Neurol.* **22**, 643.

Kruger, L., and Maxwell, D. S. (1966). *Amer. J. Anat.* **118**, 411.

Kreutzberg, G. W. (1966). *Acta Neuropathol.* **7**, 149.

Kryspin-Exner, W. (1943). *Z. Anat. Entwicklungsgesch.* **112**, 389.

Lewis, P. D. (1968). *Brain* **91**, 721.

Ling, E. A., and Leblond, C. P. (1973). *J. Comp. Neurol.* **149**, 73.

Ling, E. A., Paterson, J., Privat, A., Mori, S., and Leblond, C. P. (1973). *J. Comp. Neurol.* **149**, 43.

Luse, S. (1956). *J. Biophys. Biochem. Cytol.* **2**, 531.

Malmfors, T. (1963). *J. Ultrastruct. Res.* **8**, 193.

Matthews, M. A., and Kruger, L. (1973). *J. Comp. Neurol.* **148**, 313.

Maxwell, D. S., and Kruger, L. (1965). *J. Cell Biol.* **24**, 141.

Mori, S., and Leblond, C. P. (1969a). *J. Comp. Neurol.* **135**, 57.

Mori, S., and Leblond, C. P. (1969b). *J. Comp. Neurol.* **137**, 197.

Mori, S., and Leblond, C. P. (1970). *J. Comp. Neurol.* **139**, 1.

Mugnaini, E., and Forstronen, P. F. (1967). *Z. Zellforsch. Mikrosk. Anat.* **77**, 115.

Mugnaini, E., and Walberg, F. (1964). *Ergeb. Anat. Entwicklungsgesch.* **37**, 194.

Munger, B. L. (1958). *Amer. J. Anat.* **103**, 1.

Nissl, F. (1892). *Allg. Z. Psychiat. Psych.-Gerichtl. Med.* **48**, 197.

Nissl, F. (1894). *Zentralbl. Nervenheilk. Psychiat.* **17**, 337.

Opalski, A. (1934). *Z. Gesamte Neurol. Psychiat.* **149**, 221.

Palay, S. L. (1958). *In* "Biology of Neuroglia" (W. F. Windle, ed.), pp. 24–38. Thomas, Springfield, Illinois.

Palay, S. L., McGee-Russel, S. M., Spencer, G., and Grillo, M. A. (1962). *J. Cell Biol.* **12**, 385.

Paterson, J. A., Privat, A., and Leblond, C. P. (1972). *Anat. Rec.* **172**, 380.

Paterson, J. A., Privat, A., Ling, E. A., and Leblond, C. P. (1973). *J. Comp. Neurol.* **149**, 83.

Penfield, W. (1932). "Cytology and Cellular Pathology of the Nervous System." Harper (Hoeber), New York.

Peters, A. (1964). *J. Anat.* **98**, 125.

Peters, A., and Vaughn, J. E. (1967). *J. Cell Biol.* **32**, 113.

Peters, A., Palay, S. L., and Webster, H. de F. (1970). "The Fine Structure of the Nervous System," p. 129. Harper (Hoeber), New York.

Petersen, K. U. (1969). *Z. Zellforsch. Mikrosk. Anat.* **100**, 616.

Phillips, D. E. (1973). *Z. Zellforsch. Mikrosk. Anat.* **140**, 145.

Privat, A. (1970a). *Anat. Rec.* **166**, 364.

Privat, A. (1970b). *Int. Congr. Neuropathol., 6th, Paris* p. 447.

Privat, A. (1972). Thèse doc. Biol., Univ. Paris p. 58.

Privat, A. (1973). *Anat. Rec.* **175**, 416.

Privat, A., and Leblond, C. P. (1972). *J. Comp. Neurol.* **146**, 277.

Privat, A., Robain, O., and Mandel, P. (1972). *Acta Neuropathol.* **21**, 282.

Pruijs, W. M. (1927). *Z. Gesamte Neurol. Psychiat.* **108**, 298.

Rakic, P. (1971). *J. Comp. Neurol.* **141**, 283.

Rakic, P., and Sidman, R. L. (1973). *J. Comp. Neurol.* **152**, 103.

Ramón y Cajal, S. (1909). "Histologie du système nerveux de l'homme et des vertébrés." Maloine, Paris.

Ramón y Cajal, S. (1913). *Trab. Lab. Invest. Biol., Univ. Madrid* **II**, 255.

Rapos, M., and Bakos, J. (1959). *Z. Mikrosk.-Anat. Forsch.* **65**, 211.

Reese, T. S., and Karnowsky, M. J. (1967). *J. Cell Biol.* **34**, 207.

Roussy, G., and L'Hermitte, J. (1930). *Rev. Neurol.* **37**.

Russel, G. V. (1962). *Tex. Rep. Biol. Med.* **20**, 338.
Rydberg, E. (1932). *Acta Pathol. Microbiol. Scand., Suppl.* **10**, 1.
Schaper, A. (1897). *Science* **5**, 430.
Schultz, R. L. (1964). *J. Comp. Neurol.* **122**, 281.
Sidman, R. L., and Hayes, R. (1965). *J. Neuropathol. Exp. Neurol.* **24**, 172.
Sjöstrand, J. (1965). *Z. Zellforsch. Mikrosk. Anat.* **68**, 481.
Sjöstrand, J. (1966). *Acta Physiol. Scand., Suppl.* **270**, 19.
Sjöstrand, J. (1971). *Exp. Neurol.* **30**, 178.
Skoff, R. P., and Vaughn, J. E. (1971). *J. Comp. Neurol.* **141**, 133.
Smart, I. (1961). *J. Comp. Neurol.* **116**, 325.
Stensaas, L. J., and Gilson, B. C. (1972). *Z. Zellforsch. Mikrosk. Anat.* **132**, 297.
Stensaas, L. J., and Reichert, W. H. (1971). *Z. Zellforsch. Mikrosk. Anat.* **119**, 147.
Stensaas, L. J., and Stensaas, S. (1968). *Z. Zellforsch. Mikrosk. Anat.* **86**, 104.
Stenwig, A. E. (1972). *J. Neuropathol. Exp. Neurol.* **31**, 696.
Takano, I. (1964). *Okajimas Folia Anat. Jap.* **40**, 1.
Torvik, A. (1972). *J. Neuropathol. Exp. Neurol.* **31**, 132.
Torvik, A., and Skjorten, F. (1971). *Acta Neuropathol.* **17**, 265.
Vaughn, J. E. (1969). *Z. Zellforsch. Mikrosk. Anat.* **94**, 293.
van Crevel, H., and Verhaart, W. J. C. (1963). *J. Anat.* **97**, 451.
Vaughn, J. E., and Pease, D. C. (1967). *J. Comp. Neurol.* **131**, 143.
Vaughn, J. E., and Pease, D. C. (1970). *J. Comp. Neurol.* **140**, 207.
Vaughn, J. E., and Peters, A. (1968). *J. Comp. Neurol.* **133**, 269.
Vaughn, J. E., Hinds, P. L., and Skoff, R. P. (1970). *J. Comp. Neurol.* **140**, 175.
Virchow, R. (1854). *Arch. Pathol. Anat. Physiol. Klin. Med.* **6**, 135.
Watson, W. E. (1965). *J. Physiol. (London)* **180**, 741.
Wendell-Smith, C. P., Blunt, M. J., and Baldwin, F. (1966). *J. Comp. Neurol.* **127**, 219.
Wolf, J. (1965). *Z. Zellforsch. Mikrosk. Anat.* **66**, 811.
Wong-Riley, M. T. T. (1972). *J. Comp. Neurol.* **144**, 61.
Yasuzumi, G., Tsubo, I., Sugihara, R., and Nakai, Y. (1964). *J. Ultrastruct. Res.* **11**, 213.

Three-Dimensional Reconstruction from Serial Sections

RANDLE W. WARE

California Institute of Technology
Pasadena, California

AND

VINCENT LOPRESTI

Columbia University, New York, New York

I. Introduction

The rapidly expanding study of the nervous system, from the lowest invertebrates to the mammals, has led to increased interest in improved methods for visualizing and quantitating the structure and relationships of its fundamental elements, the neurons. The problems are severe because of (1) the inhomogeneous nature of the tissue, and it is the inhomogeneities that are responsible for its function and must be elucidated; (2) the combination of extent, density, and fineness of nerve fibers; (3) the fact that statistical methods as used in the field of stereology, while perhaps useful for characterizing various properties, such as fiber and bouton structure and density, or nature and frequency of connections between various cell types, are obviously incapable of providing the information that is essential for determining the specific connectivities of a network. There is in addition the realization that it is not possible to identify reliably a profile simply by the appearance of its cytoplasm, particularly at a synapse, so it seems inescapable that it has to be followed for some distance if one hopes to make a firm statement about its origin. Furthermore, it must be assumed that simply designating a contact a synapse is not sufficient to describe its possible analog functioning. Some synapses occur on large branches where their activity may be readily integrated into the total cellular activity, whereas others occur on long fine processes and are therefore relatively isolated. Thus information about the environment around the synapse is necessary in order even to begin to make a guess as to its possible significance. Our present knowledge of the nervous system has come largely from specialized neurological techniques for light microscopy, which have not necessitated the brute-force procedures of serial section reconstruction. The Golgi methods have proved invaluable in overcoming the problems of density and extent in describing the structure of individual elements and the overall organization of nervous tissue, although by the random nature of their staining they have not given detailed connectivity information except by inference. Similarly, degeneration techniques have allowed the gross following of pathways that are otherwise unobservable to their terminations. Various procedures utilizing anterograde and retrograde axonal flow permit ever more localized regions of the nervous system to be "tagged," so that afferent and efferent fibers can be separated from the bulk of the surrounding tissue. Newer techniques of intracellular injection, an ideal replacement for the capricious Golgi method when cells of interest can be identified and impaled with microelectrodes, permit total specificity of staining to be obtained. When used in multiple injections, they allow regions of connectivity to be located at the whole-mount level and, if desired, studied by both light and electron microscopy.

Sjostrand (1958) showed that the graphical and model-building reconstruction techniques of the old German anatomists can be usefully applied to neural tissue at the level of the electron microscope, and was immediately followed by others. By a suitable choice of material it was Trujillo-Cenóz (1965a,b) who, like His working a century earlier with embryos, used them to give a unified look at a system that had previously been known only by its parts, this time not with an organism as a whole but a piece of nervous processing which was separate both functionally and anatomically. It seems clear that connectivity information can be obtained by the use of serial section analysis. Not so clear is exactly how to proceed tactically. In some invertebrates in which neurons are sufficiently short, complete electron microscope analysis is possible (Sections III,A,2, and IV,A). In other invertebrates and vertebrates in which neurons extend over large distances one has to be more clever, first using combinations of light microscope techniques such as intracellular injection, Golgi staining, or degeneration to locate a region of interest, and then using serial section electron microscope analysis once the appropriate regions are isolated (Sections, II,C,2 and III,A,1, and 2). The matter of creating a reconstruction is also a problem. Traditional methods, while adequate for simple structures, are prohibitively time-consuming when the volume of tissue is large, so that hundreds or thousands of sections are needed, and the items to be reconstructed intermingle extensively. Thus they seem more suitable for presentation of results than for investigation. Of the methods available, serial section cinematography (Section II,A,3) seems to offer an attractive alternative for studies requiring many sections and in which the laborious work of the more common techniques tends to discourage research.

It is the purpose of this article to review some of the methods of three-dimensional reconstruction from serial sections that have been successfully used in the study of many different types of tissue, and to present enough of the details arrived at by their use to show the kinds of information obtained, which could not have been deduced from the study of individual sections alone. Finally, we describe some of the newer attempts at reconstruction being developed specifically for the investigation of nervous tissue.

II. Vertebrates

A. Historical

Dissection, or semiordered sequential destruction, has been the rather straightforward method forced upon biologists for seeing inside their chosen materials. Assembling information about the structure back together again into a presentable form has been more of a problem. Originally, one was dependent purely on artistic talent for conveying the insights gained to colleagues far from the operating table (Fig. 1). Those reconstructions that were the most successful and instructive were drawn from the point of view of an actual observer, using the techniques of geometric perspective and proportion to give a sense of reality to the finished drawings, and were of necessity what are now called artist's interpretations. The old German anatomical journals up to about 1920 contain many fine examples of what the cooperation between scientist and artist could accomplish in spreading detailed biological knowledge from the specialist to the nonspecialist. With the advent of suitable microtomes, or "cutting engines," in the late eighteenth to early nineteenth centuries, some precision was introduced into the dissection procedure, and it became both desirable and possible to produce much

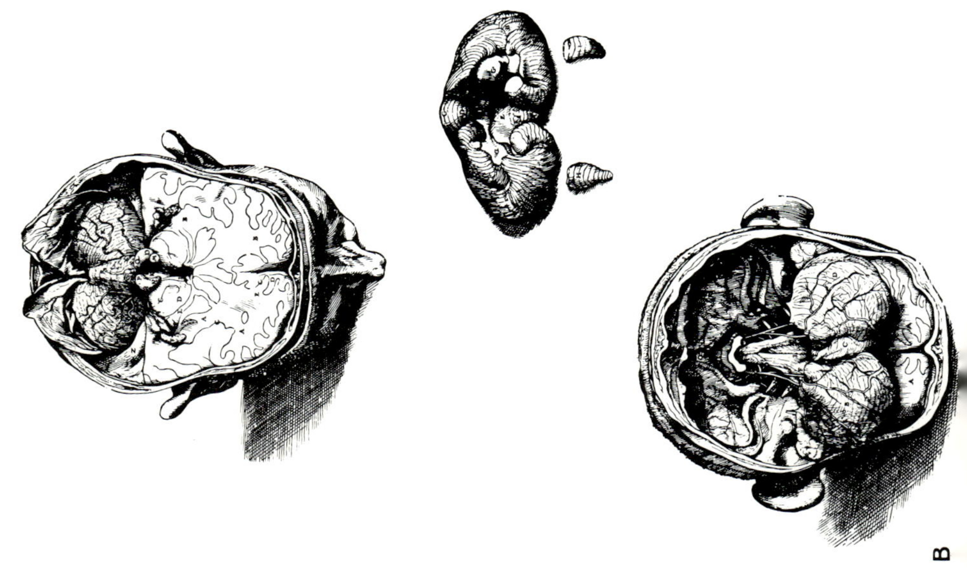
B

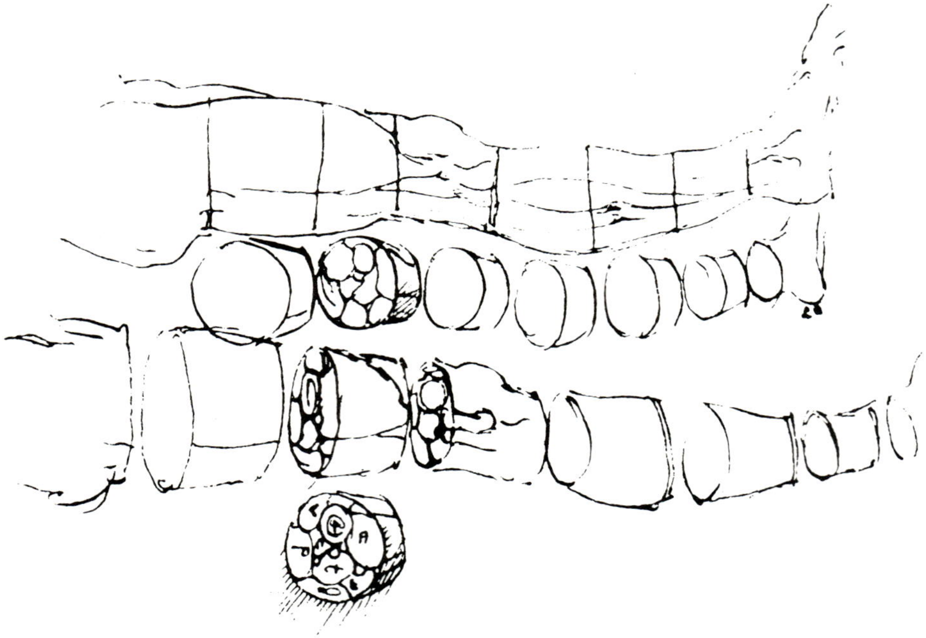

more scientifically accurate reconstructions than could be made from observations carried out strictly on whole material. To do this required relying in the first analysis more on mechanistic than artistic procedures.

The event that made possible a serious consideration of quantitative three-dimensional reconstruction was the discovery by Klebs in 1869 that with wax embedding sections of uniform thickness could be reproducibly obtained from microtomes already available. The German anatomists immediately seized upon this technique and within a few years developed the principal methods of reconstruction which, but for slight variations, are used today. The conceptual usefulness, and not just the novelty, of these methods for giving biologists their first real quantitative insight into structures that previously had been known only from dissection or from imagination liberally applied to single sections can be appreciated from the number of papers presented on the subject in the period 1880–1887 and the often excruciating detail used by investigators in describing their procedures to one another. It might be kept in mind, however, that in spite of the rapidity with which the development of techniques as a whole took place they were seldom, obvious though they may seem today, the work of a single investigator.

1. *Graphical Methods*

The first type of reconstruction to be developed, used by the famous German anatomist His (1880) in the study of human embryos, was graphical, and oddly enough is not what today would be considered the most obvious one. Rather than a full-face drawing of the object in the direction of sectioning, it produces a reconstruction viewed perpendicular to this direction. The method is simply to parallel project the outlines of objects in the section onto a straight line, in the same direction for each section, and then to separate these line projections by a distance corresponding to the magnified section thickness for the final reconstruction (Fig. 2). This method proves useful as long as the surfaces so reconstructed are not unduly complicated and has the advantage that, with artistic interpolation, a single series can be used to produce a reconstructed image from any direction perpendicular to the direction of the

Fig. 1. (A) The right leg of a man, measured and then cut into sections which show the parts of his anatomy. The cross section of a leg is a system of anatomical representation still used today. It is suggested that Leonardo's procedure derives from a device developed by the theorists of perspective who, in order to make the complex forms of the human body constructed according to Brunelleschi's *costruzione legitima,* had to prepare a series of cross sections taken at various levels. (Courtesy of University of California at Los Angeles Art Library, Belt collection.) (B) Anatomical illustrations of Vesalius. (Courtesy UCLA Medical History Library.)

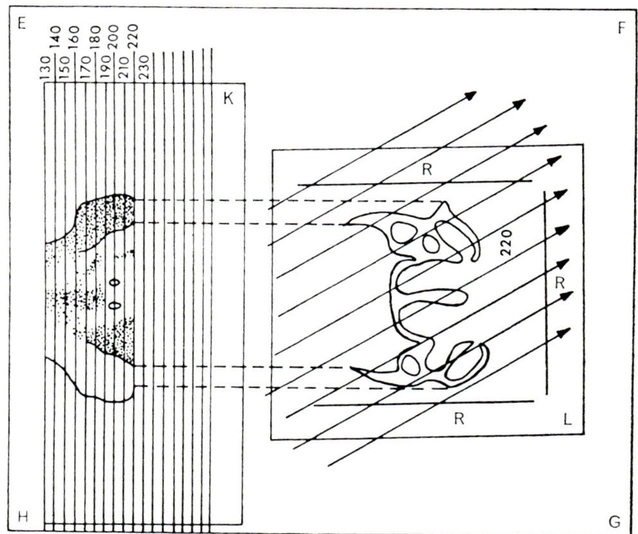

FIG. 2. Procedure for projecting required points from the tracing onto grid lines, with interpolative shading added. (From Gaunt, 1971.)

sectioning. Examples of these results obtained by this method are shown in the original drawings and reconstructions of His reproduced in Fig. 3. With the aid of such graphical reconstructions, His produced free-hand clay models of human embryos, drawings of which can still be found in medical texts today (Goss, 1966).

In 1886, Kastschenko (1886, 1887) pointed out a major difficulty with this method in its original description by His, namely, that it made no provision for aligning sections unambiguously with respect to one another, but depended instead on the experience and insights of the investigator. Thus it was at best useless, and at worst gave incorrect results, in the hands of the untrained. To remedy this he introduced what became known as the external marker technique, that of introducing some simple objects near to but outside the specimen along its entire length that can be used to align the planes of successive sections. In Kastschenko's case these markers were the precisely trimmed edges of the paraffin block that contained the specimen, specially coated with powdered carbon so that the section's edge remained visible even after the paraffin was removed in subsequent histological processing. At the same time he introduced for the first time the technique of superimposing outlines, with hidden line removal, traced from a number of sections to give a contour drawing of the object as if seen from face on. This procedure

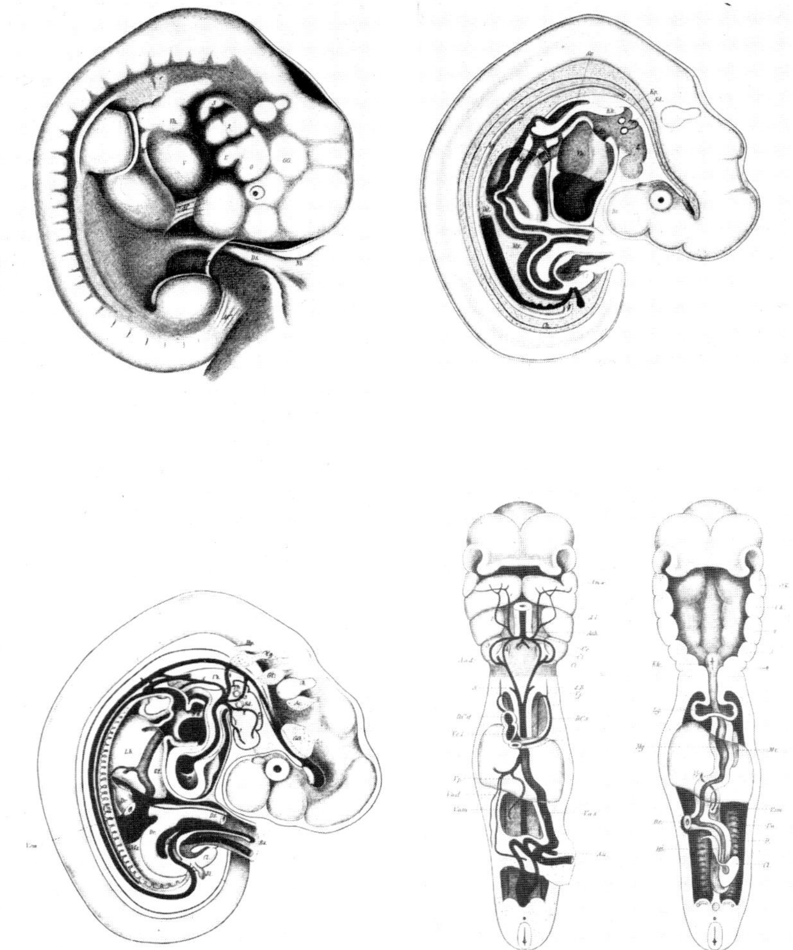

FIG. 3. (A) Upper left: Embryo B, surface view. Upper right: Embryo B, showing intestine, heart, aorta, and central nervous system. Lower left: Profile reconstruction of embryo B, showing the head ganglia and nerves and vessel system. The dotted lines within the hemispheres denote the border of the intermediate brain. The heart is opened to expose the ending of the veins in the auricle and the origin of the aorta in the ventricle. The anterior half of the right auricle is removed to show the aorta bulb. The intestine is shown interrupted to show the process of the vena parietalis sinistra. Lower right: Frontal reconstruction of the same embryo. (B) [p. 332] Sections used to produce the reconstructions in (A). (From His, 1880, courtesy of the UCLA Medical History Library.)

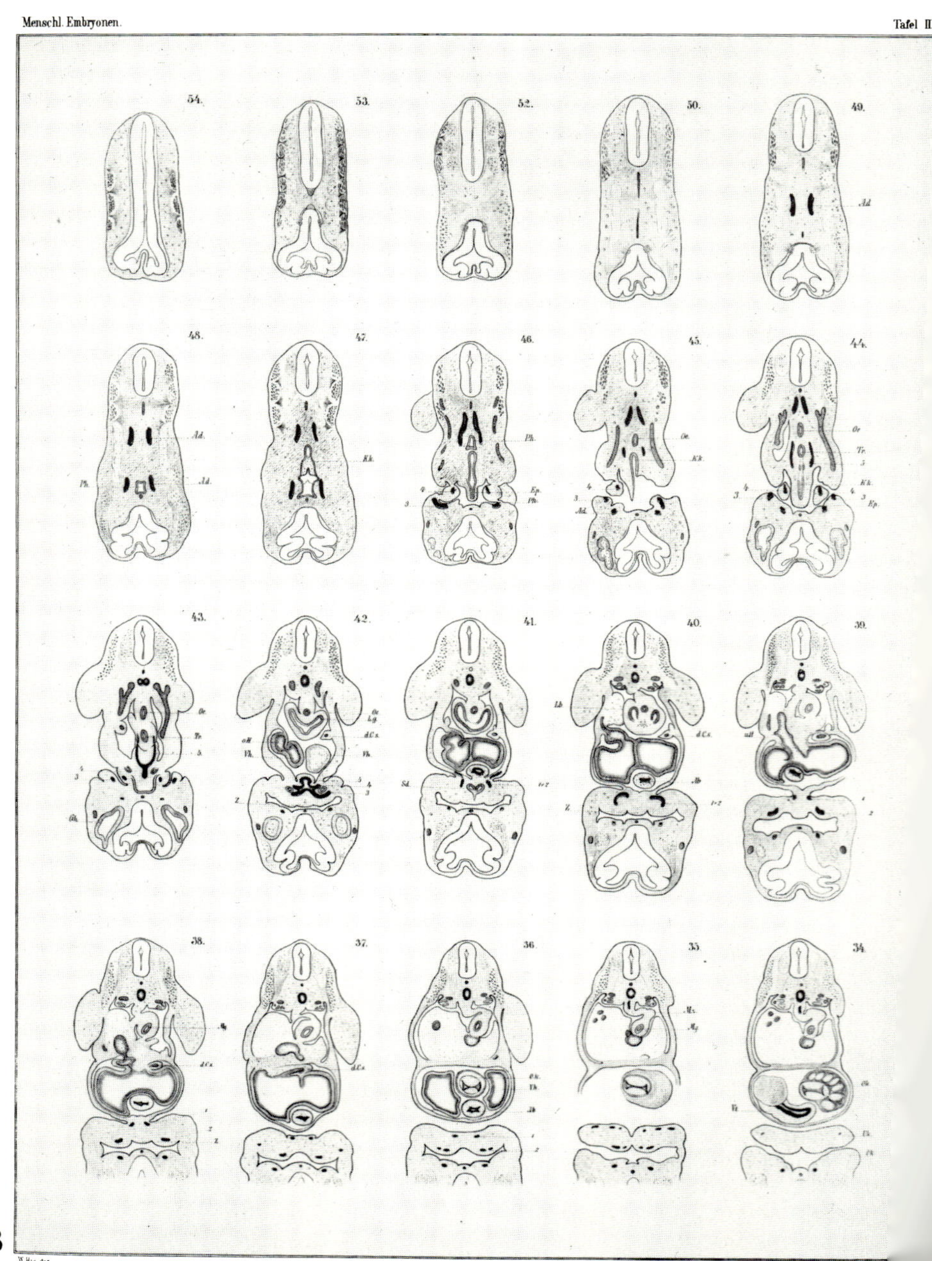

Fig. 3B. See page 331 for legend.

gives satisfactory results as long as not too many sections are superimposed or not too much detail is included in any one drawing. It has the advantages that it is fast, requiring no additional graphical projections or manipulations, and independent of knowledge of section thickness.

Nearly 25 years elapsed before a workable graphical method was presented that freed workers from the limitations imposed by the initial choice of a direction for sectioning, perhaps because one of the leaders in the field of model reconstruction had said that it was not possible except with models (Strasser, 1886). In 1911, Odhner demonstrated that contour drawings could in fact be produced in a view oblique to the direction of sectioning. The method simply requires displacing successive micrographs with respect to one another uniformly along a straight line before drawing the superimposed contours on the final reconstruction, although to produce the correct mathematical foreshortening each drawing of the original section must be redrawn (Fig. 4). This is most conveniently accomplished by placing a square grid over the original and transferring the intersections of lines of the section and lines of the grid onto a second grid of parallelograms. Alternatively, as proposed by Dixon and Howarth (1958), foreshortening can be produced by photographing the tracings of the micrographs at the desired angle and then using these photographs to construct the final tracing. Reconstructions produced in this way from electron micrographs have been used to provide remarkable illustrations for the works of Poritsky (1969), Mitchell and Thaemert (1965), and Thaemert (1966, 1970).

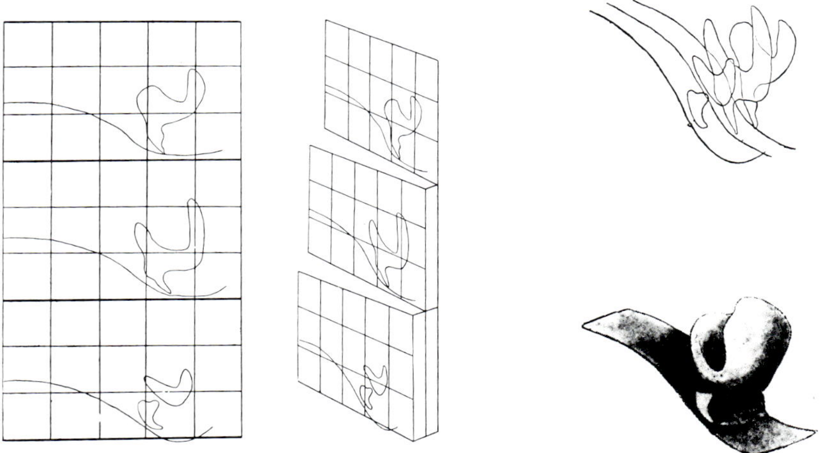

FIG. 4. Odhner's graphic procedure for oblique reconstruction. (From Odhner, 1911.)

The art of simple oblique or full-face contour drawing was brought to its highest point by Krieg (1949a,b), who was not an artist, in his illustrations of the brains of the rat, monkey, and man. The documentation of his lifelong work of tracing fiber pathways in the monkey brain following about 200 assorted lesions (Krieg, 1966) consists of figures that are unsurpassed in clarity and detail by the best medical illustrations. The only addition he made to Odhner's method was occasionally to add partial perspective, accomplished by constructing his parallelogram grids with vanishing points (Fig. 5). As pointed out by Poritsky (1969), however, this in itself is not sufficient to produce true perspective, which would

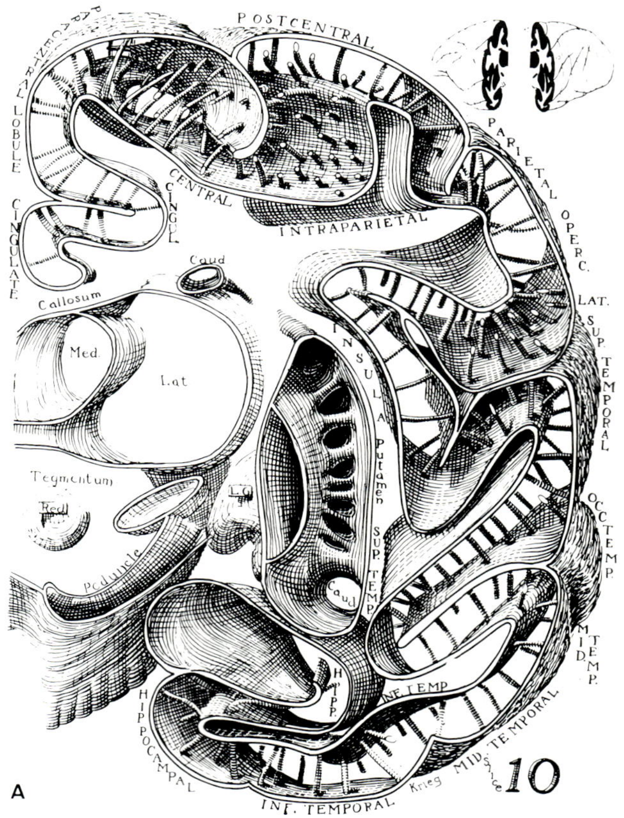

FIG. 5. (A) The tenth of 18 4-mm slices through the monkey brain made by the simple full-face contour tracing technique from 40-μm sections taken through the slice. (From Krieg, 1949a.) (B) Outer edge of all sections used in perspective reconstruction, in position. (From Krieg, 1949b.)

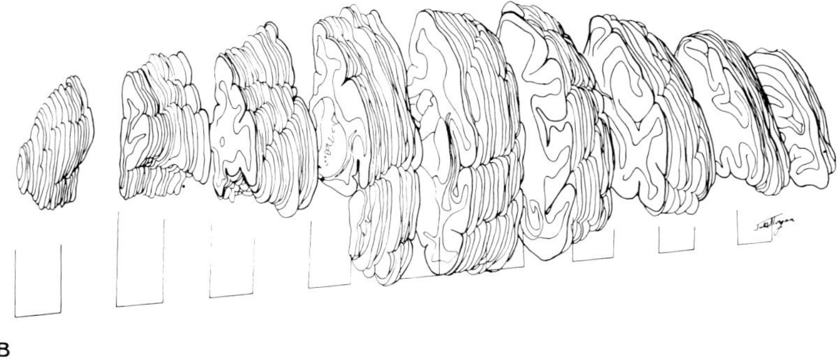

FIG. 5B. See facing page for legend.

require a tilting of the section planes with respect to one another as well.

A method analogous to that of His for producing projective drawings of cylindrical objects cut perpendicular to their long axis has recently been described by Thompson and Gottlieb (1972). They state that a view produced by their method "compares to those of other reconstruction techniques in the same way that a mercator projection map of the earth compares to a photograph taken from space." The procedure is simply to center a sheet of polar coordinate graph paper over the axis of the micrographs. The angular position and extent of each object within the specimen is plotted along the abscissa of the final drawing at an ordinate value corresponding to the section number. Reconstructions produced in this way were used to follow the time course of development of pupal imaginal tissues in *Drosophila*.

2. *Model Building*

Shortly after His described the method for projective graphical reconstruction, Born (1883) published a procedure for constructing solid models from wax plates cast of the appropriate thickness. Like His, he made no explicit provision for accurately aligning sequential sections and relied instead on familiarity with the material. Only after Kastschenko described his method for external markers did he incorporate it into his model building (Born, 1888). Initially, wax plates were made by pouring a calculated amount of molten wax onto hot water which was then allowed to cool slowly. Microscope images through a camera lucida were traced onto paper and the hardened wax plate simultaneously by means of a pencil and carbon paper. Drawings of some of the models made by Born using this technique are shown in Fig. 6. Three years later Strasser (1886, 1887b) introduced the procedure of embedding both sheets of cardboard and the original drawing in the wax plate which was then

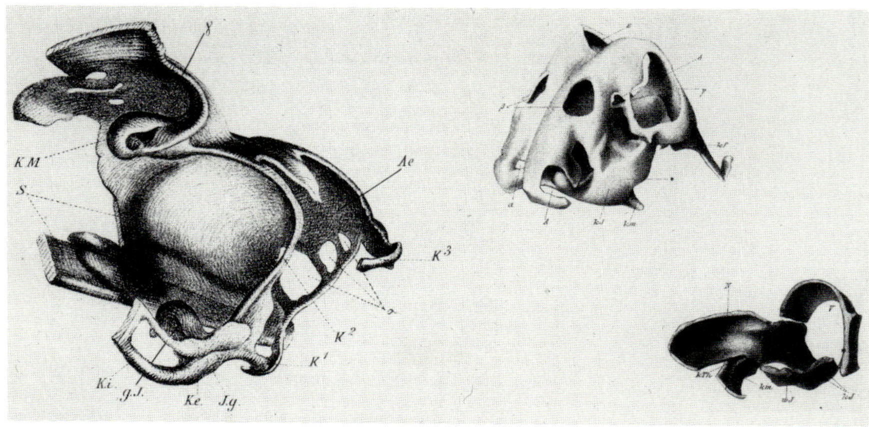

Fig. 6. Left: Model of the cartilage nasal skeleton of an embryo of *Tropidonotus natrix*. Seen from beneath and to the side (Born, 1882). Center: Model of the cartilage skeleton of the ethmoidal region of an embryo of *Lacerta vivipara* shortly before emergence from the egg. Seen from anterior and to the side. Right: Model of the cartilage skeleton of a part of the nasal cavity of *Platydactylus muralis*. Seen from behind.

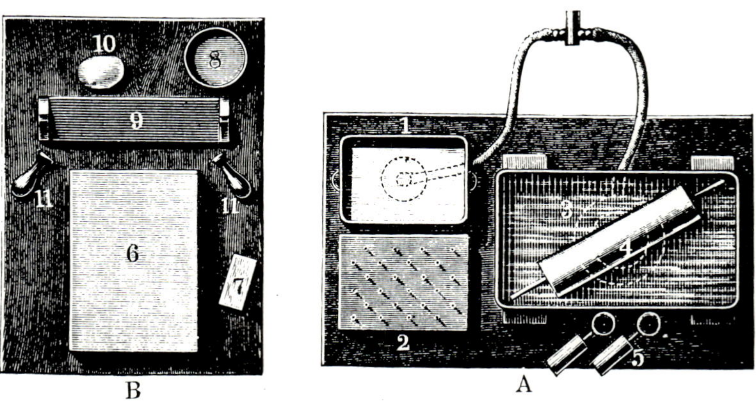

Fig. 7. Strasser's apparatus for rolling wax sheets to the desired thickness. The roller (4), which rests in a trough of boiling water (3), is transported with handles (5) to a holder (9). The softened wax sheet, with the embedded micrograph tracing and sufficient cardboard to give the approximate thickness desired, is placed on the lithographer's stone (6). Guides of the desired thickness (not shown) are placed parallel to 6, the roller is held with handles (11) and used to flatten the sheet to the thickness of the guides. (Woodcut from Strasser, 1887b.)

rolled to the required thickness to decrease its fragility (Fig. 7). Disassemblable models were made by leaving thin cross-bridges of wax between parts as the tracings were cut from the wax plates and then severing these with a hot spatula after the model was completed by fastening all plates together with heat or turpentine. Strasser also mentioned that, with proper illumination, as many as four plates of clear wax with color inscribed tracings could be stacked to give an easily made three-dimensional contour reconstruction of the sectioned material.

Since then only slight modifications have been introduced into model-building techniques (Sack, 1966; Moore and Hayden, 1963), largely in the form of the materials used for the plates. Born (1888), for example, analyzed merits of the various types of wax available at the time ranging from the 4 marks/kg preferred yellow beeswax to the 0.3 marks/kg less satisfactory resin–wax mixture. Strasser (1887a) suggested but did not use stacked glass plates with contours drawn on their surfaces. His (1904) used stacked celluloid sheets to form the basis for illustrations in one of his texts, but at low magnifications and always using his method of projective reconstruction. Rolshoven (1937) made an amusing variation on this by mounting two sets of celluloid sheets on separate frames, one set with external and one with internal structures, so that they could be interleaved at will. Kerr (1902) traced his structures on frosted glass sheets which were then cleared in stacking by the addition of oil between the plates. Pedler and Tilly (1966) described the construction of a pantograph with a hot wire which could cut out traced structures from plastic foam. Several interesting methods have been tried to fill in the discontinuities caused by the thickness of whatever material is used to construct the sheets. Lapin (1927), for example, constructed his sheets of clear gelatin, cut out the items of interest, and then filled the resulting holes with gelatin of different colors. Jorgensen (1971a) used Plexiglas sheets to reconstruct portal tracts in liver tissue. After tracing vessels and bile ducts on the sheets, he drilled them out and filled them in the completed model with different-colored fluids. One of the significant types of modifications for large-scale production of a single model uses the original wax model only as the form for a mold which is subsequently injected with a polymerizable plastic (Blechschmidt, 1954).

It is probably safe to say that model building, in spite of its often striking products, has consistently retained second place to graphical reconstruction except for some educational purposes, as in medicine, where it is necessary to appreciate subjectively the space-filling properties of an object. This is due partly to the requirements imposed by mass communication, but mainly because models are time-consuming to construct, incapable of clearly displaying structures within structures (especially tubes or fiber tracts), and not as suitable for quantitative measurements as are properly constructed two-dimensional projections.

3. *Serial Section Cinematography*

The German anatomists were not long in developing yet another means of reconstruction, that of serial section cinematography, the photographing of aligned sequential sections onto movie film. This method has enjoyed less popularity than others over the years, probably because of the large number of uniform, high-quality sections necessary for its successful application and the difficulty in obtaining exact alignment of the sections. The first report of this technique was by Reicher (1907a) on two films he made, one from 1235 and another from 1060 serial sections of the human brain, and presented at the Dresden Neurological Congress. His procedure for alignment of the sections (Reicher, 1907b), tedious and not very accurate, was to project each brain slice at the same magnification as it appeared on the film onto an intermediate screen covered with a grid of different-colored lines, center it using a high-powered magnifying glass, and swing the screen out and the camera in and photograph. He expected to achieve better results by using celloidin embedding followed by mounting of each section at the same relative position on a glass slide using either the section edge or embedded external markers for reference, and went so far as to have a special apparatus constructed for positioning the sections accurately. The films produced by the crude alignment procedure had some annoying jumps from one frame to the next, but were demonstrative enough for him to remark, "Nerve fibers cross over one another before our eyes, the brain nerves originate from their nuclei, traverse the brain, and emerge from it. In short, the difficulty of comprehending the spatial course of fiber systems and pathways is eliminated, and within a few minutes a continuous, visual picture of the organization of fiber pathways is displayed. Further investigations will reveal whether this method is simply a didactic aid or a new method for research" (Reicher, 1907a). Unfortunately, he never continued these investigations, as he soon became fascinated with the more ordinary forms of microcinematography (Reicher, 1910).

The problem of registering the sections with themselves and in proper relation to the sprocket holes of the film has been solved in several quite different ways since then. Perhaps the least successful method, at least in terms of awkwardness and frequency of use, was mentioned by Widakowich (1907) in the same year as Reicher's presentation, and later by Bush (1952) in a more automated form. The procedure, clearly shown in Bush's figure (Fig. 8), is to mount individual paraffin sections directly on cinefilm. It was claimed that this reduces problems of section compression and distortion, since the section is supported in its final position

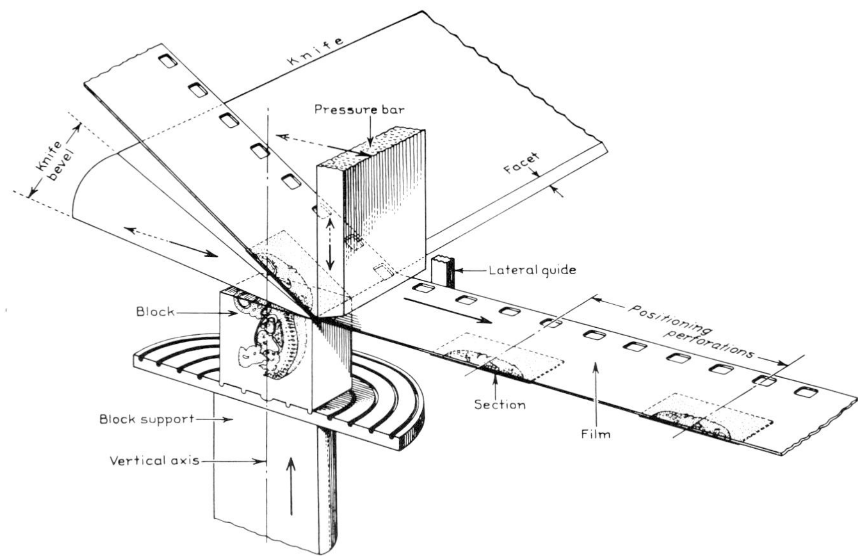

FIG. 8. Scheme for automatically applying serial sections to 35-mm perforated cellulose acetate film. This diagrammatic median view shows the block support carrying an embedded embryo, with the knife completing a cut with the pressure bar still in the functioning position. Forward motion is indicated by a solid arrow on the film, and reciprocal motions are indicated by solid-broken-line arrows. The lateral film guide moves with the pressure bar but is carried on a separate mount. (From Bush, 1952.)

on the film during the entire slicing operation, of particular advantage during the sectioning of frozen material. The sections were processed for histological examination on the film substrate and could then be shown either directly or after photographic copying of the original.

Still a different means of avoiding the alignment problem was described by Hegre (1951), and has been used in variations many times since. The crucial development in his case was that of a block surface staining method, described in 1946 (Hegre, 1946) and a later improved one in 1967 (Hegre, 1967), which permitted the exposed surface of a cut block to be stained with sufficient contrast for photography by reflected light. By using an automated device which sliced off a section, applied stain to the remaining block, and photographed this surface, he eliminated problems such as alignment and differential section distortion. The first use he made of this procedure was an attempt to refute some claims of the opponents of the myogenic theory of heart conduction by studying the atrioventricular conduction system of the human heart at 6-μm intervals (Read *et al.*, 1952). It enabled him to follow the course of the bundle of

His as it leaves the atrioventricular node, crosses the septum, and then bifurcates and, in his words, to "demonstrate more clearly than any other method the cardiac conduction system." He later made unpublished cinematographic observations on lung tissue and spinal cord and created an educational film of a 9-mm pig embryo (Hegre, 1966).

The next mention of this method was by Postlethwait (1962; Postlethwait *et al.*, 1964), who used it to examine plant vascularization through the transition zone between the root and shoot of seedlings. After an independent rediscovery of the virtues of cinematography, Zimmerman and Tomlinson (1965) began a long series of publications with a type of observation made more than once (Elias, 1971a): "Diagrams purporting to represent the course of vascular bundles in palm stems are reproduced frequently in textbooks. These seem to be based less on direct observation than on figures copied from earlier publications." Since the material in question can have diameters up to 0.5 m and over 20,000 vascular bundles, this is perhaps understandable. Zimmerman simplified the problem by using a smaller palm, *Raphis*, with a stem diameter of 2 cm and with 1000 bundles, and ultimately designed a microtome capable of handling specimens of arbitrary length (Zimmerman and Tomlinson, 1972).

Cinematographic reconstruction of human brains using surface photography is currently being carried out by Livingston (personal communication) at the University of California at San Diego. Surface staining was found to produce "sections" of nonuniform color, and so was given up in favor of unstained brain embedded in paraffin or bulk staining with Stöhr's picrocarmine procedure. Special microtomes 50 times as large as the ones described by Postlethwait (1962) were constructed from lathe beds, one for paraffin-embedded material and one for frozen brains prepared with water-based stains. Forty-eight educational films of supposedly normal brains, each of which proved to have lesions which had been overlooked clinically and at post-mortem examination, were made with the former device and eight with the latter.

These methods are fundamentally macrophotographic in nature, which limits the amount of information recorded. Serial sections are, however, made for use in the microscope, so it was natural for biologists to extend the principles of serial section cinematography to that level of resolution. The first procedure that did this and at the same time avoided the problems of alignment by using not real but optical sections was described by Saxl (1927), who made a cinematographic record during a through-focus pass in light microscopy of a transparent thick (greater than the depth of focus of the objective) section. Elias (1935) used this method to document the absence of anastamoses in epidermal melanophores of the tadpole *Bombinator pachypus.* His procedure was to photograph 0.5-μm optical sections four times each on cinefilm focusing once into, and then out of, the specimen. This enabled him to produce a film loop which, along with a suitable stop-motion projector, could simulate the through-focus procedure of a microscope but with a projected image. Drawings were made of each "section" and assembled into graphical or physical models by the methods described.

Peacock and Price (1932) explicitly rejected the practicability of Widakowich's method and described the first procedure for photographing an arbitrary enlargement of *true* sections by means of a compound microscope. To do this they mounted, horizontally and in series, a light source, condenser, microscope, 16-mm cinecamera which could be swung into and out of the optic axis of the system, and a large vertical board onto which the microscope could project its image. During the photography a section was positioned on the microscope stage so that its projected image aligned with a tracing of the previous section, the camera was swung into position, and the image photographed. Apparently, the only use they made of this device, using 400 to 800 6-μm sections, was to show that, whereas in single sections the nest cells in malignant tumors usually appeared as isolated clumps of keratinized cells separated from each other and from the surface of the tumor, in a completed reconstruction they were seen to be in direct continuity with each other and with the surface of the tumor. Basically this same procedure was described again by Lugg (1972), who produced color cinefilms of up to 2000 sections through traumatized spinal cord preparations.

The problem with this tracing method of alignment for quantitative work is that it drastically underutilizes the inherent resolution of the system in the perpendicular direction. Whereas ordinary cinematographic film can have a resolving power of 50 lines/mm and microfilm up to 400 lines/mm, and with pin registration cinefilm can be reproducibly positioned to 40 lines/mm, the interplane resolution is limited by the ability of the operator to align an imperfect tracing of a section with a real one, which in its final film image may have details too fine to record by hand except with extreme persistence. In addition, the construction of even a crude drawing for each section is time-consuming, especially in series hundreds of sections in length. To overcome both of these objections, Zimmerman and Tomlinson (1966) devised a method, shown in Fig. 9, by which two compound microscopes could be optically coupled, enabling the operator and the camera to look through either or both simultaneously. In this scheme the crucial element is a beam-splitting device used in reverse so that it combines the two optical images. Each microscope is illuminated independently, and the photographic procedure is

FIG. 9. Optical shuttle device described originally in Zimmerman and Tomlinson (1966). (Zimmerman, original photograph.)

to place two successive sections on alternate microscopes, illuminate and photograph one, illuminate both so that the second can be aligned with the first, and then illuminate and photograph only the second. By this back-and-forth procedure a cinefilm can be created in which each section is aligned directly with the previous one without the necessity of an intermediate drawing. Using this procedure for fine resolution and the surface photography method described for an overall view, Zimmerman revolutionized the study of the vasculature of monocotyledonous plants, notoriously difficult because of its apparent complexity and lack of order. In their first article, (Zimmerman and Tomlinson, 1965) they stated, "The vascular system of *Raphis* has been resolved by an essentially dynamic process, which is reflected in the subsequent description which refers to the distribution of vascular bundles by a dynamic terminology." With their cine procedure they demonstrated conclusively that all bundles behave essentially alike, all bundles maintain their individuality and proceed indefinitely up the stem, and all bundles in the central, uncrowded part of the stem trace out a shallow helical path, rotating clockwise when followed upward. A few appropriately chosen sequences were sufficient to show a "previously unrecognized pattern which we believe to be fundamental for monocotyledonous stems as a whole . . . and the developmental principle which underlies this structural pattern" (Zimmerman and Tomlinson, 1972). With the cooperation of the Institut für den Wissenschaftlichen Film in Göttingen, Zimmerman created an educational film for public distribution showing some of his results (Zimmerman, 1971).

Serial section cinematography at the level of the electron microscope was first described by Ware (1971) and Levinthal and Ware (1972). Their procedure had the additional advantage that the items photographed on the cinefilm were not sections themselves, but negative transparency photographs of them, thus permitting the use of a single machine for serial section photographs produced from any source. A schematic diagram of a similar apparatus (Ware *et al.*, 1974) is shown in Fig. 10A.

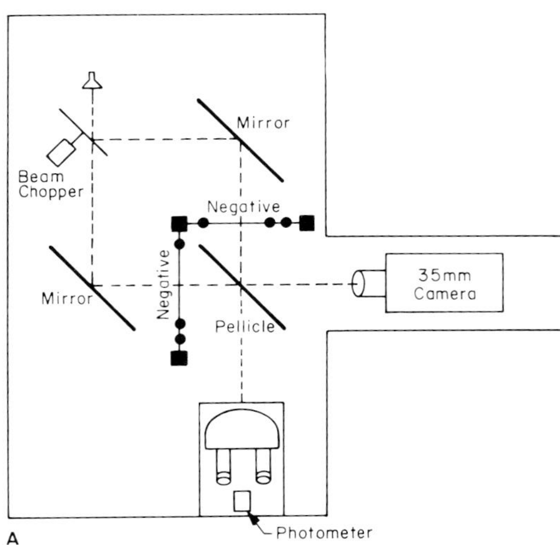

FIG. 10. (A) Schematic drawing of the cine apparatus described in the text. (B) Photograph of the same apparatus. (Unpublished illustrations.)

FIG. 10B. See page 343 for legend.

The basic idea is that negatives from two sequential sections are held vertically and at right angles in carriages capable of translation and rotation. By means of a pellicle, both can be seen through either the observation or camera port. Rather than being illuminated simultaneously during alignment, they are alternately illuminated at a frequency just below fusion by means of a beam chopper, a modification of which was also introduced independently by Zimmerman and Tomlinson (1968) into their optical shuttle device. Alignment is thus achieved not by the awkward procedure of matching up corresponding points on the two negatives, but by minimizing the subjective sensation of motion in the region of interest. "In the region of interest" is meant literally, since unlike thick sections, thin sections are subject to numerous types of relative distortions which cannot be controlled. First, there is compression during the cutting of even pure plastic. Adding a specimen which is not of exactly the same hardness as the embedding medium compounds the problem. To some extent this can be minimized by "flattening" the sections as they float in the trough with chloroform vapor or a locally applied pulse of heat, but even then sections can remain differentially distorted at the levels of resolution obtained with the electron microscope. Second, there are tensions in the plane of the sections during the final moments as they dry down on the supporting film. In our laboratories we have found that these can be minimized by dipping the grid just prior to collection of the ribbon into a very dilute solution of a nonionic detergent such as Triton X. Third, and perhaps most important, the electron beam causes considerable heating and shrinking of the section locally, especially when it is highly focused. This can be reduced by means of a cold trap but cannot be entirely eliminated especially since, in series work, ribbons are usually placed on a supporting film of fairly large area, and regions of interest are likely to be very far from the firm support offered by the grid. Thus any alignment must be a compromise, and "blinking" successive sections is the best way to achieve it. For reasons of convenience, a master is made on microfilm with only one frame per section. Depending on the use to which the film is to be put, it can then be duplicated by a competent optical house in any format, such as forward and back passages through the specimen, multiple frames per section, fade-ins, and so on. With this type of machine movies are routinely created at the rate of 60 to 80 frames per hour. Since serial sections are made at the rate of 300 per hour and can be easily photographed at the rate of 400 per day on a modified Zeiss electron microscope, this leaves most of the investigators time available for analysis of the reconstructed material.

The ultimate uses of cinefilms have been less well documented than the means of their production. Reicher (1907a) simply showed them as movies at a neurological conference where, as judged by his rapid desertion of the idea, they apparently were not well received. Livingston (personal communication) has distributed his films to television and for educational purposes and is currently working with the Jet Propulsion Laboratories of the California Institute of Technology to use computer analysis of the movies for studies in quantitative and comparative morphology. Using three optically filtered images from each section, he has differentiated brain structures in the cinemorphology films of unstained human brain. Zimmerman, like Elias (1935), used the film as a research tool by employing a stop-motion rear surface projection system to present aligned micrographs one at a time on a table where measurements were made and courses plotted for individual vascular bundles and fibers. Cinematography at the level of the electron microscope is, first, simply a more convenient storage device for scanning many hundreds of electron micrographs than a stack of prints. Relationships among component parts are easily remembered when the relevant micrographs are viewed rapidly in aligned position, thus obviating the need for construction of models, solid or graphical, except for final presentation. It was unexpectedly discovered in addition that individual nerve fibers in dense neuropil which seem to be lost in a slow scan can often be found using the sense of motion felt when the film is

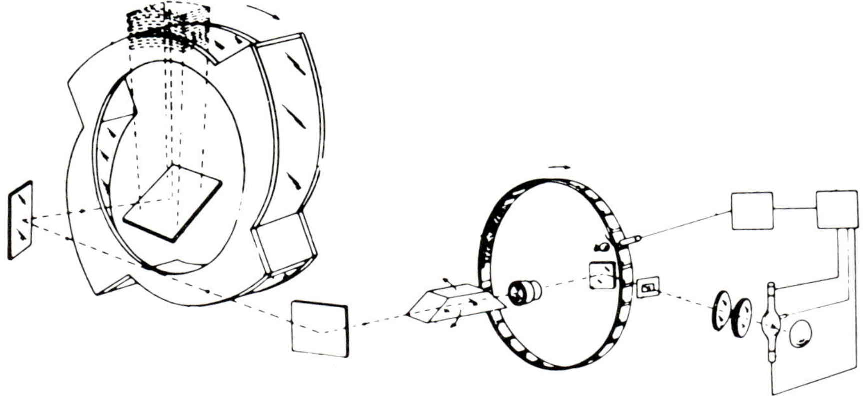

FIG. 11. Schematic diagram. (From de Montebello, 1967.)

projected at a rapid rate. In order to make use of the full resolution and contrast range present on the microfilm micrographs, Ware *et al.* (1974) have abandoned the projection method and instead look directly at the film plane with a zoom dissecting microscope. The way these films are employed in transferring three-dimensional information to a computer is described in Section IV,A. A novel method of using a cinefilm or its equivalent to produce a spatial, rather than temporal, reconstruction has been described by de Montebello (1967, 1969) and is shown diagrammatically in Fig. 11. Briefly, a filmstrip of aligned sequential micrographs is mounted on a rotatable circular transparent drum. As each frame reaches a certain position, a strobe illuminating light is triggered which, through a system of lenses and mirrors, projects the frame onto the inside of a curved transmitting screen forming the spiral arm of another rotating drum. Looking from outside the second drum, a viewer thus sees the projected image of each frame at a different height. If the rotations are sufficiently rapid and properly timed, his perception is of a static reconstruction of the sectioned object.

B. Applications in Medicine

Interestingly enough, serial section reconstruction techniques have not been consistently used in medicine to elucidate either basic structures or deviations from them in diseased material. Some investigations have simply given a verification of structures previously known, but a few have provided fundamentally new information which could only have been obtained from a systematic study of serial sections. Perhaps the most famous example of the perpetration of error resulting from the unimaginative use of single sections by usually high authorities was brought to light by Elias (1948, 1949). Since the work of Gerlach in 1849 it had been accepted, except by a few doubters who were openly ridiculed by more prestigious workers, that the basic structural elements of the liver are cylindrical, that is, that the liver is composed of cords of cells surrounded by blood spaces called sinusoids. This theory was maintained for nearly 100 years until Elias sought new documentary evidence for inclusion in an instructional filmstrip prepared but for the final illustrations according to the widely held ideas. It was only during the preliminary efforts to construct a wax plate model for this film that he realized that no matter how the liver was cut one saw only longitudinal sections of the "cords." Thus he concluded that the liver could not consist of columnar arrangements of cells, but must be made of curved plates, sheets, or laminae, one cell thick. He went on to construct wax plate models, both from drawings of serial sections and from optical section cinematography (Elias, 1935), to show that this is in fact the case and that the laminae are not discrete but anastamose with one another and enclose between them sinusoidal spaces which form a continuous labyrinth.

In 1967, Takahashi and Hayama (1967) presented evidence for the morphological nature of a well-known class of liver disorders, the so-called intrahepatic cholestasis, for which no clear histological abnormality could be found by single-section analysis. This disorder is characterized by jaundice of obstructive character but is not accompanied by mechanical obstruction of gross bile ducts and had been regarded as showing a discrepancy between the function and structure of the liver in regard to bile secretion. In their description of previous analyses these investigators point out that the original hypothesis of the nature of the disorder (advanced in 1937), that it might be attributed to cholangiolar damage in the junctional area, became discredited when needle biopsy consistently failed to show any severe histological changes in the liver. Subsequently, several hypotheses were presented, none dealing with the structure and continuity of the bile duct itself.

To investigate a possible direct morphological correlate of the disorder, these investigators selected three cases of intrahepatic cholestasis and two cases of serum hepatitis. In all cases there was no pathological process that might have caused ductular obstruction. Their method was to form graphical and plate models from 200 to 300 6-μm serial sections of a randomly selected 200-μm square portal area and to reconstruct the architecture of the cholangiolar system. Figure 12A shows their reconstruction of the bile duct system from a normal liver. It is highly irregular, and anastamoses of ductules show that topologically speaking it is not a "tree." The main interlobular duct, however, provides a more-or-less

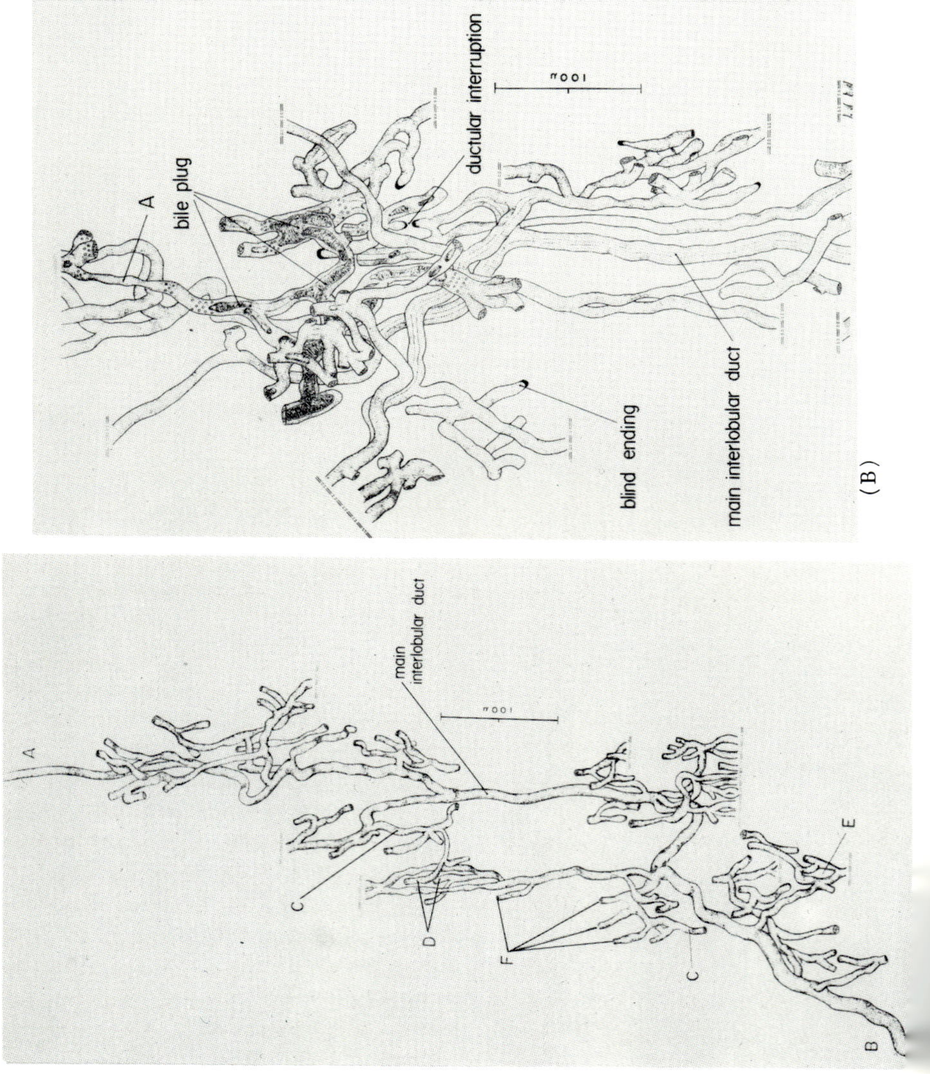

A
bile plug
ductular interruption
blind ending
main interlobular duct
A
main interlobular duct
C
D
F
C
E
B

(A) (B)

direct path for transport of bile. Figure 12B shows a reconstruction from one of their cases of cholestasis. It can be seen that bile plugs fill ductular lumina, many ductules end blindly, and the plexus has no connection with the main interlobular duct. The other two cases of cholestasis show the same derangements, but to a larger degree. An interesting observation they made was that in both cases of fatal viral hepatitis exactly the same type of disruption of the cholangiolar architecture was present. They concluded with the observation that the pattern of derangement suggests random and indiscriminate lesions affecting the cholangiolar system and a subsequent ductular regeneration which, although present, is insufficient to restore normal bile flow. In a subsequent study (Oh-I *et al.*, 1969), the same technique was used in cases of congenital bile duct atresia to show an intrahepatic bile duct system composed of an extremely complex network of proliferating ductules which fuse to form a labyrinth structure. Sometimes the main duct is still present, but often it is replaced by meandering ductules.

In other liver studies, Kelty *et al.* (1950) sought the source of the intrahepatic portal obstructions existing within the cirrhotic liver, which lead to increased portal tension. This seemed of interest since, whereas most emphasis had been placed on the nature of the scar tissue and its contractions, it was frequently noted that little or no correlation existed between the amount of fibrous tissue in a cirrhotic liver and the increase in portal pressure. By constructing a glass plate model of regenerating nodules near venous structures, these investigators showed that compression of large veins and displacement of small ones followed exactly the shape of contacting nodules. In contrast, regions about large vessels which consisted mainly of connective tissue without any adjacent nodules did not show any compression. Their conclusion was that the source of portal hypertension must be the compression of vessels by the pressure of growth and expansion of the regenerative nodules.

An example of the same type of faulty thinking that led to incorrect ideas of normal liver structure was pointed out by Jorgensen (1971b) in the case of congenital hepatic fibrosis. Several investigators had noted the presence of longitudinally cut bile ducts encircling vessels, but had not investigated them further. Jorgensen submitted one such case to serial

---

FIG. 12. (A) Graphic reconstruction of the normal cholangiolar area. A to B, Main interlobular duct; C, lateral branches running in reverse direction; a pair of ductules from D follow entirely different directions in their courses; E, anastamosis between ductules; F, junction of ductules with the limiting plate. (B) Cholangiolar area in case 3. In the upper half of the figure a complicated ductular plexus with numerous bile plugs is demonstrated. This is isolated from communication with the main interlobular duct. (From Takahashi and Hayama, 1967.)

reconstruction and found, as expected, that the "ducts" were actually sheets surrounding the veins. He related this to the embryology of the liver, in which the intrahepatic duct system starts out as a duct plate which is later perforated by connective tissue into true ducts, and suggested some sort of mechanism of regression arrest as the source of the peculiar duct system in congenital hepatic fibrosis. Still another example of this type of erroneous deduction from single two-dimensional micrographs was pointed out by Elias (1971a) in regard to the development of the human testis. The gonads of 15 to 25-mm male embryos are characterized by the presence, in sections, of long branching stripes which for 70 years had been identified as cords. A commonsense approach would identify them as sheets, yet this is not the terminology used even in recent texts. In 1971, Elias (1971b) reported that as a result of quantitative stereological methods supported by reconstruction from serial sections he was able to make the proper identification of these folds as nonpolarized pseudostratified epithelium embedded in mesenchyme. At the 30 to 35-mm stage the sheets begin to break up into cylindrical rods. The transformation is complete only at the 110-mm stage.

At the level of the electron microscope it was not until 1970 that Thaemert (1970, 1973) resolved nearly a decade of controversy about the nature and extent of the autonomic innervation of the atrioventricular node by actually following nerve fiber processes over a large part of their length. The problem was that several reports on the ultrastructure of the atrioventricular node using single-section analysis prior to 1970 had indicated only a small number of neuromuscular contacts within the two cardiac nodes of many animals, with a single exception. In addition, some of the single-section analyses differed as to the structures of the nodal cells themselves, some seeming to show uniformity of cell type and others indicating the cells could be grouped into various rigid classifications. The source of these discrepancies, as Thaemert discovered, is that there is a gradation of both properties from the anterior to the posterior poles of the node. The degree of contact between autonomic fibers and the nodal cells ranged from very light at the anterior pole to very heavy at the posterior pole. Nearly all the cells of the latter region were shown to contain one or more nerve processes with numerous swellings containing only clear vesicles (Fig. 13). There was a corresponding gradient, but in the opposite direction, in the degree of contact between the nodal cells themselves as judged by the presence of maculae occludentes, ranging from high among the smaller, globular cells of the anterior region to low among the larger, more highly branched cells of the posterior region. On this basis, Thaemert made the suggestion that the posterior nodal

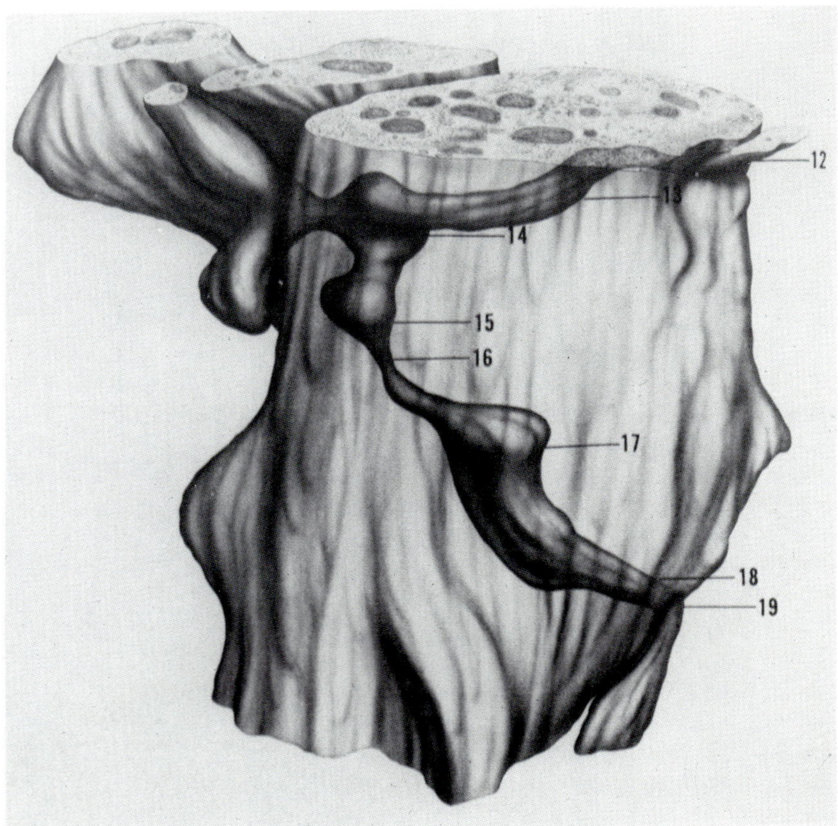

FIG. 13. This three-dimensional illustration constructed from 85 sections shows the position of a vesiculated nerve process within a sarcolemma-lined tunnel inside a nodal cell. From right to left, a nerve process enters the nodal cell to become varicose while passing through the cell transversely and leaves the tunnel on the other side. However, before it does so, an irregularly beaded branch is emitted from a varicose junction and passes diagonally downward to emerge from the cell below. Numbers indicate levels illustrated by electron micrographs. (From Thaemert, 1970.)

cells have a regulative function, whereas those more anterior are specialized for conduction.

A series of studies involving the structural interrelationships of the vascular, tubular, and autonomic nerve components of the juxtaglomerular apparatus (JGA) has been carried out by Barajas and co-workers using serial sections prepared for electron microscopy (Barajas and Latta,

1963; Barajas, 1970, 1971; Muller and Barajas, 1972; Barajas and Muller, 1973). The JGA can be defined as the area at the hilus of the glomerulus in the kidney where a portion of the distal convoluted tubule enters into intimate relationship with the vascular pole of the glomerulus, which is composed of afferent and efferent arterioles (see Fig. 14, top). The most important function associated with this area is control of the glomerular filtration rate. This is effected both by vasoconstriction and dilation, and especially by control of the ionic permeability of the distal tubule by renin. This substance is released by granular cells located in both the wall of afferent and efferent arterioles near the hilus and in the nearby mesangial region. In both cases the granular cells are interspersed with typical smooth muscle cells. Various physiological observations led to a hypothesis that the granular cells might be baroreceptors and release renin upon sensing changes in arterial blood pressure, and also to an alternative mechanism proposing that renin release is a function of certain changes in the distal tubule, which are somehow transmitted to the granular cells.

Barajas took the approach that the precise anatomical relations of the vascular and tubular components must first be elucidated in order to decide upon a reasonable hypothesis, and that for structures of such complexity this had to be done at resolutions available only with the electron microscope. Clay models based on preliminary observations of light micrographs of 1- to 2-μm serial sections had indicated extensive areas of contact between the distal tubule and the efferent arteriole in the region of the hilus (Barajas and Latta, 1963). In analyzing series of up to 500 thin sections electron microscopically, this initial observation was confirmed. It was found in every series through a JGA region that an extensive area of contact occurs between the distal tubule and both efferent arteriole and cells of the mesangial region. This contact is characterized by cytoplasmic extensions of distal tubular cells projecting to the arteriole where fusion of the basement membranes of both elements occurs. A contact region can also be seen between the tubule and afferent or efferent arterioles where the basement membranes are apposed but not fused. Composite drawings showing these contact areas were made from montages of each section (Fig. 14). These contacts can be seen as possibly representing the morphological correlate of the observed physiological relationship between distal tubule volume and renin secretion. In a subsequent article (Barajas, 1971), it was reported that the contacts of distal tubules with granular cells are always of the nonfused type. More importantly, a study of the autonomic nerve components of the JGA was initiated both histochemically and by serial section electron microscopy. In following the adrenergic and cholinergic autonomic nerve

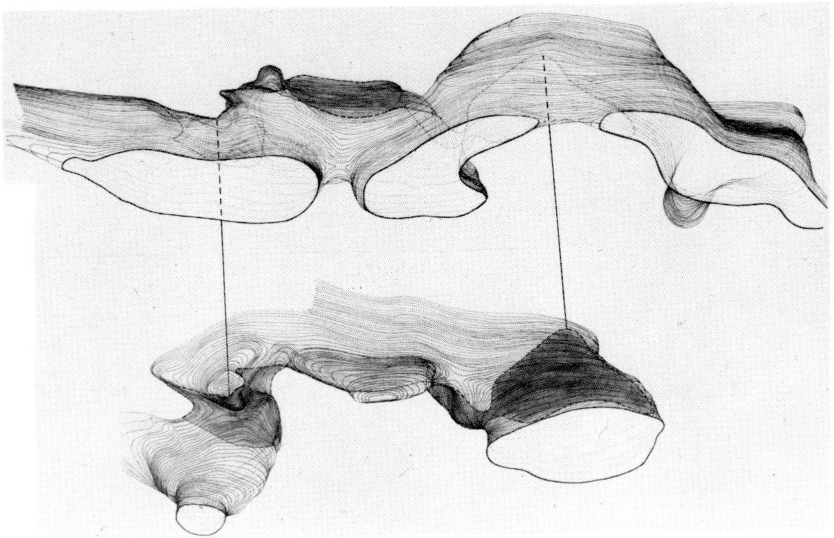

FIG. 14. Above: Separate three-dimensional representation of the surface of the vascular component. The afferent arteriole (in lower part of lower frontal plane of section) joins with the mesangial region (lower frontal plane of section) at this plane of sectioning. The afferent arteriole joins with the mesangial region and the hilar portion of the efferent arteriole (in upper part of lower plane of section) in the lower midportion of the figure. The efferent arteriole is the middle plane of section. The upper plane of section is the capillary region. The area in which contact occurs is circumscribed by dotted lines, and the contact areas on the vascular component that face the viewer are darkly shaded. Those on the opposite side are left unshaded. Below: The surface of the tubular component represented as a three-dimensional drawing. Contact areas are circumscribed by dotted lines. The frontal portion of the contact area on the tubular component that faces the viewer is more darkly shaded than the areas on the opposite side. Contour tracings are from $2\frac{1}{2} \times 5$ ft montages. Dotted lines indicate path of separation of these two apposed structures. (From Barajas, 1970.)

supply to the arteriolar smooth muscle, it was observed that some of the axons in this bundle make contact with both proximal and distal tubular cells by means of varicosities resembling the *en passant* synaptic contacts with the smooth muscle (Muller and Barajas, 1972). As noted in previous electron microscope studies, both granular and typical smooth muscle cells receive innervation. Histochemical investigation showed that adrenergic and cholinergic nerves have a similar distribution in this respect. In a more detailed quantitative study of a series of 398 sections (Barajas and Muller, 1973), it was found that less than one third of the cells of efferent and afferent arterioles were innervated by the autonomic axons in the region studied, while of 30 cells in the mesangial region only 3 were innervated, 2 of which contacted the distal tubule. In all innervated granular cells, the presynaptic swellings contained dense core vesicles of the adrenergic type. No varicosities with clear vesicles were found near granular cells. Furthermore, all cells received contacts from on the average of three to four separate axons, raising the possibility of a graded neural control of granule cell secretion. The most important observation was made by following autonomic axons over long distances. In this way the same axon was seen to contact by *en passant* swellings both granular and smooth muscle cells of afferent and efferent arterioles, as well as both proximal and distal tubule cells. These anatomical relationships thus provide a basis for multiple effects in the control of glomerular filtration rate, and furthermore point out the possibility that these several mechanisms may be integrated and coordinated by virtue of the innervation of the various elements by the same autonomic axons.

C. Nervous System

1. *Spinal Motoneurons*

As emphasized by Sherrington (1892), the spinal motoneurons can be viewed as the final common pathway for the integration of sensory stimuli with the multitude of influences from more central areas. This concept of the motoneuron as an integratory machine at the single-cell level has been extensively studied both physiologically and anatomically. A few attempts have been made to obtain anatomical information related to their function by three-dimensional reconstruction. These investigations have concentrated on the quantitative and qualitative synaptology of spinal motoneurons, that is, on a fairly complete description of the types, relative proportions, and distribution of the afferent synapse population to well-defined areas of these cells. Although a significant body of information exists from single-section statistical analysis of synaptology, the overall picture generated from serial sections has provided additional topographical information not otherwise available.

A series of investigations reported on by Elfvin was concerned with the fiber patterns and synaptology of autonomic motoneurons in the superior cervical and inferior mesenteric sympathetic ganglia of the cat. These ganglia are arranged as an interconnected chain external to the spinal cord proper. They are organized in such a way that preganglionic fibers originating in the spinal cord have synaptic input to the intrinsic motoneurons of the ganglia. These postganglionic neurons in turn send impulses to visceral smooth musculature. The preganglionic fibers are a cholinergic system, while most of the postganglionic neurons are adrenergic. In an early study, Elfvin (1963a,b) used very short series of thin sections (20 to 40 sections each) to describe for the first time both the ultrastructure and synaptic interrelationships among the preganglionic afferent terminals and the perikaryon and dendrites of selected ganglion motoneurons. The short lengths of the series made it necessary in most cases to infer the axonal or dendritic nature of any fiber from ultrastructural criteria, since it was not possible to trace processes to their cells of origin. Elfvin studied large numbers of synapses, both terminal and *en passant*, between preganglionic axons and the dendrites and perikarya of ganglion cells. These axons were seen to wind around the dendrites, making conventional synapses at several points. It was not possible in the short series used to determine whether one preganglionic axon served several ganglion motoneurons or whether one ganglion cell received synapses from a multiplicity of preganglionic axons.

In a later more detailed study with series of 300 to 500 sections through the inferior mesenteric ganglion, Elfvin (1971a,b,c) better characterized the possible synaptic relationships among the neuronal elements. The data from each section were traced on clear plastic sheets from which composite three-dimensional drawings and summary wiring diagrams were made (Fig. 15). The rationale for this extended anatomical study came from experiments on curarized ganglia (e.g., Libet and Tosaka, 1970), which brought to light an intrinsic inhibitory neuronal system in the sympathetic ganglia which is apparently mediated by monoamine transmitters. Histochemical staining of the ganglionic neurons for monoamines had revealed that these adrenergic neurons displayed a characteristic fluorescence primarily as a crownlike pattern around the cell body. In his ultrastructural study, Elfvin showed that each ganglion cell possesses a large number of very short, thin perikaryal processes containing dense-core vesicles of a size typically found in adrenergic nerve terminals. These so-called accessory dendrites seemed to be a likely source of the crown of monoamine fluorescence and were usually identifiable as being extensions of the cell body only by following them through a long series of sections. Elfvin found that most preganglionic fibers synapse onto both standard and accessory dendrites, but hardly at all onto the ganglion cell body. More importantly, he found contacts with synaptic densifications between pairs of accessory dendrites of adjacent cells and between accessory and standard dendritic branches near almost every preganglionic axon synapse onto one or both of the dendritic elements. Dendrodendritic contacts of this type were usually absent where there was no preganglionic axon synapse. These findings, combined with the presence of monoaminelike vesicles close to the dense dendritic contact point, supported the possibility that the contacts represent the intrinsic inhibitory

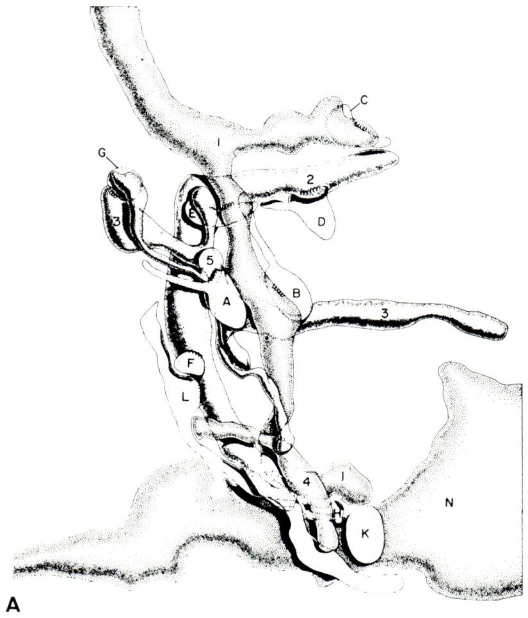

FIG. 15. (A) Reconstruction illustrating the relationships of fibers near a ganglion cell body (N). 1–5, Ganglion cell fibers; A–L, preganglionic fibers. Synapses are indicated by a spiral. (From Elfvin, 1963a.) (B) Schematic representation which illustrates the contact relationships between pre- and postganglionic neurons in the sympathetic ganglia of the cat. The elongated parts of the preganglionic fibers indicate synaptic regions. The various types of contacts are: a, axodendritic on large dendrite; b, axosomatic; c, axodendritic on accessory dendrite; d, axoaxonic; e, dendrodendritic between accessory dendrite and large dendrite; f, dendrosomatic; g, dendrodendritic between large dendrites; h, dendrodendritic between narrow branchlet from one dendrite and a large dendrite. (From Elfvin, 1971c.)

system in the ganglion and that this system operates between dendrites of adjacent ganglion cells.

In what is probably the most detailed study of its kind to date, Conradi and Skoglund (Conradi, 1969a,b,c,d; Conradi and Skoglund, 1969a,b) made a two- and three-dimensional analysis of the population of synaptic boutons contacting the dendrites, perikaryon, and axon initial segment of lumbosacral spinal α-motoneurons in adult and newborn cats. The aim of these investigations was to provide a morphological correlate to the spinal reflex system of dorsal root innervation to the motoneurons in the context of the considerable physiological changes occurring during the postnatal development of this system.

Six distinct types of bouton could be identified on the basis of charac-

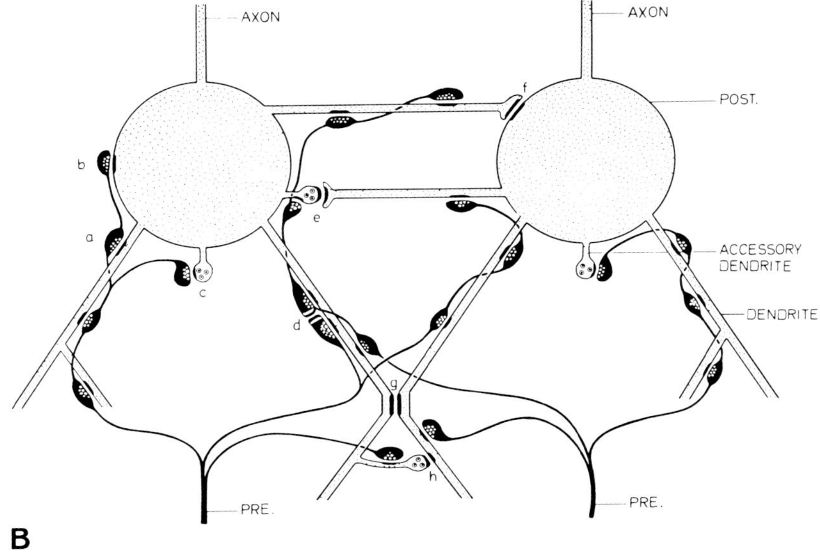

FIG. 15B. See facing page for legend.

teristic size and ultrastructure. In a statistical analysis conducted using either equally spaced single sections or all single sections covering a small representative volume, these investigators estimated the frequency of occurrence of each bouton type and the percentage of the motoneuron surface covered by boutons. It was found that 50% of the perikaryal surface and that of thin dendritic branches is covered by terminal boutons, while the figure for the proximal dendrite surface is 70%. In all cases the remainder of the neuronal surface was found to be covered with astrocytic glial processes. In addition, a nonrandom distribution of bouton types was found. For example, type-F boutons are found primarily on the perikaryon, with a constant decrease in frequency of occurrence further out on the dendrites. The reverse distribution occurs for type S. Type T is found exclusively on thin dendritic branches, type C on perikaryon and proximal dendrites, and type M on proximal dendrites. Thus each region of the motoneuron was found to possess its own characteristic bouton spectrum, supporting the validity of the classifications.

A complete three-dimensional reconstruction was made first simply to check the validity of the sampling methods used in the single-section analysis, and second to obtain information on the topographic array of boutons on the motoneuron surface. To accomplish this a series of 750 sequential thin sections was made, 220 of which completely encompassed a proximal motoneuron dendrite in longitudinal section. For quantitative measurements drawings of the dendrite surface and associated boutons were traced from montages of every few sections, and the true percentage of the dendritic surface covered by boutons, both combined and of each separate type, was calculated. The resulting values were within 5% of those obtained from the statistical analysis of systematically chosen single sections, indicating that a complete three-dimensional reconstruction is not necessary to obtain reliable sampling statistics. The complete reconstructions, however, showed that most individual boutons established more than one synaptic contact, an observation not made in the single-section studies.

In order to represent the topographical information derived from serial sections, Conradi used what he called the mantle surface method for longitudinally cut cylinders, which is in effect a flattening of the dendrite into one plane so that the three-dimensional surface is represented in only two dimensions. In this method the area of the spread-out cylinder is outlined on a transparent sheet of paper. Each section through the dendrite is represented as a pair of parallel longitudinal lines, the uppermost section as a line at the center of the area, and the lowest section as two lines at either edge. The transparent sheet is then placed over each micrograph in sequence, and the outlines of the dendrite in the micrograph aligned with their corresponding lines on the sheet. Lengths of apposition of boutons, as well as other parameters, are entered along each line representing the cross section of the dendrite. The result of such a reconstruction for the adult cat and kitten is shown in Fig. 16. This representation clearly illustrates the clustering of the F-type boutons at the axon hillock, while showing that the other types seem to be randomly distributed along the surface.

Three-dimensional wax models of a motoneuron initial segment were also made (Fig. 16). In the region of the axon hillock and initial segment, it was shown that there were only S-, C-, and F-type boutons present on the hillock, but rarely the T-type and never the M-type. No presynaptic structures at all were found on the initial segment of adult neurons. In experiments in which dorsal rhizotomy was performed, all M-type boutons and some S-type boutons disappeared from all areas of the motoneuron 3 days after cutting the dorsal roots. This allowed tentative identification of at least the M-type as the terminal of a dorsal root afferent. The characteristic flattened synaptic vesicles in the F-type suggested that they may be inhibitory in function (Bodian, 1966a,b). Taken together these observations indicate that all excitatory dorsal root synapses occur at the proximal dendritic level, a result suggested by the physiological studies of Rall *et al.* (1967), while at the axon hillock, the site of action potential generation, the inhibitory terminals outnumber the excitatory terminals by 50%. An additional finding of the degeneration experiment was that an M-type bouton consistently has a P-type apposed to its surface. This was taken to indicate a presynaptic inhibitory effect of P on M, an inhibition of dorsal root activity by small axon terminals which has been well documented physiologically (Frank and Fuortes, 1957; Eccles, 1964).

The two- and three-dimensional representations of motoneurons in neonatal kittens permitted comparisons with adult animals (Fig. 16). It was found that at birth the proximal dendrites have a smaller overall percentage of their surface covered by boutons than in the adult, and that the reverse was the case for the cell body. Large areas of dendrite were devoid of synaptic contacts, a situation never found in the adult and, more importantly, all adult types of boutons were present in the neonate except the M-type. The first appearance of M on proximal dendrites at 3 weeks was correlated with the known development of sensory-motor coordination at that time.

An earlier attempt to map the topographic surface distribution of presynaptic boutons was carried out by Karlsson (1966b), and provides an interesting comparison with the work of Conradi. Karlsson reconstructed two lateral geniculate nucleus interneurons from serial sections by electron microscopy and computed the percentage of perikaryon and proximal dendrite surface covered by boutons, although he made no attempt to identify the origins of any of the boutons. He found that the density of bouton covering of neuritic spines was an order of magnitude higher than on the shafts of the neurites and two orders of magnitude higher

---

FIG. 16. [pp. 360, 361] (A) Left: Surface map of the proximal portion of an adult cat dendrite. The region included ranges from the cell body (proximal) to the junction at the first branch point. The horizontal row of numbers refers to the section number in the series. Center and right: Wax model of an axon hillock (not painted) and initial segment (stippled with black), reconstructed from serial sections and seen from the front and behind. A cell body is at the bottom of the figure. Note the compact myelin body just outside the beginning of the myelin (upper black region). Boutons on the axon hillock are also included. Black, F-type; white stippled, S-type. (B) Left: Surface map of the proximal portion of a neonatal kitten dendrite. The region included ranges from the cell body (proximal) to the junction at the first branch point. Center and right: Two aspects of a wax reconstruction of the axon hillock (not painted) and the initial axon segment (stippled) in the newborn stage. A cell body is at the bottom of the figure. Note that the initial segment is curved and gives off a collateral just proximal to the beginning of the myelin (black). Both the axon hillock and initial segment are covered by boutons of the F-type (black), S-type (stippled), and C-type (white and separately indicated). (From Conradi, 1969a,b,c,d.)

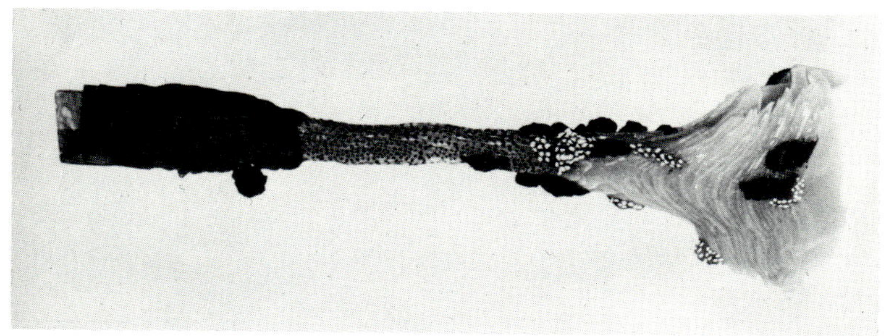

A

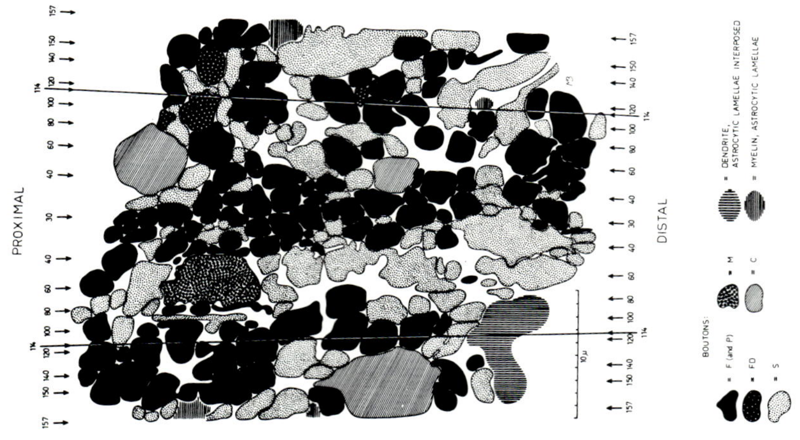
PROXIMAL
DISTAL
BOUTONS:
= F (and P)
= FD
= S
= M
= C
= DENDRITE, ASTROCYTIC LAMELLAE INTERPOSED
= MYELIN, ASTROCYTIC LAMELLAE
10 µ

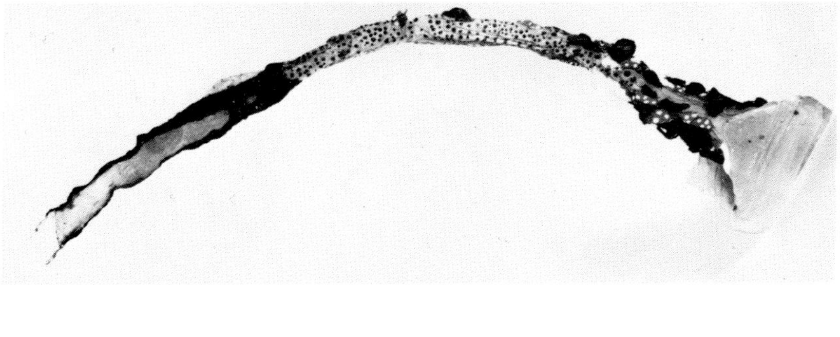

B

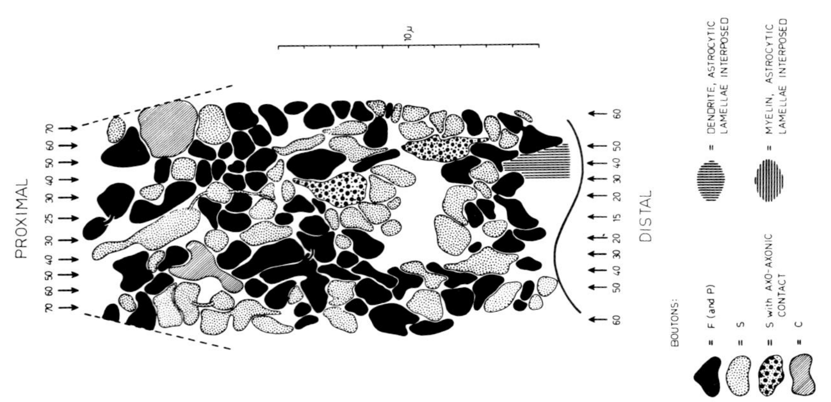

FIG. 16A and B. See page 359 for legend.

than on the perikaryal surface. The surface covering of boutons averaged for all parts of the neurons studied was only 15%, and almost all the remaining 85% of the surface was covered by glia. The large difference in bouton covering between this interneuron and the 50–70% bouton cover found in Conradi's study of spinal motoneurons seems to emphasize the characterization of the spinal motoneuron as the convergence point and final integratory pathway for a multitude of neuronal influences on the periphery.

Spinal motoneurons have also been reconstructed at the light microscope level using intracellular injection techniques. In fact, the first report of an intracellular staining technique for vertebrate neurons that resulted in the staining of the perikaryon and all processes was by Globus *et al.* (1968) in just such a study. They injected tritiated glycine iontophoretically into cat spinal motoneurons by means of triple-barreled electrodes. Spinal cords were then embedded in Paraplast, sectioned at 6 μm, and exposed for autoradiography. Most of the cells injected were found to have heavy, uniform label throughout the soma and dendrites, the latter being traced as far as 600 μm from the nucleus. Axons were somewhat more difficult to follow than dendrites because of their smaller diameters, which even in 6-μm sections probably resulted in considerable absorption of radioactivity. Nevertheless, in some cases they could be followed as far as 1300 μm from the nucleus, a distance corresponding to that expected from normal rates of axonal flow. Dendritic geometry and dimensions were found to be in agreement with results using the standard Golgi procedure, implying that the method gave results as reliable as this classic technique. In 1970, Barrett and Graubard (1970) used the Procion technique of Stretton and Kravitz (1968) to inject the same type of cell, but made only projection reconstructions of dendritic arborizations from 100-μm sections. They were, however, able to follow the Procion marker from a single cell in as many as eight sequential sections. Both axons and dendrites were heavily stained. Axons were traced for up to 1000 μm into the ventral white matter, and dendrites were consistently stained 400 μm from the soma, occasionally as far as 600 μm. They qualified the Procion technique with the statement: "The distant fine dendritic branches faded out gradually, as if they were not stained intensely enough to make their full length visible." Yet in those cells considered "intensely stained," summed lengths of dendrites and surface areas of visible dendrites agreed with similar measurements from Golgi-prepared material.

Knowledge of exact cell geometries obtained from the injection method in motoneurons has been used to evaluate experimentally the unit membrane resistant R_M. Lux *et al.* (1970) used dendritic branches reconstructed from automicroradiographs of 6-μm paraffin serial sections of neurons that had been injected with tritiated glycine to determine those geometric parameters necessary to represent the dendrite tree mathematically by the equivalent-cylinder method of Rall (1959). In seven neurons they used these parameters along with the total cell impedance measured in the same cell to arrive at an average R_M of 2750 ohm-cm^2. Barrett and Crill (1971) used the simpler Procion injection technique to determine the same quantity in 10 neurons. They claimed that in their preparations one of the conditions necessary for using the equivalent-cylinder model, namely, that $\Sigma d_i^{3/2}$ (where d_i are the diameters of dendrites emerging from a branch point) is a constant, was not satisfied. Therefore they adopted the alternate procedure of independently computing the impedance of each dendritic segment and summing them appropriately to obtain a calculated value of R_N, the total cell resistance. Since R_N is a simple function of R_M, the desired quantity, they determined a value for R_M as that which gave the observed R_N. Their value was somewhat lower than that of Lux *et al.*, and they found that if they used the equivalent-cylinder simplification on their neurons they consistently obtained the higher value.

2. *Retina*

By far the largest amount of serial section reconstruction at the electron microscope level has been done on the vertebrate retina. In fact, it was Sjostrand (1958), working on the receptor cell terminals of the guinea pig retina, who was the first to establish by example that serial section reconstruction with ultrathin sections was both a feasible and necessary extension of ordinary electron microscope analysis. The reasons for the popularity of retinal study are easy to understand. The sense of sight has occupied the minds of philosophers and physicians for longer than any other as being the key input to the human mind. Anatomically, the retina is a small piece of brain, well separated from the remainder of the mass, composed of relatively short fibers which are highly compacted and, moreover, which are arranged in a highly systematic fashion as shown first by the classic silver impregnation studies of Cajal and more recently by means of electrophysiological recordings. For a thorough review of the current knowledge of retinal structure, the reader is referred to the excellent and readable historical article by Stell (1972). What is presented here are some of the crucial studies using extensive serial reconstruction that have formed our present ideas regarding the detailed structure of the vertebrate retina. It is helpful conceptually to divide the synapses of the retina into three categories: (1) first order, in the outer plexiform layer, involving only interactions among receptor cells; (2) second order, also in the outer plexiform layer, involving synapses of bipolars and horizontals with receptors; (3) third order, in the inner plexiform layer, involving synapses among bipolars, amacrines, and ganglion cells, those responsible for the retinal output to the brain. In Fig. 17 are shown schematically all the major types of cellular interconnections that have been found to date in the vertebrate retina, with the exception of the first-order junctions. Historically, these were the first to be elucidated by serial section analysis in the often quoted article by Sjostrand (1958) on the receptor terminals of the guinea pig "pure rod" retina, in which the first truly three-dimensional reconstruc-

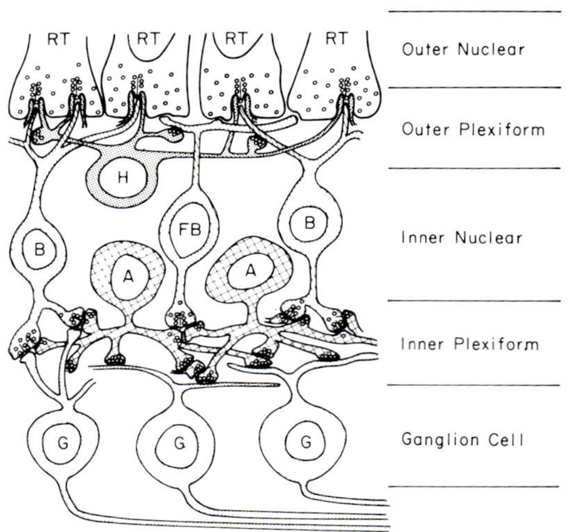

Fig. 17. Summary diagram of the arrangements of synaptic contacts found in vertebrate retinas. In the outer plexiform layer, processes from bipolar (B) and horizontal (H) cells penetrate into invaginations in the receptor terminals (RT) and terminate near the synaptic ribbons of the receptor. The processes of flat bipolar cells (FB) make superficial contacts on the bases of some receptor terminals. Horizontal cells make conventional synaptic contacts onto bipolar dendrites and other horizontal cell processes (not shown). Since horizontal cells usually extend further laterally in the outer plexiform layer than do bipolar dendrites, distant receptors can presumably influence bipolar cells via the horizontal cells. In the inner plexiform layer, two basic synaptic pathways are suggested. Bipolar terminals may contact one ganglion cell dendrite and one amacrine process at ribbon synapses (left) or two amacrine cell (A) processes (right). When the latter arrangement predominates in a retina, numerous conventional synapses between amacrine processes (serial synapses) are observed, and the ganglion cells (G) are contacted mainly by amacrine processes (right). Amacrine processes in all retinas make synapses of the conventional type back onto bipolar terminals (reciprocal synapses). Only conelike receptor terminals are shown for simplicity. (Modified from Dowling, 1970.)

tion of a rod synaptic ending (spherule) was presented. In this article it was reported that series of up to 40 250 A unstained sections of methacrylate-embedded retinas were analyzed to determine the structure of a typical rod spherule. These series were not of sufficient length to give unequivocal identification of many terminating processes which came from distant cell bodies, but they were long enough to show clearly that the synaptic endings of nearby cones (pedicles) give off fine processes which contact the spherules at their vitread poles, near where bipolar cell dendrites synapse. Each spherule, it was determined, receives a process from three or four pedicles from different directions, some adjacent and some up to 10 μm (three receptor rows) away. Sjostrand (1965) later clarified that these are not typical synaptic junctions, but that there are membrane darkenings in the region of apposition, and synaptic vesicles, although not clustered, are present on either side of the contact region. Missotten (1965), in a monograph published 7 years later, which describes the ultrastructure of the human retina as determined by extensive serial sectioning of Epon- rather than methacrylate-embedded material, found an identical result, adding that pedicle–spherule contacts are very close, about 100 A, and often show a desmosomelike darkening. In 1965 and 1969, Sjostrand (1965, 1969) extended his original observations on guinea pig receptors by analyzing longer series of sections (up to 300). Then he showed, in addition to lateral connections between pedicles and spherules, that the pedicles themselves are connected by mutual processes. Moreover, he found that a single pedicle process can contact both a spherule and another pedicle, in which case the process contacts the spherule not in the usual vitread region but on its lateral surface. Only one pedicle so terminates on any spherule, the others establishing contacts at the vitread pole where the processes from bipolar and horizontal cells enter the synaptic body invaginations. These more typical pedicle processes were seen to contact spherules of up to two receptor cells. In one series of 98 sections he constructed a complete connectivity map of these processes covering 22 spherules and 12 pedicles (Fig. 18). In this reconstruction there was no sign of any systematic arrangement of these contacts, although "over a certain limited area the long processes from the pedicles appear to be preferentially oriented in one particular direction" (Sjostrand, 1969). Missotten (1965) likewise showed in the human retina that cones are connected with each other and with many spherules by means of lateral extensions (6 to 12) of the pedicles, and that the contacts are at the region of the synaptic hillus on the spherules. Although he showed that a single pedicle extension can contact several spherules, he does not mention a process contacting both a pedicle and spherule, and states

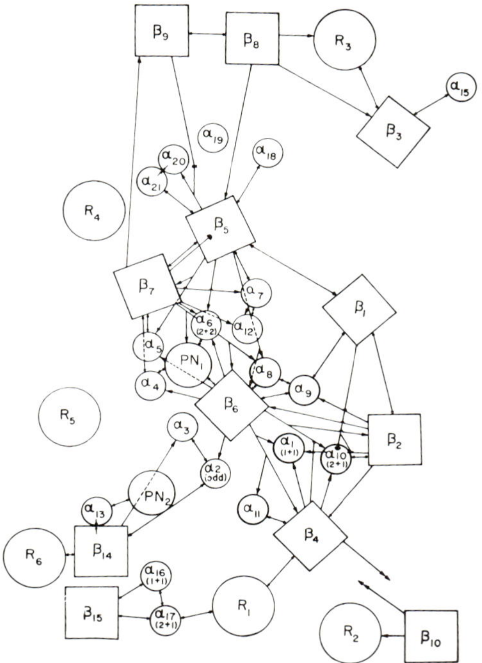

FIG. 18. Schematic presentation of lateral interreceptor contacts in a fair-sized area of the outer plexiform layer. Only the central part of the scheme is complete with respect to such connections. (From Sjostrand, 1965.)

that many spherules have no pedicle contacts at all. Lasansky (1971) demonstrated similar interreceptor contacts in the presumed all-cone retina of the turtle. With a procedure first used extensively by Kolb (1970), he decreased by a significant amount the length of the series of ultrathin sections necessary to prove his point by remounting and re-sectioning whole identified cells which had been impregnated with silver by the method of Golgi as modified to preserve ultrastructure. Thus from the light microscope he was able to determine that there are two types of cone pedicles, one sending 6 to 10 short (10 to 15-μm) extensions in all horizontal directions, and one sending out a single long (30- to 40-μm) extension which travels in different directions in neighboring cones. This single process subsequently branches near its terminus, giving rise to an equivalent number of dendritic endings. By serially sectioning only the endings, he determined that they abut against the lateral processes (horizontal cells) at the dyads (similar to triads but without the central bipolar dendrite) of other cone pedicles. He deduced that these terminations correspond to the dark-cytoplasm terminations near dyads in conventionally fixed material, and on the basis of this identification concluded that the pedicle extensions make "additional specialized contacts" of an *en passant* nature in the outer plexiform layer with other unidentified processes. Although it is clear that this technique of examining only "marked" cells cannot give a complete connectivity diagram of the type found by Sjostrand (1969), it provides a very valuable and fast method of describing the nature of and participants in the various types of synaptic structures observed.

Still another large effort to determine the exact nature of contacts between receptor cells was undertaken by Nilsson (1964) in the retina of the frog. In this case no processes from pedicles were found, and Nilsson defined interreceptor contacts as "direct relationships between synaptic bodies (terminals) over large areas without interposition of glial processes." No membrane darkenings, attachment zones, or synaptic vesicles were claimed to accompany these contacts, only a light gap 40–170-A wide. These results were obtained by scanning an area covering 215 receptor cells; 650 serial sections over a vertical distance of 30 μm were used to identify the synaptic bodies in the synaptic layer with specific cell types at the inner segment level. It was found that (1) each terminal of a red rod is in contact with three other terminals, those of another red rod, a single cone, and the principal member of a double cone; (2) each terminal of a single cone and each terminal of the principal member of a double cone is in contact with the endings of three red rods; (3) the synaptic endings of green rods and the accessory elements of double cones make no direct contacts with terminals of other receptors.

Complete understanding of the second-order synapses was made a little more difficult because of the seemingly great array of synapse configurations and complexities observed in single-section analysis. Systematic three-dimensional reconstructions of terminals ultimately revealed a fundamental unity in synaptic organization, and in addition showed exactly how each type of postsynaptic cell entered the synaptic complex. The basic structure of rod spherules was worked out by Sjostrand (1958, 1965, 1969) in the guinea pig, Missotten (1965) in the human, and Kolb (1970) in the monkey. Complete reconstructions of cone pedicles were given by Missotten (1965) in the human, and Pedler and Tilly (1965) in the monkey, and a more extensive connectivity analysis was provided by Kolb (1970). The rod spherule is simpler, containing only one invagination and far fewer invaginating synaptic fibers. In the guinea pig, Sjostrand found three distinct types of spherules (Fig. 19). Although his series were up to 300 sections in length, he was not able to identify

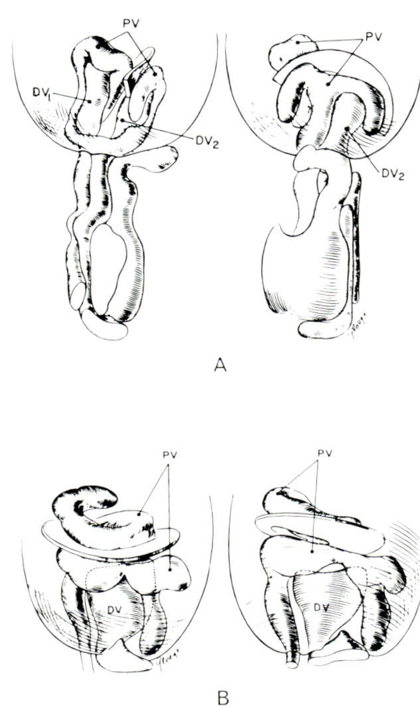

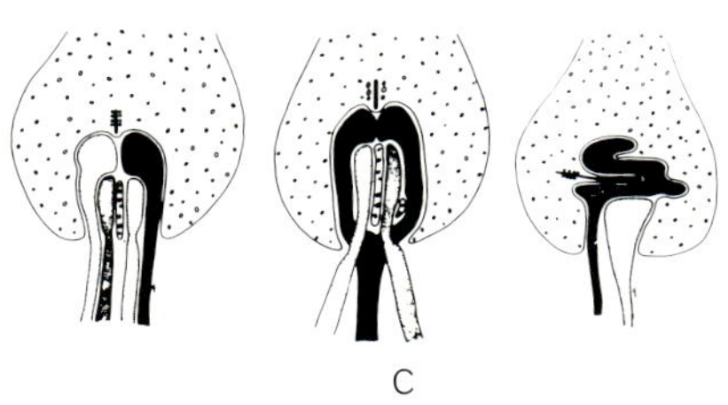

FIG. 19. (A) Three-dimensional reconstruction of synaptic contacts of alpha type (rod) receptor cells, the 1 + 2 type. Proximal vacuoles (PV) are branches of a single process entering through the vitread pole of the synaptic body. Two other processes enter independently and end as distal vacuoles (DV_1 and DV_2) in contact with the proximal vacuoles. (B) Three-dimensional reconstruction of an alpha type (rod) receptor cell synaptic body of the 1 + 1 type. Proximal vacuoles (PV) are formed from a process entering the vitread pole of the synaptic body and assuming a spiral course. A single process of a different type forms a single distal vacuole (DV). (C) Types of two-receptor cells. Left: 2 + 2 type, with two separate processes forming a pair of proximal vacuoles and two other separate processes forming a pair of distal vacuoles. Center: 1 + 2 type, with one process contributing to both proximal vacuoles, and two separate processes forming a pair of distal vacuoles. Right: 1 + 1 type, with one process forming the proximal vacuoles and another contributing a single distal vacuole. (From Sjostrand, 1969.)

the cells giving rise to the lateral, more deeply penetrating dendrites (proximal vacuoles). The central dendrites (distal vacuoles) were repeatedly identified as originating from rod bipolar cells, although it was not clear whether they came from the same or different bipolars. Missotten found a larger variation in the number of fibers invaginating into the spherules, from two to seven. Two to three of these correspond to Sjostrand's proximal vacuoles, and in agreement with him he found that they occupy a larger volume and a lateral position within the invagination, have a complicated form, and often contain small vesicles or other inclusions. Although his series were at most 70 sections long, he states that these processes are terminal extensions of the horizontal cell telodendria. The remaining invaginating processes, from one to five, similarly are smaller and penetrate less deeply and were traced to rod bipolar cells. Thus each spherule can in principle be linked to from one to five different rod bipolars. In disagreement with Cajal (1933) and Polyak (1941), he found no case in which the rod bipolar dendrites synapsed onto the sides of cone pedicles or rod spherules. In her analysis of the monkey retina, Kolb (1970) avoided the necessity of having to take excessively long series of sections by cutting only the dendritic terminals of selected Golgi-impregnated cells. These cells were cut out of plastic-embedded retina and remounted for thin sectioning. By complete analysis of the dendritic spread (15-μm diameter) of a small bipolar, she determined that it contacted 33 rod spherules, by no means all in its dendritic field, and no pedicles. In addition, only a single dendrite was found to penetrate each spherule as one of the central processes. Since, like Missotten, she found two or more central processes per spherule, she concluded that the unstained processes she observed must come from rod bipolars with overlapping fields. By serial sectioning of Golgi-impregnated horizontal cell axons, she further showed, like Missotten, that they went only to rod spherules and formed the lateral processes of the invaginations. In all cases a spherule had only one of its lateral processes stained, implying that it received fibers from two horizontal cells, a conclusion Sjostrand and Missotten were unable to make conclusively even from their extensive series.

The cone pedicle has been less studied in detail, and although the basic features described by Sjostrand (1958) concerning its extensions to rods and other cones were confirmed in the pigeon (Cohen, 1963) and gray squirrel (Cohen, 1964) by Cohen, in the gecko by Pedler and Tansley (1963), and in the alligator by Kalberer and Pedler (1963), there was either no or conflicting information concerning the identities of the processes taking part in the synapses. Missotten (1965) was the first to make a serial section analysis of the cone pedicle. In agreement with previous single-section work, two types of synapses were found, one in which the postsynaptic dendrite only slightly indents the pedicle (several hundred per cone), and one that resembles the synapse of the rod in having an invagination lined with two large, deeply penetrating processes (but without vesicles) and a marked presynaptic ribbon, but unlike it in that it contains only one other process inserted between them. This unit, on the order of 25 per cone, was termed the synaptic triad. Unfortunately, the unanimity concerning the identity of the processes contacting spherules is not to be found in the case of cone pedicles. The conclusions of Missotten were that (1) flat bipolars form the slightly indented contacts on several cones; (2) midget bipolars form some of the central processes of triads of a single cone; (3) horizontal cell dendrites form some of the central elements of the triads in several cones; and (4) horizontal telodendria are "unequivocally" the lateral processes in the pedicle. Kolb (1970), however, in a more recent and well-documented work, made the correspondingly different conclusions as a result of extensive serial sectioning of Golgi-impregnated monkey retina neurons that (1) a flat bipolar forms about 25 of the slightly indented contacts on each of about 6 cones, all the cones in its receptive field being contacted, and a new cell type, a flat midget bipolar, forms about twice as many slightly indented contacts on only one cone. These midget contacts are invariably paired at the base of the invaginations; (2) the previously identified midget bipolar, now called the invaginating midget, forms all the central processes of the triads in a single cone; (3) horizontal cell dendrites invariably form lateral processes, usually one but occasionally both, of several triads in each of from 7 to 12 neighboring cones (see Fig. 20), all the cones in its receptive field being contacted. This is in fundamental agreement with the results of Lasansky (1971) and Stell (1967); (4) horizontal telodendria were never observed to contact pedicles. Her results on the horizontal cells were confirmed and extended in a later article (Boycott and Kolb, 1973). Whether these rather striking differences are due to improper identifications in Missotten's admittedly short (70 sections) series, or sampling errors compounded by the capriciousness of Golgi staining, has yet to be resolved; but it does indicate that if serial sectioning of unmarked cells alone is to succeed, it must clearly be a much larger effort than has been attempted so far.

Patterns of intercellular connectivity forming third-order synapses in the inner plexiform layer are much less well understood than in the outer, largely because the cellular processes there are much longer. Thus the short series reported on by Missotten (1965) in his monograph proved useless for undertaking a serious connectivity analysis of the inner plexi-

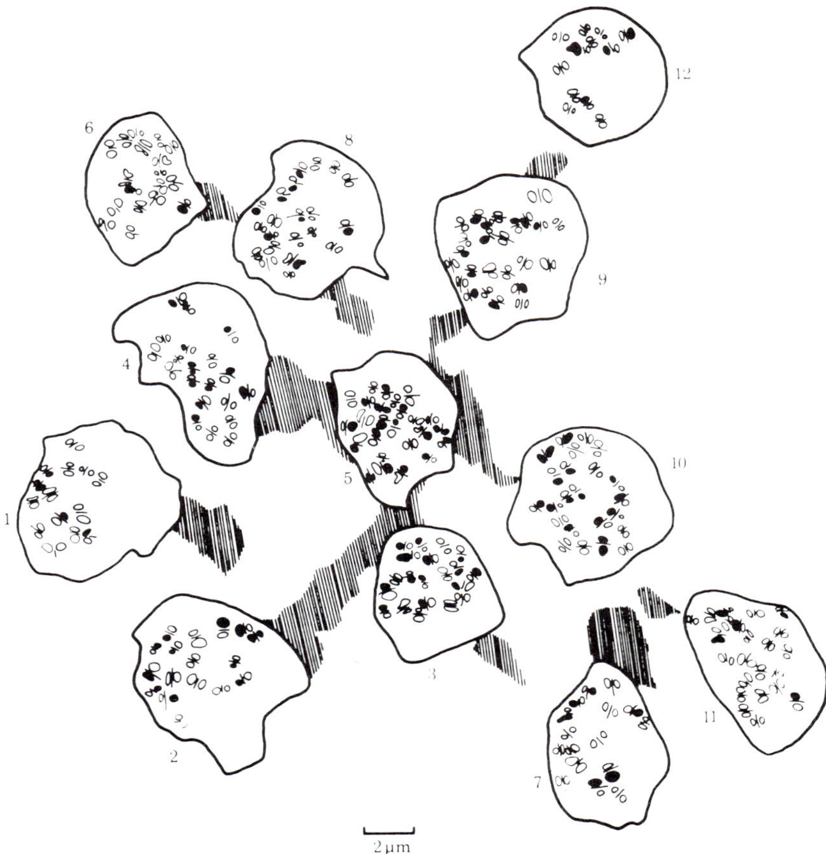

FIG. 20. Tracing from electron micrographs of a type-B horizontal cell in flat section. The cell body lies below pedicle 5, and some of the main dendrites are represented by the shaded areas. The number of triads seen for each pedicle are indicated by the ribbon and two lateral elements; the contributions of dendritic terminals from the horizontal cell are indicated by the black lateral elements. (From Kolb, 1970.)

form layer, and so far no one has attempted using the Golgi method with anything like the devotion of Kolb (1970) in her study of receptor synapses to map out the receptive fields and type and number of contacts engaged in by even the cells of primary interest, the ganglion cells. Extensive serial section reconstruction has so far been limited to determining the nature and distribution of synapses on bipolar telodendria. The first analysis of this type was reported by Allen (1969), in which he reconstructed the entire axonal surface of a bipolar of unidentified type in the human retina from 67 consecutive sections. He identified two types of efferent synapses in the bipolar telodendria: ribbon synapses, consisting of a characteristic synaptic ribbon within the bipolar cell with clustered vesicles, and conventional synapses, those generally observed with an accumulation of vesicles at the darkening of the synaptic membrane. In contrast to the first type, it was found to have only one postsynaptic fiber. The ribbon synapse was shown to consist not just of the two postsynaptic elements usually identified in single-section analyses, but up to four. Moreover, neighboring ribbon synapses often share postsynaptic fibers, with a single fiber taking part in as many as four synapses. Its characteristic ribbon was shown to be a disclike structure similar to that observed in rod spherules and described by Sjostrand (1958). Further complexity is added since each synaptic ribbon examined was associated with a complex of individual efferent and afferent synapses. No pattern was found to characterize these complex synaptic areas, except that they resulted "from a fairly simple interlocking system of repetitive pairs of triples joined by a common member, and, at fairly regular intervals, with a single member joining smaller groupings into larger complexes." Figure 21 shows his completely reconstructed bipolar axon and

---

FIG. 21. [pp. 373–375] (A) Three dimensional drawing of primary retinal bipolar telodendrion, perikaryon, and proximal segments of dendrites. Reconstruction was made from montages of each of 67 consecutive serial sections believed to include almost the entire cell. The lighter area over the nucleus represents the cut edge. The Golgi apparatus (lower edge of nucleus) obscures centrioles which lie nearby. The mitochondria are accurately represented in number, shape, and distribution within the cell. Toward the upper edge of nucleus (shown as semitransparent) there is an oval nucleolus; several scattered mitochondria in cytoplasm surrounding the nucleus on opposite side of cell are shown. The cell processes shown as surrounding the telodendrion are those that form synaptic contact with the cell or are glial and in extensive intimate association. The approximate size, orientation, and location of the synaptic ribbons are shown (synaptic ribbon in distal axon shown at arrow). Synaptic vesicles are also shown in the locations where they occurred, but are not drawn precisely to scale. (B–D) Topographic projection of the surface of each of the boutons of the bipolar telodendrion, depicting areas of contact of the various surrounding processes (identified by number). Each ribbon synaptic complex is represented by a black circle and a primed number. The simple efferent synapses are shown in squares, each identified by a primed number as well. The projection was made from montages of micrographs printed at ×46,000 and was drawn to the same scale. The boutons of the telodendrion exist in different planes; each is shown as it appeared in the sequence of the numbered serial sections of the series (1 to 67 shown as dotted lines on scale at bottom). The orientation of the boutons with respect to the main cell body is shown within the enclosed segments of the outline drawing of the entire cell. The stippled small areas indicate sites at which boutons are connected. In most cases glial processes were not numbered separately, but all glial contacts are identified. (From Allen, 1969.)

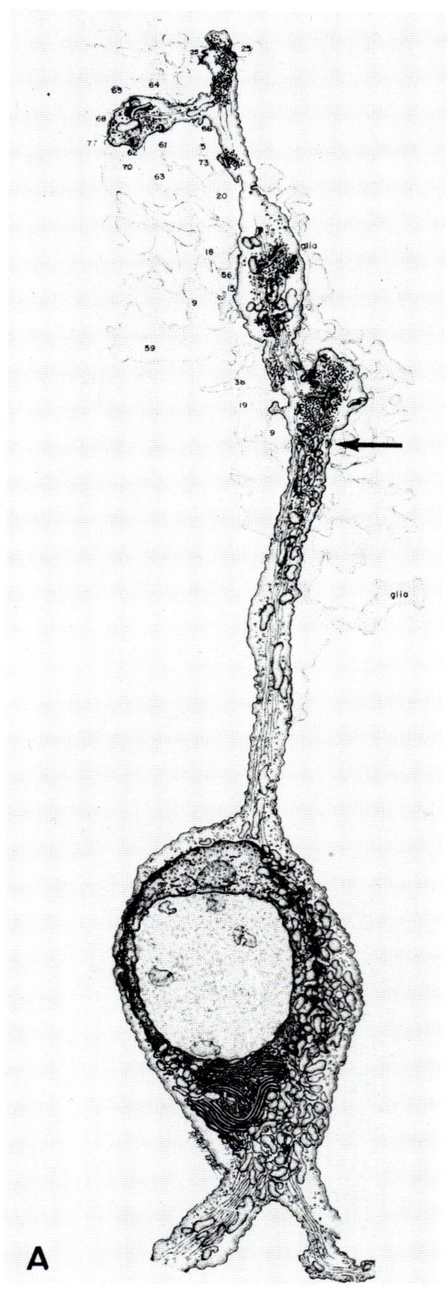
25
23
65
64
68
66
77
62
61
5
70
73
63
20
glia
18
56
15
9
59
38
19
9
glia
A

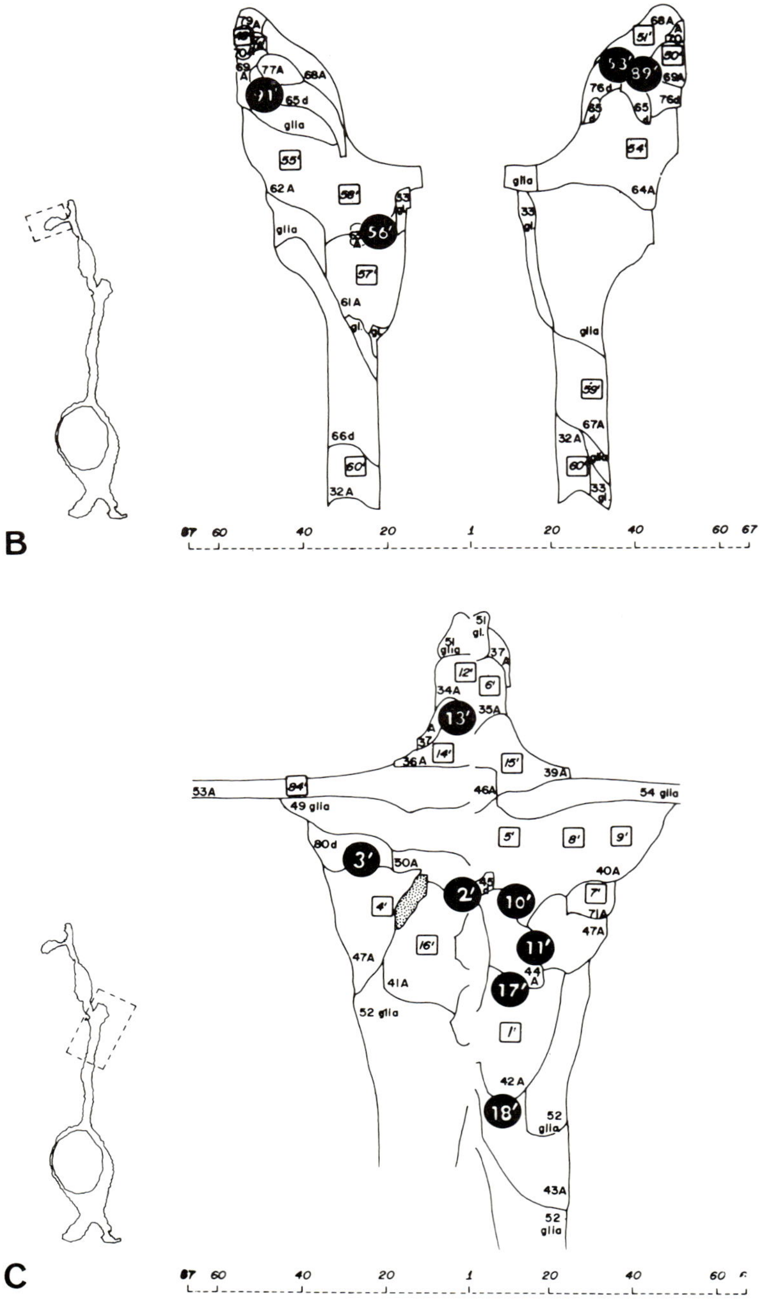

B

C

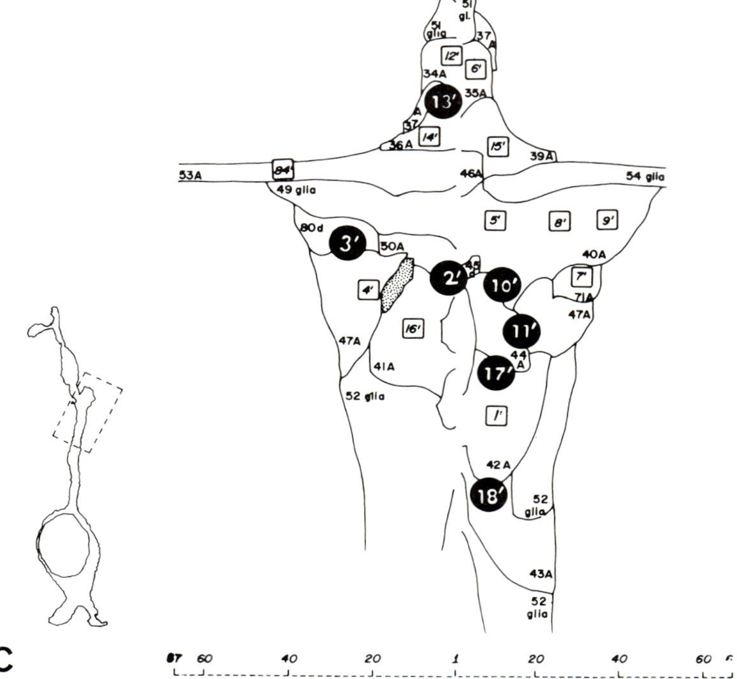

Fig. 21B and C. See page 372 for legend.

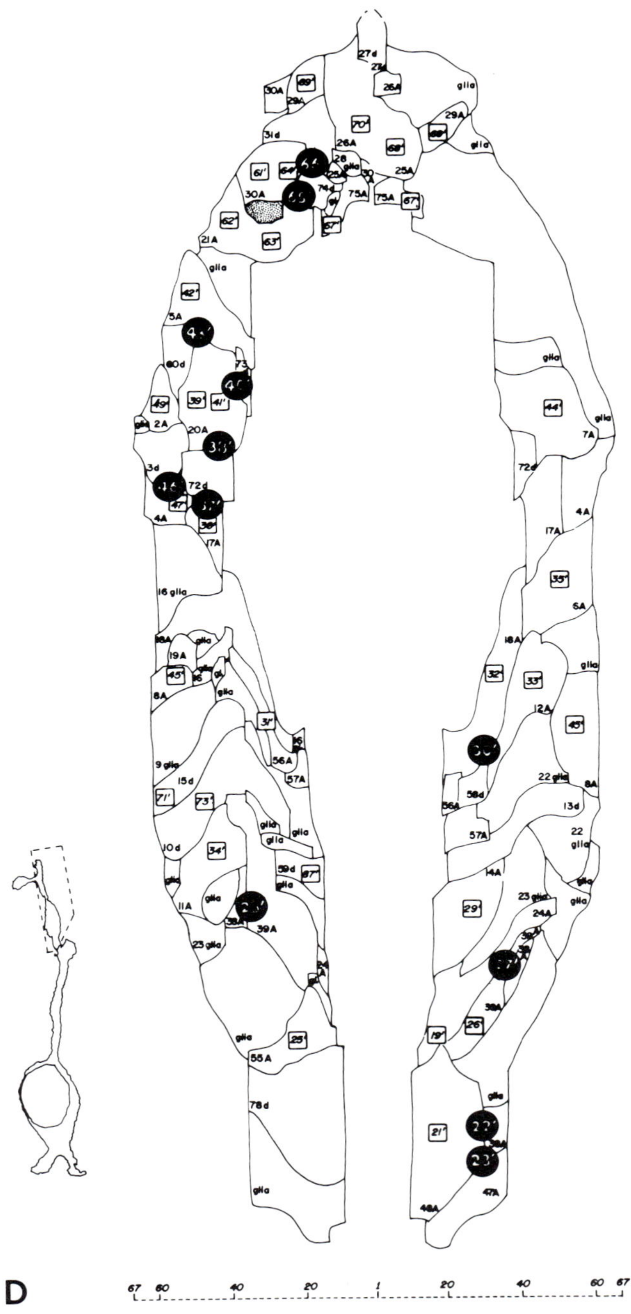

D

FIG. 21D. See page 372 for legend.

the topographic surface maps of synapses derived from it. A more recent article by Witkovsky and Stell (1973) reports the same type of reconstruction of a larger number of identified mud puppy bipolars, but over only part of their axons. These investigators too found that complete sectioning of a ribbon synapse reveals considerably more complexity than had been determined by single-section analysis (e.g., Dowling and Boycott, 1966). A slight variation from the observation on human bipolars is that some presynaptic ribbons are associated with no postsynaptic element and that 30% have only one. Whereas Allen (1969) and Foos and Miyamasu (1973) found that roughly 50% of human amacrine cell synapses are reciprocal, Witkovsky and Stell found only 5–25% are so involved in the mud puppy. This is subject to their qualification that only a fraction of any given bipolar axon was studied. By calculating the synaptic surface density from their own surface maps and from the results of Dubin (1970) they made the interesting observation that, whereas in all vertebrate retinas the density of ribbon synapses is nearly the same, the density of amacrine conventional synapses varies widely. Since it has also been shown that experience can alter the number of these synapses but not the ribbon type (Fisher, 1972; Fifkova, 1972), these investigators considered the interesting possibility that amacrine conventional synapses are "floating," while ribbon synapses are genetically hard-wired into the retina and not subject to modification.

3. *Cortex*

Approaches to the determination of fiber pathways and synaptic patterns at higher levels of the vertebrate central nervous system have generally not made use of the kinds of extensive serial section analysis described for spinal and retinal neurons because of the large distances involved. The most useful techniques for central nervous system pathway reconstruction have relied on specific silver impregnation methods for degenerating axons and presynaptic terminals (Nauta and Gygax, 1954; Fink and Heimer, 1964; Colonnier, 1964). These methods involve making lesions at the sites of location of the presynaptic perikarya and fixing the tissue some defined period of time subsequent to the operation. Serial paraffin sections are normally taken at thicknesses of from 10 to as many as hundreds of microns, but a frequently used staining procedure involves the silver impregnation of every fifth to tenth section and staining for cells in sections adjacent to each silver impregnated one. In this way degenerating fiber pathways can be followed with respect to the groupings of cellular landmarks both in transit and at their final destinations. Serial section analysis in which every section is used has been shown to be useful in regions like the cortical and precortical areas which show an easily discernible laminar organization either anatomically or physiologically and for which there is a known topographic projection of the more peripheral neurons onto the more central ones. Thus various brainstem nuclei, the visual and somatosensory cortices, and the olfactory bulb are amenable to such investigations.

An example of how the mapping of degenerating terminals can be used to demonstrate a fundamental architectural principle in the cortex has been provided by Hubel and Wiesel (1972). There have been large numbers of studies concerning both the physiological and anatomical organization of the mammalian visual pathways. It is now accepted that the retinal cells project in topographic order onto the thalamic relay neurons of the lateral geniculate bodies. The geniculate of each side receives input from the ipsilateral temporal half-retina and the contralateral nasal half-retina. These inputs remain segregated at the geniculate in the sense that the nucleus is organized into several discrete layers, each layer receiving the retinal ganglion cell terminals of either ipsilateral or contralateral origin but not both. In turn, the geniculate neurons convey their impulses to the visual cortex in the occipital lobe of the brain. This cortical area is also stratified horizontally into alternate cell-sparse and cell-rich layers. Hubel and Wiesel (1962) demonstrated by physiological recording in the cat that the cortical cells in layer IV, to which the afferent geniculate terminals mainly project, "have the simplest properties and show the least intermingling of inputs from the two eyes." Binocularly driven "complex" visual cortical neurons are located in layers above or below layer IV. The visual cortex is also partitioned vertically into two overlapping but independent systems of columns. In the first of these, cells with a similar orientation preference are grouped together. In the second system cells are grouped according to eye dominance, the cells of one column mostly favoring the left eye, those of the next the right. Layer IV, whose cells are strictly monocular, is thus divided into a mosaic of left eye and right eye patches (Hubel and Wiesel, 1963). Thus physiological recordings indicate that the separation of left eye and right eye inputs at the geniculate layers seems to be continued as vertical column separation in the geniculocortical projection, at least in cortical layer IV. The fairly clear-cut anatomical layers of the geniculate, especially in the monkey, correlate well with the physiological layering, but such is not obviously the case in vertical columns of layer IV of the visual cortex.

In a series of experiments in the macaque, Hubel and Wiesel (1969, 1972) attempted to provide an anatomical demonstration of the cortical ocular dominance columns. This was accomplished by making small lesions in one or a few geniculate layers, and in each case recording the locations of degenerating geniculate terminals in the cortex. Unlike most other degeneration studies the degree of spatial resolution needed in this work required that the whole area containing degenerating terminals be carefully reconstructed from serial sections so that the specific spatial patterns of degeneration resulting from geniculate lesions of varying extent could be compared. Lesions were made by lowering an electrode into the geniculate and heating the tip when it had reached the desired layer as determined by recording geniculate responses to retinal stimulation. To eliminate uncertainty in the identification of any single stained dot as a degenerating terminal, only cortical areas where the stained images occurred in clusters were considered.

What was desired was an *en face* view of layer IV of the cortex as if flattened into a plane. Because of the folded nature of the cortex such a view was obtained most easily by sectioning perpendicular to this direction, and then graphically straightening the lines of degeneration in layer IV as they followed the fold. The cortex was sectioned serially at 30 μm in the coronal plane, and all sections were stained for degenerating axons. Each grouping of degenerating terminals in cortical layer IV was traced on graph paper, using a new line for the micrograph of each successive section. Blood vessels were used in successive sections as alignment aids. When the geniculate lesions involved two layers with different eye input, the pattern of cortical degeneration appeared as a continuous band or bands in layer IV interrupted only at the ends. By contrast, when a large lesion was made in one geniculate layer, only a regular pattern of terminal degeneration occurred. Bands of degeneration were interrupted at regular intervals by segments of about equal extent where no degenerating terminals were seen (Fig. 22). This distinct repeating pattern of bands and interbands was taken to represent the anatomical correlate of alternating eye inputs to layer IV. These investigators proposed that an orthogonal arrangement of these cortical ocular dominance columns with the second vertical column system of orientation preference of cortical neurons would be consistent with this and other data (Fig. 23). A study of the postsynaptic neurons involved in these cortical ocular dominance columns is currently being undertaken by Sterling (personal communication). It is known from single-section analysis that axons from the geniculate synapse heavily onto spines and dendrites in cortical layer IV. To identify the types of postsynaptic cells (e.g., pyramidal or stellate) and the locations of their cell bodies, serial section cinematography with electron micrographs is being used to follow fibers postsynaptic to degenerating geniculate fibers.

The olfactory bulb has recently been the object of studies combining silver impregnation of both normal and degenerating material with

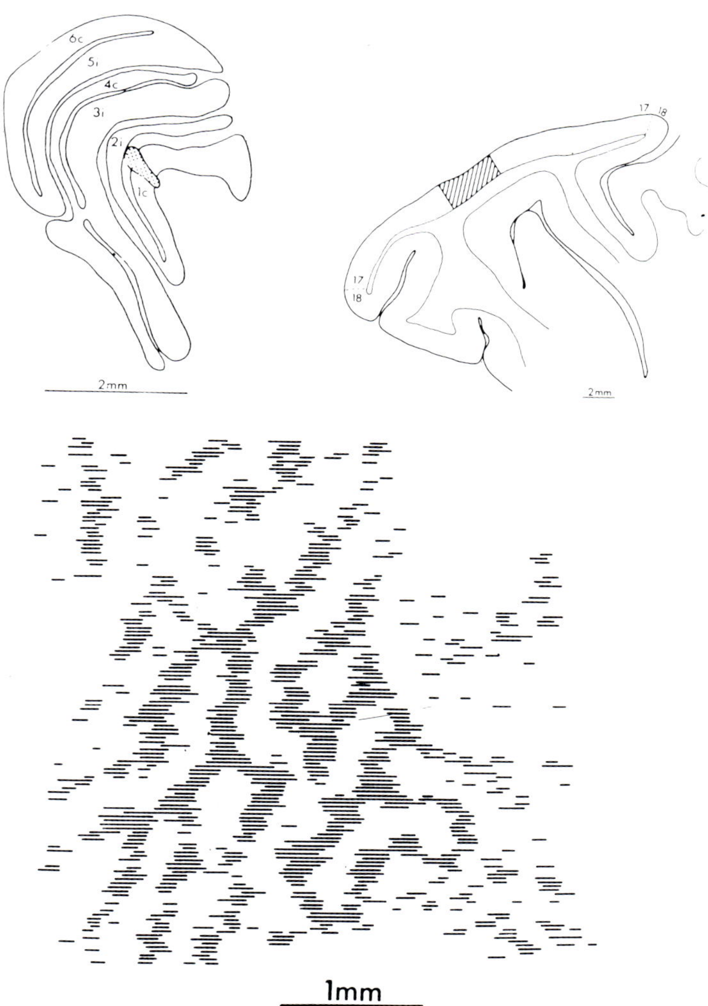

FIG. 22. Monkey 41. Upper left: Tracing of lateral geniculate body from coronal section showing the extent of a lesion in the most ventral layer. Upper right: Site of degeneration (indicated by shading) in striate cortex. Parasagittal section about 15 mm from midline. Left is posterior. Lower: Reconstruction of region showing terminal degeneration in layer IVb. Each horizontal line segment represents the degeneration in a single band of a single section. Adjacent 30-μm sections were plotted on graph paper. Curved sections of the gyrus were straightened graphically. Anterior is to the right, medial is up. (From Hubel and Wiesel, 1972.)

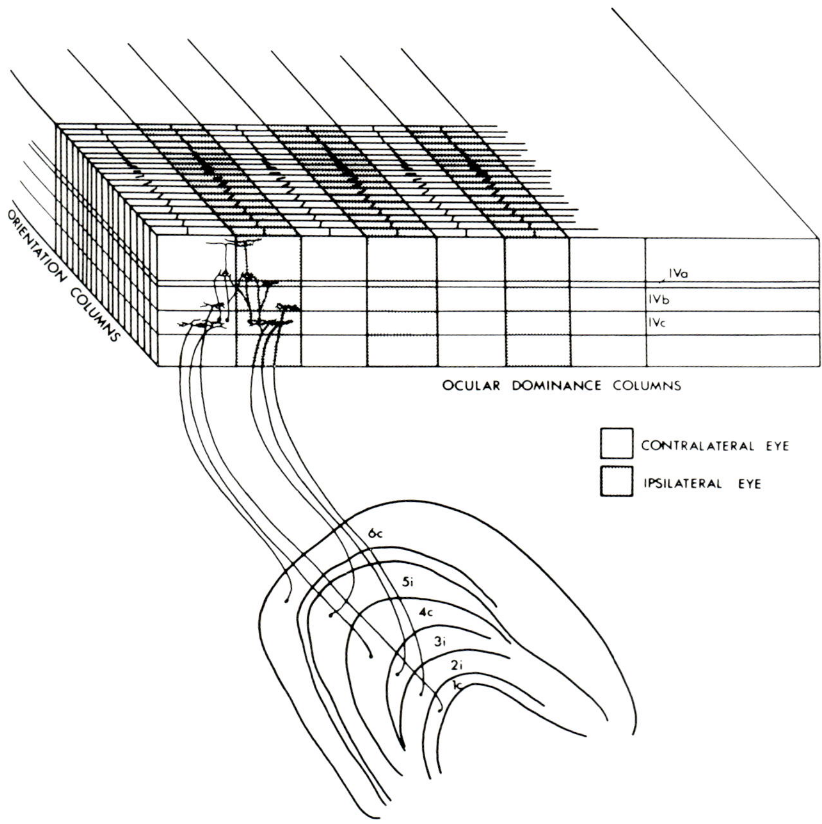

FIG. 23. Schematic diagram showing a possible relationship between ocular dominance columns and orientation columns, assuming the orientation columns are also long, narrow, parallel slabs and that their arrangement is very orderly. Note that the width of these orientation slabs is much less than that of the ocular dominance columns. A complex cell in an upper layer is shown receiving input from two neighboring ocular dominance columns, but from the same orientation column. (From Hubel and Wiesel, 1972.)

serial section electron microscopy. A large number of reports at both light and electron microscope levels had previously demonstrated the well-defined laminar organization in the bulb. Incoming olfactory nerves with cell bodies in the nasal mucosa terminate in synaptic glomeruli, mainly on the dendrites of mitral and tufted cells. The axons of the latter cells form the secondary olfactory tracts, conveying the output of the bulb more centrally. In this peripheral layer of the bulb, there is also an intrinsic interneuron present, the periglomerular cell. In deeper layers of the bulb, two types of interneurons are found: the granule cells, which participate in an extraglomerular reciprocal synapse with mitral and tufted cells, and a short-axon cell about which little is known. The granule cells show a distinctive morphology in Golgi impregnations, having what appears to be two dendritic systems, one ramifying into the external plexiform layer beneath the glomeruli, and the other with a less complex branching into the ventricular or deeper margins of the bulb. These cells have no clearly distinguishable axonal processes, and in addition are the only cells in the bulb that show a system of dendritic spines and varicosities.

Physiological studies had demonstrated the existence of inhibitory mechanisms acting on mitral and tufted cells, which can be activated by stimulation of three different groups of central afferents to the bulb (centrifugal fibers) and which are apparently mediated by interneurons. In the absence of detailed synaptological information on this system, Price and Powell (1970a,b,c,d) undertook a combination of silver staining and serial section electron microscope studies in the rat designed both to clarify certain features of granule cell morphology and synaptic relations with mitral cells and to establish the sites of synaptic termination of the centrifugal fiber systems. Series of up to 150 sections were made and wax models built to demonstrate the three-dimensional synaptic relationships. In following the transition from the granule cell perikaryon to the initial portions of both groups of processes emanating from it, these investigators could find no evidence of axon initial segment ultrastructure. It was confirmed in detailed observation of more distal processes that this intrinsic neuron resembles the retinal amacrine cell in having no demonstrable axon. Spines were found over the whole neuronal surface, and differences in ultrastructure were observed between the spines and varicosities of deep dendrites and the so-called gemmules (Rall *et al.*, 1966) of the external dendrites, which participate in the reciprocal synapses. In serial sections it was possible to determine that all spines received at least one synapse. Those synapses in which granule cell spines, gemmules, or varicosities are the postsynaptic elements are always of the same morphology and are called type A for convenience. A second morphological type of synapse, type B, in which the granule cell was also postsynaptic, showed a complementary distribution to the first, being found almost exclusively on dendritic shafts and perikarya. The more detailed serial section analysis of gemmules showed that the reciprocal synapse between granule cell gemmules and the dendrites (and perikarya) of mitral and tufted cells consisted of an A type in which the granule cell is postsynaptic and a C type in which it is presynaptic to the mitral cell, the latter type differing from B only in average synaptic vesicle size (Fig. 24).

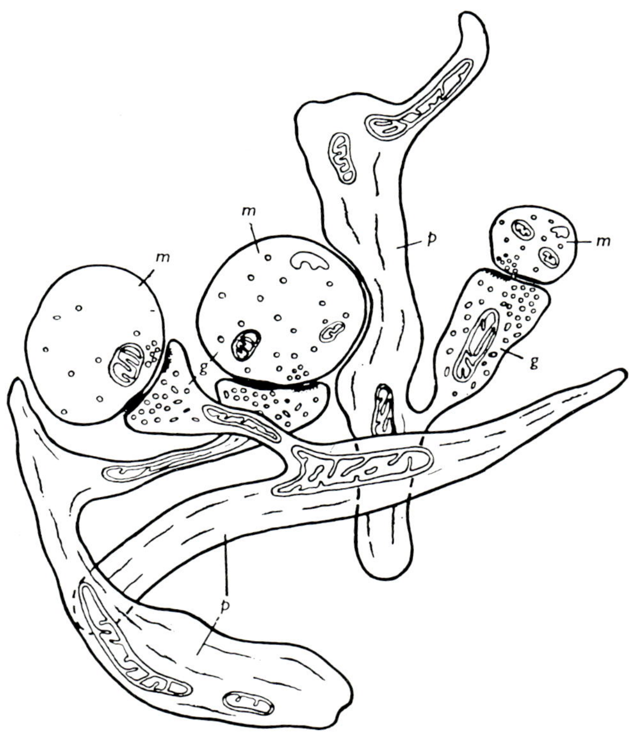

FIG. 24. Illustration of the relationships of peripheral processes, gemmules, and mitral cell dendrites. Reconstruction from 14 serial sections. g, Gemmule; m, mitral cell dendrite; p, peripheral process. (From Price and Powell, 1970a.)

In experiments in which lesions were separately made in each of the three centrifugal pathways, a good correlation was made with the synaptology of unoperated animals, suggesting a tentative "wiring diagram" for this region of the olfactory bulb. These investigators emphasized that "for the identification of the postsynaptic profiles contacted by degenerating terminals, the use of serial sections proved invaluable." They found that the degenerating terminals of all three centrifugal pathways ended on the dendritic spines, gemmules, and varicosities of granule cells by way of type-A synapses, corresponding perfectly with the distribution of this type in unoperated animals. Moreover, in simultaneous lesions of all three pathways just outside the bulb, no degeneration was seen in terminals that formed type-B synapses onto granule cells, and many type-A terminals had likewise not degenerated. The conclusion drawn was that these intact terminals must come from intrinsic neurons. The suggestion was made on the basis of the known type-A contribution of mitral cells to the reciprocal synapses that the remaining undegenerated type-A synapses should derive from mitral and tufted cell collaterals, leaving the short-axon cell as the possible intrinsic source of the undegenerated type-B terminals to granule cell dendritic shafts and perikarya. A further clarification was the finding that, other than in the glomeruli, mitral and tufted cells seemed to be postsynaptic at reciprocal synapses but nowhere else within the bulb. Thus the serial section-derived synaptology was not only consistent with but also elaborated on the background of physiological knowledge by suggesting that the granule cell subserves the role of final common path through which both centrifugal and intrinsic influences act on the output cells. A more recent report (Pinching and Powell, 1972) has clarified some aspects of the distribution of centrifugal fiber terminals in the bulb, and a serial section study of the glomerular region (Pinching and Powell, 1971a,b,c) has been valuable in confirming synaptic interrelations in this area.

White (1972) used large numbers of series of about 60 sections each to classify types of dendrodendritic synapses in the glomerular region of the rat. The dendritic nature of a process was inferred from ultrastructural criteria and, because of the short lengths of the series, none of the dendrites could be traced back to the cells of the origin. White observed synapses between dendrites that received afferent input from presumed olfactory axons and others that did not, but never directly between two dendrites that were both postsynaptic to olfactory axons. The various dendrodendritic contacts encountered led White to postulate that the synapses between dendrites with afferent input and those without might occur between dendrites of mitral and tufted cells and those of the periglomerular interneurons, respectively, as this scheme would then be indicative of a similar type of interneuron-mediated inhibition as definitively found in deeper layers of the bulb by Price and Powell (1970a,b,c,d).

Observations made from serial section electron microscopy on the olfactory bulb of the cat (Willey, 1973) have correlated well with those of the aforementioned investigators on the rat. Although he did not trace dendrites to their cell bodies of origin, Willey described similar dendrodendritic reciprocal synapses. An unusual finding was the occurrence of reciprocal synapses between dendrites, presumably of interneuronal origin, and the perikaryon of mitral cells which were easily identifiable by their large size and characteristic ultrastructure. The mitral cell contribution to this reciprocal synapse consisted of a cluster of vesicles apposed to a membrane densification occurring within the perikaryal cytoplasm itself.

In an investigation into developmental interactions in the rat olfactory bulb, Hinds (1972a,b) used Golgi staining preliminary to a serial section electron microscope study to define stages in the initial differentiation of mitral cells. The silver impregnations indicated that mitral cell morphological differentiation could be staged on the basis of cell body position and orientation in the neural tube and by the nature and orientation of cell processes. Hinds then used series of up to 250 thin sections to attempt to correlate the ultrastructure of developing processes with the stages assigned from light microscopy. He defined the stage at which the mitral cells of any region could be considered to possess a true axonal process. An unexpected finding was the extent of precocious olfactory nerve penetration into the developing bulb demonstrable in serial sections. Of significance was the observation that these axonal mitral cells seemed to occur only in areas of the bulb into which olfactory axons had previously penetrated, leading to the suggestion of an inductive influence of these growing axons on the differentiation of mitral cells, their eventual primary synaptic sites.

III. Invertebrates and Cell Structure

A. Invertebrate Nervous Systems

Serial section reconstruction in invertebrates has been confined almost exclusively to the nervous system. Invertebrates are attractive neurophysiologically and neuroanatomically because they provide examples of simple yet highly integrated behavior controlled by small, well-delimited nervous systems with a tremendously decreased number of neurons compared to vertebrates. In the field of embryology, it is also becoming apparent that general principles of neurogenesis can be more easily studied at the detailed single-cell level in invertebrates. Much of the neuronal reconstruction work has been done with the light microscope to determine the geometry of individual neurons, using thick (10- to 15-μm) serial sections following the intracellular injection of dyes. The analyses of long series of thinner sections, both at the light and electron microscope levels, for the purpose of determining interneuronal connectivities have concentrated mainly on the arthropod visual system.

1. *Light Microscope Studies*

The idea of marking cells with vital dyes dates back to early embryological studies of cellular migration in gastrulation. The need for specifically intraneuronal dyes arose from a desire to mark for subsequent identification the particular cells in a network from which physiological recordings had been made. Kerkut and Walker (1962) were the first to introduce such a technique, one that involved the iontophoresis of ferrocyanide from the recording microelectrode and its subsequent reaction with saturated ferric chloride to precipitate Prussian blue dye in the neuron. The injected cell appeared to retain its normal electrical properties, and the dye often marked a segment of axon that could then be traced through several successive histological sections. The technique was later used by Sandeman (1969) to correlate physiology and gross axonal anatomy of the giant motoneurons involved in the crab eye withdrawal reflex. Other early iontophoretic intracellular staining techniques that never achieved widespread acceptance involved the use of such dyes as fast green and methylene blue. Thomas and Wilson (1965, 1966) first used these to identify the general location of Renshaw cell bodies in the cat spinal cord by light microscopy, but made no mention of the dyes' ability to stain processes. Kato *et al.* (1968) showed that cat spinal motoneurons could be identified in 40-μm frozen sections after injection of fast green, and subsequently processed for electron microscopy. This stain penetrated for an undisclosed distance into finer cell processes, but was always blocked at the axon hillock whether injected on the soma or axon side. Since the only aim of these studies was to mark the soma position or main axonal courses of cells from which intracellular electrophysiological recordings had been made, the required resolution was minimal, and the general aspects of neuronal morphology were available from whole-mount viewing.

It was not until the fluorescent dye Procion yellow was introduced by Stretton and Kravitz (1968) that intracellular staining achieved widespread use in the study of neuronal geometry by light microscopy. These investigators were interested in finding a technique that would allow visualization of the finest branches of fibers in a complex neuropil while changing neither the resting potential nor the characteristics of the action potential of the cell. The possibility of making a complete reconstruction of the branching pattern of injected neurons at the light microscope level with a greatly reduced number of sections in comparison to what would be required for electron microscopy was explored in the lobster abdominal ganglion. Ten-micrometer serial sections were made of injected cells in different animals which were thought to be homologous on the basis of physiological recordings and soma position. The outlines of the injected neuron in each section were traced from the micrographs and the tracings aligned so that a two-dimensional projection of the cell could be made (Fig. 25). The homology of the injected cells was supported by the striking similarity of their branching patterns in terms of the number and relative positions of individual main branches, although

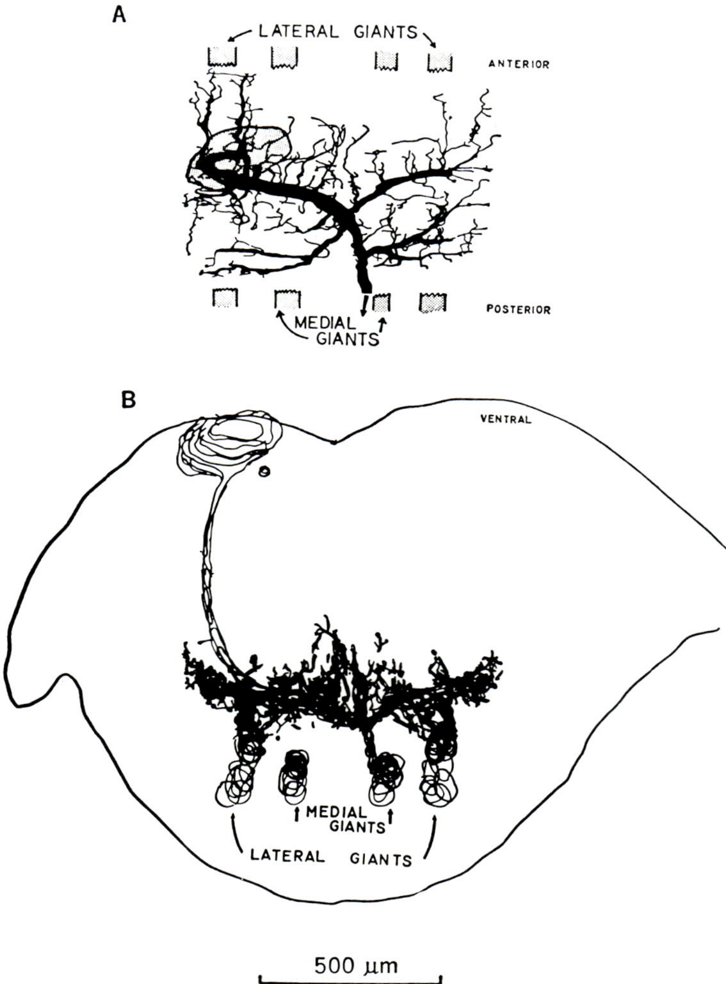

FIG. 25. (A) Plan (horizontal projection) of cell I_2, an inhibitory motoneuron to fast flexor muscles in the lobster second abdominal ganglion. The cell body is stippled; the axon continues in the direction of the arrow to the third root, where it leaves the ganglionic chain to innervate fast flexor muscles. The position of the medial and lateral giant fibers is indicated at the two ends of the ganglion. (B) Front projection of I_2. The outlines of the medial and lateral giant fibers have been drawn at 100-μm intervals. (From Stretton and Kravitz, 1968.)

some microheterogeneity was found in the finer branches. The finest branches filled by the dye were on the order of 1 μm in diameter, and it was impossible to determine whether these branches were in fact neuronal terminations or whether smaller-diameter subdivisions existed but were not filled with dye.

Since its introduction, serial section reconstruction of Procion-injected cells has served to identify muscle terminations of specific motoneurons (e.g., Stuart, 1970; Davis, 1970), to compare the reproducibility of axonal pathways and more complex branching patterns between homologous neurons in different ganglia (e.g., Kravitz *et al.*, 1968; Selverston and Kennedy, 1969; Wilkins and Larrimer, 1971), and in general to verify anatomically connectivity patterns assumed from physiological data. Bentley (1970) reconstructed the main axonal pathways of 40 locust flight motoneurons, certain of which acted synergistically with one another, to infer that this action is mediated directly rather than by means of interneurons. The most complete series of investigations of this type came from the work of Kennedy and co-workers in a detailed study of the relationship of command interneurons to the coordinated output of motoneurons in each segmental ganglion of the crayfish. These motoneurons control the fast flexion of abdominal segments in swimming behavior. The physiology of the abdominal cord was studied by Hughes and Wiersma (1960a,b), who found that continuous intersegmental giant fiber pathways (command interneurons) are present in the cord, some of which are excitatory and some inhibitory. Two of these giant fiber systems were extensively studied by the Kennedy group, which differentiated electrophysiologically between a medial and a lateral pair of giant fibers on the basis of their sensory input and the patterns of motor output they evoked (Evoy and Kennedy, 1967; Kennedy *et al.*, 1967, 1969). Kennedy *et al.* (1969) reconstructed various motoneurons in the third abdominal ganglion after Procion injection, all of which were shown to receive input from both of the giant fiber systems, in order to determine anatomically the connectivities of each motor cell with the ipsi- and contralateral giant fibers. In all cases the anatomical predictions of connectivity (Fig. 26) were borne out by subsequent recordings from the appropriate motoneurons while the giant fibers were stimulated separately. Larrimer *et al.* (1971) injected the homologous flexor motoneurons in the sixth abdominal ganglion, which were previously shown to be activated only by medial giant fibers. As expected, they showed processes going to the medial but not the lateral giant fibers.

A similar anatomical connectivity analysis of segmental swimmeret motoneurons was made by Davis (1970) in the lobster. Abrupt endings of Procion-injected motoneuron dendrites on axons in the ganglion were

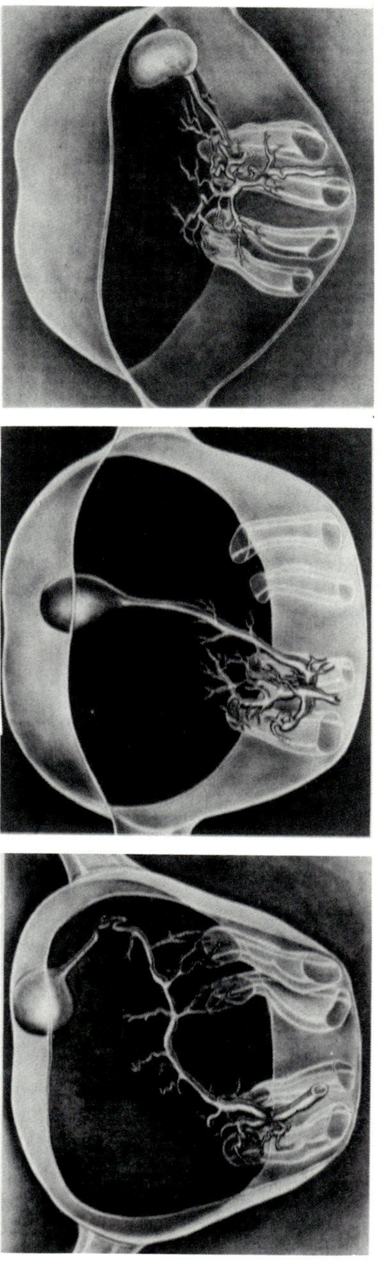

Fig. 26. Morphology of flexor motoneurons F_3, F_5, and F_9. Reconstructions are artist's interpretations made from stacks of tracings from serial sections. (From Selverston and Kennedy, 1969.)

taken to indicate functional synaptic coupling when viewed on the background of the crayfish studies mentioned. This view was further supported by comparison of a connectivity map for one of the motoneurons made using serial section reconstruction of a dye-injected neuron with an input map prepared from the data of Wiersma and Hughes (1961) on the basis of data containing only those neurons known physiologically to mediate inputs to the swimmeret motoneurons. This comparison showed that there were axons on the input map that occupied positions corresponding to the fibers upon which the swimmeret motoneuron processes terminated on the morphologically constructed connectivity map.

Studies such as these have made it clear that at least in some cases serial reconstruction of Procion-injected neurons can reliably demonstrate contacts between cells known to be physiologically coupled. That the technique can be misleading in terms of the numbers and distribution of synapses is illustrated by the studies of Nichols and Purves (1970) and Purves and McMahan (1972). Nichols and Purves studied the receptive fields of leech sensory neurons mediating three different types of sensory stimuli both by physiological recording and Procion yellow injection in order to demonstrate the connections of these sensory neurons to two large motoneurons controlling contralateral abdominal contractions. All sensory cells had many small axonal branches in close association with the processes of both motoneurons, suggesting that the connections of sensory and motor cells occur between multiple small processes rather than at a single large junction. The arborizations of the sensory cells are so elaborate that these investigators concluded the exact sites of synaptic contact could not be demonstrated light microscopically. Purves and McMahan (1972) injected the motoneurons in order to mark them for electron microscopy. In short series of thin sections examined electron microscopically, they found that the main process of the cell gave off some long branches up to 150 μm in length, but that these were greatly outnumbered by short branches of one to a few μm. These short branches, which were not noted by Nichols and Purves (1970), in fact were seen to receive the majority of synapses onto the cell with only occasional synapses occurring on the main processes and longer branches. Thus it seems that light microscope examination of Procion yellow–injected cells has served admirably for showing the pathways of larger fibers. These pathways undoubtedly indicate the regions of synaptic fields of the injected neurons, although whether one can infer the precise locations of synapses from the light microscope studies remains to be shown. Fortunately, there are now coming into use several electron-dense dyes which have the same unusual penetrating power of Procion yellow (Christensen, 1973; Gillette and Pomeranz, 1973). It can be expected that they will be used to isolate likely synaptic regions by light microscopy for ultimate verification and fine-structural study by electron microscopy in ways analogous to those used by Kolb (1970) with Golgi-stained material (Section II,C,2).

2. *Electron Microscope Studies*

Aside from the large and continually growing body of three-dimensional information derived from dye injection techniques, there has been a smaller number of attempts at complete serial section reconstruction of invertebrate sensory-central pathways using the more common staining methods for light and electron microscopy. Not surprisingly, these studies have been largely confined to structures showing a periodic repeating arrangement. The visual systems of arthropods are especially suited to such analysis, since the compound eye is composed of a series of regularly repeating light-receptive units which project onto a similar array of repeating synaptic units in the optic ganglia. The ganglia are serially arranged and provide "a natural distinction of neatly separated levels of information processing, each joined to the next by an intricate but

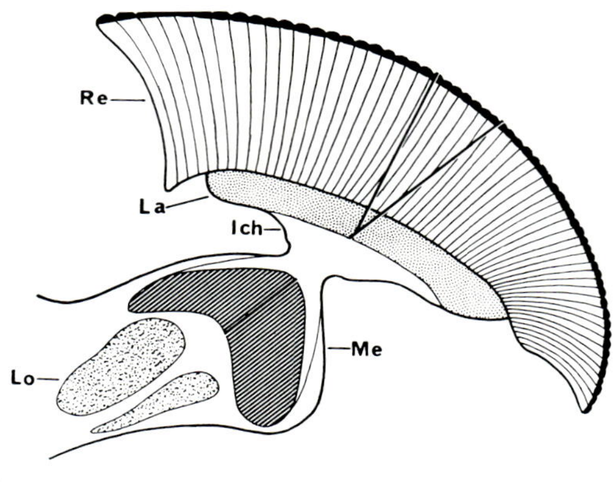

FIG. 27. (A) Survey schematic of optic system of dipteran flies, tangential section. Just below the retina of ommatidia (Re) lies a thin lamina (La), the location of the optic cartridges and the site of first visual integration. The lamina is connected to the medulla (Me) by the intermediate chiasm (Ich) or optic nerve. The lobula (Lo) is divided into two unequal neuropilic masses termed by Cajal the ovoidal and lamina ganglia. (From Trujillo-Cenóz, 1972.) (B) Higher magnification schematic view of the projection of visual axes upon the lamina and medulla in dipteran flies, horizontal section. In the retina three rhabdomere axes are shown out of the seven that each ommatidium contains. The interweaving of the retinular axons ending in the lamina results in a summation of like axes BBB, CCC, and DDD upon the second-order neurons of the lamina. The projection through the chiasma upon the medulla results in an exactly reversed anatomical representation of that in the visual field and lamina. (From Horridge and Meinertzhagen, 1970b.)

orderly system of fibers projecting the output of one level onto the input of the following one" (Braitenberg, 1967).

Studies of this type have been mainly carried out on the visual system of dipteran flies, a schematic of which is shown in Fig. 27. The retina consists of hundreds of units called ommatidia. Each ommatidium has one lens which focuses light onto eight photoreceptor cells called retinulae, two stacked exactly on top of one another and six surrounding these in a circle. The photoreceptive area, or rhabdomere, of each retinular cell is separate from all others, and therefore, except for one of the central two, each receives light from a different direction in space. Axons from the six outer retinular cells interweave in a complex manner and then terminate in the first synaptic layer, or lamina, in structures called optic cartridges (Cajal and Sanchez, 1915). The course of the two central retinular axons is different and is described later. The lamina is connected by the optic chiasm to the second region of optic processing, the medulla, and the medulla projects onto the remaining two optic ganglia and thence to the brain in a manner that has not been investigated.

Cajal and Sanchez (1915) long ago recognized by means of ordinary Golgi staining that the short retinular axons terminate on the second-order lamina neurons in repeated cylindrical units termed optic cartridges,

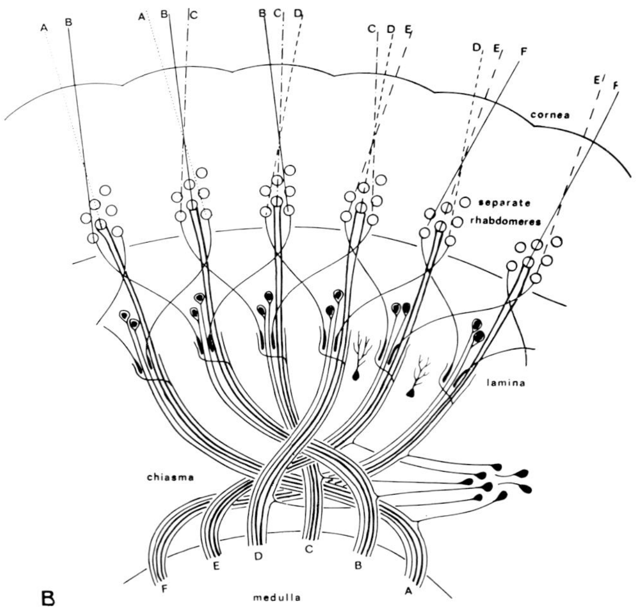

FIG. 27B. See facing page for legend.

but detailed information on the functional structure of the cartridges was not obtained until systematic electron microscope studies were made. Trujillo-Cenóz and Melamed extended the observations of Cajal by using short (40- to 50-section) series of electron microscope sections through lamina cartridges of *Sarcophaga* and *Lucilia* (Trujillo-Cenóz and Melamed, 1963; Trujillo-Cenóz, 1965a,b). In these, as in their later studies, three-dimensional representations of the data were made in the form of models built from balsa wood, cardboard, and wax. The basic structure of optic cartridges was shown to be consistent with most observations of Cajal, with six retinular fibers (called terminals within the cartridge) arranged in a circle around the axons of two intrinsic lamina neurons. A third intrinsic neuron L_3, corresponding to Cajal's type II lamina cell, was also identified, but its relationships with other fibers were not elucidated. A fourth intrinsic neuron L_4, which formed connections among the neighboring cartridges, was identified much later by Strausfeld and Braitenberg (1970). Boschek (1971) used serial section electron microscopy to work out partial connectivities of L_3 and L_4. Wax models of the synaptic area of the cartridge (Trujillo-Cenóz, 1965a) conclusively showed that each of the two postsynaptic lamina cells sends out many small branches to each of the six presynaptic terminals throughout the length of the cartridge (Fig. 28). As a result both lamina cells can be considered postsynaptic to all six terminals, indicating the important convergence of input in the cartridge. It was further shown that fibers of more central origin, the so-called centrifugal or nervous bag fibers, are important elements of lamina cartridges. They were found to be presynaptic to each retinula cell axon in the course of their weaving throughout the cartridge. A thorough description of the different glial types was also provided, especially pointing out the separation of cartridges from one another by cells corresponding to the epithelial glia of Cajal. Certain buttonlike projections into retinula terminals, which in single sections had been interpreted as neuronal synapses (Pedler and Goodland, 1965), were clearly shown to be of glial origin in the serial section study and were termed capitate projections.

This elucidation of the basic cartridge structure and connectivity can be seen as having provided an important foundation for the work of Braitenberg (1966, 1967) in Musca, who addressed himself to the larger problem of the retina-lamina projection pattern. The complex interweaving of the six short retinular axons at the base of the retina as described by Cajal and Sanchez was seen in these more systematic studies to represent a very well-ordered projection of the visual field onto the lamina cartridges. Braitenberg deduced the pattern of this projection by reconstructing the paths of retinular fibers between the base of the retina and the neighboring end of the optic cartridges using 10-μm serial light micro-

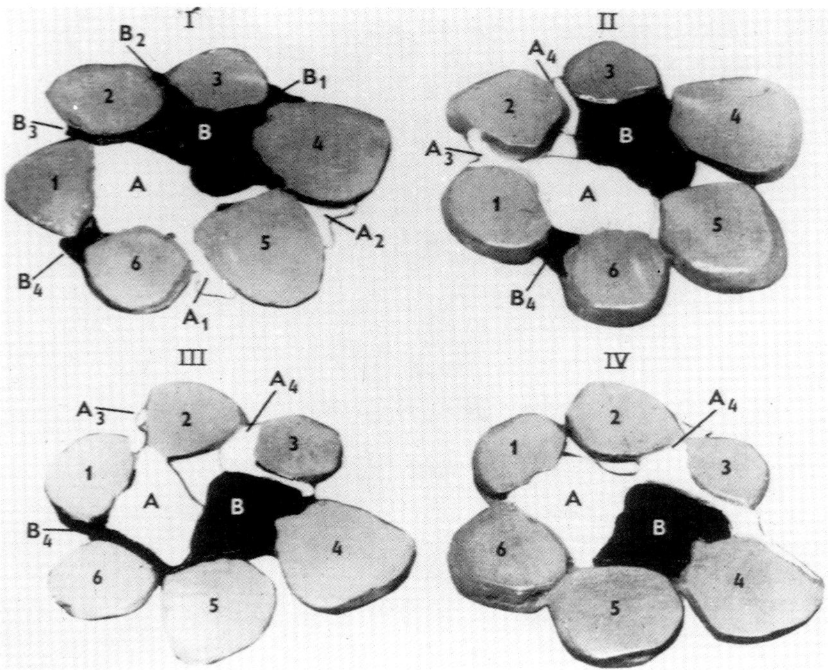

FIG. 28. These models represent 30 serial sections of about 1000 A each, covering a small portion of an optical cartridge. The centrifugal fibers and glial projections have been omitted in these models. Capital letters with a subindex indicate the prolongations of the second-order fibers. Model I covers the first 9 sections; model II covers the following 7 sections, and models III and IV the last 14 sections. (From Trujillo-Cenóz, 1965a.)

scope sections of silver-stained material. He described the arrangement of retinular cell rhabdomeres within an ommatidium as an asymmetric, roughly circular one similar to the petals on a flower. The arrangement was identical in every ommatidium, so that an unambiguous rotational numbering sequence (after Dietrich, 1909) could be assigned to the eight retinular cells of every ommatidium in the eye. In following the retinular axons through successive sections in transit to the lamina, it was found that the rotational sequence was maintained as the bundle of fibers from each ommatidium went through a turn of 180°. Following this rotation, Braitenberg found complex interweaving which consists of fibers with corresponding numbers in neighboring ommatidia departing their respective bundles in the same direction. The net result of this at the lamina is that the flowerlike arrangement of rhabdomeres of an ommatidium is reflected in a similar pattern in the array of cartridges in the lamina reached by the fibers of any one ommatidium, but rotated by 180°. Thus the six short axons from the six outer retinular cells of an ommatidium project to six different cartridges, and neighboring cells of the ommatidium project to neighboring cartridges (Fig. 29). Braitenberg reasoned that this arrangement made sense only if the angle between the optic axes of neighboring retinular cells in one ommatidium was the same as the angle between the central axes of neighboring ommatidia. The validation of this point by Kirschfeld (1967) led to the conclusion that the receptor cells from six different ommatidia whose axons converge on one cartridge all receive light from one direction in space. In other words, each retinular cell of a given ommatidium receives light from a slightly different direction but, because of the regular crossing-over of the fibers before they enter the lamina, each laminar cartridge receives light from only one direction. The limitations of resolution in the 10-μm silver-stained sections did not allow a firm conclusion about central retinular

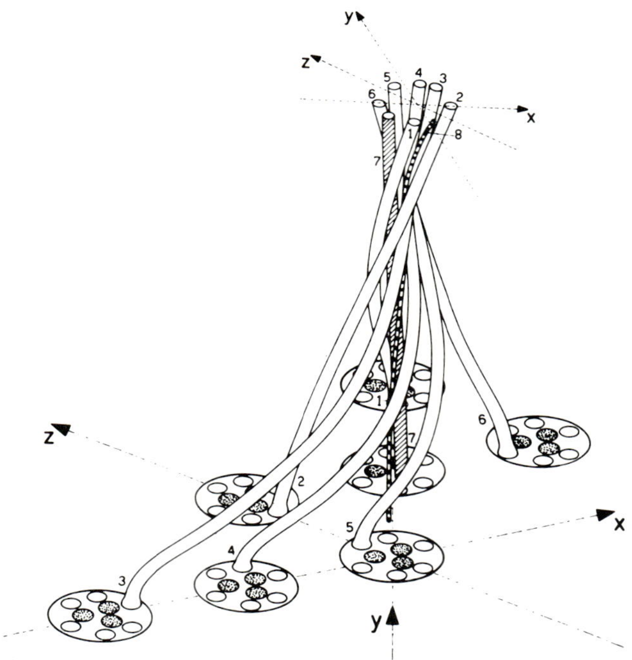

FIG. 29. Diagrammatic perspective view of the distribution of fibers arising from one ommatidium onto the lamina. Note the cyclic order of fibers 1 to 6 in the cartridges. The relation of fiber 7 to the cartridge is purposely undefined in this drawing. (From Braitenberg, 1967.)

fibers 7 and 8 from each ommatidium, which were lost in tracing through the series at the level of the lamina cartridges.

The paths of these two fibers were described by Trujillo-Cenóz and Melamed (1966a) as a result of the better resolution obtained by serial section electron microscopy in a study of the detailed three-dimensional structure of the ommatidium. These reconstructions were extremely useful both in suggesting possible schemes for the funneling of light through the dioptric apparatus and in describing the positional relationships between the superior and inferior central cells of the ommatidium (retinular cells 7 and 8), whose axons were seen to bypass the optic cartridges completely and enter into the lamina-medulla chiasm. This report and a shorter one (Trujillo-Cenóz and Melamed, 1966a,b) were also concerned with the projection pattern of the short retinular axons onto the cartridges, and overlapped in both scope and time with Braitenberg's work. The conclusions reached were basically the same, except for slight differences in the geometric descriptions of the pattern.

The observations on the superior and inferior central cells of the ommatidium were strengthened and extended in analyses of series up to 400 thin sections in length (Melamed and Trujillo-Cenóz, 1968). It was found that the microvilli comprising the rhabdomeres of these two cells were oriented in perpendicular directions. The extension of the series of sections to include the lamina-medulla chiasm and the neighboring regions of the medulla made it possible for them to show that the long retinular fibers of the two central cells not only traversed the chiasm along with the lamina cell axons and centrifugal fibers, but that they actually terminated on intrinsic medulla neurons in varicosities containing synaptic ribbons. As a consequence of the work of Waterman (1966) regarding the dichroic nature of light absorption by rhabdomere visual pigment and of the recognition of the common optic axis of the two cells, the conclusion was reached that the superior and inferior central cells could represent an intraommatidial polarized light analyzer, and that information about the plane of polarization would thus be available only at the level of the medulla. In an attempt to identify the synaptic interaction of the central retinular fibers, Campos-Ortega and Strausfeld (1972) made lesions in individual ommatidia in order to kill the superior and inferior central cells. The degenerating axons were silver-stained so that they could be followed to their terminations at different levels of the medulla. Once the location of these terminals was identified, the proper regions of nondegenerated preparations were prepared for serial section electron microscopy to show the presence of synaptic specializations throughout the extent of these long retinular fibers in the medulla.

In continuing attempts to elucidate the elements involved in the lamina-medulla projection, serial section electron microscopy was combined with serial section light microscopy (Trujillo-Cenóz, 1969) and with silver staining (Trujillo-Cenóz, 1970). In the former work light microscopy was used to make a tentative identification of the thick fibers of the medulla as the terminal segments of the pair of lamina cell axons postsynaptic to the short retinular terminals in the optic cartridges. With electron microscopy these thick fibers were shown to make numerous synapses of a peculiar ribbon type onto branches of the plexus of horizontal fibers where they terminate in the first synaptic stratum of the medulla. Golgi impregnations were used in the latter work to identify the centrifugal (nervous bag) fibers of lamina cartridges as specifically originating from cell bodies in the medulla. Each centrifugal fiber was seen to arborize terminally in the lamina, where the branches climb up the short retinular axons to which they are presynaptic. This connectivity scheme suggests that the centrifugal termini represent a control system feeding back the cartridge output to the medulla onto the retinular axon terminals in the optic cartridges. A schematic representation of all these relationships within an optic cartridge is shown in Fig. 30.

Horridge and Meinertzhagen (1970a,b) used the known structure of the cartridge and its duplicated presence within the lamina to attempt an inference of the rules of fiber growth by studying an anticipated catalog of errors in the morphology of some cartridges. For their study they used light microscope examination of 1000–2000 1-μm serial sections through the osmium-fixed *Calliphora* visual system. Laborious following of axonal profiles was carried out on printed micrographs of each section. Braitenberg's attentions had for the most part been devoted to ommatidia in the periphery of the eye. Horridge and Meinertzhagen, however, chose to follow retinular projections of 120 ommatidia centered about the eye equator, where an anterior-posterior axis divides the eye into mirror-symmetric dorsal and ventral halves and where they expected errors might be more frequent. All optic cartridges were found to receive the usual six short retinular axon terminals, except for those underlying three ommatidial rows on either side of the equator. The cartridges of the four inner rows each received eight terminals, and those of the two outer rows seven. This is as expected, since when they depart from their respective bundles at the complex interweaving four of the six short axons from an ommatidium spread toward the equator while only two spread away from it. By mapping the lamina terminations of the short retinular axons of both equatorial and more peripheral ommatidia, it was found that on the basis of predicted optic axes none of the short axons from these 120 ommatidia terminated in an incorrect lamina cartridge. The arrangement of the terminals around the two second-order lamina

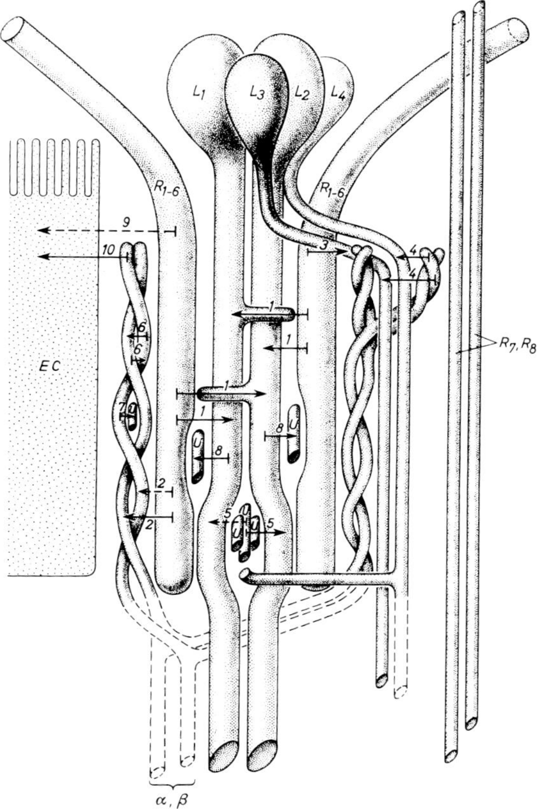

FIG. 30. A semischematic drawing of the structure and synaptic interactions of the optic cartridges. The numbers and arrows indicate synapses. The somata of the monopolar neurons have been labeled L_{1-4}. Only two of the six retinular cell axon terminals (R_{1-6}) and two of the six paired centrifugal fibers (α, β) have been shown for the sake of clarity. R_7, R_8, Axons of photoreceptors; EC, epithelial glial cells; U, unidentified fiber fragments. Broken lines have been used to indicate areas of inadequate evidence. Not to scale. (From Boschek, 1971.)

neurons within the cartridge shows a unique rotational sequence related to the positions of the retinular cells of origin within their respective ommatidia. Errors in the location of terminals among their neighbors in this rotational sequence occurred only 10 times, all but one mistake being found in cartridges close to the equator. But based on the synaptic relationships elucidated by Trujillo-Cenóz (1965a), these errors were not expected to be functionally significant.

A similar study was carried out on a fly in which regional disruptions of the complex short-axon lattice occurred as the result of a double dislocation of the eye equator as deduced from the breakdown of mirror symmetry (Meinertzhagen, 1972). By using serial sections to follow the terminations in the lamina of the short retinular axons from 144 ommatidia in the region of the dislocation, Meinertzhagen found that, contrary to the results for normal eyes, 17 short axons terminated in the wrong cartridge as deduced from the optic axes of their retinular cells. Almost all the errors were found in axons that would normally have terminated in cartridges more than one row away from the cartridge underlying the ommatidium of origin. In none of these incorrect terminations did an axon travel farther than it normally would have in making the correct termination. That is, most of the errors showed terminations a few cartridges nearer the ommatidium of origin than the correct termination site. In mapping the rotational sequence within a cartridge of incorrectly terminating axons, a conclusion was reached that had been suggested from the earlier work, namely, that in spite of occasional variation found in terminal rotational position the common feature was that all terminals lay on the side of their cartridge nearest the ommatidium of origin.

Meinertzhagen pointed out that the combined results of the serial section studies in normal and dislocated eyes can serve at least to rule out certain possibilities for the factors acting to produce the pattern in development. Certain mechanisms of sequential guidance in the growth of these axons toward the proper cartridge were considered unlikely, since an error at one point would tend to result in a chain of errors across the eye, a situation never encountered. The primary rule operating in the determination of rotational sequence of termination in a cartridge seems to involve a cessation of growth once any short axon reaches a point in the cartridge closest to its ommatidium of origin, as indicated by the fact that even axons projecting to the incorrect cartridge terminate in this way. Finally, Meinertzhagen concluded that the temporal pattern of differentiation, that is, the sequence of axonal growth, is of great importance and cannot be discovered except by studying the detailed three-dimensional anatomical relationships at different embryological stages.

The importance of this last point has been borne out in electron microscope serial section reconstructions of the developing retina–lamina connections in dipterans (Hanson, 1972; Trujillo-Cenóz and Melamed, 1973) and in the crustacean *Daphnia magna* (LoPresti *et al.*, 1973, 1974). Trujillo-Cenóz and Melamed (1973) have given a detailed description of the precocious connection between eye imaginal discs and the lamina anlage along a bundle of about 40 nerve fibers presumably originating in the larval sense organs. In late larval life this so-called optic stalk was seen to serve as the growth substrate for the retinular axons as they invade the lamina. The most significant observation made from the serial section analysis was that the cells of one ommatidium form a preliminary structure in which the eventually separate rhabdomeres are at first fused together and in which the retinular axons of these cells are initially uncrossed, projecting to the same area of the lamina. The fused receptor unit progresses to its final morphology with its separated retinular cells, while at about the same time the projection to the lamina undergoes a complex rearrangement to yield the characteristic crossed projection of the eye onto the optic cartridges. The results of Hanson (1972) are in general agreement with those of Trujillo-Cenóz and Melamed and perhaps go slightly further in emphasizing the possibility that this initial projection may serve an inductive function in terms of the lamina neurons. This point is made even more directly by LoPresti *et al.* (1973, 1974) in reconstructions of different embryonic stages in *Daphnia.*

B. Cell Structure

The polymorphic nature of mitochondria has been well documented (Novikoff, 1961). Early light microscope cinematography showed them to be "extremely variable bodies which are continually moving and changing shape in the cytoplasm. There are no definite types of mitochondria, as any one type may change into another" (Lewis and Lewis, 1914). Fusions were described, particularly in relation to the formation of *Nebenkerne* (Tahmisian *et al.*, 1956). In unicellular organisms light microscopy had revealed "long sinuous elements forming a network generally external to the chloroplast in *Chlamydomonas, Eudorina, Pandorina,* and *Volvox*" (Hollande, 1942). There were numerous reports of images of mitochondria in single sections connected to each other by narrow membrane bridges purported to represent mitochondrial fission (Diers, 1966). Limited serial section work at the level of the electron microscope showed that uniform populations of some specialized forms of mitochondria do occur (Stempak, 1965; Stephens and Bils, 1956; Christensen and Chapman, 1959; Senger and Saacke, 1970), but the earliest and most extensive work of this type (Andersson-Cedergren, 1959) on mouse motor end plates seemed to confirm the topologically simple nature of mitochondrial structure. It is only recently that complete reconstruction of specialized mitochondrial forms and some unicellular organisms revealed that mitochondria can attain complexities far greater than implied by previous investigators who described them as being "spherical to rod-shaped with occasional branching" (Lang, 1963).

In the early spermatids of insects, the mitochondria undergo an aggregation process to form a compact rounded structure called the *Nebenkern.* Andre (1959, 1962) and Pratt (1962) analyzed very short series of sections in an attempt to characterize the relationships between the mitochondria forming the structure but were unable to make significant process. Pratt (1968) later used several series of sections through entire *Nebenkerne* at different stages of formation and dissolution to make graphic reconstructions from an oblique angle. With these she showed that newly formed *Nebenkerne* are unified structures made from anastamoses of rodlike mitochondrial segments which later separate into two interlocking but unconnected ringlike units. Further transformations occur which break these interlocking rings, yielding other complex structures which ultimately give rise to separate mitochondria.

An indication that irregularly shaped mitochondria exist in normal unicellular organisms came first from two studies by Keddie and Barajas (1969, 1972), in which reconstructions were made in an attempt to find a morphological difference among species. In the first report the fraction of cell volume occupied by various organelles was estimated in *Pityrosporum orbiculaire.* Models of plastic were built, and organelle volumes measured either by plotting each section on graph paper and estimating the area or by the use of a planimeter, and then multiplying by the section thickness. In comparing the characteristics of mitochondria of this species with those of *Pityrosporum ovale*, they showed that, although the fraction of cell volume occupied by these organelles was similar in both species, the number of discrete mitochondria per cell ranged from about 20 to 40 of simple shape in *P. ovale* but only 1 to 4 of highly branched, complex shape in *P. orbiculaire.* It was concluded that in one species the mitochondria seemed to keep pace with cell volume by the formation of buds and branches, while in the other they appeared to increase in number rather than size. Barajas (personal communication) has recently stated that species difference may not be as important as phase of growth in determining exact mitochondrial structure.

Similar observations on yeast mitochondrial form have been made more recently by Hoffman and Avers (1973) in *Saccharomyces.* In nine complete series of 80 to 150 sections, these investigators report that each reconstructed cell contains only one branched giant mitochondrion. The total length of the organelle is more or less the same in different cells, although the diameter seems to vary as a function of the growth phase at which the cell is fixed. Even in budding cells it was found that branches of the giant organelle extend into the bud and preserve structural continuity until the new cell wall separates the bud from the mother cell. As noted by Keddie and Barajas (1972), classic ovoid or ellipsoidal profiles were commonly seen in single cross sections, which leads to the suspicion that previous ideas of mitochondrial size and number may in some cases be artifacts of the sectioning procedure.

Another finding of complex mitochondria was made by Schötz and co-workers (1972) during their reconstruction of the (+)-gamete of the unicellular alga *Chlamydomonas reinhardii.* This study was directed primarily at understanding the relationships and structures of the subcellular organelles, and to this end they used 80 serial sections to construct a complete three-dimensional model of 2 of the 141 cells they studied in detail (Fig. 31). The model shows the common result that *Chlamydomonas* cells are highly asymmetric. Not as clear from this particular figure is that the chloroplast is likewise asymmetric and often has several perforations through which extensions of the highly branched mitochondria (Fig. 32) penetrate. Of particular interest is that in all cases the mitochondria observed in the vicinity of the chloroplast were very closely apposed to its surface and even made deep troughlike indentations over their entire length. Although it had been previously determined from ordinary single-section analysis that the nuclear envelope spread into endoplasmic reticulum (ER) of the cytoplasm, these investigators showed that in a single cell this does not just happen occasionally. The nuclear envelope was shown to form not only swellings of ER but also vacuolelike structures in the peridictyosomal region which produce many small vesicles whose function, they guessed, is to contribute to the regeneration of the Golgi cisternae. In fact, so bound to ER are these and all other observed vacuoles that they concluded it was possible that the cytomembranes and organelles surrounded by membranes may form at least temporarily a continuous system within the cell.

Other cellular organelles have been studied less often by serial section analysis, although occasionally with interesting results. Karlsson (1966a) completely sectioned the perikarya of two interneurons in the rat lateral geniculate nucleus and made a quantitative three-dimensional analysis of all organelles present including the configurations of cytomembranes. Some Golgi complexes, each consisting of a group of flattened cisternae, were found to be continuous with others throughout the whole cell body, and branchings and fusions of different complexes were very common. A single 9 + 0 cilium was found in each of the two neurons studied, where it projected out from a typical basal body. In one neuron a basal body without a cilium was found. Karlsson implied that only by complete three-dimensional analysis is it possible to determine whether the presence of such a cilium is universal, since it was noted that out of many hundreds of other perikarya observed in single sections attachment of

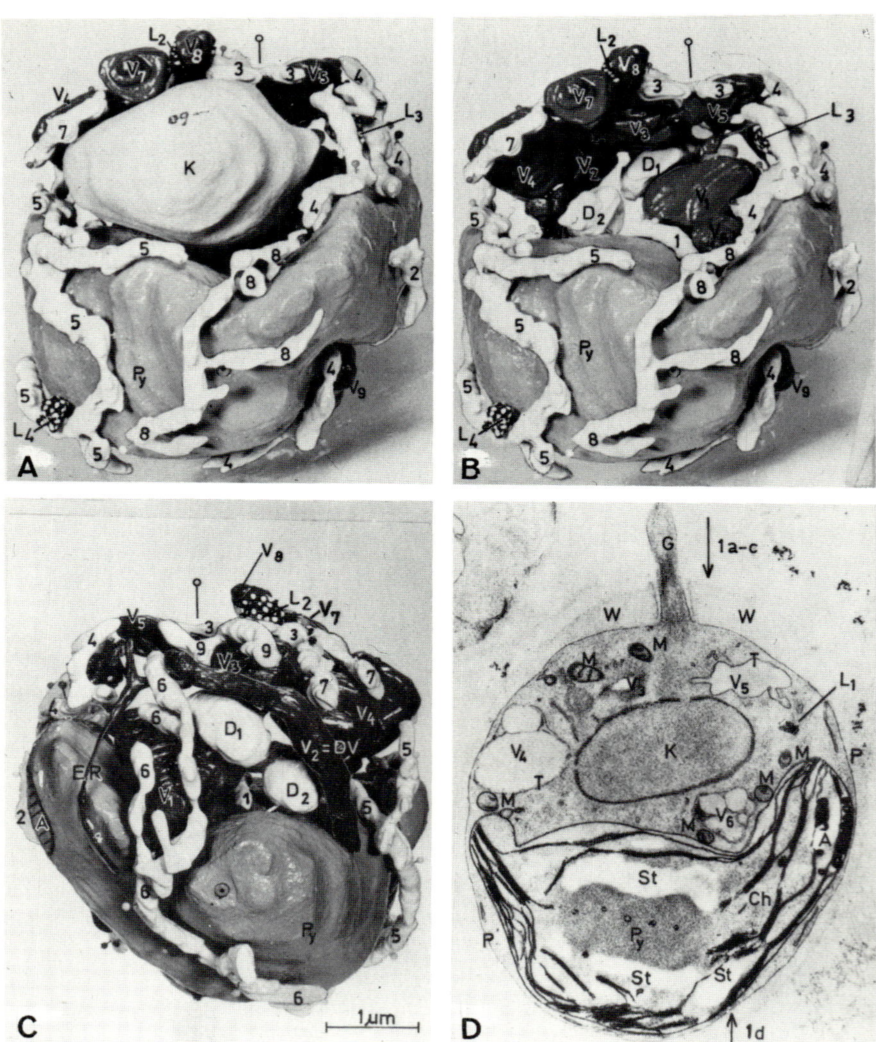

FIG. 31. (A–C) Model of the *Chlamydomonas* cell. (A and B) View from above, perpendicular to the plane of the section shown in (D). (A) Complete cell. (B) Cell after removal of the cell nucleus in order to permit an internal view. (C) View from below, perpendicular to the plane of (D). (D) A section from the series used in the construction of the model represented in (A) to (C). Fixation with potassium permanganate. The cell nucleus is cut near its lower surface so the nuclear membrane is sectioned tangentially. V, Vacuoles; K, nucleus; Py, pyrenoid; Ch, chloroplast; W, cell wall; T, tonoplast; St, starch; D, dictyosome; ER, endoplasmic reticulum; L, lipid bodies; A, eyespot; M and integers, mitochondria; P, plasmalemma. (From Schötz *et al.*, 1972.)

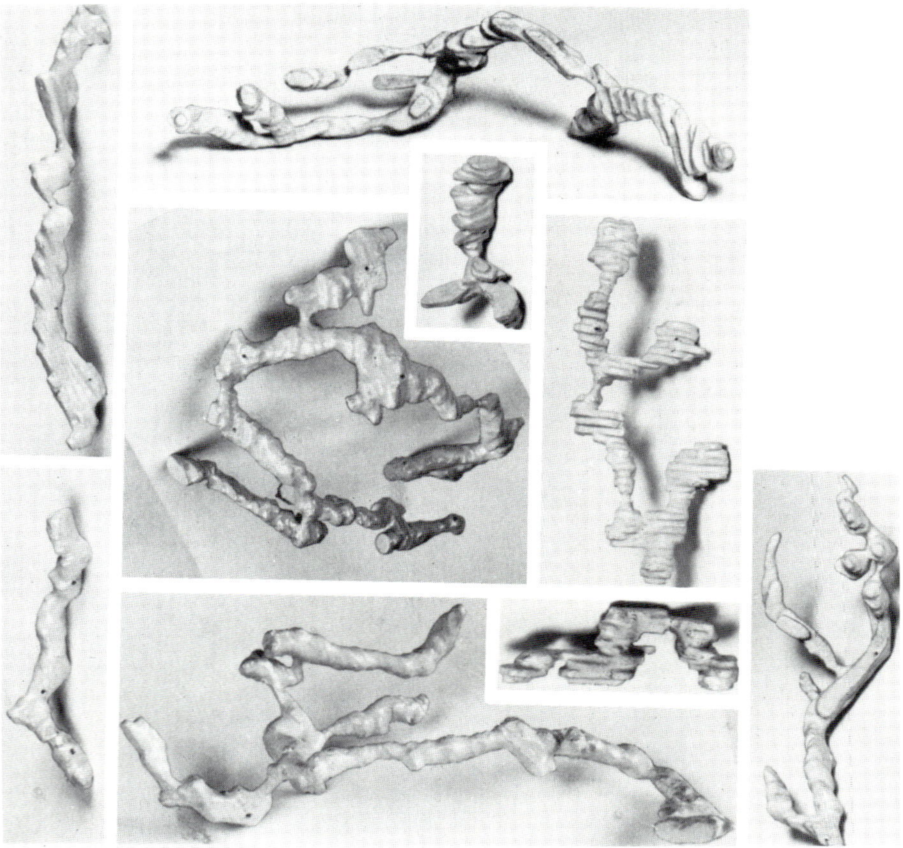

FIG. 32. Mitochondria from *Chlamydomonas*. (Original photographs, courtesy of Schötz.)

a cilium to a cell body could be seen in only one. Out of 569 mitochondria reconstructed, 179 were found to be branched, the remainder having a simple rodlike or ellipsoidal shape.

A nuclear inclusion of rodlike shape was noted by Karlsson (1966a) in only one of a number of nuclei examined carefully. These had been described as early as 1900 by light microscopists (Siegesmund *et al.*, 1964), but only in some nuclei and only some species. Willey and Schultz (1971) reviewed the more recent literature on the subject and made a systematic three-dimensional study of these structures in the cat olfactory

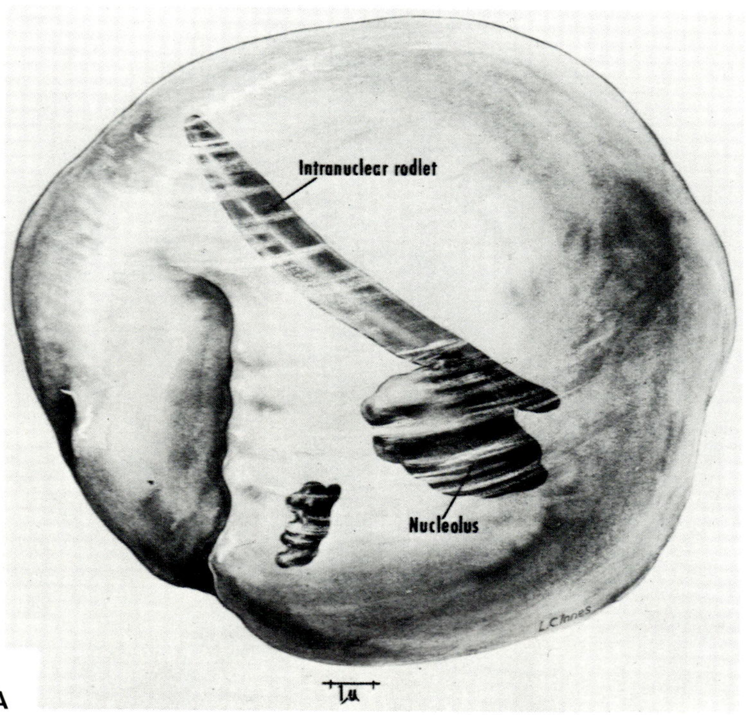

FIG. 33. (A) Three-dimensional reconstruction from serial sections of a neuron nucleus with an intranuclear rodlet. The rodlet is oriented in the nucleoplasm without any apparent relationship to the nucleolus or nuclear membrane. The nuclear membrane is folded in two places. (B) A nucleus with an intranuclear lattice. The lattice is not drawn in fine detail. Morphological continuity between the intranuclear lattice and nuclear membrane is not evident. Subsections of the nucleus are shown below to indicate the curvature of the lattice in the nucleoplasm. (From Willey and Schultz, 1971.)

system, a location where they were found to be particularly prevalent. Both the rod and the fibrillar lattice forms were observed, often in adjacent cells but never in the same cell. The rodlet (Fig. 33A) consists of 6-nm diameter filaments spaced approximately 15-nm apart arranged in a linear bundle of constant diameter with tapering at the ends. Although the rod never penetrates the nuclear membrane, it was seen to abut against it at both ends. In the olfactory bulb the rod inclusions are found in the periglomerular, internal granule, and mitral cells, whereas in the anterior olfactory nucleus and prepyriform cortex they do not appear to be unique

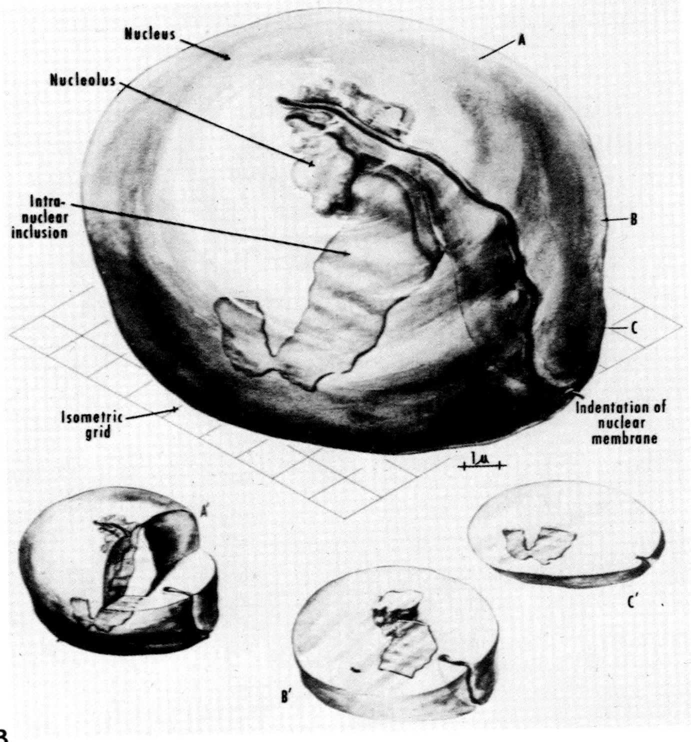

FIG. 33B. See facing page for legend.

to any particular type of neuron. The fibrillar lattice (Fig. 33B) consists of either four or eight stacked sheets of parallel 10-nm fibrils, the orientation of the fibrils in one sheet being at an angle of from 30° to 90° with that in the adjacent sheets. Unlike the rod, the lattice was never found to be in intimate contact with the nuclear membrane.

Reconstructions of the protozoan red cell parasite *Anthemosoma garnhami* have been made by Vivier and Petitprez (1972). These investigators were specifically interested in the complex vacuolar system of the parasite with regard to its involvement in the digestion of hemoglobin of the host rodent red cell. The finding was that an entire vacuolar subsystem communicates directly with the space external to the parasite cell by means of a system of micropores on its surface. These cavities are filled with host cell erythrocyte stroma. Neither complete invacuolization nor real ingestion of hemoglobin could be seen in the cytoplasm of the parasite. The conclusion was that the digestion process is extracellular, probably occurring in the externally communicating vacuolar system.

Hinds and Ruffett (1973) serially sectioned areas of developing mitral cells in the mouse olfactory bulb in an attempt to trace the bundle of axonal microtubules to its origin in the perikaryon. They found that the microtubules continue as a discrete bundle of unchanging diameter to a termination on one side of the nucleus, while the remainder of the perikaryal cytoplasm is almost devoid of tubules. This fact in conjunction with other observations led them to suggest that the direction in which the axon extends may be determined within the perikaryon by the orientation of the bundle of microtubules. A recent attempt to reconstruct the microtubules composing the mammalian mitotic spindle combined whole-mount viewing by high-voltage electron microscopy with the analysis of serial sections by computer graphics (McIntosh *et al.*, 1973). The combined results are taken as being consistent with the sliding of tubules over one another at anaphase.

A question recently posed by Teichberg and Holtzman (1973) involves the structural continuity of smooth endoplasmic reticulum (SER) throughout the length of nerve axons. The cytochemical data of these investigators indicates that adrenergic vesicles may arise at least in part from the budding of membrane-bound vesicles from SER. They suggest that, if it is continuous, this membrane system could represent another possible axonal transport system for specialized purposes, such as direct transport of newly synthesized transmitter from perikaryal rough endoplasmic reticulum (RER) through axonal SER to nerve terminals where it is packaged into vesicles and released by budding. Teichberg (personal communication) demonstrated continuity of SER through a short distance of axoplasm (1 μm). Other observations in the developing axons of *Daphnia* embryos has shown that a continuous somewhat branched membrane system exists through 4–5 μm at the distal end of the fiber with no termination at either end of the series (LoPresti, unpublished observations). Karlsson (1966b) found in his study of two completely reconstructed rat lateral geniculate interneuron perikarya that all proximal branches had cytomembranes continuous with those of the cell body. These observations raise the possibility of complete continuity from distal processes to the cell body in both growing and mature neurons.

Chromosome structure has always been of great interest in cytological research but, in terms of its substructural complexity and the dynamic changes in structure and function involved in progression through the phases of the cell cycle, it is still the least well understood of cellular organelles. There are three levels of chromosome organization that might be amenable to serial section reconstruction, and at all three there is cyclic variation in structure. The first of these involves the substructure of a single chromatin fiber in terms of its relative content and arrangement of DNA, protein, and RNA components. The second is that of the organization of nucleoprotein chromatin fibers in the interphase and metaphase configurations. Finally, there is the suprachromosomal organization, including such problems as the relationships between individual or pairs of chromosomes during meiotic prophase or the relationship of the whole set of chromosomes to the spindle microtubules. The most obvious problem in conducting such studies has been the inability to interpret consistently at the macromolecular level the patterns of electron density seen in cross sections through fixed chromatin fibers. DuPraw (1970) has pointed out that the images obtained from high-resolution electron microscopy of very thin sections are dependent not only on the method of chromosome extraction and fixation but also vary considerably with respect to the embedding medium used. One is also confronted by the practical limits of resolution in plastic-embedded material, and further by the nonuniform distortion problems caused in successive sections by the high beam currents required at magnifications of over ×100,000. This last problem can to some extent be overcome by the use of image-intensification systems, but generally it has been found preferable for very high-resolution work to use modifications of whole-mount techniques (DuPraw, 1968, 1970). As a result, it is only at the suprachromosomal level of organization that serial section analysis has been attempted, and this in the specific area of the structure and function of synaptinemal complexes at the first meiotic prophase.

The first study of this type involving serial sectioning of whole nuclei was carried out by Wettstein and Sotelo (1967). Previously, Sparvoli *et al.* (1965) had used small numbers of serial sections to examine the structure of *Tradescantia* chromosomes, and Bernhard (1966) had compared the nucleolar structure of normal and pathological cells by serial drawings. Wettstein and Sotelo reconstructed in drawings two whole nuclei of spermatocytes in midpachytene of first meiotic prophase in order to examine the extent of synaptinemal complexes between homologous chromosomes. The ultrastructure of the complex had been previously described as early as 1960 from single sections (Sotelo and Trujillo-Cenóz, 1960). These investigators photographed every section at high magnification and found that the synaptinemal complexes are similar in all pairs of homologs, both in terms of the medial pairing region and in the constant spacing of the lateral elements. In all autosomes the complex was as long as the bivalent itself, that is, the homologs seemed paired throughout their whole length. In the sex vesicle, a term used to describe the region of nucleus containing the nucleolus and sex chromosomes, it was found that the X–Y complex is surrounded by nucleolar material and in contact with it through all sections. These investigators emphasized that a random section through the X–Y synaptinemal complex would allow its identification as a pair of autosomal homologs, whereas in actuality the X–Y pair is the only one with a complex limited to only one area of its association and is unambiguously identifiable in serial sections.

A series of investigations by Solari and co-workers has provided many useful observations in the area of X–Y pairing in spermatocytes. The synaptinemal complex between the X–Y pair in mice was reconstructed serially at different stages from zygotene to diplotene (Solari, 1969a,b). The sex vesicle becomes attached to the nuclear membrane for the first time at zygotene, but a synaptinemal complex is not visible until pachytene. At this stage two independent electron-dense cores are visible inside the sex vesicle, a shorter core which does not change in appearance from zygotene to pachytene, and a larger core which contains the synaptinemal complex. There are three attachment points on the nuclear membrane, one from each core and the third in common. The complex at the free end of the long core was seen to have disappeared by late pachytene and, contrary to the observations of Wettstein and Sotelo, the X–Y complex seemed to be free of the nucleolus except at midpachytene. Solari proposed that the sex vesicle be renamed the heterochromatic X–Y pair on the basis of this evidence, the pair showing approximately the same degree of condensation throughout meiotic prophase I and contrasting sharply with the autosomes. The sentiments of Wettstein and Sotelo were echoed by Solari in citing an earlier study (Solari, 1964) in which the X–Y synaptinemal complex was not revealed in single sections.

The main implication of such a complex in X–Y meiotic pairing is a region of homology between these two chromosomes at the molecular level which has also been demonstrated by various molecular hybridization studies. Solari's observations in the mouse were for the most part identical in studies of rat (Urena and Solari, 1970) and human (Solari and Tres, 1970) spermatocytes. These findings were represented in models made by tracing each section on clear plastic sheets, spacing them properly, and drilling guiding holes so that the model could be assembled and disassembled easily (Solari, 1970). By measuring the cores in each plastic sheet, the length of each electron-dense core was obtained for all three species. In the rat and human X–Y pairs, it was found that, although close associations exist, the pair seems to be independent of nucleolar material at all stages.

The correlation between the appearance of synaptinemal complexes and functional meiotic pairing was strengthened by Moens. He first serially sectioned the bouquet arrangement of meiotic chromosomes in *Locusta* (Moens, 1969) and determined that synapsis was initiated near the nuclear membrane at both the centromeric and noncentromeric ends of the chromosome. In this species, as well as in the reconstructions of the chromosome complement of a small protist (Moens and Perkins, 1969), it was found that all synaptinemal complexes became attached to the nuclear membrane at both ends of the chromosome pairs. In a study of meiotic chromosome pairing in natural and artificial *Lillium* polyploids, Moens (1970) demonstrated that in an autotetraploid switches of pairing partners within each set of four homologs was reflected in switches between synaptinemal complexes. In an allotriploid the synaptinemal complex showed abnormal lateral elements, as would be expected if these complexes are reflections of functional pairing at the crossover stages of meiosis.

Although it is generally thought that the aforementioned work indicates an involvement of synaptinemal complexes in structurally binding homologous chromosomes together during pachytene, the relationship between these complexes and pachytene crossing-over with resultant chiasma formation is less clear. In a serial section study of the synaptinemal complex in the achiasmatic meiosis of mantid spermatogenesis, Gassner (1969) showed that unlike the achiasmatic meiosis in some dipterans in which synaptinemal complexes and pairing of homologs are absent there are definitely normal-looking complexes present between homologs in mantids. Contrary to the case in most species, in which as already described complexes are present throughout pachytene but disappear before the end of prophase I, Gassner noted that normal complexes are present until disjunction of homologs in anaphase I, and remnants can still be seen at telophase on each half-bivalent. Unfortunately, unlike the situation in *Drosophila* males, in which a lack of crossing-over has been demonstrated genetically, such is not the case in mantids, and the possibility of chiasma occurrence was not ruled out, since chiasmata are normally only demonstrable in the repulsion of homologs after synaptinemal complex breakdown in diplotene of metaphase I. The persistence of the complexes in this case did not allow a definite conclusion as to the presence of chiasmata.

Stack (1973) has taken a similar approach in studying both triploid and tetraploid varieties of *Allium*. Over 100 serial thin sections were required to produce three-dimensional reconstructions of whole nuclei in the form of drawings. The advantages of this genus are the more-or-less complete synapsis in the triploid, as well as the completeness of the achiasmatic condition. This latter point is demonstrated by the fact that, unlike in the mantids, the synaptinemal complexes do disappear by diplotene, with the result that homologs completely separate at this time, there being no chiasmata to hold them together. Stack found that the complexes of the triploid are indistinguishable from those of the chiasmatic tetraploid. This evidence in combination with that of others based on different methods led Stack to conclude that synaptinemal complexes are involved in structurally binding homologs in register so that crossing-over and resultant chiasma formation may then occur in such a way that these chiasmata would keep the homologs together upon breakdown of the complexes at late pachytene–early diplotene. It appears from the available data that alternative hypotheses of some more direct involvement of the synaptinemal complexes in chiasma formation are not favored.

Gillies (1972) serially reconstructed the pachytene nucleus of *Neurospora*, and found that all seven bivalents showed synaptinemal complexing and that six were attached to the nuclear membrane at both ends, the satellite end of the seventh or nucleolar chromosome pair showing a free end. In addition, Gillies found that the values obtained for chromosome lengths obtained from serial section electron microscopy were smaller than those obtained from light microscope measurements. All the reasons for this are not clear, but it may be that these lengths are a product of some of the fixation and embedding artifacts mentioned above. Although the studies described here make it evident that there are certainly some quite useful applications of serial section reconstruction for the study of whole chromosomes, the limitations of the technique for both quantitative measurements and substructural reconstructions are for the present perhaps insuperable.

IV. Computer Methods

In making the transition from two to three dimensions, two problems are encountered in using the information available. The first is strictly the graphical one, that is, how to present a three-dimensional structure on paper, the traditional two-dimensional means of communication, and how adequately to represent relationships of structures that may be in intimate contact. As has been discussed in detail, this has usually been solved by various types of interpretive drawings, each one individually made, or by photographs of models constructed out of solid material. The second deals with the fact that one often wants to do something with a structure once one has it by some sort of data analysis, mathematical modeling, or sequential presentation in time. The points representing the structure thus take on a quantitative, not just a descriptive, significance. Consequently, one would like not only to have large amounts of three-dimensional data available in a systematic form, but in addition to be able to use them in various computations.

To aid in the solution of both these problems, recourse is now being made by several groups to the speed and facility offered by digital computers, especially computers equipped with graphics terminals. In these cases the computer is used in all stages of the reconstruction process, including collecting primary data from two-dimensional single-section micrographs, storing and analyzing them, and creating a visual representation of the reconstructed material. So far the techniques have been developed mainly for use in conjunction with neurological studies requiring accurate morphology, and can be divided into two categories relating to the degree of automation used in collecting the primary data. The examples mentioned here are only beginnings and in some cases are unpublished.

A. Semiautomatic

There has been an independent realization among many workers hoping to use the advantages of the computer in reconstruction work that the computer can serve as a useful tool without fundamentally new programming concepts. Therefore problems of pattern recognition and section interpretation have been left to the human operator, and the computer relegated, insofar as data collection is concerned, to the status of a notebook. This comes from entirely practical, not philosophical, reasoning. First, digitization of a single picture at resolutions available with microscopes, and particularly the electron microscope, requires an enormous amount of storage space. Second, computation with stored pictures is very lengthy on even the largest of computers presently available. But these objections aside, the fact remains that machine interpretation of pictures at the conceptual level has not reached the complexity that can be achieved almost instantly by the human brain. The area which at present seems most approachable from a machine point of view is that of interpretive line tracing, that is, analysis of structures that are, or can be considered to be, lines on a clear background (see Section IV,B). Some attempts at this have been made (Garvey *et al.*, 1972) in studying Golgi-impregnated neurons, but have had to be temporarily abandoned (Wann *et al.*, 1973) in favor of an interactive approach using a human as the prime interpreter. This has been necessary for example at branch points, in places where imperfect staining results in a discontinuous image, or in cases, not infrequent, in which the desired image is surrounded by an equally dense background to the extent that only an experienced microscopist can use those subtle clues available to extract it. To circumvent these difficulties one might think of having a human investigator abstract contour information, say, from individual micrographs and using a computer to process the line tracings; but then one eliminates the possibility of human reinterpretation of the original micrographs at a later time based on new information or information gained from analysis of several micrographs in a series. The difficulties are not of course insoluble, but for the moment it has seemed advantageous to many investigators to begin to use the capabilities of small computers which are readily available to handle the already solvable problems of data management and collection.

One example of the interactive use of a computer graphics system has been implemented at the University of California at San Diego by Selverston and co-workers (personal communication) for recording information at the level of the light microscope with the intent of obtaining exact cell geometry parameters so that biophysical properties can be determined for identified cells. These properties and the cell architecture can then be used to predict the flow of synaptic currents within the cell. This is of interest since the ganglion being studied, the stomatogastric ganglion of the lobster, is *not* characterized by constancy of cell body and fiber position from animal to animal. In spite of this all measured physiological interactions between pairs of the 30 cells are extremely reproducible by such criteria as amplitude and specificity of connections, as might be expected since ganglia in different animals must be functionally quite similar. So the question arises how the network compensates for the fact that homologous components in different ganglia may not be constructed in exactly the same way. One possibility is that alterations in impedance due to length changes can be offset by changing certain cell shape parameters. The determination of cell shape and the location of contact points between pairs of functionally related cells can suggest how this functional equivalency is obtained.

To obtain electronic parameters from geometrical measurements of a multibranched neuronal tree, two procedures developed by Rall (1959) have been used. Under certain circumstances (Section II,C,1), it is possible to represent the entire dendritic tree by an equivalent cylinder of axoplasm. Calculations involving one cylinder or simple groups of them can be made analytically. Another method that has been employed is to lump parts of the tree into compartments with specified capacitances and conductances. This model generates a system of coupled differential equations which can only be solved numerically. What these treatments have in common is that they abstract information from the real geometrical configuration and work with the abstracted model. In the case of the lobster stomatogastric ganglion, in which homologous cells in different ganglia are functionally alike but may be geometrically dissimilar, not just in relative lengths and diameters but perhaps also in their branching patterns, it would be interesting to see if it is possible to obtain a simplified representation for these cells which would yield consistent input–output characteristics from animal to animal.

The voltage–current relationships for a neuronal cell reconstructed from serial sections can be computed if simplifying assumptions are made. As for any passive electrical system, the input and output voltages relate to the input and output currents through equations of the form

$$V_1 = RI_1 - QI_2$$
$$V_2 = QI_1 - R'I_2$$

where the voltages and currents may be taken at two arbitrary points (e.g., at two points where electrodes have been placed), and the input impedances R and R' and the transfer impedance Q are functions only of the material between those two points and of the complex frequency. The processes of the cell are treated as being made up of many small cylinders joined together. For a single such cylinder impedances r and q may be obtained by evaluating the Laplace-transformed solution of the cable equation

$$\lambda^2 \frac{d^2V}{dx^2} = V + \tau \frac{dV}{dt}$$

at $x = 0$ and $x = L$. This yields for the time-independent case

$$r = r' = r_i \coth L$$
$$q = r_i/\sinh L$$

where r_i is the axoplasm resistivity per unit area. Then R, R', and Q for an arbitrary network can be obtained by combining the rs and qs of these subunit cylinders using a straightforward application of Kirchhoff's laws iteratively on a computer. The input impedance R is experimentally verifiable with a single electrode in the soma. Although it is difficult to place a second electrode into processes in the neuropil, computing the impedances between two arbitrary points is useful in order to compare identified cells in different animals. The reconstructed cells may be represented in two-dimensional diagrams as an aid in picking out interesting pairs of points (Fig. 34). Many such pairs are tried on different but homologous cells to obtain tapered cylinder models whose dimensions are computed from the exact steady-state impedances. These elements may be selected so as to combine in a way that will result in an approximate matching of the frequency dependence of the exact solution.

FIG. 34. (A) Computer drawing of a reconstructed PVG cell. (B) Computerized flattening of the same cell into two dimensions. (Original illustrations, courtesy of S. Glasser.)

The procedure for obtaining the shape parameters from cobalt- or Procion-injected material consists in rear-projecting color micrographs of sequential 10-μm sections of injected neurons one at a time onto a large screen (Fig. 35). Along two edges of the screen is a sonic sensing device (Graf Pen) which, by time delays, is capable of measuring the position of the tip of a sound-emitting pen when positioned in the plane of the detectors. In this way coordinate information in the form of contours is stored in a PDP 11/45 computer as the operator moves the pen over the projected micrograph. Display of the recorded information is immediately presented on the screen of a Vector General graphics unit driven by the 11/45. Appropriate information (ganglion contour, cell body, and fiber profiles) is recorded for each section independently without regard for proper alignment of the sections with respect to one another. Two succeeding operations are used to convert the individually stored section images into a complete three-dimensional reconstruction. The first consists of displaying each consecutive pair of sections so that by a visual best fit criterion the image of section 2 can be rotated and translated on

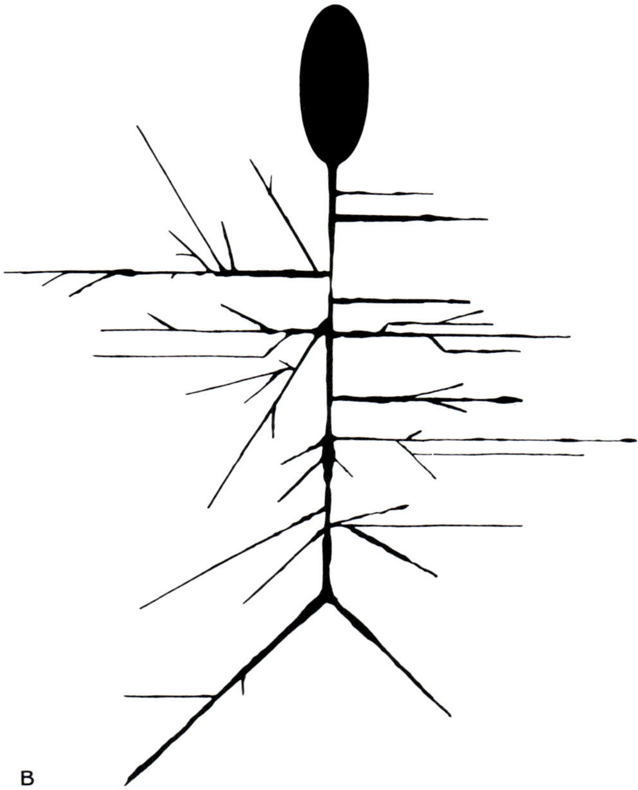

FIG. 34B. See facing page for legend.

the graphics display so that it aligns with section 1, then section 3 with section 2, and so on throughout the specimen. This places a series of aligned, but as yet unconnected, sections in the computer memory. The next step consists of forming connections between the profiles of the nerve fibers seen in the individual sections, and this utilizes the full real-time continuous three-dimensional rotating capacities of the Vector General unit. Two or more section images are displayed and rotated by whatever amount seems convenient to the operator, so that the tracings are seen at an oblique view. By means of a cursor on the screen which can be positioned in three dimensions a fiber stump in one section is located and identified, and this position is keyed to the computer. The continuation of the stump is similarly located and keyed on the next section image, thus completing the connectivity of the fiber in three dimensions.

A completely traced PVG cell is shown in Fig. 34. In Fig. 34B the topography of the cell shown in Fig. 34A has been totally distorted to give an informative two-dimensional projection, since cells whose topography is quite different in different ganglia appear much more similar in appearance when displayed this way. The topology and fiber diameters, however, are accurately reproduced.

The first approach of this type was reported by Levinthal and Ware (1972) using an Adage Graphics Terminal 50 display computer, with the difference that the alignment of sections was carried out on the micrographs themselves rather than on abstracted images of them. This is accomplished prior to data collection by the cinematographic procedure described in Section II,A,3. A further difference is that feedback from the computer to the operator is increased by optically superimposing the computer screen and section image. In this way the operator can see exactly the coordinate information being stored in the computer as it is being recorded, because its display is viewed on top of the original micrograph. The purpose of these two features is to permit all three dimensions to be treated fundamentally the same, since during the data recording a single structure is followed and traced to its completion wherever it might go in the movie. A movable spot on the screen indi-

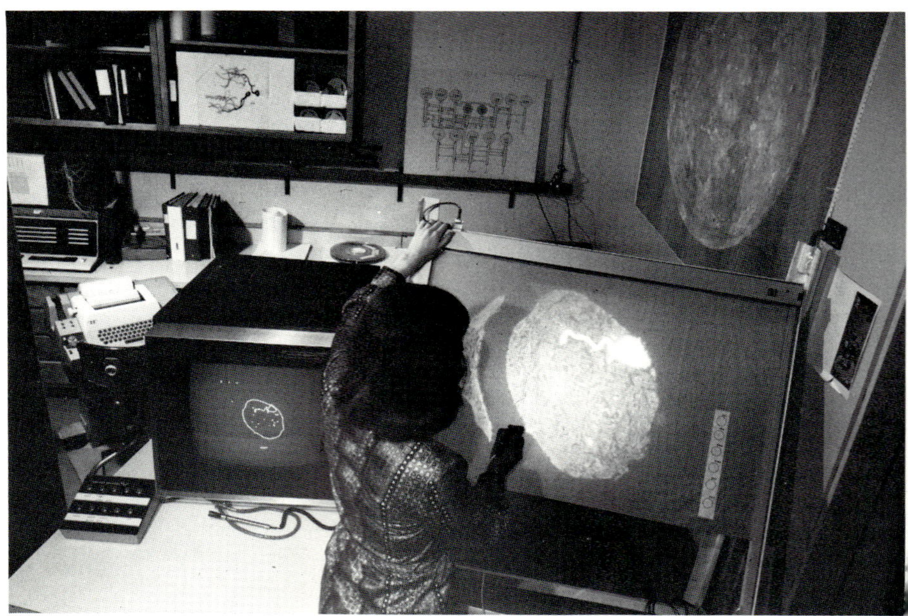

FIG. 35. Computer recording system used by Al Selverston at the University of California at San Diego. See text for explanation. (Original photograph.)

cates the x-y position in the micrograph, and the computer controls the z position by operating the projector. By selectively displaying the tracings of one or two previously recorded sections on top of an untraced micrograph, the operator can decide exactly where the object is to be continued. This is of particular advantage in nervous tissue in which nerve profiles may appear quite similar in dense neuropil and may change rapidly from section to section even in very thin-section electron micrographs. In addition, it allows a smooth trace of the contour to be formed.

The system described was programmed to function in one of two modes, a contour mode in which the outlines of structures on chosen sequential sections were traced to the desired degree of accuracy in order to give a realistic three-dimensional appearance to the object reconstructed, and a "tree" mode in which the neurons were recorded as stick figures defined by an ordered sequence of discrete points, or nodes. In the latter case provision was made for associating with each point a topological modifier to designate it as a root (cell body), terminus (end of fiber), branch point, continuation (intermediate) point, or synapse location with a pre- or postdesignation. Branch points were placed sequentially, as they were encountered, on a special list, the "push-down" list, so that when a terminus was recorded the computer could return the operator to the most recently encountered such point for further tracing. When this particular return operation was carried out, the computer displayed on top of the returned-to micrograph the already traced input and output nodes so that the operator could see immediately where tracing was to be continued. A topological specification of the neuron was made by disregarding the continuation points and reconstructing a "reduced" tree (Fig. 36).

Computer quantitation of the structures traced has four main purposes in this system. First, there is the simple graphical one, that of displaying, singly or in combination, the items that have been traced. With the particular graphics system used, this can be done either statically from any direction of view or dynamically by providing a real-time rotation of the object in space. Second, Ware (1971) has described a graph theoretical method of representing the topological structure of any arbitrary tree in a unique, canonical form. This enables one to compare the topology of trees, independent of their topography, either visually or by computational methods (Fig. 37). Third, homologous structures in different animals or similar structures in the same animal can be directly compared topographically either visually or computationally. Fourth, and perhaps ultimately most useful, synapse information from each individual neuron included during the recording can be used to form an intercellular connectivity map of a completely traced ganglion, a procedure prohibitively time-consuming if attempted by hand.

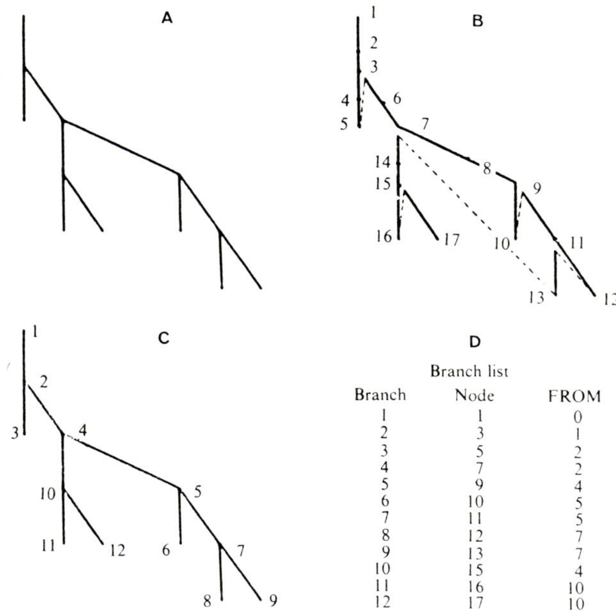

Branch list

Branch	Node	FROM
1	1	0
2	3	1
3	5	2
4	7	2
5	9	4
6	10	5
7	11	5
8	12	7
9	13	7
10	15	4
11	16	10
12	17	10

FIG. 36. Illustration of the principle used in tracing a branched structure. (A) A tree. (B) A possible tracing sequence. Broken line: return from a terminus to the most recently recorded unsatisfied branch point. (C) Numbering of branch points only. (D) Topological specification of the tree. (From Levinthal and Ware, 1972.)

The AGT-50 system with electron microscope cinematographic reconstruction of the central nervous system of the rotifer *Asplanchna brightwelli* was used in a study designed to determine the degree of identity among the nervous systems of different animals of the same species. Possible causes for differences can be divided into specific ones, such as genetic or environmental effects, and nonspecific ones, meaning random fluctuations that occur not as definite responses to any particular influence but simply because development at the fine-structural level may not follow a strictly controlled sequence of events. The rotifer was considered suitable for this study, since it reproduces by diploid parthenogenesis. Therefore by cloning a single parent animal it is possible to eliminate genetic and differential environmental effects as the source of any differences observed among members of the clone. In addition the rotifer brain possesses only about 200 cells and has a very simple organization of its neuropil so that complete reconstruction is possible. The systematic study of 13 females (Ware, 1971) showed first of all that within any one animal there is, with a few exceptions and except on the midline where the

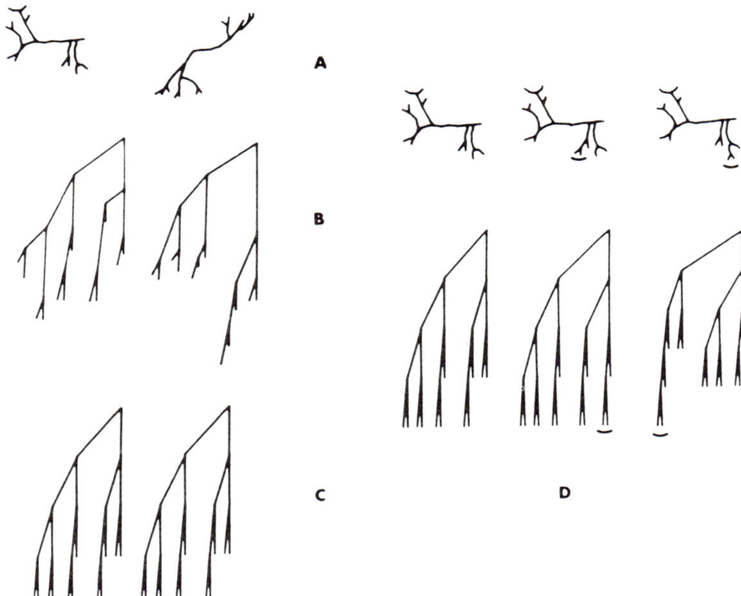

FIG. 37. Illustration of the formation of topological maps. The cells of (A) have been converted into maps in (B) and (C). The topologies are the same; only the topographies have been changed to detect the coincidence. In (C) all lengths are arbitrary. In (B) the same manipulation has been done and the x distances are arbitrary, but the y distances are the contour lengths which, in the original structures, separate the branch points. The second type of conversion can be carried out in a straightforward manner by means of the corresponding FROM list only, whereas the first requires all the three-dimensional information available about the paths between the nodes. In (D) one of the trees has been slightly altered to demonstrate the effect on the graph. (From Levinthal and Ware, 1972.)

processes of many bilaterally paired cells meet, perfect bilateral symmetry in cell body position and type, shape of processes within the neuropil, and location of identified synapses. Even seemingly random invaginations of one cell body into another are reproduced in mirror image on the two sides of the midline. Second, the comparison of many animals showed that the exceptions to bilateral symmetry seen in any one animal are exactly reproduced in different animals. All cell bodies could be unambiguously matched, since nearly every one has some distinctive characteristic that differentiates it from its neighbors. In the neuropil every identified synapse and characteristic organization of fibers examined was found in all animals in which they were studied. Initial studies on the embryology of the rotifer were reported by Seldon (1972, personal communication) using the same procedures. By working backward from the adult he found that every brain neuron could be followed to a stage where no neuropil was present. His results indicate that the only apparent anastomosing of nerve cells observed in the adult, between two bilaterally paired motoneurons receiving very strong synaptic input from two bilaterally paired sense organs, may be real in that the binucleate structure of the adult seems to be formed from two neurons which are embryologically distinct. The studies of Horridge and Meinertzhagen (1970a,b) mentioned in Section III,A,2 demonstrated the absence of connectivity errors in the construction of the repeated units of the dipteran compound eye in which, because of their large number, it might be expected that the system as a whole could function with a small amount of error in the construction of some of its elements. The findings in the rotifer brain, in which each fiber is distinct and there is no redundancy in information processing, show that the developmental process in this much more primitive animal can be just as precise to the extent that each fiber reaches the exact location necessary for the carrying out of its unique function.

The AGT-50 system, with hardware and software improvements designed to increase the number of tracing and display options as described by Levinthal *et al.* (1974), has been applied to an analysis of both the final connectivity and the development of specific retina–lamina connections in the small crustacean *D. magna.* A computer illustration made from a low-magnification movie through the eye and optic ganglion is shown in Fig. 38A. Using the offspring of females, which reproduce by diploid parthenogenesis, it was shown in series of up to 1500 thin sections that each of the 22 fused ommatidia projects to a type of optic cartridge composed of five postsynaptic lamina neurons (Macagno *et al.*, 1973). In a detailed analysis of the direct projection pattern of the same identified ommatidia onto their cartridges in different animals, a unique and reproducible connectivity was demonstrated between the eight retinular cells of the ommatidium and the five postsynaptic lamina neurons. Three-dimensional tree representations of the synaptic terminal branching pattern of the same retinular axons in several different animals show that there is some variation in the number and topographical distribution of the branches, although the connectivity pattern itself remains identical. A synaptological study of this region showed that the number of total synaptic areas in each retinular terminal is large in all cases but varies from as few as 20 to as many as over 100 between the same terminal and postsynaptic lamina cell in different animals. This suggests the possibility that, whereas connectivities are strictly controlled developmentally, the strength and exact morphology of synaptic interaction may be subject to other influences.

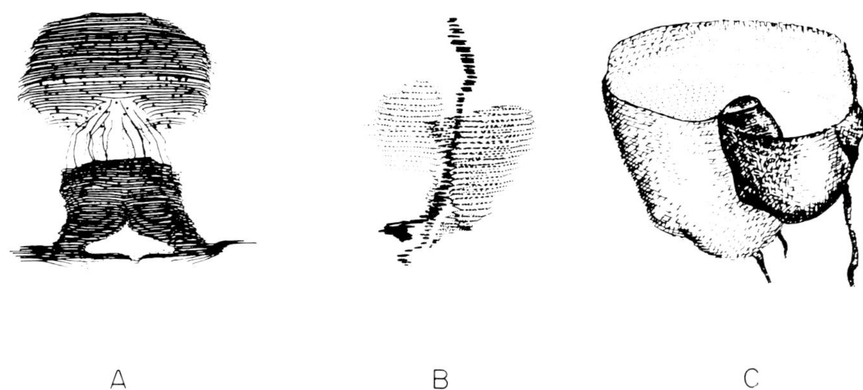

FIG. 38. *Daphnia.* (A) Low-magnification computer reconstruction of the eye, optic fibers, and optic lamina in an adult *Daphnia.* (From Macagno *et al.*, 1973.) (B) Higher-magnification reconstruction of lamina cells 1 and 3 of the same developing cartridge shown in relation to the lead optic fiber. It can be seen that cell 1 has a well-defined axon, while cell 3 appears to be in the process of initial axon outgrowth. (From LoPresti *et al.*, 1973.) (C) Even higher magnification reconstruction of the anterior half of a lamina neuroblast undergoing the gliallike reaction with its lead optic fiber (not shown). (From LoPresti *et al.*, 1973.)

Reconstructions were also made from series of 500 to 800 sections of all 22 differentiating ommatidia and their optic cartridges at five different embryonic stages spanning 10 hours of a 50-hour gestation. LoPresti *et al.* (1973) reported in these studies that the spatial array of ommatidia mapping one to one on the similar array of cartridges could be explained by successive waves of ommatidial differentiation near the midline of the eye followed by mutual lateral migration patterns of these ommatidia and the initially undifferentiated lamina cells to which their axons project. The vacated midline positions are then occupied by new undifferentiated cells in both eye and lamina, at which point the next wave of differentiation ensues. This process is repeated until all ommatidia are differentiated. Study of the same ommatidia and developing cartridges at the different embryonic stages showed that ommatidial differentiation involves precocious axon proliferation by only one of the eight retinular cells of the ommatidium. This "lead fiber" was shown to grow close to the midplane of the lamina successively over the surface of five undifferentiated lamina neuroblasts as it proceeds from the base of the eye posteriorly. Three-dimensional reconstructions of these five cells at higher magnifications revealed that after initial contact each neuroblast undergoes a transient gliallike reaction with the lead fiber. A computer illustration of part of a lamina cell in this configuration is shown in Fig. 38C. As the whole developing cartridge migrates laterally, the five lamina neuroblasts were seen to proliferate their axons in the order in which they had first been contacted in the midplane, and therefore in the order in which they underwent the gliallike reaction with the lead fiber. An illustration of lamina cells 1 and 3 of a cartridge are shown with the lead optic fiber in Fig. 38C. A serial section study at higher magnifications (LoPresti *et al.*, 1974) revealed the presence of transiently formed gap junctions between the lead fiber and lamina neuroblast at this characteristic interaction but nowhere else, a fact supporting the notion that the growing retinular axon successively exerts an inductive effect on the differentiation of the five eventual postsynaptic cells of its optic cartridge.

Still another computer graphics installation of this general type has been set up at Caltech for the study of the nematode by Russell and coworkers (Ware *et al.*, 1975), and of the catfish retina by Naka (see Section IV,B). In both cases reconstruction is being carried out by the cinematographic procedure described (Section II,A,3), and data are recorded using a computer system made up of a PDP 11/45 with a Vector General–based graphics output. The attractiveness of the nematode for detailed morphological examination is that mutations are easily induced so that one can hope to locate the neurological correlate of genetically determined behavioral alterations. The nematode, like the rotifer, possesses a very small central nervous system composed of about 200 cells, although it is somewhat less compact. The sensory system consists of organs on the snout which communicate with the central nervous system, called the nerve ring because it completely encircles the axially located esophagus, by means of six symmetrically placed nerve bundles passing through the anterior end of the animal. The nerve cells surrounding the nerve ring are shown in Fig. 39. Those that have fibers in the six anterioraly directed nerves are marked according to whether their fibers are in the dorsal, ventral, or lateral bundles. Constancy of structure from animal to animal does not seem to be as complete as in the rotifer. For example, in one animal out of five studied in a particular region a very characteristic glial cell was seen to be missing. In another case a normally bipolar cell was seen in one instance to be monopolar, although the single fiber divided some microns from the soma and sent its two branches to regions corresponding to those occupied by the two unbranched fibers of the normal bipolar cell. Motor output to the 32 identified head muscle cells is easily studied completely within the nerve ring and nearby cords since, unlike other animals, nematodes have muscle cells which send processes to the central nervous system rather than vice versa. One of the immediate results of tracing the ciliated cells of the sense organs to the region of muscle synapsis in the nerve ring has been the identification of an evolutionarily primitive sensory motor cell, that is, a sensory cell that makes direct synaptic contact with a muscle.

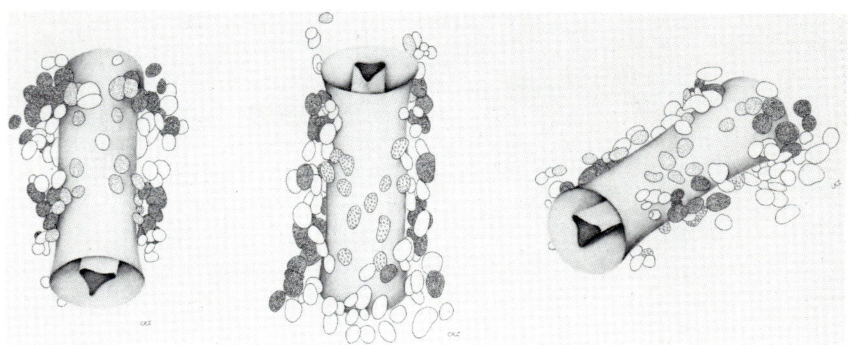

FIG. 39. Reconstruction of the nerve ring region of the nematode *Caenorhabditis elegans* made from serial electron micrographs showing three views of the same animal (left to right: from above, from below, from the side). Nerve and glial nuclei are shown clustered about the cylindrical esophagus. The nerve ring fits into the slightly anterior region where few cell nuclei are present. Cell identification: dorsal sensory, light dot; ventral sensory, x; lateral sensory, dark dot; chemosensory (amphidial), circles. (Original illustration.)

Quantitation of various neuronal parameters which can be easily observed in Golgi-stained material is considered desirable by many investigators. Those parameters studied are ones such as the extent of dendritic spread, total dendritic length, degree (order) of branching, orientation of the dendritic tree, and density of dendritic spines. These are relevant first of all just as a means of neuronal classification. In addition, they are often used to correlate alterations in neuron structure with alterations in behavior, for example, as a function of age of the animal particularly during critical developmental periods, or following specific sensory input deprivation, lesions of input cells, or extended periods of enforced immobility. Two computer graphics systems have recently been set up to speed the collection and analysis of such data, one at the University of California at Los Angeles (Lindsay, personal communication) and another at Washington University in St. Louis (Woolsey *et al.*, 1972; Wann *et al.*, 1973). A primitive precursor of these systems was first described many years earlier by Glaser and Van der Loos (1965), using analog computation with no graphics output.

The system described by Woolsey *et al.* (1972) and Wann *et al.* (1973) is made for use with thick sections, although no provision is made for tracing in several sections sequentially. In this case coordinate information is directly available to the computer (a PDP 12), since it controls the x-y position of the stage of the microscope used for observation, as well as the z position, of the microscope focus, by means of stepping motors. The procedure for tracing a neuron consists in the operator moving the microscope stage and focus, through the computer, so that desired points along a neuron branch are brought sequentially into coincidence with the center of the field of view. When a point is so centered the computer is keyed by the operator to record its coordinates within the specimen and any additional data designations desired (branch point, terminus, etc.). An illustration of the type of reconstruction obtained is shown in Fig. 40.

The system at the University of California at Los Angeles is also designed for use with thick (100–200 μm) sections and includes a provision for tracing through sequential sections. It consists of a manually operated microscope which displays the object on a closed-circuit television screen and a PDP 8 for data storage. Two perpendicular edges

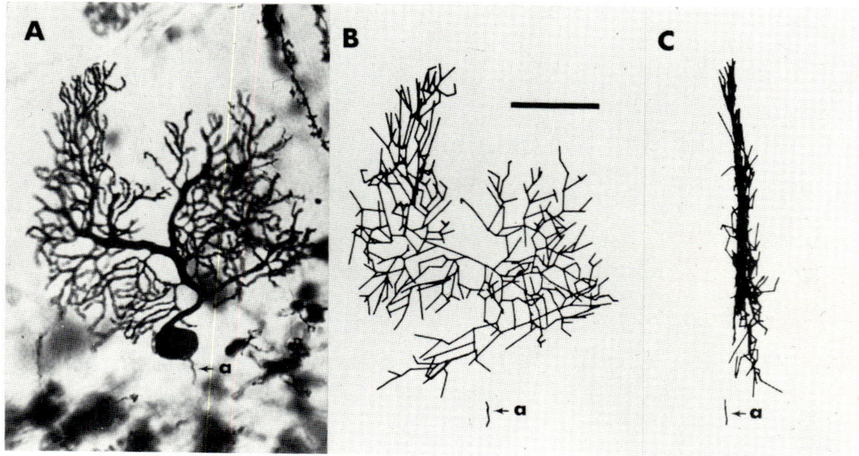

FIG. 40. (A) Photomicrograph of a Purkinje cell from a parasagittal section of the cerebellum of a mouse. Golgi–Cox method, 150-μm-thick section. The number of data points recorded from this cell was 553, and the time required to enter them was about 2 hours. The total three-dimensional length of the dendritic tree is 4435 μm, the number of dendritic segments is 414, and the highest order of bifurcation is the twenty-third. (B) Photograph of the computer oscilloscope display generated from the data logged from the Purkinje cell shown in (A). The cell body is not depicted. Observe the excellent agreement between the appearance of this reconstruction and the actual biological specimen. (C) Photograph of the computer-generated oscilloscope display from the Purkinje cell shown in (A). A coordinate transformation of 90° about the vertical axis has been applied to the data displayed in (B). Note the planar distribution of the Purkinje cell dendrite which is viewed as if sectioned parallel to the long axis of a cerebellar folium. All figures are reproduced at the same magnification. Bar, 50 μm; a, axon. (From Woolsey, 1973.)

of the television monitor are equipped with linear arrays of sonic sensors (Graf Pen), and x-y coordinate information is obtained by placing a sonic pen on the screen at the point of interest. The z coordinate is determined by the computer from a potentiometer on the microscope fine focus adjust. As in the other systems described one of a number of data categories can be associated with each point along with its coordinates.

Development of computerized reconstruction techniques specifically designed for purposes of medical illustration and instruction has been described by Gott (Gott *et al.*, 1969; Willey *et al.*, 1973). Procedures for transferring coordinate information from tracings of micrographs of sequential sections to the computer are somewhat the same as have been described although noninteractive. The significant difference is that the computer drawings of these data are made with perspective and hidden line removal. This feature is usually sufficient to enable a trained medical illustrator to complete his work with only a few views from different angles of the reconstructed object (Fig. 41). In a particularly interesting example of how a computer-produced image, although crude by artistic standards, can be instructive, Gott, in conjunction with the Cardiovascular Research Laboratory, Loma Linda University School of Med-

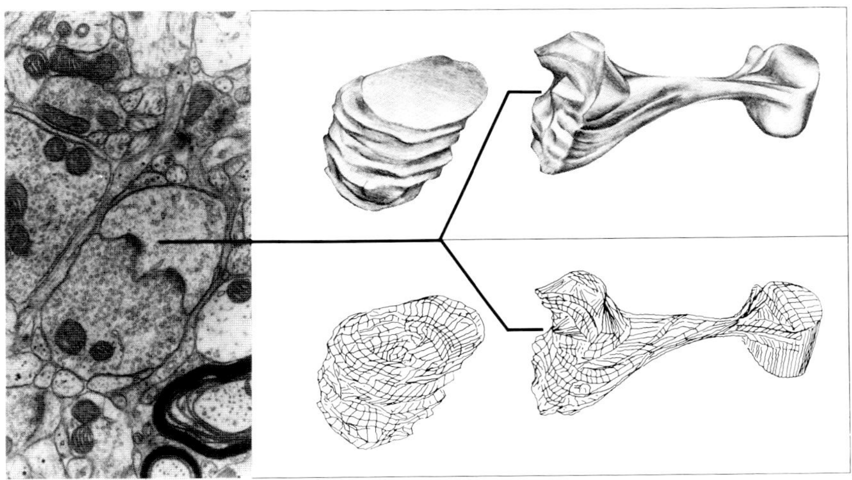

FIG. 41. The nerve junction drawing shown above was prepared using the Aerospace Corporation's computer graphics technology. A series of thin tissue slices was photographed in a Loma Linda University electron microscope at ×31,800 magnification. The upper micrograph identifies the synapse outline, while the lower line drawing shows cues added to a computer reassembly of many such sections. The finished rendering was first traced in outline by the medical illustrator (for accuracy and speed), and then shaded for detail. (Gott, original illustration.)

icine, has produced a computer animation film of a beating human heart (Gott, 1970). A three-dimensional computer model of the heart and its four chambers in diastolic position was made by means of a through-focus series of sections produced by an x-ray polytome of a postmortem heart at 5-mm intervals. Animation for each chamber was prepared using the results of quantitative analysis of cineangiograms. Cineangiography data were provided by high-speed motion pictures obtained from x-ray image–amplified pictures of the heart pumping blood containing a blood-soluble radiopaque material. The measured changes in chamber dimensions were plotted against time, and ultrasonic techniques provided an independent determination of the same spatial data. Construction of the final, time-dependent computer model did not require a separate reconstruction from serial sections for each position, but rather simply a scaling of the initial diastolic image using these quantitative measurements. Proper phasing of the motion of each chamber was obtained both from cineangiography and the electrocardiogram of a healthy patient with a heartbeat rate of 60 per minute. A time sequence of the resulting reconstructions is shown in Fig. 42.

B. Automatic

A computer-based system at Carnegie-Mellon University for the automatic analysis of serial light microscope micrographs of intracellularly stained neurons has been described by Reddy *et al.* (1973). This differs from the ones mentioned in that it not only has the interactive tracing capabilities described, but in addition is being designed to work with internal images of micrographs that have been digitized and image-processed. This permits the alignment of sections and the following of nerve processes through the sections to be accomplished largely by the computer. The object of this analysis is to provide substantially the same types of quantitative information sought by Selverston and co-workers, that is, dendritic diameters along their entire length and exact topography and topology of the neuron, although more rapidly and with much less human intervention. These data can then be related to physiological or developmental aspects of the nervous system or permit the morphological comparison of homologous neurons in different ganglia which have been studied and compared electrophysiologically prior to embedding and sectioning.

The center of Reddy's system is a PDP 10 computer with 192K words of core storage. The large core size is necessary for the analysis of the digitized micrographs, each of which has its intensity measured over a grid of 1000 × 1000 points. With an average of 250 sections per ganglion (lobster abdominal ganglion), it can be seen that the storage capabilities

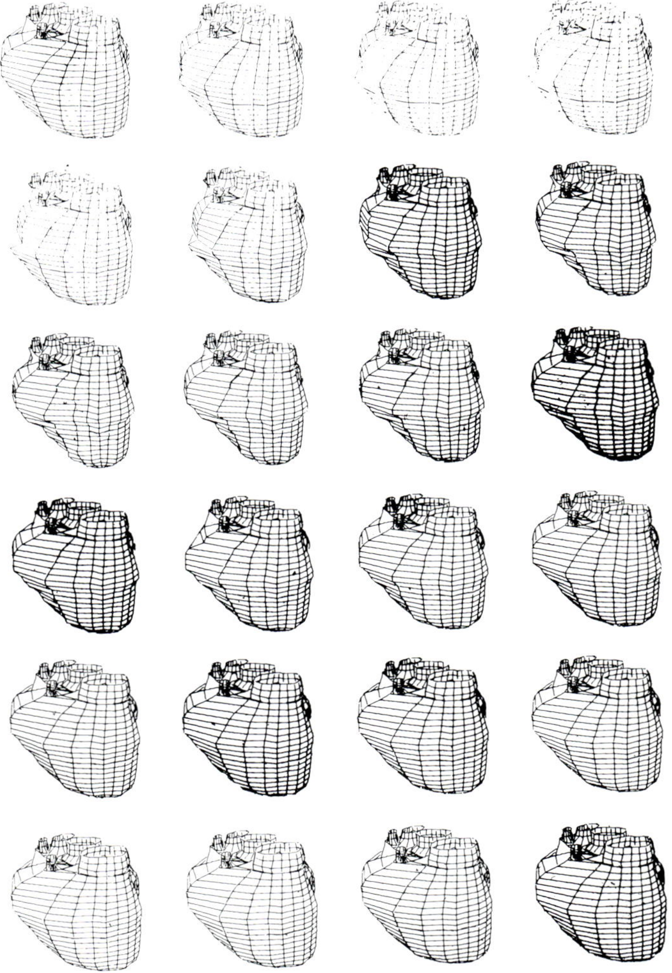

FIG. 42. Complete heart cycle. (From Gott *et al.*, 1969.)

must be large. To help ease this problem, image-compression programs have been developed that reduce permanent storage requirements by a factor of 10. Conceptually, there are two main steps in creating a reconstructed neuron. The first is to obtain a computer image of all nerve profiles as they appear on each section through the ganglion. The second is to have the computer form the proper connections of these individual profiles between the sections. In the semiautomated use of this system an

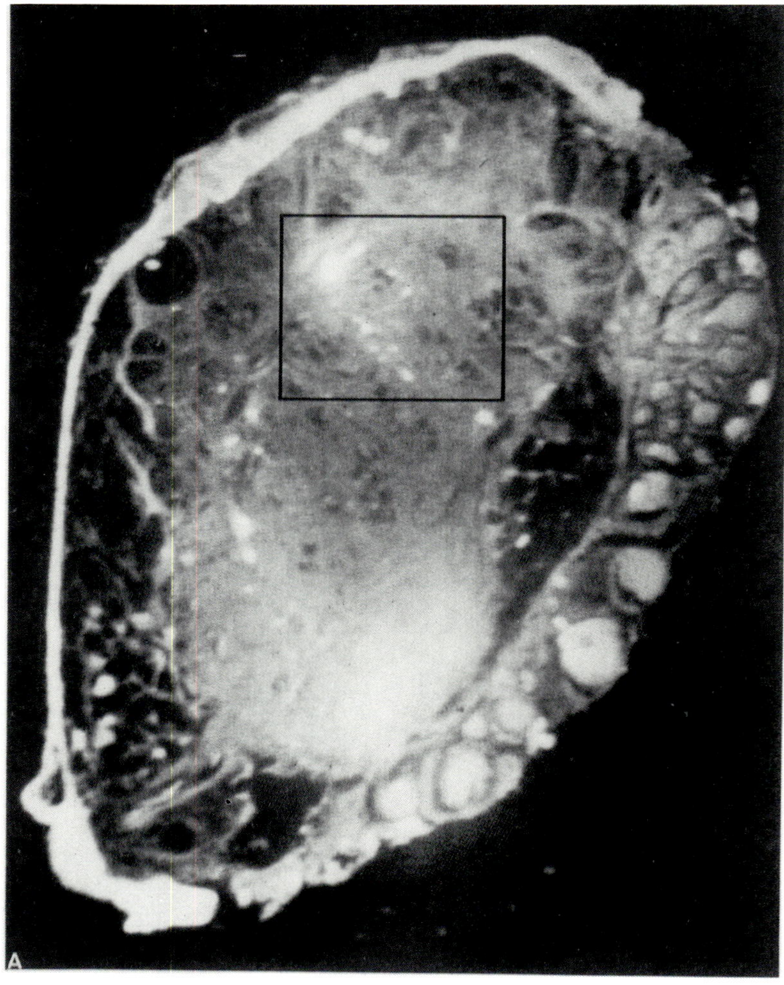

FIG. 43. (A) Fluorescence micrograph of a section containing branches of a power stroke swimmeret motoneuron which has been injected with the dye Procion yellow. (B) Computer image of that portion of the ganglion outlined in (A), printed on the Xerox Graphic Printer. (C) Dendritic profiles obtained by thresholding and edge detection on the computer image of (B). (From Reddy *et al.*, 1973.)

operator traces the contours of each item of interest by methods similar to those already discussed and labels them in each section for further use by the connectivity programs. In the automated state a commercially available image dissector (Information International, Inc.) measures the

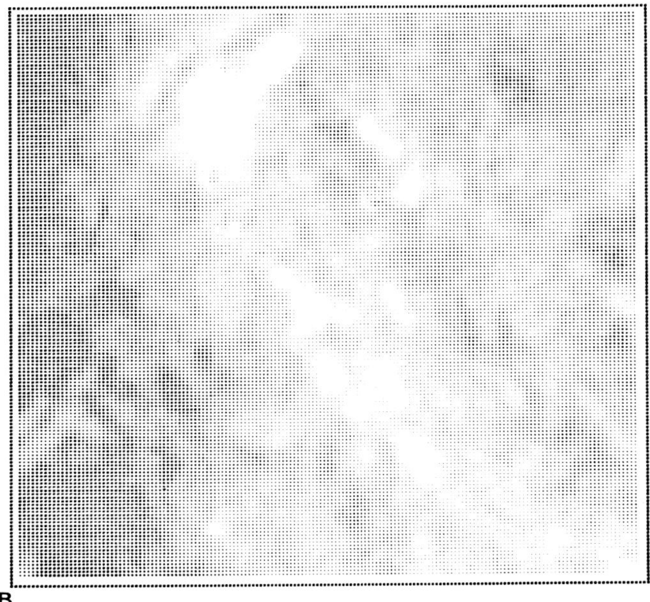

B

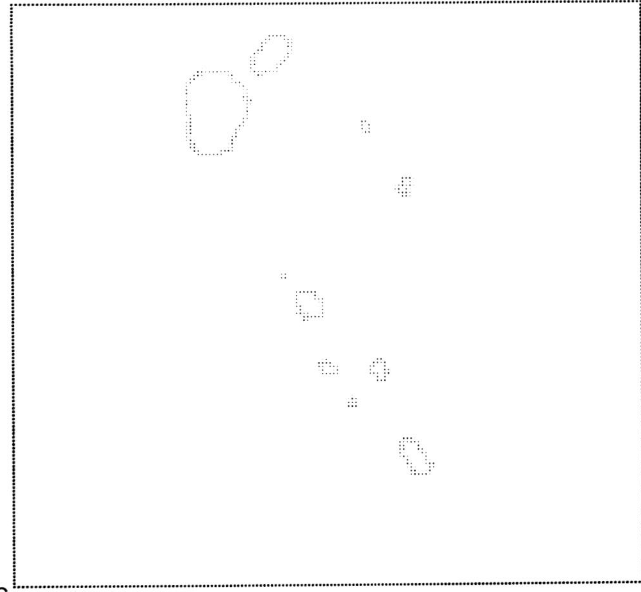

C

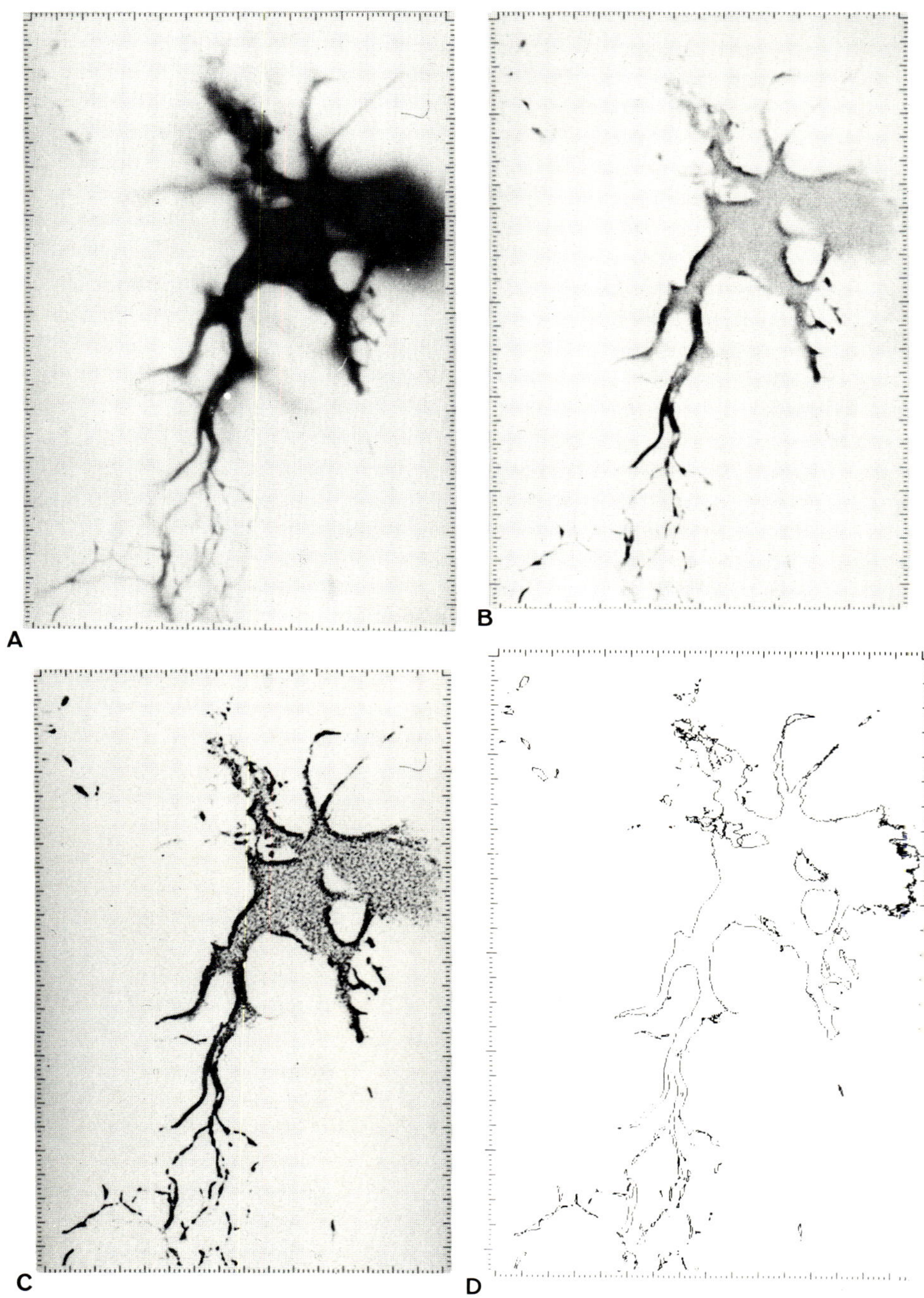
A
B
C
D

optical density of each of 10^6 points over the micrograph of each section and stores them. These optical densities of course cannot be used by the connectivity programs, since they do not constitute a machine-recognizable image by themselves. First, each set of micrograph densities must be image-processed and analyzed to abstract nerve profiles for each micrograph. Standard concepts such as thresholding and edge detection by differentiation are used to convert the densities into these contours. The results are shown in Fig. 43. Once profiles are available for each section, the assembly of sections can proceed. This requires locating dendritic profiles which are in some sense near each other in adjacent sections and logically connecting them in a machine-recognizable way. As the tree is built from pieces in this way, branch points are identified and recorded and diameters are associated with each dendritic segment. The graphical display of the reconstructed neuronal structure is, first, an end in itself. Second, it enables an operator to modify the computer-generated reconstruction or to aid actively in the connectivity formation process when a logical impasse is reached by the machine. Display problems dealing with the generation of transparent surfaces, the elimination of hidden lines, and the manipulation of the image at will to see selected portions at high magnification or bring other objects into the field of view are also being studied.

Another automatic system is in operation at the Jet Propulsion Laboratories (JPL) of the California Institute of Technology (Weinstein and Castleman, 1971; Castleman *et al.*, 1974) for use with Golgi-stained neurons in the catfish retina. The principal differences are: (1) It is set up to work with both histological and optical sections, although only optical sectioning has been fully implemented; and (2) its output can be either a genuine tree structure, with branch points and fiber diameters appropriately cataloged, or simply a complete gray-level display of the tissue presented from any angle of view. The difference between these two is that in the first case the computer "knows" what it is displaying topologically in that machine intelligence has been used to process nerve profiles on each section and connect them between sections. In the second case a series of intensities is displayed without any machine internal knowledge of what structures might be represented.

FIG. 44. (A) Digitized optical section of a Golgi-stained catfish retina neuron. (B) Same optical section "deblurred" by the method described in the text. Note that objects that appear on many sections (large fibers, cell nucleus) tend to have their intensity reduced, while objects that occur only in the section undergoing correction are enhanced. (C) The image of (B) to which edge enhancement as described in the text has been applied. (D) The outlining procedure described in the text applied to (B). (Castleman, original illustration.)

Optical sections are digitized directly through a light microscope over a raster of 1000 × 1000 points. This is done using a high-numerical-aperture objective so that optical sections can be usefully taken every micron. The first stage of the reconstruction consists of image-processing these individual section images. This itself consists of two steps. The first, unique to the JPL system, is to subtract out of every section image the image of both neighboring sections which have been computationally "blurred" to correct for the fact that, in the section being operated upon, they appear slightly out of focus. The results of this residual image removal applied to an original optical section shown in Fig. 44A are shown in Fig. 44B. The second step, which is the only one applied to histological sections, is to mathematically enhance the appearance of edges in each corrected section. This is accomplished in a manner exactly analogous to the process of lateral inhibition in animal visual systems. The results of this operation applied to Fig. 44B are shown in Fig. 44C. A third type of image-processing may or may not be carried out, depending on the type of final reconstruction desired, and is used with both optical and histological sections. This is to eliminate the gray-level information of the picture and replace it simply by outlines of structures present. The procedure consists of representing all intensities above a certain threshold level as 1 and all others by 0. The resulting two-level silhouette image is reduced to an outline by discarding interior points simply by setting all 1s to 0s if they are surrounded on all four sides by 1s. This type of outlining applied to the image-processed section in Fig. 44B is shown in Fig. 44D. Note that this procedure does not use any kind of image analysis. The computer does not know anything about the connectivities, continuities, locations, or even the existences of any structures present.

The second stage in producing a simple visual reconstruction consists of stacking the image-processed section images, rotating the block of them by the desired angle, and projecting the block onto a single plane. An example of a neuron reconstructed in this manner is shown in a complete gray-level stereopair in Fig. 45. The same procedure can also be carried out with the outline section images, thereby giving a skeleton view of the neuron.

This system is currently being used by Naka (personal communication) in a combination of neurophysiological and histological approaches to the classification of neurons in the catfish retina. The traditional method of examining a fixed retina is to take sections 50- to 100-μm thick perpendicular to the plane of the retina. All existing schemes of neuron classification, based largely on Golgi-stained material, depend on images of neurons seen in this plane. The problem is that the dendritic spread on many neurons in the catfish retina may be 500 to 800

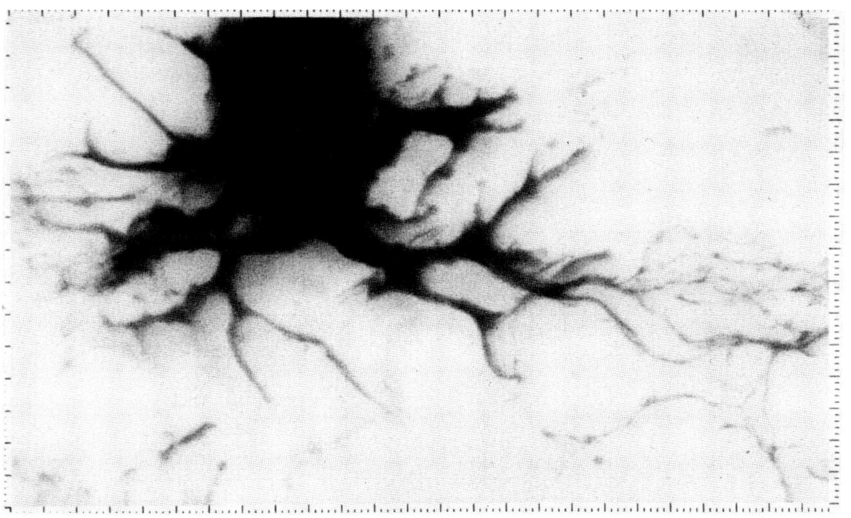

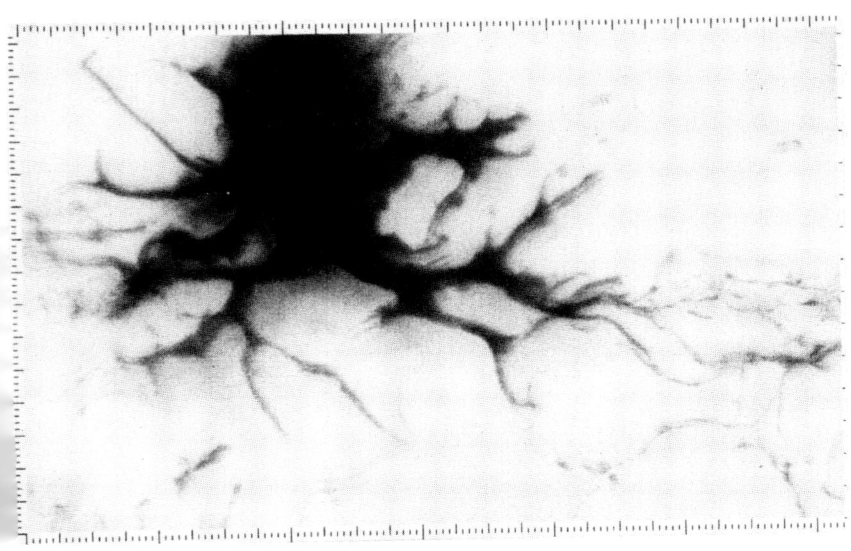

FIG. 45. Stereo pair of the same neuron shown in Fig. 44 produced by computer tilting of the planes of a stack of optical sections. (Castleman, original illustration.)

μm in the horizontal direction, whereas the dendritic height is only 30 to 50 μm vertically. In this case, then, the logical choice for a direction of sectioning is horizontally, so that all the neurons can be contained within a single section. Using the JPL system Naka cuts the retina in a horizontal plane and reconstructs neuron images as seen in a perpendicular plane, thus allowing direct visual comparison with standard cell descriptions. The reason for cutting the specimen in such a way as to obtain an entire Golgi or intracellularly stained neuron within a single thick section is that the block can then be remounted and cut for further examination by electron microscope cinematography. This ultimately will provide a picture of a single neuron based on electrophysiological recordings, overall morphological type, and characterization of the types and distribution of synapses on its dendritic tree.

V. Conclusion

Reconstruction of objects from serial sections by standard techniques unfortunately requires a large amount of effort which is not directly related to the purpose of the investigation. Historically, its first applications were in embryology, where it enabled workers to see systematically for the first time the spatial relationships of developing organs and systems. Its application in medicine and biology has permitted an understanding of both normal and abnormal tissues which was impossible with the ordinary histological techniques used in conjunction with single-section analysis. With the advent of the electron microscope, it has enabled us to understand the morphology of some of the smallest things resolvable by that instrument. Since it is one of the few techniques we have for making a detailed examination of fine structure, it must be hoped that with proper systematization it can be used to progress from the modeling of objects that can be observed in under 100 sections to the morphological analysis, by whatever means, of those nonmolecular intercellular relationships that are responsible for the integrative capacities of a system of cells.

References

Allen, A. (1969). *In* "The Retina: Morphology Function and Clinical Characteristics" (B. R. Straatsma, M. O. Hall, R. A. Allen, and F. Crescitelli, eds.), UCLA Forum in Medical Sciences, No. 8, pp. 101–143. Univ. of California Press, Berkeley.

Andersson-Cedergren, E. (1959). *J. Ultrastruct. Res., Suppl.* **1**, 1.

Andre, J. (1959). *Ann. Sci. Nat., Zool. Biol. Anim.* **12**, 283.

Andre, J. (1962). *J. Ultrastruct. Res., Suppl.* 3, 1.

Barajas, L. (1970). *J. Ultrastruct. Res.* **33**, 116.

Barajas, L. (1971). *Science* **172**, 485.

Barajas, L., and Latta, H. (1963). *Lab. Invest.* **12**, 257.

Barajas, L., and Muller, J. (1973). *J. Ultrastruct. Res.* **43**, 107.
Barrett, J. N., and Crill, W. E. (1971). *Brain Res.* **28**, 556.
Barrett, J. N., and Graubard, K. (1970). *Brain Res.* **18**, 565.
Bentley, D. R. (1970). *J. Insect Physiol.* **16**, 905.
Bernhard, W. (1966). *J. Nat. Cancer Inst., Monogr.* **23**, 13.
Blechschmidt, E. (1954). *Z. Anat. Entwicklungsgesch.* **118**, 170.
Bodian, D. (1966a). *Science* **151**, 1093.
Bodian, D. (1966b). *Bull. Johns Hopkins Hosp.* **119**, 16.
Born, G. (1879). *Morphol. Jahrb.* **5**, 62.
Born, G. (1882). *Morphol. Jahrb.* **8**, 188.
Born, G. (1883). *Arch. Mikrosk. Anat.* **22**, 584.
Born, G. (1888). *Z. Wiss. Mikrosk.* **5**, 433.
Boschek, C. B. (1971). *Z. Zellforsch. Mikrosk. Anat.* **118**, 369.
Boycott, B. B., and Kolb, H. (1973). *J. Comp. Neurol.* **148**, 115.
Braitenberg, V. (1966). *Z. Vergl. Physiol.* **50**, 212.
Braitenberg, V. (1967). *Exp. Brain Res.* **3**, 271.
Bush, V. (1952). *Science* **115**, 649.
Cajal, S. R. (1933). *Trab. Lab. Invest. Biol., Madrid* **28**, Appendix, 1.
Cajal, S. R., and Sanchez, D. (1915). *Trab. Lab. Invest. Biol., Madrid* **13**, 1.
Campos-Ortega, J. A., and Strausfeld, N. J. (1972). *Z. Zellforsch. Mikrosk. Anat.* **124**, 561.
Castleman, K. R., Fuchs, H., and Schantz, M. J. (1974). *IEEE Trans. Circuit Theory* (in press).
Christensen, A. K., and Chapman, G. B. (1959). *Exp. Cell Res.* **18**, 576.
Christensen, B. N. (1973). *Science* **182**, 1255.
Cohen, A. I. (1963). *Exp. Eye Res.* **2**, 88.
Cohen, A. I. (1964). *Invest. Ophthalmol.* **3**, 198.
Colonnier, M. (1964). *J. Anat.* **98**, 47.
Conradi, S. (1969a). *Acta Physiol. Scand., Suppl.* **332**, 5.
Conradi, S. (1969b). *Acta Physiol. Scand., Suppl.* **332**, 49.
Conradi, S. (1969c). *Acta Physiol. Scand., Suppl.* **332**, 65.
Conradi, S. (1969d). *Acta Physiol. Scand., Suppl.* **332**, 85.
Conradi, S., and Skoglund, S. (1969a). *Acta Physiol. Scand., Suppl.* **333**, 5.
Conradi, S., and Skoglund, S. (1969b). *Acta Physiol. Scand., Suppl.* **333**, 53.
Davis, W. J. (1970). *Science* **168**, 1358.
de Montebello, R. L. (1967). *Int. Congr. Stereol. Proc., 2nd, Chicago* pp. 306–307.
de Montebello, R. L. (1969). *Ann. N. Y. Acad. Sci.* **157**, 487.
Diers, L. (1966). *J. Cell Biol.* **28**, 527.
Dietrich, W. (1909). *Z. Wiss. Zool.* **92**, 465.
Dixon, A. D., and Howarth, P. (1958). *J. Anat.* **92**, 162.
Dowling, J. E. (1970). *Invest. Ophthalmol.* **9**, 655.
Dowling, J. E., and Boycott, B. B. (1966). *Proc. Roy. Soc. Ser. B* **166**, 80.
Dublin, M. W. (1970). *J. Comp. Neurol.* **140**, 479.
DuPraw, E. J. (1968). "Cell and Molecular Biology." Academic Press, New York.
DuPraw, E. J. (1970). "DNA and Chromosomes." Holt, New York.
Eccles, J. C. (1964). "The Physiology of Synapses" Springer-Verlag, Berlin and New York.
Elfvin, L.-G. (1963a). *J. Ultrastruct. Res.* **8**, 403.
Elfvin, L.-G. (1963b). *J. Ultrastruct. Res.* **8**, 441.

Elfvin, L.-G. (1971a). *J. Ultrastruct. Res.* **37**, 411.
Elfvin, L.-G. (1971b). *J. Ultrastruct. Res.* **37**, 426.
Elfvin, L.-G. (1971c). *J. Ultrastruct. Res.* **37**, 432.
Elias, H. (1935). *Z. Wiss. Mikrosk.* **52**, 424.
Elias, H. (1948). *Amer. J. Anat.* **84**, 311.
Elias, H. (1949). *Amer. J. Anat.* **85**, 379.
Elias, H. (1971a). *Science* **174**, 993.
Elias, H. (1971b). *Anat. Rec.* **169**, 310.
Evoy, W. H., and Kennedy, D. (1967). *J. Exp. Zool.* **165**, 223.
Fifkova, E. (1972). *Exp. Neurol.* **35**, 458.
Fink, R. P., and Heimer, L. (1967). *Brain Res.* **4**, 367.
Fisher, L. J. (1972). *Nature (London)* **235**, 391.
Foos, R. Y., and Miyamasu, W. (1973). *J. Comp. Neurol.* **147**, 447.
Frank, K., and Fuortes, M. G. F. (1957). *Fed. Proc., Fed. Amer. Soc. Exp. Biol.* **16**, 39.
Garvey, C., Young, J., Simon, W., and Coleman, P. D. (1972). *Anat. Rec.* **172**, 314.
Gassner, G. (1969). *Chromosoma* **26**, 22.
Gaunt, W. A. (1971). "Microreconstruction." Pitman, London.
Gillette, R., and Pomeranz, B. (1973). *Science* **182**, 1256.
Gillies, C. B. (1972). *Chromosoma* **36**, 119.
Glaser, E. M., and Van der Loos, H. (1965). *IEEE Trans. Bio-Med. Eng.* **12**, 22.
Globus, A., Lux, H. D., and Schubert, P. (1968). *Brain Res.* **11**, 440.
Goss, C. M. (1966). "Gray's Anatomy." Lea & Febiger, Philadelphia, Pennsylvania.
Gott, A. H. (1970). "Teaching Heart Function by Computer Animation." Aerospace Corp., El Segundo, California.
Gott, A. H., Kubert, B. R., Bowyer, A. F., and Nevatt, G. W. (1969). *AFIPS Conf. Proc., Spring Joint Comput. Conf., Boston* **34**, 637.
Hanson, T. E. (1972). *Annu. Rep. Div. Biol., Calif. Inst. Technol.*, p. 41.
Hegre, E. S. (1946). *Stain Technol.* **21**, 161.
Hegre, E. S. (1951). *Va. J. Sci.* **2**, 10.
Hegre, E. S. (1966). "The Anatomy of the 9 mm Pig Embryo as Seen in Serial Sections" (16 mm educational film in color). Commonwealth Motion Pictures, Richmond, Virginia.
Hegre, E. S. (1967). *Int. Congr. Stereol., Proc., 2nd, Chicago* p. 293.
Hinds, J. W. (1972a). *J. Comp. Neurol.* **146**, 233.
Hinds, J. W. (1972b). *J. Comp. Neurol.* **146**, 253.
Hinds, J. W., and Ruffett, T. L. (1973). *J. Comp. Neurol.* **151**, 281.
His, W. (1880). "Anatomie menschlicher Embryonen." Vogel, Leipzig.
His, W. (1904). "Die Entwicklung des menschlichen Gehirns." Hirzel, Leipzig.
Hoffman, H.-P., and Avers, C. J. (1973). *Science* **181**, 749.
Hollande, A. (1942). *Arch. Zool. Exp. Gen.* **83**, 1.
Horridge, G. A., and Meinertzhagen, I. A. (1970a). *Proc. Roy. Soc. Ser. B* **175**, 69.
Horridge, G. A., and Meinertzhagen, I. A. (1970b). *Z. Vergl. Physiol.* **66**, 369.
Hubel, D. H., and Wiesel, T. N. (1962). *J. Physiol. (London)* **160**, 106.
Hubel, D. H., and Wiesel, T. N. (1963). *J. Physiol. (London)* **165**, 559.
Hubel, D. H., and Wiesel, T. N. (1969). *Nature (London)* **221**, 747.
Hubel, D. H., and Wiesel, T. N. (1972). *J. Comp. Neurol.* **146**, 421.
Hughes, G. M., and Wiersma, C. A. G. (1960a). *J. Exp. Biol.* **37**, 291.
Hughes, G. M., and Wiersma, C. A. G. (1960b). *J. Exp. Biol.* **37**, 657.
Jorgensen, M. (1971a). *Acta Pathol. Microbiol. Scand. A* **79**, 298.

Jorgensen, M. (1971b). *Acta Pathol. Microbiol. Scand. A* **79,** 303.
Kalberer, M., and Pedler, C. (1963). *Vision Res.* **3,** 323.
Karlsson, U. (1966a). *J. Ultrastruct. Res.* **16,** 429.
Karlsson, U. (1966b). *J. Ultrastruct. Res.* **16,** 482.
Kastschenko, N. (1886). *Arch. Anat. Physiol. Anat. Abt.* p. 388.
Kastschenko, N. (1887). *Anat. Anz.* **2,** 426.
Kato, M., Frijimori, B., and Hirata, Y. (1968). *Brain Res.* **9,** 390.
Keddie, F. M., and Barajas, L. (1969). *J. Ultrastruct. Res.* **29,** 260.
Keddie, F. M., and Barajas, L. (1972). *Int. J. Dermatol.* **11,** 40.
Kelty, R. H., Baggenstoss, A. H., and Butt, H. R. (1950). *Gastroenterology* **15,** 285.
Kennedy, D., Evoy, W. H., Dane, B., and Hannawalt, J. T. (1967). *J. Exp. Zool.* **165,** 239.
Kennedy, D., Selverston, A. I., and Remler, M. P. (1969). *Science* **164,** 1488.
Kerkut, G. A., and Walker, R. J. (1962). *Stain Technol.* **37,** 216.
Kerr, G. R. (1902). *Quart. J. Microsc. Sci.* **45,** 1.
Kirschfeld, K. (1967). *Exp. Brain Res.* **3,** 248.
Kolb, H. (1970). *Phil. Trans. Roy. Soc. London, Ser. B* **258,** 261.
Kravitz, E. A., Stretton, A. O. W., Alvarez, J., and Furshpan, E. J. (1968). *Fed. Proc., Fed. Amer. Soc. Exp. Biol.* **27,** 749.
Krieg, W. J. S. (1949a). *J. Comp. Neurol.* **91,** 1.
Krieg, W. J. S. (1949b). *J. Comp. Neurol.* **91,** 39.
Krieg, W. J. S. (1966). "Functional Neuroanatomy." Univ. of Illinois Press, Urbana.
Lang, N. J. (1963). *Amer. J. Bot.* **50,** 280.
Lapin, M. W. (1927). *Z. Wiss. Mikrosk.* **44,** 134.
Larrimer, J., Eggleston, A., Masukawa, L. M., and Kennedy, D. (1971). *J. Exp. Biol.* **54,** 391.
Lasansky, A. (1971). *Phil. Trans. Roy. Soc. London, Ser. B* **262,** 365.
Levinthal, C., and Ware, R. W. (1972). *Nature (London)* **236,** 207.
Levinthal, C., Macagno, E. R., and Tountas, C. (1974). *Fed. Proc., Fed. Amer. Soc. Exp. Biol.* (in press).
Lewis, M. R., and Lewis, W. H. (1914). *Amer. J. Anat.* **17,** 339.
Libet, B., and Tosaka, T. J. (1970). *Proc. Nat. Acad. Sci. U. S.* **67,** 667.
LoPresti, V., Macagno, E. R., and Levinthal, C. (1973). *Proc. Nat. Acad. Sci. U. S.* **70,** 433.
LoPresti, V., Macagno, E. R., and Levinthal, C. (1974). *Proc. Nat. Acad. Sci. U. S.* **71,** 1098.
Lugg, M. M. (1972). *J. Amer. Med. Technol.* **34,** 23.
Lux, H. D., Schubert, P., and Kreutzberg, G. W. (1970). *In* "Excitatory Synaptic Mechanisms" (P. Andersen and K. S. Jansen, eds.), p. 189. Universitetsforlaget, Oslo.
Macagno, E. R., LoPresti, V., and Levinthal, C. (1973). *Proc. Nat. Acad. Sci. U. S.* **70,** 59.
McIntosh, J. R., Siskin, J., Ross, B. M., and Vanderslice, K. D. (1973). *J. Cell Biol.* **59,** 208a.
Meinertzhagen, I. A. (1972). *Brain Res.* **41,** 39.
Melamed, J., and Trujillo-Cenóz, O. (1968). *J. Ultrastruct. Res.* **21,** 313.
Missotten, L. (1965). "The Ultrastructure of the Human Retina." Arscia, Brussels.
Mitchell, H. C., and Thaemert, J. C. (1965). *Science* **148,** 1480.
Moens, P. B. (1969). *Chromosoma* **28,** 1.
Moens, P. B. (1970). *J. Cell Sci.* **7,** 55.

Moens, P. B., and Perkins, F. O. (1969). *Science* **166**, 1289.
Moore, A. B., and Hayden, J. (1963). *Stain Technol.* **38**, 351.
Muller, J., and Barajas, L. (1972). *J. Ultrastruct. Res.* **41**, 533.
Nauta, W. J. H., and Gygax, P. A. (1954). *Stain Technol.* **29**, 91.
Nichols, J. G., and Purves, D. (1970). *J. Physiol.* (*London*) **209**, 647.
Nilsson, S. E. G. (1964). *J. Ultrastruct. Res.* **10**, 390.
Novikoff, A. B. (1961). *In* "The Cell" (J. Brachet and A. E. Mirsky, eds.), Vol. 2, pp. 299–421. Academic Press, New York.
Odhner, N. (1911). *Anat. Anz.* **39**, 273.
Oh-I, D., Kasai, M., and Takahashi, T. (1969). *Tohoku J. Exp. Med.* **99**, 129.
Peacock, P. R., and Price, L. W. (1932). *J. Roy. Microsc. Soc.* **52**, 265.
Pedler, C., and Goodland, H. (1965). *J. Roy. Microsc. Soc.* **84**, 161.
Pedler, C., and Tansley, K. (1963). *Exp. Eye Res.* **2**, 39.
Pedler, C., and Tilly, R. (1965). *In* "The Structure of the Eye" (J. Rohen, ed.), 2nd Symp., Wiesbaden, p. 29. Schattauer, Stuttgart.
Pedler, C., and Tilly, R. (1966). *J. Roy. Microsc. Soc.* **86**, 189.
Pinching, A. J., and Powell, T. P. S. (1971a). *J. Cell Sci.* **9**, 305.
Pinching, A. J., and Powell, T. P. S. (1971b). *J. Cell Sci.* **9**, 347.
Pinching, A. J., and Powell, T. P. S. (1971c). *J. Cell Sci.* **9**, 379.
Pinching, A. J., and Powell, T. P. S. (1972). *J. Cell Sci.* **10**, 621.
Polyak, S. L. (1941). "The Retina." Univ. of Chicago Press, Chicago, Illinois.
Poritsky, R. (1969). *J. Comp. Neurol.* **135**, 423.
Postlethwait, S. N. (1962). *Turtox News* **40**, 98.
Postlethwait, S. N., Mills, R., and Lohmann, K. B. (1964). *J. SMPTE Soc. Motion Pict. Telev. Eng.* **73**, 629.
Pratt, S. A. (1962). M. S. Thesis, Univ. of Rochester, Rochester, New York.
Pratt, S. A. (1968). *J. Morphol.* **126**, 31.
Price, J. L., and Powell, T. P. S. (1970a). *J. Cell Sci.* **7**, 91.
Price, J. L., and Powell, T. P. S. (1970b). *J. Cell Sci.* **7**, 125.
Price, J. L., and Powell, T. P. S. (1970c). *J. Cell Sci.* **7**, 157.
Price, J. L., and Powell, T. P. S. (1970d). *J. Cell Sci.* **7**, 631.
Purves, D., and McMahan, U. J. (1972). *J. Cell Biol.* **55**, 205.
Rall, W. (1959). *Exp. Neurol.* **1**, 491.
Rall, W., Shepherd, G. M., Reese, T. S., and Brightman, M. W. (1966). *Exp. Neurol.* **14**, 44.
Rall, W., Burke, R. E., Smith, T. G., Nelson, P. G., and Frank, K. (1967). *J. Neurophysiol.* **30**, 1169.
Read, J. L., Hegre, E. S., and Russi, S. (1952). *Circulation* **7**, 42.
Reddy, D. R., Davis, W. J., Ohlander, R. B., and Bihary, D. J. (1973). *In* "Intracellular Staining in Neurobiology" (S. D. Kater and C. Nicholson, eds.), p. 227. Springer-Verlag, Berlin and New York.
Reicher, K. (1907a). *Neurol. Zentralbl.* **26**, 496.
Reicher, K. (1907b). *Ges. Deut. Naturforsch. Arzte* **79**, 235.
Reicher, K. (1910). *Berlin Klin. Wochenschr.* **47**, 484.
Rolshoven, E. (1937). *Z. Wiss. Mikrosk.* **54**, 328.
Sack, W. O. (1966). *Anat. Rec.* **154**, 233.
Sandeman, D. C. (1969). *J. Exp. Biol.* **50**, 87.
Saxl. (1927). *Photogr. Korresp.* **63**.
Schötz, F. (1972). *Planta* **102**, 152.

Schötz, F., Bathlet, H., Arnold, C. G., and Schimmer, O. (1972). *Protoplasma* **75**, 229.
Seldon, L. (1972). M.S. Thesis, Mass. Inst. of Technol., Cambridge, Massachusetts.
Selverston, A. I., and Kennedy, D. (1969). *Endeavour* **28**, 107.
Senger, P. L., and Saacke, R. G. (1970). *J. Cell Biol.* **46**, 405.
Sherrington, C. S. (1892). *J. Physiol. (London)* **13**, 621.
Siegesmund, K. A., Dutta, C. R., and Fox, C. A. (1964). *J. Anat.* **98**, 93.
Sjostrand, F. S. (1958). *J. Ultrastruct. Res.* **2**, 122.
Sjostrand, F. C. (1965). *Colour Vision: Physiol. Exp. Psychol., 1964 Ciba Found. Symp.* p. 110.
Sjostrand, F. S. (1969). *In* "The Retina: Morphology, Function and Clinical Characteristics" (B. R. Straatsma, M. O. Hall, R. A. Allen, and F. Crescitelli, eds.), UCLA Forum in Medical Sciences, No. 8, p. 63. Univ. of California Press, Berkeley.
Solari, A. J. (1964). *Exp. Cell Res.* **36**, 160.
Solari, A. J. (1969a). *J. Ultrastruct. Res.* **27**, 289.
Solari, A. J. (1969b). *Genetics* **61**, Suppl., 113.
Solari, A. J. (1970). *Chromosoma* **29**, 217.
Solari, A. J., and Tres, L. (1970). *J. Cell Biol.* **45**, 43.
Sotelo, J. R., and Trujillo-Cenóz, O. (1960). *Z. Zellforsch. Mikrosk. Anat.* **51**, 243.
Sparvoli, E., Gay, H., and Kaufman, B. P. (1965). *Chromosoma* **16**, 415.
Stack, S. (1973). *J. Cell Sci.* **13**, 83.
Stell, W. K. (1967). *Amer. J. Anat.* **121**, 401.
Stell, W. K. (1972). *In* "Handbook of Sensory Physiology" (M. G. F. Fuortes, ed.), Vol. VII/2, p. 111. Springer-Verlag, Berlin and New York.
Stempak, J. G. (1965). *Anat. Rec.* **151**, 420.
Stephens, R. J., and Bils, R. F. (1965). *J. Cell Biol.* **24**, 500.
Strasser, H. (1886). *Z. Wiss. Mikrosk.* **3**, 179.
Strasser, H. (1887a). *Z. Wiss. Mikrosk.* **4**, 168.
Strasser, H. (1887b). *Z. Wiss. Mikrosk.* **4**, 330.
Strausfeld, N. J., and Braitenberg, V. (1970). *Z. Vergl. Physiol.* **70**, 95.
Stretton, A. O. W., and Kravitz, E. A. (1968). *Science* **162**, 132.
Stuart, A. E. (1970). *J. Physiol. (London)* **209**, 627.
Tahmisian, T. N., Powers, E. L., and Devine, R. L. (1956). *J. Biophys. Biochem. Cytol.* **2**, Suppl. 4, 325.
Takahashi, T., and Hayama, T. (1967). *Tohoku J. Exp. Med.* **91**, 149.
Teichberg, S., and Holtzman, E. (1973). *J. Cell Biol.* **57**, 88.
Thaemert, J. C. (1966). *J. Cell Biol.* **28**, 37.
Thaemert, J. C. (1970). *Amer. J. Anat.* **128**, 239.
Thaemert, J. C. (1973). *Amer. J. Anat.* **136**, 43.
Thomas, R. C., and Wilson, V. J. (1965). *Nature (London)* **206**, 211.
Thomas, R. C., and Wilson, V. J. (1966). *Science* **151**, 1538.
Thompson, C. F., and Gottlieb, F. J. (1972). *Mikroskopie* **28**, 125.
Trujillo-Cenóz, O. (1965a). *J. Ultrastruct. Res.* **13**, 1.
Trujillo-Cenóz, O. (1965b). *Cold Spring Harbor Symp. Quant. Biol.* **30**, 371.
Trujillo-Cenóz, . (1969). *J. Ultrastruct. Res.* **27**, 533.
Trujillo-Cenóz, O. (1970). *Z. Zellforsch. Mikrosk. Anat.* **110**, 336.
Trujillo-Cenóz, O. (1972). *In* "Handbook of Sensory Physiology" (M. G. F. Fuortes, ed.), Vol. VII/2, p. 5. Springer-Verlag, Berlin and New York.
Trujillo-Cenóz, O., and Melamed, J. (1963). *Z. Zellforsch. Mikrosk. Anat.* **59**, 71.

Trujillo-Cenóz, O., and Melamed, J. (1966a). *Funct. Organ. Compound Eye, Proc. Int. Symp., Stockholm, 1965* pp. 339–361.

Trujillo-Cenóz, O., and Melamed, J. (1966b). *J. Ultrastruct. Res.* **16**, 395.

Trujillo-Cenóz, O., and Melamed, J. (1973). *J. Ultrastruct. Res.* **42**, 554.

Urena, F., and Solari, A. J. (1970). *Chromosoma* **30**, 258.

Vivier, E., and Petitprez, A. (1972). *J. Ultrastruct. Res.* **41**, 219.

Wann, D. F., Woolsey, T. A., Dierker, M. L., and Cowan, W. M. (1973). *IEEE Trans. Bio-Med. Eng.* **20**, 233.

Ware, R. W. (1971). Ph.D. Thesis, Mass. Inst. of Technol., Cambridge, Massachusetts.

Ware, R. W., Clark, D., and Russell, R. (1975). In preparation.

Waterman, T. H. (1966). *Funct. Organ. Compound Eye, Proc. Int. Symp., Stockholm, 1965* p. 493.

Weinstein, M., and Castleman, K. R. (1971). *Proc. Soc. Photo-opt. Inst. Eng., Biomed.* **26**, 131.

Wettstein, R., and Sotelo, J. R. (1967). *J. Microsc. (Paris)* **6**, 557.

White, E. L. (1972). *Brain Res.* **37**, 69.

Widakowich, V. (1907). *Zentralbl. Physiol.* **21**, 784.

Wiersma, C. A. G., and Hughes, G. M. (1961). *J. Comp. Neurol.* **116**, 209.

Wilkins, L. A., and Larrimer, J. (1971). *Amer. Zool.* **11**, 674.

Willey, T. J. (1973). *J. Comp. Neurol.* **152**, 211.

Willey, T. J., and Schultz, R. L. (1971). *Brain Res.* **29**, 31.

Willey, T. J., Schultz, R. L., and Gott, A. H. (1973). *IEEE Trans. Bio-Med. Eng.* **20**, 288.

Witkovsky, P., and Stell, W. K. (1973). *J. Comp. Neurol.* **150**, 147.

Woolsey, T. A. (1973). *In* "Intracellular Staining in Neurobiology" (S. D. Kater and C. Nicholson, eds.), p. 278. Springer-Verlag, Berlin and New York.

Woolsey, T. A., Wann, D. F., Cowan, W. M., Dierker, M. L., and Shinn, C. M. (1972). *Soc. Neurosci., Annu. Meet., 2nd.*

Zimmerman, M. H. (1971). "Dicotyledonous Wood Structure," Film No. E 1735/1971. Inst. Wis. Film, Göttingen.

Zimmerman, M. H., and Tomlinson, P. B. (1965). *J. Arnold Arboretum, Harvard Univ.* **46**, 160.

Zimmerman, M. H., and Tomlinson, P. B. (1966). *Science* **152**, 72.

Zimmerman, M. H., and Tomlinson, P. B. (1968). *Amer. J. Bot.* **55**, 1100.

Zimmerman, M. H., and Tomlinson, P. B. (1972). *Bot. Gaz.* **133**, 141.

Subject Index

V

Contents of Previous Volumes

Volume 4

Volume 5

Volume 6

Volume 7

Volume 8

Volume 14

Volume 15

Volume 16

Volume 17

Volume 18

Volume 19

Volume 20

Volume 21

Volume 22

Volume 23

Volume 24

Volume 25

Volume 26

Volume 27

Volume 28

Volume 29

Volume 30

Volume 31

Volume 32

Volume 33

Volume 34

Volume 35

Volume 36

Volume 37

Volume 38

Volume 39

Errata

INTERNATIONAL REVIEW OF CYTOLOGY
Volume 39

Pages 368-377:

Illustration shown as Fig. 22 belongs to legend No. 21

Illustration shown as Fig. 23 belongs to legend No. 22

Illustration shown as Fig. 24 belongs to legend No. 23

Illustration shown as Fig. 25 belongs to legend No. 24

Illustration shown as Fig. 26 belongs to legend No. 25

Illustration shown as Fig. 30 belongs to legend No. 26

Illustration shown as Fig. 29 belongs to legend No. 27

Illustration shown as Fig. 21 belongs to legend No. 28

Illustration shown as Fig. 27 belongs to legend No. 29

Illustration shown as Fig. 28 belongs to legend No. 30

B
C
D
E
F
G
H
I
J